Contemporary Ergonomics 2003

Also available from Taylor & Francis

Biomechanics of Upper Limbs
A. Freivalds
Hbk: 0-7484-40926-2
Pbk: 0-7484-40827-0

Designing Pleasurable Products
P. Jordan
Now also available in paperback: 0-415-29887-3

E-commerce Usability
D. Travis
Pbk: 0-415-25834-0

Evidence-based Patient Handling
S. Hignett *et al.*
Hbk: 0-415-24631-8
Pbk: 0-415-24632-6

Focus Groups
Edited by J. Langford *et al.*
Pbk: 0-415-26208-9

Human Factors in Lighting – 2nd ed.
P. Boyce
Hbk: 0-7484-0949-1
Pbk: 0-7484-0950-5

Human Factors Methods for Design
C. Nemeth
Hbk: 0-415-29798-2
Pbk: 0-415-29799-0

Human Thermal Environments – 2nd ed.
K. Parsons
Hbk: 0-415-23792-0
Pbk: 0-415-23793-9

Increasing Productivity and Profit
M. Oxenburgh *et al.*
Hbk, with CD-ROM: 0-415-24331-9

Introduction to Ergonomics
R.S. Bridger
Hbk: 0-415-27377-3
Pbk: 0-415-27378-1

Working Postures and Movements
Edited by N. Delleman *et al.*
Hbk: 0-415-27908-9
Pbk: 0-415-27809-7

Contemporary Ergonomics 2003

Edited by

Paul T. McCabe

WS Atkins, UK

THE **Ergonomics society**

First published 2003
by Taylor & Francis
11 New Fetter Lane, London EC4P 4EE

Simultaneously published in the USA and Canada
by Taylor & Francis Inc,
29 West 35th Street, New York, NY 10001

Taylor & Francis is an imprint of the Taylor & Francis Group

Mental workload of the railway signaller
L Pickup, J Wilson, T Clarke

Baseline ergonomic assessment of signalling control facilities
J Capitelli, F O'Curry, M Stearn, J Wood

The role of communication failures in engineering work
P Murphy, JR Wilson

This book is produced from camera-ready copy supplied by the Editor
Printed and bound in Great Britain by TJ International Ltd, Padstow, Cornwall

British Library Cataloguing in Publication Data
A catalogue record for this book is available from the British Library

Library of Congress Cataloging in Publication Data
A catalogue record has been requested

ISBN 0-415-30994-8

CONTENTS

OCCUPATIONAL HEALTH AND SAFETY

SLIPS, TRIPS AND FALLS

FATIGUE

DISPLAY SCREEN EQUIPMENT

VISION

APPLIED PHYSIOLOGY

AIR TRAFFIC CONTROL

DRIVING

RAIL

MOTORCYCLE ERGONOMICS

DESIGN

PREFACE

Contemporary Ergonomics 2003 is the proceedings of the Annual Conference of the Ergonomics Society, held in April 2003 at Heriot-Watt University, Edinburgh, UK. The conference is a major international event for ergonomists and human factors specialists, and attracts contributions from around the world.

Papers are chosen by a selection panel from abstracts submitted in the autumn of the previous year and the selected papers are published in *Contemporary Ergonomics*. Papers are submitted as camera ready copy prior to the conference. Each author is responsible for the presentation of their paper. Details of the submission procedure may be obtained from the Ergonomics Society.

The Ergonomics Society is the professional body for ergonomists and human factors specialists based in the United Kingdom. It also attracts members throughout the world and is affiliated to the International Ergonomics Association. It provides recognition of competence of its members through its Professional Register. For further details contact:

The Ergonomics Society
Devonshire House
Devonshire Square
Loughborough
Leicestershire
LE11 3DW
UK

Tel: (+44) 1509 234 904
Fax: (+44) 1509 235 666

Email: ergsoc@ergonomics.org.uk
Web page: http://www.ergonomics.org.uk

MANUAL HANDLING

DEVELOPMENT OF MANUAL HANDLING ASSESSMENT CHARTS (MAC) FOR HEALTH AND SAFETY INSPECTORS

SC Monnington[1], CJ Quarrie[2], ADJ Pinder[3] and LA Morris[2]

[1]*Health and Safety Executive, Government Buildings, Ty Glas, Llanishen, Cardiff, CF14 5SH, UK*

[2]*Health and Safety Executive, Magdalen House, Stanley Precinct, Bootle, Merseyside, L20 3QZ, UK*

[3] *Health and Safety Laboratory, Broad Lane, Sheffield, S3 7HQ, UK*

A tool has been developed to aid inspectors of health and safety when assessing manual handling operations. It is based on flow charts that guide a non-specialist through assessments of the most important risk factors for single person lifting, carrying and team lifting operations. A 'traffic light' system indicates levels of risk for each factor assessed. The tool was trialled with user groups and reviewed by an ergonomist to ensure consistency with existing HSE guidance and the published literature. It was found to be easy to learn and quick to use. Additional work ensured that the tool was usable and reliable and benchmarked it against existing methods for assessing manual handling. The notebook-format final version has a score sheet and guidance on prioritising tasks for examination. The tool still requires validation against injury data.

Introduction

Prevention and control of work related musculoskeletal disorders (WRMSD) is one of the eight Priority Programmes of the Health and Safety Commission (HSC). Success in this area is vital if the targets of a 20% reduction in the incidence of work-related ill health and a 30% reduction in the number of work days lost due to work-related ill health (HSC, 2000) are to be achieved by 2010.

Health and Safety Executive (HSE) and Local Authority (LA) inspectors will play an important part in preventing WRMSDs. As well as enforcing health and safety law, they provide advice on risk factors and control measures on a wide range of health and safety issues and therefore need to be able to quickly identify high risk activities. However, it can be difficult to establish the degree of risk in relation to WRMSD because of the lack of quantitative exposure-response relationships and the wide range of risk factors.

The Ergonomics Section of the Health and Safety Laboratory (HSL) was commissioned by the Human Factors Unit of HSE's Health Directorate to produce a tool for UK health and safety inspectors to use when assessing manual handling operations. To ensure the tool was suitable for use by the target users, criteria were specified that the tool should be very quick and easy to use (e.g., few pages and intuitive design); it must link with other published information on manual handling, particularly the HSE guidance on the Manual Handling Operations Regulations, 1992 (HSE, 1998); and it should intuitively indicate good manual handling practice. A number of existing assessment tools were considered, but none fully met the design criteria. This led to the development of the Manual handling Assessment Charts (MAC), a new flow chart-based tool for assessing manual handling operations involving lifting, carrying and team lifting. A pilot version was made available to HSE and some LA inspectors in April 2001. At this stage Morris (2001) published a brief description of the charts and their development.

This study outlines the conceptualisation and initial development of the MAC which was the basis for further usability, reliability and benchmarking work (Care *et al.*, 2002, Tapley and Buckle, 2003, Pinder 2003) prior to the formal release of the final version of the MAC to all HSE and LA inspectors in November 2002.

Conceptualisation of the MAC and selection of risk factors

The initial concept was a flow chart fitting on one side of A4 paper that would allow a user to consider a series of risk factors in a simple logical way. A "traffic light" system was used to grade risks within each factor, with green representing good practice and low risk, and red representing high risk and poor practice. To enable prioritisation of remedial action, weighted numerical scores were linked to each risk grade so that an aggregated score could be obtained. This score could also be used to highlight improvements following changes to the operation. Charts were drafted for assessment of lifting, carrying and team handling operations.

An ergonomist's approach to examining manual handling in the field formed the basis for the approach used in the tool. To ensure that the risk factors used in the charts were related to those described in HSE (1998), the risk factors in Schedule 1 of HSE (1998) were considered and the factors most frequently assessed as being problematic when observing manual handling were listed. It was considered that load-related issues, such as bulk or sharp edges, would not be found on a regular basis and that these factors could be implicitly dealt with by other factors. Thus, a bulky load could impact on the low back moment of force, while sharp edges would relate to hand coupling. The listed factors were then ranked in order of observation/importance based on several field visits and prominent reviews of epidemiological research on work-related back disorders (Bernard, 1997; Op De Beeck and Hermans, 2000). Importantly, there is a consensus that physical risk factors associated with an increased risk of low back disorders and disability are heavy physical work; lifting and handling of loads; and awkward movements and postures (e.g., bending and twisting).

The selected factors were examined closely to develop severity categories for the individual risks. To do this, the manual handling literature was reviewed for suitable indications of risk. In addition, footage of manual handling activities was reviewed with the aim of picking out commonalities or other postural indicators that would be useful as user-friendly definitions for the categories. It was clear that for carrying and team handling there would be common aspects with lifting and that this should be preserved, while additional factors could be added in as appropriate (e.g., co-ordination for team handling and obstacles encountered while carrying).

Load weight and frequency

The aim was to develop a graphical approach by which an assessment could be made based on the weight of the load and the repetition rate. The source of the psychophysical data underlying the graphs was Snook and Ciriello (1991). For simplicity, the graphs and guidelines were developed to be applicable to both males and females. The boundary between the green and amber zones is based on the maximum weights that Snook and Ciriello (1991) found acceptable to average females. The boundary between the amber and red zones is the load/frequency combination acceptable to average males. The boundary between the red and purple zones is the load/frequency combination acceptable to 90th percentile males, until the load reaches 50 kg. The 50 kg upper limit of the red zone was selected as being twice the 25 kg maximum figure in the risk filter in Appendix 1 of HSE (1998). Inclusion of the purple zone was considered important by the project team, to indicate a level of risk higher than the red zone due to such loads being acceptable to very few industrial workers.

Hand distance from low back

The further the hands supporting the load are away from the low back the greater the torque at the low back being counteracted by the muscles and ligaments that support it. This relates directly to postures adopted during an operation and the bulk of the load, in particular the dimension in the sagittal plane away from the body. Lifting capability decreases with increases in horizontal distance, specifically in relation to the spinal stresses experienced (Mital and Kromodihardjo, 1986) and acceptable weights selected (Snook and Ciriello, 1991). Increased low back moment is one of several indicators of an increased risk that a job has a high incidence of low back disorders (Marras *et al.*, 1995). HSE (1998) illustrates the issue in detail (paragraph 54 to 55).

Vertical lift region

This factor was developed to identify high risk aspects of an operation related specifically to forward bending of the back and reaching with the arms. Studies have shown that combinations of high bending and compressive stresses can be damaging to the intervertebral discs (Adams and Hutton, 1985).

The choice of risk grade definition was undertaken to enable the user to easily relate the position of the hands holding the load with anatomical landmarks on the operative undertaking the operation. Hence the adoption of floor level, knee height, elbow height and head height as easily discernible landmarks separating the risk grades. The green zone was selected to relate well with the guidelines provided in HSE (1998), depicted as better practice. The red zone relates to extreme vertical positions (above head height and floor level or below) that require the adoption of awkward postures.

Trunk twisting / sideways bending

Biomechanical analysis indicates that asymmetrical lifting activities give rise to greater low back stress (spinal compression) than symmetrical lifting (Mital and Kromodihardjo, 1986). Also, epidemiological studies have indicated that asymmetric motion factors are risk factors for low back disorders (Marras *et al.*, 1995).

It was considered that the best approach was to separate the trunk asymmetry into twisting and sideways bending to aid consistent scoring of potentially complex situations of trunk movement and posture. Therefore, the amber score is given if either twisting or sideways bending is present, while red is given if both are seen in the same operation.

Other factors
The working environment can give rise to postural constraints that reduce the flexibility of operatives to vary their posture or impose added stress to the musculoskeletal system. Ridd (1985) demonstrated a reduction in lifting capability with a reduction in headroom. Mital (1986b) found that carrying in narrow passages reduced the carrying capacity.

The quality of the grip on a load during manual handling is important. Studies have shown that, with a good grip (i.e., handles), 4% to 30% more load can be handled than in less favourable grip situations (Garg and Saxena, 1980; Snook and Ciriello, 1991). Similarly, a more suitable grip affords less opportunity for fatigue of the forearm muscles and therefore reduces the risk of the load being dropped during the operation.

The grip afforded at the feet has clear implications in terms of how force is applied and thus the assurance of grip. It was considered important that selected factors related to the risk of slipping or tripping be included (e.g., contamination on the floor).

Environmental factors (poor lighting, strong air movements and extremes of temperatures) were also included in each chart.

Conformity to design criteria

As part of a planned development programme, constructive comments and opinion were sought from the target users of the charts. At this stage the aim was to examine the prototype charts from the point of view of the design criteria and so produce a draft tool for further evaluation. A peer-review by an experienced ergonomist examined technical aspects of the charts.

Two discussion group sessions were undertaken involving five operational inspectors with the purpose of familiarising them with the manual handling charts. During the three hours of each session the basic principles of the charts were described and instruction for their use provided. The inspectors then used the charts to evaluate seven manual operations (three lifting, two carrying and two team handling) shown on video. After each viewing, each inspector filled out the appropriate chart and totalled up the score. The scores were discussed and compared. Any discrepancies were identified with respect to possible problems experienced due to ease of use rather than as an exercise in reliability and consistency of scoring. Issues pertaining to the use of the tools, such as training provision, information needed to help inspectors effectively use the tools in the field and systems of support, were also discussed.

The peer-review was carried out by an ergonomist at HSL experienced in manual handling issues. The review considered the technical validity of the selected risk factors; how the charts related to the risk filter in Appendix 1 of HSE (1998); how individual factors should be scored and how total scores should be interpreted; and the general layout of the charts.

Findings

Inspectors found the MAC quick and easy to use. The basic idea of the tool and the layout using the flow chart was very well received. The flow chart format of two pages per chart was found to be very easy to follow and apply. The consensus from the user trials was that the charts helped to improve consistency and competence when inspecting workplaces where manual handling operations occur. Inspectors found that the links between the charts and existing HSE information were helpful. One inspector suggested that a simple one page 'aide-memoire' of the factors and how to make an assessment would be useful for on-site use. This approach was subsequently drafted for evaluation.

The peer-review found the charts to be a valid approach for assessing manual handling. The risk factors selected for inclusion were found to be justified. The "traffic light" approach was considered attractive and could be linked to the low, medium and high risk classes used in the example assessment checklist in Appendix 2 of HSE (1998). Since the quality of the existing epidemiological data does not permit accurate dose-response relationships between exposure to the risk factors and incidence of low back pain to be drawn up, the weightings of the factors must be seen as preliminary and no attempt has been made to link total scores to the taking of enforcement action by inspectors, especially since they must consider a range of other factors before doing so.

Inspectors emphasised that they are very busy and that any training requirement for the MAC needed to take account of this. Clearly, a system that required minimal training would be most beneficial. They acknowledged the advantages of structured training provided by a facilitator particularly with regard to ensuring consistency of scoring and assessment. However, opinion on the format of training was diverse. Given the absence of a consensus, it was considered prudent to include investigations of the impact of training on use and reliability of scoring during the evaluation.

The discussion groups and peer-review considered the total scores to be useful for ranking tasks in order of priority for change and a useful way of assessing improvements when changes had been implemented. The weighting of scores in the factors seeks to indicate which factors are of greater importance. However, the key assessment would not be the overall score but the risk grades for individual factors. Leading on from this, the peer-review highlighted the need for a longer term study of the validity of the charts in relation to injuries incurred (an idea that several inspectors favoured).

Several improvements to the layout of the charts were made based on suggestions made by inspectors. There was general agreement that a smaller 'pocket' size version would be most useful. This format was subsequently released and subjected to a usability evaluation. A separate score sheet that could be used to record information and be held on file was also considered to be useful. As a result of an observation in the peer-review that proactive inspection of every manual handling operation at a location would be very time-consuming, prompts to aid selection of operations requiring assessment were added onto the score sheet: 'Task is identified as high risk in company risk assessments'; 'Task has a history of manual handling incidents (e.g., RIDDOR reports)'; 'Task has a reputation of being physically demanding or high risk'; 'Employees performing the task present signs that they are finding the task arduous (e.g., sweating profusely)'.

After initial development, the indications were that the MAC conformed to the design criteria. Further development work was therefore undertaken to ensure the finalised product was suitably usable, reliable and related well to existing assessment tools (Quarrie, 2002, Care *et al.*, 2002, Tapley and Buckle, 2003, Pinder, 2003).

Acknowledgements

The authors wish to acknowledge the support given by colleagues in the MAC project team, user groups and in HSE policy and operational directorates.

References

Adams, M.A. and Hutton, W.C. 1985, The effect of posture on the lumbar spine, *Journal of Bone and Joint Surgery*, **67B**, 625-629

Bernard, B. 1997, *Musculoskeletal disorders and workplace factors. A critical review of epidemiologic evidence for work-related musculoskeletal disorders of the neck, upper extremity, and low back,* (US DHHS, Cincinnati, Ohio), Pub. No. 97- 141

Care, B., Quarrie, C. and Monnington, S.C. 2002, *Testing and improving the usability of the Manual handling Assessment Chart (MAC),* (Health and Safety Executive, Bootle). Internal report

Garg, A. and Saxena, U. 1980, Container characteristics and maximum acceptable weight of lift, *Human Factors*, **22**, 487-495

Health and Safety Executive 1998, *Manual Handling Operations Regulations 1992. Guidance on Regulations* (Sudbury, Suffolk: HSE Books), L23, Second edition

Health and Safety Commission 2000, *Securing Health Together: A long term occupational health strategy for England, Scotland and Wales*, HSE Books, MISC 225

Marras, W.S., Lavender, S.A., Leurgans, S.E., Fathallah, F.A., Ferguson, S.A., Allread, W.G. and Rajulu, S.L. 1995, Biomechanical risk factors for occupationally related low back disorders, *Ergonomics*, **38**, 377-410

Mital, A. 1986a, Prediction models for psychophysical lifting capabilities and the resulting physiological responses for work shifts of varied durations, *Journal of Safety Research*, **17**, 155-163

Mital, A. 1986b, Subjective estimates of load carriage in confined and open spaces. In W. Karwowski (ed.) *Trends in Ergonomics/Human Factors III* (Amsterdam: North-Holland), 827-833

Mital, A. and Kromodihardjo, S. 1986, Kinetic analysis of manual lifting activities: Part II - Biomechanical analysis of task variables, *International Journal of Industrial Ergonomics*, **1**, 91-101

Morris, L. 2001, MSDs - Development of a practical workplace risk assessment tools. *The Ergonomist*, No. 373, July 2001, 4-5. Also available on the Internet at: http://www.ergonomics.org.uk/resources/newsinfo/hsenews.htm#Development

Op De Beeck, R. and Hermans, V. 2000, *Research on work related low back disorders* (Office for Official Publications of the European Communities, Luxembourg)

Pinder, A.D.J. 2003, Benchmarking of health and safety inspectors' Manual handling Assessment Charts (MAC). In P.T. McCabe (ed), *Contemporary Ergonomics 2003*, (Taylor & Francis, London)

Quarrie, C., 2002, Manual Handling Assessment Charts to be used by HSE inspectors. *The Ergonomist*, No. 390, December 2002, 1-2. Also available on the Internet at: http://www.ergonomics.org.uk/resources/newsinfo/hsenews.htm#Usage

Ridd, J.E. 1985, Spatial restraints and intra-abdominal pressure, *Ergonomics*, **28**, 149-166

Snook, S.H. and Ciriello, V.M. 1991, The design of manual handling tasks: Revised tables of maximum acceptable weights and forces, *Ergonomics*, **34**, 1197-1213

Tapley, S and Buckle, P. 2003, Reliability of Manual handling Assessment Charts (MAC) developed for regulatory inspectors in the United Kingdom. In P.T. McCabe (ed), *Contemporary Ergonomics 2003*, (Taylor & Francis, London)

RELIABILITY OF MANUAL HANDLING ASSESSMENT CHARTS (MAC) DEVELOPED FOR REGULATORY INSPECTORS IN THE UNITED KINGDOM

SE Tapley[1] and Peter Buckle[2]

[1] *Health and Safety Executive (HDD2)*
Magdalen House
Trinity Road
Bootle L20 3QZ

[2] *Robens Centre for Health Ergonomics,*
EIHMS,
University of Surrey,
Guildford GU2 7TE

A manual handling assessment tool (MAC) has been developed by the HSE for use by regulatory inspectors in the UK. The MAC was designed to assist inspectors in initial identification, assessment and evaluation of significant physical risk factors for manual handling injury. Inter and intra- reliability of the tool has been assessed using videotapes of manual handling tasks. Levels of agreement and lack of differences in scoring tasks using the MAC tool would indicate that regulatory inspectors were able to reliably identify the risk factors.

Introduction

Prevention and control of WRMSD is one of the priority programmes in the Health and Safety Commission's (HSC) strategic plan selected to help meet targets set out in the Revitalising Health and Safety (RHS) strategy document (HSC 2000). This strategy sets national targets for the health and safety system including a target to reduce the number of working days lost per 100 000 workers from work related injury and ill health by 30% by 2010 and to reduce the incidence rate of cases of work related ill health by 20% by 2010.

Health and Safety Executive (HSE) and local authority inspectors play an important part in preventing WRMSD. As well as enforcing health and safety law, they provide advice on risk factors and control measures on a wide range of health and safety issues. They therefore need to be able to quickly identify high-risk activities. However, it can be particularly difficult to establish the degree of risk in relation to WRMSD because of the lack of quantitative exposure-response relationships, the wide range of risk factors and the interactions between them.

Many methods are available for use by ergonomists and practitioners when assessing exposure to risks associated with WRMSD's (Li and Buckle 1999a). Often, these have been developed for a particular research purpose and are based on the experts' view of what occupational risk factors should be considered and how they should be measured. This has resulted in methods that are frequently so sophisticated that only researchers or well-trained analysts are able to use them (Li and Buckle 1999b).

A range of existing methods was reviewed to examine their feasibility for use by regulatory inspectors (Monnington et al 2002). None of the tools reviewed met the needs of inspectors. Another consideration was that none of the tools reviewed had been validated in relation to injury risks. It was decided that a new manual handling tool targeted specifically at regulatory inspectors should be developed.

A series of manual handling assessment charts (MAC) were then developed. These consider three types of manual handling - lifting, team lifting by up to four persons, and carrying. The MAC was designed to assist regulatory inspectors in the initial identification, assessment and evaluation of the significant physical risk factors for manual handling injuries. MAC uses a flow chart approach and requires the inspector to work through a sequence of physical risk factors.

Table 1. Risk factors in the MAC charts

Load	Taken from the graph that accompanies the charts
Hand distance from low back	
Vertical lift distance	
Twist/asymmetry	
Postural constraints	Whether the operator is physically constrained whilst carrying out the operation
Grip on the load	
Environmental factors	Temperature, airflow and lighting
Carry distance	
Obstacles en route	Steps, slopes, ladders
Floor condition	Whether dry,wet or contaminated
Communication and coordination	During team handling

Individual factors such as age, physical fitness and strength are not included on the chart but are considered in the application of the criteria and are also recorded on the scoring sheet. A colour band and numerical score is allocated for each risk factor and these are added together to give a cumulative risk ranking score to indicate level of risk.

The aim of this study was to assess the inter and intra-reliability of the MAC when used by regulatory inspectors.

Method

Reliability assessment was carried out by surveying 10% of HSE field inspectors who regularly deal with manual handling issues. The sample group consisted of four different groups:

i) General inspectors who were classed as non-expert who received a briefing on how to use the MAC before the survey (briefed)

ii) General inspectors who received the briefing after the survey (non-briefed)

iii) Expert ergonomists who received a briefing on how to use the MAC before the survey (expert/briefed)

iv) General inspectors for the test-retest.

The sample group viewed four different videotaped tasks taken during reactive inspection visits (the criterion for use in the scoring exercise was that the clips clearly demonstrated the risk factors in each of the three tools) and made an initial scoring assessment using the charts. Tasks one and two were lifting tasks, task three was a carrying operation and task four was team handling. Survey data were analysed to assess the levels of agreement and differences in assessment by the groups within the whole sample. Intra -reliability was assessed using a test-retest approach.

Information that would have been elicited by questioning and observation during a site visit was given in a verbal protocol at the start of the scoring exercise; this included the load weight and repetition rate, carrying distance, communication and coordination, floor surface and other environmental factors.

The data were analysed using SPSS. The levels of agreement on scoring risk factors within each sample group were tested using the Kendal coefficient. Difference in scoring between experts and non-experts and between briefed and non- briefed inspectors were assessed using the Mann-Whitney test. Differences in scoring by inspectors carrying out scoring (Test 1) and then repeating the scoring (Test 2) six months later were evaluated using the Wilcoxon test. Each risk factor on the chart has a Red, Amber or Green scoring choice, the percentage scoring frequencies for these choices for each task by each group were computed.

Statistical significance levels of $p<0.05$ were set as being appropriate for such an assessment tool. This was also in line with other studies of this type.

Results

Levels of agreement in scoring all risk factors for each task by briefed, non-briefed, expert and non-expert sample groups are summarised in table 2. Results were computed using Kendal coefficient of concordance. A w result may range from between 0 (no agreement) to 1 (complete agreement).

Table 2. Overall level of agreement by all inspectors

Task	Sample 1 (n=50) Briefed /non experts	Sample 2 (n=22) Non briefed	Sample 3(n=5) Experts
1	$w = 0.91$	$w = 0.93$	$w = 0.89$
2	$w = 0.93$	$w = 0.99$	$w = 0.77$
3	$w = 0.89$	$w = 0.97$	$w = 0.93$
4	$w = 0.78$	$w = 0.82$	$w = 0.77$

w = strong agreement at >0.9
w = good agreement at >0.7

Assessment of the results indicates that there is a good to strong association of agreement between inspectors identifying the risk factors as all w-values exceeded 0.7 (minimum observed was 0.77). It is not possible to make direct comparisons between groups, however, the generally high level of similarity of range of scoring and percentage frequency of scoring for the groups would add weight to this finding allowing comment to be made that the level of agreement is good.

These results indicate the range of scoring as well as the percentage frequency and demonstrate the similarities in ranges and frequencies among the groups.

Difference in scoring between experts and non-experts

No statistically significant difference in scoring between the two groups were found other than for estimates of the vertical lift distance in Task 2 and the communication and coordination during the team lift in Task 4. The information for this last risk factor was given in the verbal protocol and so it would have been expected for there to be no difference in scoring, however this was not the case.

Difference in scoring between briefed and non-briefed inspectors.

The results showed that there were no statistically significant differences for the majority of the factors (as listed in table1.) Statistically significant differences were found for estimates of the Hand Distance from Low Back and Vertical Lift Distance in task 1, Grip on load in task 2 and for Vertical Lift Distance and Communications and Coordination in task 4.

Differences between test 1 and test 2 scoring. Test/retest

There was no statistically significant difference ($p< 0.05$) in scoring for any of the risk factors in any of the tasks, although Communication in Task 4 was of concern.

Percentage frequency and range of scores

These results indicated similarities in ranges and frequencies among all the groups for the risk factors. There were some variations in task 2 for the Hand distance form low back, vertical lift distance, postural constraints and grip on load. There were also some differences for communications, grip and vertical lift distance in task 4.

Discussion

There are good to strong levels of association in scoring across all the tasks and most of the risk factors with minimal statistically significant differences in scoring. However there are some discrepancies and there may be several reasons for these discrepancies.

The scoring exercise was carried out using videotaped material Denis et al (2002) commented that one of the major factors affecting observer reliability could be the difficulty of the videotaped material and potential loss of stereoscopic vision. Qualitative comments from inspectors indicated that the hand distance from low back was the main risk factor that inspectors found difficult to visualise and hard to assess from video footage. This may be improved in an inspection situation, as inspectors would be able to move around the task. It could be commented though that use of video taped material allows consistency in this type of study. Further qualitative comments about the grip risk factor showed that it had not been appreciated how important the grip on a load was during a manual handling operation. There was a significant difference between briefed and non-briefed inspectors for this factor in task 2.

Potential divided attention and timing of delivery of the scoring survey may have also affected scores. Divided attention caused by the sample groups trying to assimilate a new tool, listen to instructions and carrying out a scoring exercise simultaneously may have led to the sample groups not being able to pay full attention to what they were doing. This may not affect the ability to identify risk factors but is an issue that needs to be considered when inspecting. It also has ramifications for training as the aim would be to train inspectors to identify the risk factors easily thus freeing capacity to deal with novel settings for the risk factors.

It was noted that many of the discrepancies were in task 4. The scoring for this task was carried out just before lunch and it is postulated that the discrepancies may be related to tiredness and hunger. If this were to be the case, then it has ramifications for the inspection process as inspectors may miss risk factors when hungry and tired. However the flow chart approach of the MAC is designed to help alleviate those potential problems by leading inspectors through the risk factors logically.

Issues around the design of the score sheet that accompanies the MAC tool appeared to have had an impact on the scoring. There was some confusion among the sample groups, especially when scoring task 3, as the score sheet did not run in the same order or use the same terminology as the MAC charts.

The final inspector sample size of 77 is approximately 10% of HSE's operational field inspectors who inspect manual handling issues on a regular basis. This sample size is considered to be representative and for the findings to be extrapolated across all field inspectors. The expert sample (n=5) is small but is approximately 25% of HSE ergonomists and so can also be said to be representative. The test/retest sample group (n=8) is small and so results should be treated with some caution. However the frequency of scoring is very similar across the tasks to the other groups and this can increase confidence in the findings of the study.

The choice of subjects was largely dictated by operational and time constraints. This resulted in the use of a sample of geographical convenience, namely that of the same operating division as the researcher. However it can be argued that this sample was purposive as it was drawn from operational HSE inspectors, which was the population of interest.

A number of estimates of reliability were used. Levels of agreement alone are not a strong indication of reliability. The use of other methods such as percentage agreements in scoring, levels of difference in scoring and test/retest results has enabled a more robust view of reliability to be presented. Limitations in the power of the study resulting from the difference in size of the sample groups for experts and non-experts do remain. A method may be reliable but not valid and there is much debate regarding the reliability or validity of ergonomic methods across diverse applications and contexts (Wilson 1990; Stanton and Young, 1998; Li and Buckle, 1999a.) Li and Buckle (1999a) argue that this lack of evidence of reliability and validity may be due, in part, to a lack of motivation on behalf of those developing and using the methods to ensure they meet the development criteria set. A view that Wilson (1990) appears to agree with when commenting 'the reliability and validity of a method may well be application specific'. In this study, the MAC tool is considered to demonstrate construct validity by ensuring the risk factors identified in the tool are backed by sound scientific data (Monnington et al 2002).

Conclusion

Levels of agreement found would indicate that inspectors are identifying the factors with reasonable levels of inter- and intra- reliability. This would suggest that the tool is reliable for its purpose of assisting inspectors to identify manual handling tasks that carry significant risks for musculoskeletal disorders.

There are a plethora of assessment tools available for practitioners to use when assessing manual handling tasks; many of these tools have been context or use specific. However, it could be argued that to expect anything less would be against ergonomic principles as each assessment task is different, ergonomists seek constantly to dispel the myth that 'one size fits all' and this will surely apply to assessment tools as well.

The MAC tool, whilst designed for health and safety inspectors, would appear to be an addition to this battery of tests as this study has fund it to be a reliable assessment tool which can help practitioners to identify if and where further assessment is required.

References

Health and Safety Commission 2000, *Securing Health Together: A long-term occupational health strategy for England, Scotland and Wales*, (HSE Books, Sudbury, Suffolk), MISC 225

Li, G. and Buckle, P. 1999a, Current techniques for assessing physical exposure to work-related musculoskeletal risks, with emphasis on posture-based methods, *Ergonomics*, **42**, 674-695

Li, G. and Buckle, P. 1999b, *Evaluating Change in Exposure to Risk for Musculoskeletal Disorders - a Practical Tool*, (HSE Books, Sudbury, Suffolk), CRR 251/1999

Monnington, S.C., Pinder, A.D.J. and Quarrie, C. 2002, *Development of an inspection tool for manual handling risk assessment*, (Sheffield: Health and Safety Laboratory), HSL Internal Report ERG/02/20

Monnington, S.C., Pinder, A.D.J. and Quarrie, C. 2003, Development of Manual handling Assessment Charts (MAC) for Health and Safety Inspectors in the UK. In McCabe, P.T. (ed), *Contemporary Ergonomics 2003*, (Taylor & Francis, London)

Denis D, Lortie M, Bruxelles M. 2002, Impact of observers experience and training in reliability of observations for manual handling task. *Ergonomics* 45 (6) 441-454

Stanton N and Young M. 1998, Is utility in the mind of the beholder? A study of ergonomic methods. *Applied Ergonomics* 29 (1) 41-54

Wilson, J. 1990, A framework for ergonomics methodology in Wilson JR and Corlett EN (Eds) Evaluation of Human Work (Taylor Francis: London) 2nd edition p.120

BENCHMARKING OF HEALTH AND SAFETY INSPECTORS' MANUAL HANDLING ASSESSMENT CHARTS (MAC)

ADJ Pinder

Health and Safety Laboratory
Broad Lane
Sheffield, S3 7HQ, UK

The Manual handling Assessment Charts (MAC) have been developed to aid health and safety inspectors visiting premises where manual handling operations are carried out. When benchmarked against four other methods for assessing risk from manual handling, it was ranked as one of the most appropriate methods and one of the easiest to use. However, the rankings of severity of five tasks were random, suggesting that the tools assess risk in different ways and so cannot be compared easily. The MAC assesses multiple factors against criteria set for each factor, but does not assign an overall risk to the task. It therefore has a sounder basis than methods that combine scores from multiple factors to assign an overall risk, especially when this includes risk to both the low-back and the upper limbs since these body areas are differentially affected by different work tasks.

Introduction

Prevention and control of work related musculoskeletal disorders (WRMSDs) is one of the Health and Safety Commission's (HSC) priorities. Success in this area is vital if the targets in *Securing Health Together* (HSC, 2000) of reducing by 2010 the incidence of work-related ill-health by 20% and reducing the number of work days lost due to work-related ill-health by 30% are to be achieved.

HSL and HSE have developed jointly a risk assessment tool for HSE and Local Authority inspectors to use when visiting premises where manual handling operations are carried out by the workforce (Monnington *et al.*, 2003). This is known as the Manual handling Assessment Charts (MAC). These consider three types of manual handling - lifting, team lifting by up to four persons, and carrying. For each of these, a flowchart guides the user through an assessment and provides indications of levels of risk, using a "traffic light" system to assign colour codes and associated numerical values to each factor considered. This study reports the benchmarking of the MAC by comparing it with existing published tools used for risk assessment of manual handling operations.

Methods

Four existing assessment methods were selected for detailed comparison with the MAC. These were the 1991 NIOSH Revised Lifting Equation (Waters *et al.*, 1994); the Ovako Working posture Analysis System (OWAS) (Karhu *et al.*, 1977); the Quick Exposure Check (QEC) (Li and Buckle, 1999b); and Rapid Entire Body Assessment (REBA) (Hignett and McAtamney, 2000). A small number of ergonomists experienced in assessing industrial manual handling tasks used the five methods to assess five single-person manual handling tasks (Table 1). Video clips were used to ensure that the information presented to each expert was consistent. Information was given on weights and frequencies and durations of handling where these could not be observed from the video.

Table 1. Tasks used for benchmarking of the risk assessment tools

Task	Description	Weight
A	Depalletise compost bale, transfer to hopper, slice packaging and tip contents into hopper.	50 kg
B	Remove welded mesh from welding machine and transfer to stack behind operator	15 kg
C	Pick cases of fireworks from pallets onto mobile conveyor	25 kg
D	Pick cases of fireworks from pallets onto mobile conveyor	15 kg
E	Load wire bobbins onto rack of feed spindles	20 kg

First, they used a four-point scale, based on the OWAS Action Categories, to give a personal expert opinion of the severity of the risk of injury from each task/urgency of the need for action to reduce it. They then scored each task using the five assessment tools. Finally, they ranked the five risk assessment methods for ease of use and suitability.

With the inclusion of the Expert Opinions of each task, six assessments were made of each task (Table 2). Of these, REBA and the QEC arrived at a score which was then converted into a recommendation for action. This is done on a four- (QEC) or five-point (REBA) ordinal scale, based on that originally used by OWAS. The QEC total scores were converted into Action Levels using the score boundaries recommended by Li (2002) which were based upon the assessment of a variety of tasks using both the QEC and RULA (McAtamney and Corlett, 1993) and comparison of the exposure scores.

Table 2. Types of output from the different risk assessment tools

Expert opinion	MAC Total	QEC		REBA		OWAS Action Category	NIOSH Lifting Index
		Total	Action Level	Score	Action Level		
Ordinal 4 point scale	Ordinal integer scale	Integer scale	Ordinal 4 point scale	Ordinal scale	Ordinal 5 point scale	Ordinal 4 point scale	Ratio scale

Direct comparison of the scores provided by the different tools was not possible due to the different types of data returned by the tools (Table 2). Therefore, the non-parametric Friedman test was used to compare the rank orders of the scores using the χR^2 statistic.

Results

Four ergonomists completed the exercise to compare the five tools. After validation of the data they were entered into a spreadsheet. It was noted that there were wide discrepancies between the scores recorded by the different individuals. This was attributed partly to a lack of previous direct experience of using some of the tools and partly to the difficulty of assessing tasks from a short video clip.

Table 3. Severity of the five tasks (1= least severe, 5 = most severe)

Task	Mean ranking
B	2.083
A	2.854
C	3.292
E	3.313
D	3.458

In order to provide background information on the benchmarking exercise the severities of the five tasks were examined by using the Friedman test to compare, across the tasks, the mean rankings of the risk scores. The results, shown in Table 3, gave a value of $\chi R^2 = 6.3$ (N = 6, C = 5), $p > 0.05$.

Ease of use and suitability of the tools

The Friedman test was used to compare the rankings of ease of use of the five tools (Table 4). This gave $\chi R^2 = 11.2$,(N = 4, C = 5), $p = 0.01$. The Friedman test does not permit pairwise comparisons, but Table 4 shows that OWAS and the MAC were the two easiest tools to use and the QEC and NIOSH equation were rated as the two most difficult to use with REBA falling between the two groups.

Table 4. Mean rankings of ease of use and suitability of the tools (1 = best, 5 = worst)

Tool	Ease of use	Suitability
MAC	1.75	1.75
NIOSH	4.25	2.25
OWAS	1.5	4.25
QEC	4.25	2.5
REBA	3.25	4.25

Assessment of risk scores

Figure 1 shows how the different tools ranked the different tasks, averaged across the four experts. There are no systematic patterns in these scores, which is reflected in the value of χR^2 of 8.4 (N = 6, C = 5), $p > 0.05$, obtained from the Friedman test. Therefore, the lack of significant differences between how the risk assessment tools rank the severity of the tasks reflects random variation in the scoring systems of the different tools.

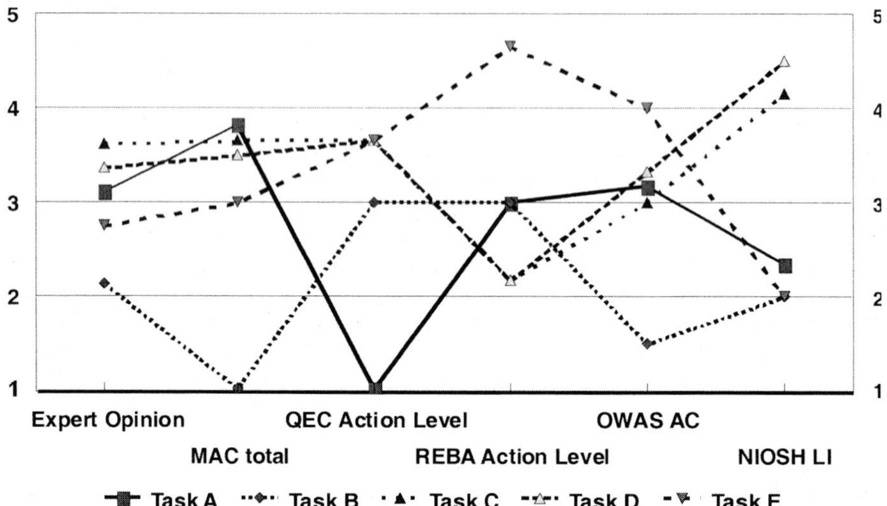

Figure 1: Mean rankings of the tasks by the tools (1= least severe, 5 = most severe)

The Friedman test was also used to compare the rankings of the suitability of the tools for assessing the risk of manual handling operations. This gave $\chi R^2 = 8.8$ (N = 4, C = 5), p = 0.05. Table 5 shows that OWAS and REBA were perceived as the least suitable, and the MAC, NIOSH equation and QEC were seen as the more suitable.

Therefore, the MAC was seen as both one of the easiest tools to use and one of the most suitable. While the NIOSH Lifting Equation and the Quick Exposure Check were seen as equally suitable, they were seen as being the hardest tools to use. OWAS and REBA were seen as the least suitable tools despite OWAS being ranked as equally easy to use as the MAC, and REBA being ranked between the MAC and the NIOSH equation.

Discussion and conclusions

Measuring exposure to musculoskeletal risk factors is both conceptually and practically complex. It can therefore be very time consuming, especially when several body parts are at risk. As a result, a number of tools have been developed to simplify the process of risk assessment and to give guidance on the urgency of remedial action. One approach is to use snapshot assessments since time sampling and weighting of measurements is time-consuming, and creates further difficulties in interpretation.

OWAS and REBA are designed for analysis of postures, often those believed to be hazardous or problematic. By contrast, the NIOSH equation is an assessment of the task without reference to the individual performing it. The MAC and the QEC assess a mixture of task parameters and postures of an individual performing the operation. The MAC does also use signs that individuals have difficulty with a task as an indication that the task is likely to be high risk and therefore worth assessing.

The benchmarking exercise showed that the MAC was one of the easiest risk assessment methods to use for assessing manual handling operations and one of the most appropriate. The benchmarking showed that there were no systematic differences in how the different methods ranked the levels of risk of the tasks studied but the way that the severity of tasks was ranked was random. It therefore appears that they assess risk in different ways and so cannot be compared easily.

While bias by the ergonomists involved in the trial cannot be ruled out, it appears that, unlike the other tools, the MAC has achieved the desirable goal of both good perceived ease of use and good perceived validity. OWAS is very rapid in use but has a very limited classification of postures. Using REBA and the QEC is more difficult because of the need to look at multiple tables. The NIOSH equation has the problem of needing accurate measurements of the parameters and involving a complex calculation which therefore needs time and care. While many of these problems can be overcome with computer software or "ready reckoners", they do reduce the acceptability of a tool. By contrast, the MAC has the advantage of only using three point scales and numerical scores that are easily added. Moreover, it has been produced in a format that is convenient for factory inspectors carrying out site visits who typically record their findings in similar size notebooks.

Despite the differences in output there are traceable links between the different tools. Thus, the MAC was developed after a review of existing methods (Monnington *et al.*, 2003). Also, the four-point scale for urgency of action that originated with OWAS reappears, via RULA, in the QEC and, as a five-point scale, in REBA. Both the NIOSH equation and the MAC make use of the same underlying psychophysical data (Snook and Ciriello, 1991) and have other factors in common, such as the grip on the load.

The status of the different tools is variable. The MAC was distributed to HSE inspectors in late 2002 and made publicly available in early 2003. Monnington *et al.* (2003), this paper and Tapley (2003) are the first reports to a scientific audience. The QEC has been published as an HSE Contract Research Report (Li and Buckle, 1999b) and presented at scientific conferences (e.g., Li and Buckle, 2000), but has not yet appeared in the peer-reviewed scientific literature. The other tools have been published in the peer-reviewed literature, REBA in 2000 (Hignett and McAtamney, 2000); OWAS as long ago as 1977 (Karhu *et al.*, 1977); and the 1991 NIOSH equation in 1993 (Waters *et al.*, 1993). OWAS has been widely used (e.g., Vedder, 1998) and the NIOSH equation was more extensively described in the *Applications Manual* by Waters *et al.* (1994).

The MAC has the advantage that it is focussed on the risk of injury to the low back from manual handling and does not attempt to score risk to the upper limb. The conceptual model underlying it assumes that risk factors, except load and frequency of handling, do not interact. The approach is to assess when the risk from an individual factor is increased and to indicate severity with a colour code and a numerical score. No attempt is made to assign meaning to the total score despite the differential weighting of the scores assigned to the factors, which should, in theory, allow levels of risk to be assigned to particular total scores and allow boundaries between risk zones to be determined. Since any real job is likely to contain several independent risks, this is a much sounder basis for risk assessment than any method that attempts to assess the overall risk of the task, especially if multiple body parts are involved.

The MAC, in common with the previously existing tools, has not yet been formally validated as predictors of risk of injury or sickness absence. HSL are currently running a project on behalf of HSE's Health Directorate to validate the 1991 NIOSH Lifting Equation in a prospective study collecting sickness absence data and relating it to the parameters of the job (Pinder, 2002). It will also be possible to use these data to test the ability of the MAC to predict work absence due to low back pain.

References

Health and Safety Commission 2000, *Securing Health Together: A long term occupational health strategy for England, Scotland and Wales*, (HSE Books, Sudbury, Suffolk), MISC 225

Hignett, S. and McAtamney, L. 2000, Rapid Entire Body Assessment (REBA), *Applied Ergonomics*, **31**, 201-205

Karhu, O., Kansi, P. and Kuorinka, I. 1977, Correcting working postures in industry: A practical method for analysis, *Applied Ergonomics*, **8**, 199-201

Li, G. 2002, *QEC Action Levels*, http://www.geocities.com/qecuk/QECActionlevels

Li, G. and Buckle, P. 1999a, Current techniques for assessing physical exposure to work-related musculoskeletal risks, with emphasis on posture-based methods, *Ergonomics*, **42**, 674-695

Li, G. and Buckle, P. 1999b, *Evaluating Change in Exposure to Risk for Musculoskeletal Disorders - a Practical Tool*, (HSE Books, Sudbury, Suffolk), CRR 251/1999

Li, G. and Buckle, P.W. 2000, Evaluating change in exposure to risk for musculoskeletal disorders - a practical tool. In *Proceedings of the IEA 2000/HFES 2000 Congress*, (The Human Factors and Ergonomics Society, Santa Monica, California), Volume 5, 407-408

McAtamney, L. and Corlett, E.N. 1993, RULA: A survey method for the investigation of work-related upper limb disorders., *Applied Ergonomics*, **24**, 91-99

Monnington, S.C., Pinder, A.D.J. and Quarrie, C. 2003, Development of Manual handling Assessment Charts (MAC) for Health and Safety Inspectors in the UK. In P.T. McCabe (ed), *Contemporary Ergonomics 2003*, (Taylor & Francis, London)

Pinder, A.D.J. 2002, Power considerations in estimating sample size for evaluation of the NIOSH lifting equation. In P.T. McCabe (ed), *Contemporary Ergonomics 2002*, (Taylor & Francis, London), 82-86

Tapley, S. 2003, Reliability of Manual handling Assessment Charts (MAC) developed for regulatory inspectors in the United Kingdom. In P.T. McCabe (ed), *Contemporary Ergonomics 2003*, (Taylor & Francis, London)

Vedder, J. 1998, Identifying postural hazards with a video-based occurrence sampling method, *International Journal of Industrial Ergonomics*, **22**, 373-380

Waters, T.R., Putz-Anderson, V., Garg, A. and Fine, L.J. 1993, Revised NIOSH equation for the design and evaluation of manual lifting tasks, *Ergonomics*, **36**, 749-776

Waters, T.R., Putz-Anderson, V. and Garg, A. 1994, *Applications Manual for the Revised NIOSH Lifting Equation*, (NIOSH, US DHHS, Cincinnati, Ohio), Report No, 94-110

DEVELOPING A MANUAL HANDLING PROCESS WITHIN AN NHS TRUST

Katrina Swain & Nigel Heaton

Human Applications
139 Ashby Road
Loughborough
Leicestershire
LE11 3AD

Manual handling injuries affect all areas of industry, with patient handling related injuries accounting for a large percentage. Compensation claims continue to increase as does enforcement action from the Health and Safety Executive (HSE). This paper presents an approach to managing manual handling issues within a National Health Service (NHS) Trust following enforcement action, identifying some of the pitfalls uncovered during the process. It will look at an organisation's response to Improvement Notices and how to integrate the practical requirements of running an NHS Trust with the legal requirements to undertake risk assessments, while examining the roles of individuals, the structures required to make the process effective and the lessons learned.

Introduction

Sixty percent of all manual handling injuries reported to the HSE (over 14,000) involve patient handling, with an estimated 3,600 nursing staff retiring each year, due to back injuries (NHS, 2002). Compensation claims for back injuries amongst nurses continue to rise, the latest 'landmark' case resulting in an award of £420,000, due to poor staffing levels and poor equipment provision (BBC, 2002). The HSE were alerted to an NHS Trust as a result of the number of reportable musculo-skeletal injuries. The Trust was served with two Improvement Notices, under the Health and Safety at Work Act 1974, section 2 and The Manual Handling Operations Regulations, 1992, for failing to undertake risk assessments and failing to train staff in the risks faced from patient handling.

External expertise was used to complement the internal structure to deal with manual handling issues. The priority was to ensure the Notices were lifted and to allow manual handling issues to continue to be appropriately managed within the hospital.

Initial information gathering and interviews identified no ownership of risks or accepted responsibility for manual handling risk management.

The time scales for the project were largely determined by the deadlines imposed by the HSE via the Improvement Notices. The first Notice, dealt with the need to conduct risk assessments, the second Notice, dealt with training.

Implementing a Manual Handling Process

Manual handling process listings were to be developed for each ward and department, processes were allocated a unique identifier to enable audit trail and cross referencing across departments. Each process was then broken down into its constituent parts providing a list of manual handling operations, again with unique identifiers. This listing then became the basis of the pre assessment prioritisation process or 'walkthrough' (Crowhurst, 1999).

The purpose of this decomposition and 'walkthrough' was to enable risk assessors to group assessments into meaningful components and to determine which operations require further assessment, by prioritising the risks.

A 'risk register' was produced for all of the risks assessed at walkthrough level. Initially the register contained a set of risks, ordered in terms of their priority (i.e. the highest levels of risk first).

On the basis of the walkthrough, assessors completed full risk assessments of the highest risks. These progressed to the point of identifying the controls necessary to reduce the risk as far as is reasonably practicable.

In some cases, walkthrough identified 'quick fixes' i.e. simple controls that would contribute to reducing the risk significantly. Where these were implemented, the walkthroughs were amended with the controls and documented accordingly, being deemed to be 'suitable and sufficient' (HSE, 1999).

Methodology

A 'top-down/bottom-up' methodology was proposed addressing policy and support at a macro level and specific operational methods at a micro level (Hignett, 2001). This involved everyone within the hospital from the Board to those working directly with patients. There were two goals:

- Where possible training would be external, accredited and validated
- A comprehensive management structure would be put in place, with local accountability, at department level and full support from the board down.

Senior management

The critical responsibilities were outlined to the board. The main teaching points from this session were to ensure that the board understood our approach to manual handling and was aware of the commitment required of the Trust if the project was to be a success. A similar session was ran for the Patient Service Managers (first line managers). At this briefing, explicit accountabilities and responsibilities were outlined, as were critical deadlines and an explanation of the effort required to comply with the Improvement Notices.

Training

Three basic training programmes were implemented (with variations).

First line managers

Two IOSH certificated training courses were delivered to the first line managers from all departments in practical issues in risk management. The main teaching points related to an explanation of modern risk management, the need to identify and prioritise health and safety risks and the roles and responsibilities of managers. The course required delegates to produce a concrete risk management plan.

Nominated risk assessors

Risk assessors were nominated for the wards and departments identified in the Improvement Notices, with remaining places allocated to other wards and departments. Risk assessors had to be experienced in their technical area and have knowledge of the requirements of the manual handling operations undertaken, knowledge of the individuals undertaking operations as well as knowledge of the loads and the working environment.

These courses comprised the bulk of the work on the project. Four types of IOSH certificated manual handling training courses were provided, tailored to the requirement of the Trust.

Risk assessors were trained to undertake either load or patient handling assessments and to achieve the requirements of the IOSH attainment level, demonstrating theoretical understanding by undertaking a written test and practical application by submitting a completed risk assessment for marking following the course. The Trust acquired 34 assessors for people handling operations and 26 for load.

Key trainers

Risk assessors who had successfully completed the above training were eligible to train as manual handling trainers (exclusively, in either a patient or load environment). These courses delivered trainers who could train others in safe handling (provided such a solution was identified on the appropriate risk assessment). The Trust acquired 26 trainers for people environments and 10 for loads.

The trainers would be responsible for delivering department or ward level training. This was training in response to specific risk assessments. It was to be delivered by trained trainers at a local level (provided the training required was within the trainer's competency). It was anticipated that significant training could be delivered on an ad-hoc basis in response to local conditions, but that all workers who require local training will receive and sign for it. Induction training would be the responsibility of the moving and handling specialist once in post.

Equipment-specific or special handling training would be outside of the competency of the trainers. Such training will be provided either by the equipment supplier or by the moving and handling specialist once in post.

Support

Working within the constraints of the project (imposed by the need to meet the terms of the Improvement Notice), a series of support days were provided to ensure that the process was followed. The support days were also intended to ensure the quality of the output of the trained risk assessors and subsequent risk assessments.

Results

The Board and a significant number of Patient Service Managers attended the briefings. It was agreed that meeting the terms of the Improvement Notice was a critical success factor for the hospital and that resources would be made available as and when required.

There was insufficient time and resource to meet the original deadline imposed by the HSE. An extension of one month was applied for and granted. The risk assessment process was complete to the satisfaction of the HSE by the end of this period.

The assessors were very busy undertaking risk assessments. This left limited time to prepare their training materials for their train the trainer module. However, the training (of trainers) programme was completed to the satisfaction of the HSE within the time limit.

Support was also provided to local risk assessors to help them to complete assessments. The Trust was required to trade-off support against training additional local resources.

The mechanics of the manual handling process (following a documented procedure and keeping records) was adopted, allowing all areas to be monitored and assessed. From a quality control point of view, being able to view directly comparable procedures between areas saved considerable time and effort.

Many of the assessors and trainers dealt with a steep learning curve and effort well, becoming fluent with their subject matter in a short period of time.

A formal manual handling policy was produced that was Trust specific with roles and responsibilities clearly identified.

Discussion

The NHS Trust had a significant number of trained risk assessors and trainers, not only enabling them to meet the requirements of the Improvement Notice, but also to continue the risk management process into the future.

The project produced trained risk assessors and trainers who took ownership of the risks in their areas and took proactive steps to identify and assess risk. The enthusiasm and motivation of the risk assessors was the most rewarding part of the whole project. They appeared to find the process empowering and felt it delivered real ownership of both problems and solutions to the ward or department level.

Local line managers now understood that manual handling was not something that Occupational Health 'do' and understood that risk assessment is a local responsibility.

From an organisational perspective, there were many lessons to be learned. The pressure on assessors was immense due to the Trust's initial refusal to request an extension from the HSE or indeed discuss the process with them. They wanted to be seen as a 'can do organisation'. Unfortunately, this meant that time scales were set by people who had no understanding of the process. Assessors worked in their own time and occasionally at home. Many perceived a lack of management support and understanding of the work they were undertaking. Indeed, in some cases it is reasonable to say that assessors succeeded in spite of rather than because of management.

Assessors were also hindered in some areas, in the early stages by the 'help' or advice of managers who did not fully understand the process or particularly the detail required. Another hindrance and frustration to assessors and managers was that they felt the goalposts were moving and that the dates set bore no relationship to the time pressures they were under. The Trust's decision to reduce the amount of external expert support to assessors in favour of more training also served to make the assessors a little isolated and 'thrown in at the deep end'.

One Year On

The project was revisited 12 months after the initial project. The Trust has undergone a major culture change with regard to the management of manual handling risk. The assessors and trainers continue in their initial work retaining ownership of risk and their own health and safety, with management supporting the process.

A manual handling specialist is now in post in an advisory and supportive role, being called upon to assist with specific problems arising from risk assessment. The major effect on the nursing staff particularly is that they feel supported enough to refuse to put themselves at risk by manually 'lifting' patients.

The organisation has undergone a steep learning curve with the work undertaken at the micro level by the assessors and trainers having an influence on macro level strategies.

Conclusion

The Health and Safety at Work act (1974) and subsequent European legislation introduced in 1993 was aimed at moving employers towards a more proactive Heath and Safety methodology in the management of risk. The Trust reacted to the enforcement action undertaken by the HSE, and through their approach highlighted many organisational management issues.

A culture change occurred very quickly in spite of the 'top down' pressure from Trust management, with an empowered group of employees who have facilitated change and safety culture within their own area.

References

Crowhurst, J., Catterall, B & Smyth, G. (2000) Managing a Manual Handling Risk Assessment Process, Contemporary Ergonomics, Taylor and Francis, London

Hignett, S. (2001) Embedding ergonomics in hospital culture: top-down and bottom-up strategies, Applied Ergonomics, vol 32, no 1, p61-69

HSE (1992) Manual Handling Operations Regulations, HMSO, London

Health and Safety at Work etc. Act 1974

NHS (2002) www.nhsplus.nhs.uk

MANUAL HANDLING IN REFUSE COLLECTION: REDUCING RISK OF MUSCULOSKELETAL DISORDERS

Andrew DJ Pinder and Edmund Milnes

Health and Safety Laboratory
Broad Lane,
Sheffield, S3 7HQ, UK

A field study of urban refuse collection was carried out to provide recommendations on reducing the musculoskeletal risks. Bag weights and collectors' heart rates were recorded. Postures and activities were filmed during refuse collection. Typical bag weights were 2 to 6 kg; less than 10% weighed over 12 kg. Stressful postures (stooping and twisting) occurred when picking up bags but durations were short. Handling heavier bags could create musculoskeletal problems for some workers, however, cardiovascular demands did not appear to be excessive. Recommendations include increased use of wheeled bins; lowering the 'rave' bar during bag rounds to reduce lifting / throwing height; team-handling of 1100 litre bins; avoidance of kerbs, steps and restricted spaces; increasing bins' wheel diameter; and reducing bin weights.

Introduction

Refuse collectors are exposed to a wide range of hazards which include the risks of musculoskeletal disorders (MSDs) that result from their manual handling activities. Poulsen *et al.* (1995) reviewed European literature to identify the range of occupational health problems that occur in refuse collectors and found that the risk of refuse collectors developing MSDs was 1.9 times higher than the rest of the working population. Within the UK there is current concern to reduce the incidence of work-related MSDs. There is also a general move towards separate collection of recyclable waste. Both of these factors could influence the working methods of refuse collectors.

Physiological factors

Refuse collection usually involves handling of bags or other loose items, and/or the handling of refuse bins, both wheeled and non-wheeled. Several studies have looked at heart rates and oxygen consumption as a means of comparing the efficiency and workload associated with these work methods. They have indicated that metabolic loads can be particularly high when transporting wheeled bins and bags up and down stairs (Friberg and Isaksson, 1974). Kemper *et al* (1990) found that the action of throwing bags into trucks consists of a far more efficient movement than lifting and emptying 6 kg non-wheeled bins into trucks. However, Frings-Dresen *et al.* (1995) reported that metabolic demands were higher during carrying and throwing bags of refuse (mean weight of 7 kg) than when handling wheelie bins. They concluded that collection of bags should increasingly be replaced by collection in wheeled containers. The actual weights have significant effects on the difficulties of handling bins and bags, and the demands are greatest when transporting loads up and down stairs and slopes. Overall, metabolic demands are greatest for non-wheeled bins, followed by bags, followed by wheeled bins.

Biomechanical Factors

Jäger *et al.* (1984) examined lower-back muscle activity during handling of various sizes of wheeled bins and found that high lumbosacral torques occurred when pulling large (1100 litre) wheeled Eurobins up kerbs. de Looze *et al.* (1995) used a biomechanical model to estimate spinal compression during a range of refuse handling tasks and found that picking up bags, throwing bags and lifting Eurobins up kerbs (even when empty) led to the highest compressions and exceeded the NIOSH low-back compression Action Limit of 3400 N. de Looze *et al.* suggested that bag collection should be discouraged due to its high task frequency, whereas handling Eurobins remained feasible due to its relatively low frequency, provided that steps are taken to reduce the spinal compression with team handling and transport across dropped kerbs. A study by Schibye *et al.* (2001) compared mechanical load on the low back and shoulders during pushing and pulling 240 litre (domestic size) wheelie bins containing loads of 25 and 50 kg. Schibye *et al* found that for these sizes of bins the overall load at L4/L5 was generally small and not affected by the differences in bin weight.

Therefore, bag handling and lifting large (1100 litre) bins up kerbs represent manual handling hazards and present refuse collectors with risks of musculoskeletal injury. Lifting and tipping non-wheeled bins into trucks is also identified as a hazard.

Psychosocial and Work Design Factors

Lund *et al.* (2001) identified several work-environment factors causing refuse collectors to leave / have to give up their jobs. These were self-rated 'extreme bending of the back', low skill discretion and low decision authority.

Ridd (2002) carried out a field study of refuse collection in two UK boroughs at about the time the study reported here was undertaken. The study reviewed a range of collection activities and identified issues which relate to the manual handling risk levels. The study also provided a good qualitative assessment of the manual handling risks and the psychosocial factors that affect the job.

Aim of this study

This study was part of an HSE project on the risks associated with refuse collection. Having identified a range of manual handling hazards from the literature, a short field study was carried out to examine the typical hazards from manual handling in urban refuse collection in the UK. The purposes were to make quantitative assessments of some of the risk factors and to make relevant recommendations on manual handling risk control strategies and to report additional findings on other risk factors and to make recommendations for their control.

Field Study Design

A Local Authority was identified which was willing to co-operate with this study and had a variety of refuse collection rounds. These included weekly collection of domestic refuse bags (including occasional non-wheeled bins); alternating weekly collection from domestic grey wheelie bins and green (recycling) wheelie bins; and trade-rounds which collected, up to daily, a mixture of trade bags, loose items and a range of sizes of wheeled bins up to the large 1100 litre 'Eurobins'.

Both the terrain, which varied from flat to steep hills, and the architecture of the properties visited affected the nature of the refuse collection task. In some locations bags and bins were left at the front of the property but in others the bags and bins were left in side or rear alleyways and therefore had to be carried / pulled some distance to the truck.

Table 1 provides basic details of the rounds that were observed. Video was taken of refuse collectors throughout the observation period in order to record postures and frequency of activities. Filming largely concentrated on activities at or near the compactor at the rear of the collection vehicle being followed, resulting in collectors often moving in and out of view as they approached and moved away from the vehicle. The video was later analysed in the laboratory using The Observer Video Pro software.

Table 1. Details of refuse collection rounds that were observed

	Round A	Round B	Round C
Type of round	Domestic bags (plus some loose items and non-wheeled round bins)	240 litre green wheelie bins for recyclable material	Trade bins (240, 360 and 1100 litres) plus green trade bags and loose items
Size of crew	5	4	2
Round start time	06:00	06:00	12:30
Round duration	8.5 - 9 hours	6.5 - 8 hours	6.5 - 7.5 hours
Start & finish of our observations	06:00 - 09:00 8 Dec	09:30 - 11:30 8 Dec	15:30-18:00 7 Dec
Weather	Dry, still, cold	Dry, still, cold	Heavy rain, wind, cold
Heart rate monitoring	on 2 collectors	on 2 collectors	Collector and 'driver-collector'

Due to some practical constraints during filming (filming from vehicle, dark conditions) the sections of video that was analysed varied in length between approximately 2 and 25 minutes. Two sections of video were analysed for Round A, one for Round B and two for Round C and the actions and postures of all the refuse collectors visible on the video were coded.

The working postures and their durations were categorised and logged according to the OVAKO Working posture Analysis System (OWAS) (Karhu *et al.*, 1981). The nature of the loads and the ways they were being handled were also coded. Heart rates were also monitored during the rounds using Polar Vantage NV™ Heart Rate Monitors. Bag and bin weights were recorded in part of Round A using a Mecmesin Force Dynamometer.

Results

Bag Weights
115 refuse bags were weighed in a housing estate during Round A. Mean weight was 6.2 kg (SD ±3.9 kg). 60% of the bags were between 2 and 6 kg. 33% of the bags were between 6 and 12 kg. Only two bags weighed over 20 kg.

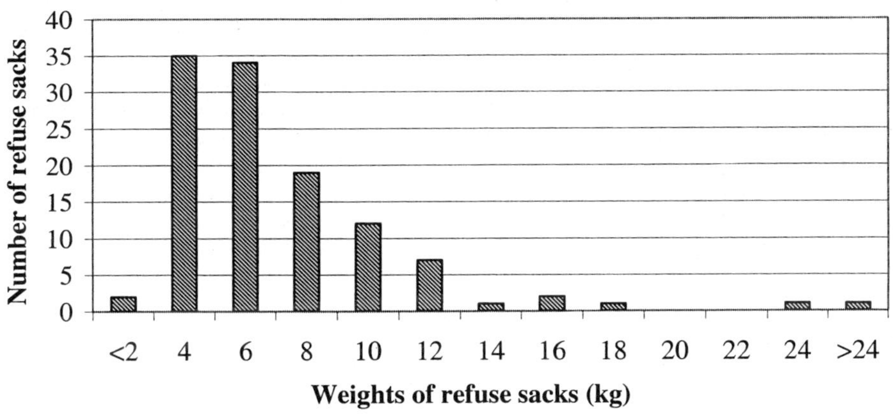

Figure 1. Weights of 115 domestic refuse sacks measured on Round A

Heart Rate Data
The heart rates were recorded from two collectors on each round every 5 seconds for total periods varying between 2 hours 32 minutes and 6 hours 30 minutes. The data indicate that the overall working heart rates were moderate, and not excessive on any of the rounds, although heart rates were slightly higher for bag rounds than bin rounds.

Table 2. Heart Rate Data Summary

Round	Overall mean heart rate *(& individual values)*	Overall mean working pulse *(& individual values)*
A (bags)	106.5 *(109, 104)*	20.5 *(21, 20)*
B (green wheelie bins)	95 *(94, 96)*	20
C (trade bins and sacks)	101 *(101, 101)*	16.5 *(14, 19)*

Posture Analysis
Refuse collectors were observed to spend between 62 and 90% of the time walking and between 10 and 36% of the time standing still on both legs. Overall, they spent between 10 and 20% of the time when actually handling refuse / bins in bent and / or twisted trunk postures. These trunk bent / twisted trunk percentages varied up to 9 % within different periods of Round A (bags) and Round C (trade). On particular rounds the mean durations of bent trunk postures ranged between 1.7 and 3 s. Mean times in twisted trunk postures ranged between 0.7 and 4.7 s. Mean times in combined twisted and bent trunk postures ranged between 0.5 and 3.9 s.

On Round A (bags) the refuse collectors were observed with one or both hands above shoulder height for between 6 and 12% of the time. On Round B (green wheelie bins) this percentage was 1% and on Round C it was between 5 and 7%. The mean durations of these hand-above-shoulder postures were typically less than 1 second and were part of dynamic throwing actions rather than controlled lifts involving quasi-static arm postures.

Additional Observations

During Round A the collectors often carried and threw two or more bags with each hand, and performed, on average, 7.1 throws per minute. They tended to hold the bags away from the body, which made walking easier (knees not hitting the bags), and reduced the risk of injury from sharp objects. However, this meant the weight of the bags was taken solely through the shoulders and arms, rather than involving some transmission of the load to the trunk. Also, the 'rave' bars inside the trucks that prevent wheelie bins falling into them during automatic tipping were around 1.5 m high. This meant that the bags and the non-wheeled round bins (110 litre, mean loaded weight 16.2 kg) that occasionally had to be lifted and emptied into the truck had to be lifted above shoulder height to be emptied. Inadequate hand holds on the bins made this task even more difficult.

On Round B (green wheelie bins) the collectors were observed to spend more time standing and waiting compared to Round A. This was due mainly to waiting for the wheelie bin lift / tip mechanism to finish operating. The operatives reported that the volumes of refuse in the green bins were less than during a grey bin collection.

A wide range of items was handled on the trade round (Round C), including bags, rolls of carpet, 240 litre and 360 litre two-wheel bins and 1100 litre four-wheel Eurobins. Collectors acknowledged that best practice was for pairs of them to move these large bins and, if possible, manoeuvre them across dropped kerbs. Considerable effort was needed to push / pull full large bins and some awkward / asymmetric trunk / hand postures were adopted. Poor access to bins was a particular problem on Round C; narrow, poorly surfaced alleys, steep slopes, obstructions and narrow doorways all increased the risks.

Discussion and recommendations

The manual handling hazards varied depending on the type of collection round. Bag rounds present risks associated with stooping and twisting postures, asymmetric lifting with often poor grips. The weights of the heaviest bags and the associated handling postures / actions were considered hazardous to at least some of the male working population due to the height to which they had to be lifted. The height of the rave bar was an important factor determining lift / throw height. Lifting round bins represents a significant manual handling risk due to excessive weight and being lifted above shoulder height. Wheeled bins appeared to present some manual handling issues when they were being pulled up / down stairs and kerbs. Obstructions and poor access were seen to contribute to manual handling risks by causing collectors to adopt awkward postures whilst applying forces.

Where possible, appropriate sized wheelie bins should replace bags and round (110 litre) bins for domestic premises (although it is recognised that where wheelie bins are impractical, bag collection will continue unless bulk collection points can be implemented). Round bins should not be lifted and tipped manually; residents should be encouraged to put only bagged refuse in them. The bags could then be taken out individually. The rave bar should be lowered as far as possible during bag rounds in order to reduce lifting height / height bags are thrown to.

The findings of this study on 1100 litre Eurobins are consistent with previous studies; team-handling should be used, especially when lifting them up kerbs and they should be maneouvered across dropped kerbs wherever possible. When moving any size bin the collectors should ensure space for movement (moving obstacles, latching doors open etc.), avoid obstacles like kerbs and steps and seek assistance from colleagues with heavier bins on stairs and slopes.

Manufacturers should be encouraged to increase wheel diameter to help reduce lifting forces when transporting bins up steps / kerbs. They should also consider using more lightweight materials. Refuse collectors should be involved in a scheme to identify locations where improvements such as resurfacing of alleys, installing dropped kerbs and ramps would be beneficial. Residents and trade proprietors should be involved in schemes to help collectors, e.g., considering where they leave their refuse for collection to reduce some of the access hazards that have been identified. Finally, collectors should be provided with Personal Protective Equipment that will effectively protect them from the risk of sharp objects in bags of refuse to enable them to grip, hold and carry the bags in an optimal way, i.e., bags not held so far away from the legs.

Bibliography

de Looze, M.P., Stassen, A.R., Markslag, A.M., Borst, M.J., Wooning, M.M. and Toussaint, H.M. 1995, Mechanical loading on the low back in three methods of refuse collecting. *Ergonomics*, **38**, 1993-2006

Frings-Dresen, M.H.W., Kemper, H.C.G., Stassen, A.R.A., Crolla, I.A.F.M. and Markslag, A.M.T. 1995, The daily work load of refuse collectors working with three different collecting methods: a field study. *Ergonomics*, **38**, 2045-2055

Jäger, M., Jordan, C., Luttmann, A. and Laurig, W. 2000, Evaluation and assessment of lumbar load during total shifts for occupational manual materials handling jobs within the Dortmund Lumbar Load Study - DOLLY. *International Journal of Industrial Ergonomics*, **25**, 553-571

Karhu, O., Harkonen, R., Sorvali, P. and Vepsalainen, P. 1981, Observing working posture in industry: Examples of OWAS application. *Applied Ergonomics*, **12**, 13-17

Kemper, H.C.G., Van Aalst, R., Leegwater, A., Maas, S. and Knibbe, J.J. 1990, The physical and physiological workload of refuse collectors. *Ergonomics*, **33**, 1471-1486

Poulsen, O.M., Breum, N.O., Ebbehoj, N., Hansen, A.M., Ivens, U.I., van Lelieveld, D., Malmros, P., Matthiasen, L., Nielsen, B.H. and Nielsen, E.M. 1995, Collection of domestic waste. Review of occupational health problems and their possible causes. *The Science of the Total Environment*, **170**, 1 19

Ridd, J. 2002, Ergonomics Appendix to SITA / GMB Back In Work Project: *'Don't Bin Your Back'*. SITA, Arundel House, 6 Portland Square, Bristol, BS2 8RR

Schibye, B., Sogaard, K., Martinsen, D. and Klausen, K. 2001, Mechanical load on the low back and shoulders during pushing and pulling of two-wheeled waste containers compared with lifting and carrying of bags and bins. *Clinical Biomechanics*, **16**, 549-559

MUSCULOSKELETAL DISORDERS

CAN ERGONOMIC MICROSCOPES REDUCE THE RISKS OF WORK-RELATED MUSCULO-SKELETAL DISORDERS?

Heather G. Gray [1] & Fiona MacMillan [2]

[1] *School of Health & Social Care, Glasgow Caledonian University, Glasgow G4 OBA*
[2] *Queen Margaret University College, Leith Campus, Edinburgh EH6 8HF*

An experimental field study was carried out to investigate the differences in upper trapezius muscle activity, overall posture, pain and perceived exertion in cytologists while scanning cervical screening slides using both standard and ergonomic microscopes. Thirteen cytologists screened eight slides with each microscope while measurements were taken. Results indicated that the ergonomic microscope was preferable to the standard microscope in terms of upper trapezius muscle activity, shoulder pain and perceived exertion. These findings have important implications for cytologists in terms of reducing potential risk factors for the development of work-related musculo-skeletal disorders.

Introduction

In 1988 the National Health Service Cervical Screening Programme (NHSCSP) was established, with the remit to reduce the incidence and mortality of cervical cancer in the UK. Nine years later, the NHSCSP published minimum ergonomic working standards for centres undertaking cervical screening work, with specific recommendations as to ergonomic microscope features (NHSCSP, 1997).

Research has revealed that microscopy work places a high level of musculo-skeletal stress (MSS) on cytologists, with symptoms increasing in prevalence and intensity over time. The main body areas affected are the neck, shoulders, back, and wrists (Rizzo *et al,* 1992, Hopper *et al,* 1997).

Several researchers have used electromyography (EMG) to measure upper trapezius (UT) muscle activity during microscopy work (Lee & Humphreys, 1985, Kreczy *et al,* 1999). Lee and Humphreys (1985) and Kreczy and colleagues (1999) all found that when their microscopists used the ergonomically designed workstations, which had angle adjustable microscope eyepieces, forearm supports and height adjustable chairs, there were significant reductions in UT activity. However, the results from these studies were drawn from small samples (3-12 subjects), with Lee's experiments performed on healthy students in a laboratory setting.

Although the use of ergonomic microscopes has been recommended by the NHSCSP (1997), to date, no research has examined their effects on cytologists in a field setting. Therefore this study aimed to examine UT muscle activity, overall posture, pain and perceived exertion in cytologists scanning cervical screening slides while operating both a standard microscope (SM), with a fixed eyepiece, and an ergonomic microscope (EM).

Subjects & Methods

Having obtained ethical approval for the study, 13 cytologists (36.1 ± 8.8 years) were recruited from one large teaching hospital in Scotland. Subjects who volunteered to participate had to fulfil specific inclusion criteria, they must have: worked with microscopes in cytology for 3 or more years; been working with microscopes for at least 10 hours per week; been able to primary screen at least 32 cervical slides per day; have completed the NHSCSP Certificate in Cytology or equivalent; have had no recent upper limb or neck fractures or injuries. The study was conducted within the Hospital's Cytology Department.

The study was a cross-sectional, two-period crossover experimental design, during which all subjects carried out cervical screening using both the SM (Leitz SM-Lux) and EM (Nikon Eclipse E400). The order of microscope presentation was randomised and counterbalanced, and there was a 20-30 minute break between them.

Throughout the study, subjects were instructed to set up their workstation and screen slides in their usual manner, using the equipment provided. The cytologists screened 8 slides with each microscope, approximately an hour per microscope; with EMG recordings taken throughout the screening of every second slide (see Figure).

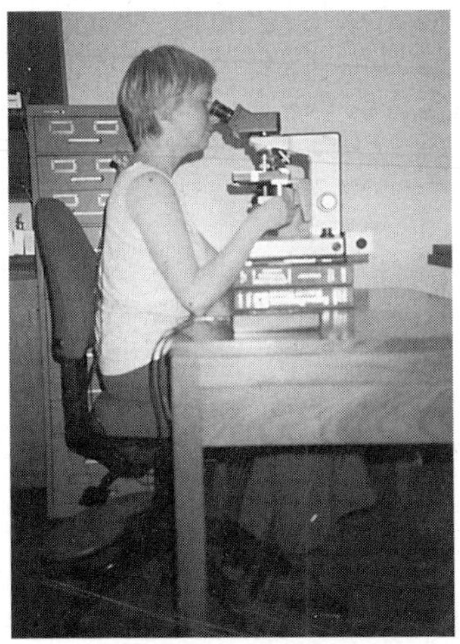

Figure: Standard and Ergonomic Microscope Workstation Set-up

Muscle activity was quantified by bilateral recordings of the UT muscles using 2 sets of integral, self-adhesive, pre-gelled, bipolar Ag/AgCl surface electrodes* (3.3cm diameter), which were connected to a pre amplifier with a gain of 375. In turn this was connected to the ME3000P8† portable EMG sampling unit resulting in a total gain of 412, a high pass filter set at 8 Hz, and a low pass filter set at 500Hz. The amplifiers had a common mode rejection ratio of 110dB. The signal was then sampled at 1000Hz, by a 12-bit analogue to digital converter.

---

* Blue Sensor, Type M-00-S, Medicotest Inc., Illinois

† ME3000P8, Mega Electronics Ltd., Kuopio, Finland

Prior to placing the EMG electrodes the skin was shaved, rubbed with an abrasive scrub, and cleaned with an ethanol wipe. Electrodes were placed symmetrically either side of the point 2 centimetres (cm) lateral to the midpoint between the 7th cervical vertebra and the most lateral point of the acromion process. The electrodes were placed on either side of this lateral point in the direction of the muscle fibres with a 3.5cm distance between the 2 electrode centres.

The EMG normalisation procedure consisted of 3 maximal voluntary contractions (MVC), with the subject sitting supported upright on a chair, and the shoulder abducted to 90° in the scapular plane. The elbow was flexed to 90° with the palm of the hand facing downwards. Three MVCs were performed on each arm, and the MVC that produced the maximal reading, measured by a hand-held dynamometer (Nicholas MMT, Model 01160, Lafayette Instrument, Lafayette Inc.), was used as the basis for the relative EMG value.

The raw EMG data for each slide was rectified, using the Multi Signal ME3000P version 3.0 software, the average EMG activity for the time period calculated and recorded in micro volts per second (μVs), and then expressed as a percentage of the MVC (%MVC).

A computer generated 10cm horizontal visual analogue scale (VAS) was used to measure pain intensity, in conjunction with a numbered Body Part Discomfort (BPD) diagram for subjects to indicate pain location. Prior to slide screening commencing, subjects were shown the BPD diagram, and asked if they had any pain. If they identified any area(s), they were then instructed to place a vertical line on the VAS at a point that they felt represented their pain intensity *at that moment*, for each body area. This process was repeated immediately after every forth slide had been screened, with subjects permitted to view their previous scores. The number of millimetres measured on the VAS provided the pain score.

A Rapid Upper Limb Assessment tool (RULA) was used to obtain an overall estimate of each subject's posture (McAtamney & Corlett 1993). In order to measure their posture, subjects were filmed with a video camcorder as they screened every second and eighth slide.

Borg's Category Scale, CR-10 (Borg 1998), was used to measure subjects' ratings of perceived exertion (RPE). The CR-10 scale was presented to them once at the end of the screening period for *each* microscope.

One-tailed two-sample *t*-tests were carried out on normally distributed data, and if data did not follow a normal distribution, non-parametric Mann Whitney U-tests were used, with the significance level (α) set at 0.05.

Results

The SM resulted in greater UT muscle activity than the EM in both shoulders for all subjects except one. The mean %MVC reduction from the SM to EM was 5.3 ± 4.4%MVC for the right shoulder, and 3.5 ± 2.0%MVC for the left, representing statistically significant reductions of 64.4% and 61.8% respectively (Table 1). Interestingly, the mean time to screen a slide with the EM (271 ± 91.5 seconds (s), range 60-592s) was 39.1s less than with the SM (310 ± 145.2s, range 62-816s).

Table 1: Upper Trapezius EMG Results

	Mean %MVC (SD)	Range %MVC	*t*-stat (*df*)	*p*-value
Right UT				
Standard Microscope	8.3 (6.0)	0.4–25.0	2.5 (11)	0.015
Ergonomic Microscope	3.0 (3.7)	0.3–15.9		
Left UT				
Standard Microscope	5.7 (4.1)	0.5–18.2	6.1 (11)	0.000
Ergonomic Microscope	2.2 (3.1)	0.2–14.4		

The area that had the highest mean and median VAS scores for the SM was the back of the neck, followed by the shoulder area (the sections on the BPD diagram representing UT), upper back, and lower back. When operating the EM the upper back was rated most painful, followed by the shoulder area, neck then lower back (Table 2). VAS scores increased over time when operating both microscopes.

Table 2: Visual Analogue Pain Scale Results

Body Area	Mean VAS Scores SM (SD)	Mean VAS Scores EM (SD)	U-value	*p*-value
Neck	27.9 (10.8)	7.5 (3.9)	9.5	0.084
Shoulders	23.4 (13.1)	11.8 (1.0)	7.5	0.043 *
Upper Back	20.6 (16.6)	6.6 (2.5)	10.0	0.098
Lower Back	13.7 (6.4)	4.5 (1.0)	18.0	0.625

* $p<0.05$ SM = standard microscope EM = ergonomic microscope

Non-parametric Mann-Whitney U-Tests were used to analyse VAS scores. The results revealed that, although there was a trend for the EM to result in less pain than the SM, there were no *statistically* significant differences between the microscopes for the neck, upper back, or lower back, but there was a statistically significant difference between microscopes for shoulder pain ($p = 0.043$).

Both microscopes yielded median RULA scores of 4, with scores ranging from 4-5 for the SM, and all EM scores were 4 (U=15.0, p=0.173). A RULA score of 4 is Action Level 2 ("investigate further"), and a score of 5 is Action Level 3 ("investigate further and change soon"). Although the median RULA scores imply that there were no postural differences between microscopes, there were actually distinct differences in joint angles (see Figure). The EM was consistent in keeping shoulder flexion below 20°, and elbow flexion between 60-100°. Whereas, the SM required all subjects to maintain >20° flexion for their left shoulder, and 23% to flex their right shoulder >20°. Furthermore, the SM required >100° right and left elbow flexion for 54% and 23% of cytologists respectively. No one operating the EM worked with wrist ulnar deviation (UD), but the SM resulted in right UD in 23% of subjects and left UD in 54% of subjects.

Finally, differences in subjects' ratings of perceived exertion (RPE) also were observed between microscopes. The mean RPE scores were fairly low, 2.4 ± 2.1 (range 0.5-8) and 0.9 ± 0.8 (range 0-2.5) for the SM and EM respectively. The mean RPE score was significantly lower when operating the EM, $t(11)=2.77$, $p=0.009$.

Discussion

The results demonstrate that the EM resulted in significantly less UT muscle activity for both shoulders. This is possibly because SM operation necessitated the following features that can result in increased UT muscle activity: unsupported forearms; a greater frequency of repetitive finger/ wrist movements; and shoulder flexion greater than 20°.

Mean %MVCs for the right UT were greater than the left for both microscopes. This is most probably because during screening, the fingers and wrist of the right hand have to move much more frequently than those of the left, resulting in increased UT activity for proximal stabilisation.

In agreement with previously mentioned microscopy researchers, being able to adjust the microscope in height and angle reduced UT EMG, as did reducing the time spent on the microscope, i.e. due to the shorter EM screening times. The fact that slides took approximately 40 seconds less time to screen when operating the EM, was probably due to it having a larger field of view, meaning subjects did not require to move the microscope stage as frequently to scan slides.

Since none of the microscopy studies to date normalised their EMG data, the results can only be compared, tentatively, to other sedentary occupations. It seems that the SM is most comparable to assembly work, with %MVCs ranging from approximately 5.5-8%MVC (e.g. Feng *et al,* 1999), and the EM to VDU work, with %MVCs ranging from 1-5%MVC (e.g. Fernstrom *et al,* 1994).

The body areas in which discomfort were reported were identical to those mentioned by other researchers and the prevalence of symptoms was most analogous to those identified by Hopper and associates (1997) who also examined UK cytologists. In this study, the mean VAS scores were fairly low, and comparable to those described by Aaras (1994), where VAS scores were 34 prior to, and 22 after, ergonomic intervention in VDU operators.

When screening slides, there was a trend for the SM to result in greater subject discomfort than the EM, however, only the shoulder area, displayed a *statistically* significant reduction in pain. It is theorised that the higher levels of discomfort when operating the SM were due to the increased, sustained UT muscle activity. Sustained contractions, and increased intramuscular pressure, result in insufficient peripheral circulation in the UT muscle, which can be a predisposing factor in the development of neck-shoulder pain (Larsson *et al,* 1995). It is postulated that this ischaemia was the primary cause of pain experienced by the cytologists. Additionally, unfavourable psychosocial factors have also been shown to result in increased perception of pain. Factors pertinent to cytology work may include: anxiety about making mistakes; time pressures and a perceived high workload; and perception of strenuous work.

According to RULA, the type of microscope did not significantly affect the cytologists' postures (p=0.173), however, on examination of the individual joint angles it seems that RULA was not sensitive enough to reveal postural differences. In this study the majority of subjects worked at Action Level 2, similar to Lomas (1998), who identified that 82% of her 11 observed cytologists were also rated at this Level.

Probably the most noticeable difference observed between the two microscopes under study was the amount of shoulder flexion required. The difference between the right and left sides was due to the asymmetrical positioning of the microscope controls. Shoulder flexion ranges during microscopy are comparable to VDU work (Vasseljen & Westgaard, 1997).

Overall, the cytologists reported low levels of physical strain when operating both microscopes, with the EM resulting in significantly less (p=0.0091) perceived exertion than the SM. Although, it should be noted that there was not always a direct correlation between UT muscle loading and RPE scores, a finding supported by Oberg and associates (1994). The mean RPE scores were similar to those found in light assembly workers, 2-5 (Mathiassen & Winkel, 1996).

Conclusions

Cytologists have been identified as an occupational group at risk of developing work-related musculo-skeletal disorders (WRMSD). However, research that has investigated interventions to reduce this risk is sparse. This study found that there were statistically significant reductions in UT muscle activity, of at least 60%, shoulder pain and perceived exertion when using the ergonomic microscope. These findings imply that cytologists would benefit considerably from using ergonomic microscopes, not only in terms of immediate reductions in muscle activity and pain, but also, long term, in the possible prevention of WRMSD and musculo-skeletal sick leave.

Aaras A. (1994) Relationship between trapezius load and the incidence of musculo-skeletal illness in the neck and shoulder, *International Journal of Industrial Ergonomics,* 14, 341-348

Borg G. 1998 *Borg's perceived exertion and pain scales*, (Human Kinetics, Leeds)

Feng Y, Grooten W, Wretenberg P. & Aborelius U.P. 1999 Effects of arm suspension in simulated assembly line work: muscular activity and posture angles, *Applied Ergonomics*, 30, 247-253

Fernstrom E, Ericson M. & Malker H. 1994 Electromyographic activity during typewriter and keyboard use, *Ergonomics*, 37, 3, 477-484

Hopper J.A, May J. & Gale A.G. 1997 Screening for cervical cancer: the role of ergonomics, in Robertson S.A. (ed.) *Contemporary Ergonomics 1997*, (Taylor & Francis, London), 38-43

Kreczy A, Kofler M. & Gschwendtner R A. 1999 Underestimated health hazard: proposal for an ergonomic microscope workstation, *The Lancet*, 354, 1701-1702

Larsson S, Cai H, Zhang Q, Larsson R. & Oberg P.A. 1995 Microcirculation in the upper trapezius muscle during sustained shoulder load in healthy women – an endurance study using percutaneous laser-Doppler flowmetry and surface electromyography, *Eur J Appl Physiol*, 70, 451-456

Lee K.S. & Humphreys L.A. 1985 Physical stress reduction of microscope operators, *Proceedings of the Human Factors Society*, 29th Annual Meeting, 789-793

Lomas S.M. 1998 An investigation into the postures adopted by the upper limbs during cervical screening, in Bridger R.S. & Charteris J. (ed) *Global Ergonomics 1998*, (Elsevier, Oxford), 353-356

Mathiassen S.E. & Winkel J. 1996 Physiological comparison of three interventions in light assembly work: reduced work pace, increased break allowance and shortened working days, *Int Arch Environ Health*, 68, 94-108

McAtamney L. & Corlett E.N. 1992 *Reducing the risks of work related upper limb disorders: a guide and methods*, Institute for Occupational Ergonomics, University of Nottingham, Nottingham, NG7 2RD

NHSCSP 1997 *Minimum ergonomics working standards for personnel engaged in the preparation on, scanning & reporting of cervical screening slides*, MDA Evaluation Report MDA/97/31, August 1997, (Medical Devices Agency, Norwich)

Oberg T, Sandsjo L. & Kadefors R. 1994 Subjective and objective evaluation of shoulder muscle fatigue, *Ergonomics*, 37, 8, 1323-1333

Vasseljen O. & Westgaard R.H. 1997 Arm and trunk posture during work in relation to shoulder and neck pain and trapezius activity, *Clinical Biomechanics*, 12, 1, 22-31

THE DEVELOPMENT OF ACTION LEVELS FOR THE 'QUICK EXPOSURE CHECK' (QEC) SYSTEM

Robert Brown[1] **and Guangyan Li**[2]

[1] *University of Sunderland, Sunderland, SR1 3SD UK*
[2] *Human Engineering Limited*
Shore House, 68 Westbury Hill, Westbury-on-Trym, Bristol, BS9 3AA UK

A preliminary study was carried out to identify an action level for the QEC system (Quick Exposure Check for the assessment of workplace risks for work-related musculoskeletal disorders). This was achieved by assessing a number of industrial tasks simultaneously using the QEC and RULA (Rapid Upper Limb Assessment) and comparing the assessment scores from both methods. The action levels of the QEC were then extracted from the corresponding RULA scores. The results suggested that a QEC score in the range of 40-49% (calculation method described in the text) could be described as equivalent to a RULA score of 3-4 (action: investigate further); similarly, a QEC score of 50-69% indicated 'investigate further and change soon' and a score of 70% or higher suggested 'investigate and change immediately'. A QEC score of less than 40% was regarded to be 'acceptable'.

Introduction

The term work-related musculoskeletal disorder (WMSD) refers to any disorder that involves the nerves, tendons, muscles and supporting structures of the body as a result of any work-related activity. There are numerous published studies which stress the 'risk factors' that are found to be associated with WMSDs. These are described as being either workplace/occupational factors, individual/personal factors (Winkel and Westgaard, 1992), or psychosocial factors (eg. Houtman et al., 1994; Theorell et al., 1991). Studies have shown that the WMSD risk associated with individual factors was small when compared with that associated with workplace exposures (Armstrong et al., 1993), and a clearer role of psychosocial factors in the development of WMSDs is still under debate (Bongers et al., 1993). A number of observational techniques have been developed since the 1970's to assess WMSD risks and these are reviewed in Li and Buckle (1999a).

The QEC system (Li and Buckle, 1999b) focuses mainly on the assessment of workplace risk factors that have been found to have a contributory role in the development of WMSDs, such as repetitive movements, force exertion, awkward postures and task duration. The tool combines the assessment of both the 'observer' and the 'worker' in its assessment outcome and indicates a risk exposure level (score) for the back, shoulder/arm, wrist/hand and the neck in relation to a particular task, and shows whether or not an ergonomic intervention proves to be effective (with decreased or increased exposure scores). However, the system is not yet able to suggest whether or not an action is needed corresponding to a particular exposure level.

The aim of the present study was to formulate action levels that may be used, in conjunction with the QEC system, to decide upon the necessity of an ergonomic intervention in the workplace.

Method

There can be different ways, in theory, to define an 'action level' or 'zone' against an exposure level assessed by such an observational tool, using epidemiological or physiological measures for example, but these may also be accompanied by issues as to what extent the measured physical workload exposures can represent a 'true' level of WMSD outcome, especially when several risk factors are presented in a combined manner.

The present study attempted to identify and use, amongst a number of tools available, an established tool with already pre-defined action limits on which to compare the scores obtained by the QEC system. Of the pen-paper based observational methods available, the RULA system (McAtamney and Corlett, 1993) was selected for this purpose. This was based on several considerations, for example, RULA is a posture based method designed for assessing the severity of postural loading; it considers the static or repetitive body movement and force exertion; it calculates assessment scores with action levels, and it is well established and widely used.

The first major hurdle to overcome was how to compare the scores obtained from QEC system, which covers individual upper body regions, with the RULA system, which uses the assessment of the upper body as a whole. This was overcome by converting the scores obtained by the QEC system for individual upper body areas into one that covered the upper body regions as a whole. The exposure level (E) is calculated as a percentage rate between the actual total exposure score X and the maximum possible total score X_{max}, using the formula: E (%)=X/X_{max} 100%

Where: X=Total score obtained for exposure to the (Back+Shoulder/arm+ Wrist/hand+Neck); X_{max}=Total maximum score for exposure to the (Back+Shoulder/arm +Wrist/hand+Neck). X_{max} is a constant value for a particular type of task, i.e., for manual handling, X_{maxMH}=176; for other tasks, X_{max}=162.

After the choice of which assessment tool to compare with the QEC system was made, it was necessary to decide upon the type and the number of tasks that were needed for this study. In total, 31 tasks were chosen from pre-recorded videotape. This was to provide an identical set of tasks that could be assessed by using both tools. The tasks were chosen to encompass as many and varied work activities as practicable, including repetitive tasks, manual handling, tasks with either static or dynamic natures, and tasks which were performed in either a standing or seated position. A list of the tasks used in this study is given in Appendix 1.

All of the tasks used in this study have received no previous ergonomic intervention, and have been filmed from a variety of angles to show the whole range of motions of the worker. This was to ensure that both systems were able to pick up the risk factors associated with the particular task in question.

In the task selection, one of the considerations taken into account was the use of any protective clothing that may have been worn by the workers such as gloves. Many of the gloves that were worn in the workplace were a general-purpose type, used by both sexes and were generally a 'one size fits all' arrangement. This meant that for many of the workers, the gloves used tended to be on the large side and often made it difficult to correctly assess the positions of the wrist. Work tasks selected with the use of such gloves were kept to a minimum, and these tasks selected were recorded from various angles and were scrutinised closely to ascertain the correct positioning of the wrists during the performance of the task.

To reduce the possible learning effects of assessing the tasks by both tools, the QEC and the RULA were used in alternate sequence and the tasks were also assessed randomly. All assessments were carried out by one ergonomist over a period of 2 days. Prior to the formal assessments, the ergonomist performed a number of self-training sessions with both methods (using pilot assessment exercises on different tasks), until he felt that he had obtained sufficient knowledge/experience in both methods and could use them almost equally well when assessing a particular task.

Results

The results obtained from the assessments of the 31 tasks are shown in Table 1.

Table 1. The results of the comparison between the QEC and RULA

Task ID	**RULA Score**	**QEC Score**				
		Percentage Total (E)	Mean	SD	Range	X/X_{max}
13A*	4	45%	46.0%	1.0	45-47%	80/176
12A*		47%				82/176
3A*	5	55%	57.0%	1.1	55-58%	96/176
1A*		57%				100/176
12C		57%				92/162
8A*		58%				102/176
9A*		58%				102/176
10A*	6	61%	65.7%	2.4	61-69%	108/176
10C		63%				102/162
13C		63%				102/162
6A*		64%				112/176
7A*		64%				112/176
11A*		65%				114/176
14A*		65%				114/176
9C		65%				106/162
5A*		67%				118/176
8C		67%				108/162
15C		68%				110/162
17C		68%				110/162
18C		68%				110/162
4C		69%				112/162
5C		69%				112/162
11C	7	70%	82.3%	8.2	70-93%	114/162
6C		73%				118/162
3C*		76%				134/176
23A*		77%				136/176
16C*		83%				146/176
14C*		86%				152/176
1C*		90%				158/176
2C*		93%				164/176
7C		93%				150/162

Note: * - Manual handling tasks.

The QEC assessment data is shown in the column entitled "Percentage Total" and is presented using the equation: E (%)=X/X_{max}100% as described earlier. A maximum possible score (X_{max}) of 162 can be obtained from the QEC score table if a particular task assessed is perceived by the observer to be of a mainly static type, these may include seated or standing tasks with or without frequent repetition and where the load/force exertion is relatively low. A maximum possible score (X_{maxMH}) of 176 indicated that the particular task assessed was perceived by the observer to be of a manual handling (including lifting, pushing/pulling and carrying loads), as indicated in Table 1.

To explain how E is calculated, consider Task 10A in Table 1 as an example. The X value (score obtained from the QEC assessment) is 108 and the X_{max} is 176. Therefore: E=(108/176) x 100% = 61%. More details on how the exposure scores are differentiated in the QEC between different types of tasks are given in Li and Buckle (1998, 1999b).

Summary and discussion

This study has attempted to develop an action level for the QEC system, so that its users can have a reference that suggests whether or not an ergonomic intervention is needed for a particular task concerned. Based on the preliminary results obtained in this study, through the assessment of 31 different tasks using both the QEC and the RULA methods, the equivalent exposure levels and recommended actions are summarised in Table 2 below:

Table 2: Summary of QEC action levels

QEC score (Percentage Total)	Action	Equivalent RULA score
< 40%	Acceptable	1 - 2
40 - 49%	Investigate further	3 - 4
50 - 69%	Investigate further and change soon	5 - 6
≥ 70%	Investigate and change immediately	≥ 7

It should be noted that the present results (Percentage Total) are used to refer to the upper body as a whole. However, the nature of the design of the QEC system was to enable rapid assessment of the exposure levels (and the exposure changes following an intervention) for various body regions, so as to identify the problematic area(s). Using an overall exposure level may 'hide' a particular 'high' score of a certain body part such as the back, since this high score can be offset by a 'low' score in another body region such as the neck, leading to a suggestion that the task is safe (although it may be potentially unsafe for the back). Therefore, it is recommended that, when an overall percentage score is found to be greater than 40%, the user refers to the original QEC score table (Li and Buckle, 1998, 1999b) and checks the individual scores of each body part so as to identify where a particular high exposure lays. To this point, the action levels identified in the present study should only be regarded as a general reference which leads to further identification of particular body areas that may need ergonomic attention.

It also needs to be pointed out that the preliminary action levels identified in the present study were based on equivalent actions recommended by RULA. Although

RULA has been widely used, the validity of the method, especially the predictive value of its assessment scores for quantifying the actual risk of WMSDs, has not yet been validated. There is also an issue about the differences in the design and application purposes of the two methods used, which may limit the value of the present results. Further studies are required to validate the relationships between the QEC assessment outcome (scores) and corresponding action levels using more sophisticated approaches.

More information on the QEC system is available at: http://www.geocities.com/qecuk/

Acknowledgement

This work was carried at the Institute for Automotive and Manufacturing Advanced Practices, University of Sunderland, as part of an MSc project.

References

Armstrong, T.J., Buckle, P., Fine, L.J., Hagberg, M., Jonsson, B., Kilbom, Å., Kuorinka, I.A.A., Silverstein, B.A., Sjogaard, G., Viikari-Juntura, E.R.A., 1993, A conceptual model for work-related neck and upper-limb musculoskeletal disorders. Scandinavian Journal of Work, Environment and Health, 19, 2, 73-84.

Bongers, P.M., de Winter, C.R., Kompier, M.A.J. and Hildebrandt, V.H., 1993, Psychosocial factors at work and musculoskeletal disease [review]. Scandinavian Journal of Work, Environment and Health, 19, 297-312.

Brown, R., 2001, Further evaluation of the Quick Exposure Check (QEC), with particular attention to the scoring system. Unpublished MSc Thesis, University of Sunderland.

Houtman, I.L.D., Bongers, P.M., Smulders, P.G.W. and Kompier, M.A.J., 1994, Psychosocial stressors at work and musculoskeletal problems. Scandinavian Journal of Work, Environment and Health, 20, 2, 139-145.

Li, G. and Buckle, P., 1998, A practical method for the assessment of work-related musculoskeletal risks - Quick Exposure Check (QEC). In: Proceedings of the Human Factors and Ergonomics Society 42nd Annual Meeting, October 5-9, Chicago, Human Factors and Ergonomics Society: 1351-1355.

Li, G. and Buckle, P., 1999a, Current techniques for assessing physical exposure to work-related musculoskeletal risks, with emphasis on posture-based methods. Ergonomics, 42, 5, 674-695.

Li, G. and Buckle, P., 1999b, Evaluating change in exposure to risk for musculoskeletal disorders – a practical tool. HSE Contract Report 251/1999, HSE Books ISBN 0 7176 1722 X.

McAtamney, L. and Corlett, E. N., 1993, RULA: a survey based method for the investigation of work related upper limb disorders. Applied Ergonomics, 24, 91-99.

Theorell, T., Harms-Ringdahl, K., Ahlberg-Hultén, G. and Westin, B., 1991, Psychosocial job factors and symptoms from the locomotor system - A multicausal analysis. Scandinavian Journal of Rehabilitation Medicine, 23, 165-173.

Winkel, J. and Westgaard, R., 1992, Occupational and individual risk factors for shoulder-neck complaints: Part II - The scientific basis (literature review) for the guide. International Journal of Industrial Ergonomics, 10, 85-104.

Appendix 1. List of tasks used during the assessment

Task ID	Task description
1A.	The manufacture of plastic plates.
3A.	The manufacture of an electric wall fan housing.
5A.	Measuring a quantity of plastic dough.
6A.	Placing the plastic dough in to a mould.
7A.	Removal of a plastic drain inspection cover from a mould.
8A.	Removal of a plastic vacuum cleaner cylinder from a mould.
9A.	The manufacture of plastic milk crates.
10A.	Removal of a plastic barrel from a blow moulding machine.
11A.	Removal of excess plastic from a plastic barrel.
12A.	Placing plastic granules in to a rotation casting machine.
13A.	The closure of a rotation casting machine.
14A.	Manufacturing plastic baths.
23A.	Disposal of excess material after the manufacture of a crankshaft.
1C.	Moving empty oil drums by hand.
2C.	Emptying bags of chemicals in to a hopper.
3C.	Loading of a delivery van.
4C.	Packing small packets of food.
5C.	Checkout operator.
6C.	Addition of parts to an engine whilst on a production line.
7C.	Placement of stacked boxes on to a production line.
8C.	A worker, checking small objects before being packed in to a box.
9C.	Placing objects on to spikes on a moving production line.
10C.	Checking small objects mounted in a box.
11C.	Poultry worker.
12C.	A forklift driver.
13C.	The packaging of poultry in to bags.
14C.	Removal of aluminium casks from an underground cellar.
15C.	Seated worker, packing small plastic tubs.
16C.	Changing a large paper roll over on a printing machine.
17C.	Packing large plastic bottles.
18C.	Making individual bricks by hand.

THE INFLUENCE OF PSYCHOSOCIAL RISK FACTORS ON ABSENCE DUE TO MUSCULOSKELETAL DISORDERS

Serena Bartys[1,2], Kim Burton[1], Ian Wright[3], Colin Mackay[4], Paul Watson[2], Chris Main[2]

[1]*Spinal Research Unit, University of Huddersfield, 30 Queen Street, Huddersfield, HD1 2SP*
[2]*Dept of Behavioural Medicine, Hope Hospital, Salford*
[3]*GlaxoSmithKline, Brentford*
[4]*Health and Safety Executive, Bootle*

The influence of psychosocial factors on absence rates due to musculoskeletal disorders remains incompletely understood; much previous research has been cross-sectional, involving a limited range of psychosocial variables. This paper reports a large prospective study into the relationship between proposed clinical and occupational psychosocial risk factors (yellow and blue 'flags') and absence rates due to musculoskeletal disorders across a multi-site, multi-task UK pharmaceutical company. Baseline data were collected from 4,637 workers, and company-recorded absence over the ensuing 15 months was documented. All of the psychosocial flags studied were individually predictive of the incidence of absence, but the relative influence of any one flag or group of flags on duration of future absence could not be distinguished, suggesting limited value for psychosocial screening.

Introduction

The significant health concern that musculoskeletal disorders (MSDs) pose in industrialised nations has led to substantial research concentrating upon work-related risk factors. The literature shows that certain jobs and certain physical workplace factors are associated with the manifold risk of experiencing MSDs, compared with working populations not exposed to these risk factors (Merllie & Paoli, 2001), though their overall effect-size seems modest (Carter & Birrell, 2000). More recently it has become clear that attention to physical workplace risk factors is not enough to protect or enhance workers' health and well-being (Westgaard and Winkel, 1997), and more attention is now being focussed on the psychosocial work environment.

Griffiths (1998) has suggested that it is the least tangible aspects of work that represent the most common threat to worker ill-health today. Indeed, it is now recognised that psychosocial workplace factors may play an important role in the recovery of MSDs (Bartys et al, 2001). In addition to the established influences from clinical psychosocial factors, such as distress and somatisation (Pincus et al, 2002), a systematic review by

Linton (2001) concluded that occupational psychosocial factors also play a significant role in future back pain problems. The factors documented were: job satisfaction, monotonous work, social support, perceived demand/control, work content, work pace, self-reported stress, perceived ability to work, perceived safety, perceived emotional effort, and attributions of cause. In addition, Bongers et al (2002) reported from a review of the literature, that high job stress and demands were consistently related to upper extremity disorders.

An explanation of how these psychosocial factors influence recovery from MSDs comes from Davis & Heaney (2000), who hypothesised that differing responses to (perceptions of) environmental factors influence how the individual may accept and cope with pain or injury. Addressing the problem of recovery from MSDs from a biopsychosocial approach acknowledges the influence of the psychosocial work environment and recognises that work can place certain constraints on the individual.

In the present study, psychosocial risk factors for delayed recovery from MSDs are referred to as 'flags', and different colours of flags represent different concepts. Yellow flags are classified as clinical psychosocial risk factors for disability; guidance for addressing these factors has come from the Accident Rehabilitation and Compensation Insurance Corporation of New Zealand (Kendall et al, 1997). The authors of that document stated that beliefs about pain and disability, fears that physical activity will be injurious, and negative perceptions of work can act as obstacles to recovery from MSDs. Blue flags have been distinguished from the yellow flags (Burton and Main, 2000), and are expressed as occupational psychosocial risk factors; blue flags are negative perceptions or attributions of the work environment, and have been proposed to be detrimental to recovery from MSDs.

In a previous study (Bartys et al, 2001), detrimental scores (cut-off points) were derived from certain psychosocial questionnaires, and individuals who exceeded these scores were considered to have their flags 'flying' and be at risk of delayed recovery from MSDs. This prospective study explores the relationship that yellow and blue flags have on absence over a 15-month period.

Methods

The first phase of this project commenced in 2000, whereby all 7,838 permanent employees of GlaxoSmithKline UK were invited to complete a booklet of questionnaires. GlaxoSmithKline is a multi-task organisation, and questionnaires were delivered to workers whose job categories were (in general terms) manual processors, research scientists, office and administration workers, field sales representatives, and management. The questionnaires were designed to collect self-report data on MSDs using an abbreviated version of the previously validated Nordic Musculoskeletal Questionnaire (Dickinson et al, 1992), and data on occupational and clinical psychosocial factors was also collected.

The following occupational psychosocial factors were measured: job satisfaction; social support; attributions of workplace causation (of back pain); sources of pressure at work; perceived exertion; and perceived control and influence at work. Two clinical psychosocial factors – psychological distress and beliefs about the inevitable consequences of musculoskeletal disorders – were also measured. In total, 4,637 workers responded to the baseline survey (59% response rate), and their absence over the previous 12 months was determined from company records.

To facilitate a prospective evaluation of the influence of these baseline data on future absence, company sickness absence data were tracked over the ensuing 15 months, and absence due to MSDs was extracted. Previously defined 'flags' calculated from some of the above psychosocial factors were used to categorise the risk of the incidence of absence and its duration. The yellow flag was psychological distress, and the blue flags were job satisfaction, social support, perceived control, attribution of workplace causation and perceptions of organisational climate.

Results

In the 15 months following the completion of the workforce survey, 214 (5%) of the responding workers took absence due to MSDs. This resulted in 267 spells of MSD-related absence and 2,461 working days lost. Chi-squared analyses of demographic variables showed that, compared to the non-absentees, there were significantly more males than females, significantly more manual workers than non-manual workers, and significantly more older workers than younger workers who took absence during the ensuing 15 months (Table 1).

Table 1. Distribution of workers who took absence, compared to those workers who did not take absence in the ensuing 15 months, categorised by sex, job type and age

Absentees (n)	Non-absentees (n)	X^2 sig. level
Male (146), Female (68)	Male (2415), Female (1871)	$P<0.05$
Manual (128), Non-manual (86)	Manual (825), Non-manual (3515)	$P<0.001$
19-40 yrs (80), 41-65 yrs (134)	19-40 yrs (2350), 41-65 yrs (1818)	$P<0.001$

It was then interesting to examine whether previously reported cross-sectional associations between the yellow and blue flags and absence due to MSDs would also emerge from the prospective study. Odds-ratios (OR) were calculated to explore the association between the psychosocial flags and incidence of future absence. Table 2 shows the previously reported cross-sectional odds ratios for absence in a previous 12-month period (Bartys et al, 2001), and then the prospective odds-ratios for the risk of taking absence in the following 15 months. The results show that broadly similar statistically significant risks were found for the incidence of future absence compared with those for reported previous absence.

Table 2. Odds-ratios for the cross-sectional and prospective associations between yellow and blue flags and absence due to MSDs

Psychosocial 'flag'	Absence previous 12m (OR)	Absence next 15m (OR)
Distress	2.4	1.6
Job satisfaction	3.2	3.2
Social support	2.9	2.4
Attribution of causation	1.8	2.1
Perceived control	2.1	1.8
Organisational climate	2.0	2.3

This workplace study also offered the opportunity to explore the influence that yellow and blue flags have on duration of future absence/time to return to work. Average duration of absence was examined for those workers who took absence due to musculoskeletal disorders in the ensuing 15 months, and it was found that the average duration of absence was 11 working days. However, the nature of musculoskeletal disorders means that there is wide range of absence durations (in this case between 1 and 119 working days), therefore it is also useful to report the median duration of absence (which was 5 working days) and the mode of that duration (which was 2 working days). These descriptives show that the majority of spells of musculoskeletal absence during this period were short-term.

Average duration of absence was then compared between workers who had zero 'flags flying', and those who had up to five 'flags flying'. It is important to note here that during this 15-month period, a trial of a psychosocial intervention for workers with MSDs was being conducted on two of the sites of GlaxoSmithKline. So, in order to reduce the risk of confounding the average duration of absence, those workers who had received the intervention during this period were excluded (n=59).

It was found that average duration of absence was significantly longer, at 13 working days, for those workers who had one or more flags flying (n=110), compared to 8 working days for those with zero flags flying (n=45) *(t=-2.63, P<0.05)* - see Figure 1.

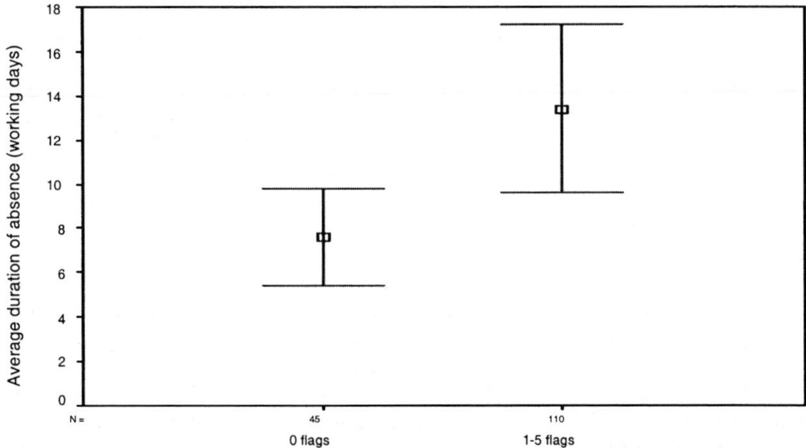

Figure 1. Average duration of absence (working days) for workers who had zero 'flags flying' compared to those workers with one or more 'flags flying'.

Although there was a clear trend for longer absence with more flags flying, i.e. 0 compared to 1, 1 compared to 2, 2 compared to 3, etc, small numbers in each of these groups meant that the differences were not statistically significant.

In order to examine whether there were different psychosocial risk factors for short and long absences, the sickness absence data were then categorised into self-certified absences (lasting up to 7 days) and medically certified absences (over 7 days). Univariate analyses using the full range of psychosocial factors listed under Methods (not just the yellow and blue flags given in Table 1) did not find any statistically significant differences between short and longer durations of absence. Further, a logistic regression analysis of the different categories of absence again using all the psychosocial factors studied, together with demographic variables (Table 1), did not reveal any statistically significant predictors for these different durations of absence.

Discussion

This study presents evidence suggesting that clinical and occupational psychosocial risk factors (yellow and blue flags) can predict the incidence of future absence. Moreover, a cumulation of these psychosocial flags was associated with longer absence. However, it was not possible to determine the relative influence of any particular flag (or indeed any of the psychosocial factors studied) on the duration of absence.

This finding may, in part, be due to the fact that the sickness absence data had a skewed distribution, with the majority of workers taking absence that lasted less than one week. This meant that a more sophisticated predictive analysis, such as linear regression, that relied on the sickness absence data being a continuous variable of normal distribution, was unhelpful. Such statistical problems with sickness absence data have also been noted by North et al (1993) whilst reporting on the Whitehall II study. The alternative approach of categorising absence into short and longer durations (self certified and medically certified) failed to reveal a predictive relationship between the psychosocial factors and duration of future absence.

Whilst the findings presented here correspond with other prospective studies that examined the influence that psychosocial factors have on the *incidence* of absence (e.g. Bigos et al, 1991), there is a surprising dearth of prospective studies that look at the influence of psychosocial factors on *duration* of absence. This study aimed to redress the balance because duration of absence, or return to work time, promises to be a more realistic target for intervention than primary prevention, and thus is worthy of further research. However, the present results imply that routine psychosocial screening in the workplace may not be useful in predicting the duration of absence due to MSDs over an extended period.

Nevertheless, it may be that the psychosocial factors that influence absence behaviour are only pertinent once absence has commenced, or pain is reported. This notion is now being explored in a controlled trial: psychosocial assessments are being made within a few days of the start of absence, and any psychosocial obstacles to recovery (yellow and blue flags) are being addressed with a cognitive-behavioural based intervention. The hypothesis is that an individually targeted psychosocial intervention, addressing specific obstacles to return to work will reduce the duration of sick leave. Results of that study will be reported in due course.

References

Bartys, S., Tillotson, M., Burton, K., Main, C., Watson, P., Wright, I. and Mackay, C. 2001, Are occupational psychosocial factors related to back pain and sickness absence? In M.A. Hanson (ed.) *Contemporary Ergonomics*, Taylor and Francis, London, 23-28.

Bigos, S.J., Battie, M.C., Spengler, D.M., Fisher, L.D., Fordyce, W.E. and Hansson, T. 1991, A prospective study of work perceptions and psychosocial factors affecting the report of back injury, *Spine*, **16**, 1-6.

Bongers, P.M., Kremer, A.M. and ter Laak, J. 2002, Are psychosocial factors risk factors for symptoms and signs of the shoulder, elbow, or hand/wrist?: a review of the epidemiological literature, *American Journal of Industrial Medicine*, **41**, 315-342.

Burton, A.K. and Main, C.J. 2000, Obstacles to recovery from work-related musculoskeletal disorders. In W. Karwowski (ed.) *International encyclopedia of ergonomics and human factors*, Taylor and Francis, London, 1542-1544.

Carter, J.T. and Birrell, L.N. 2000, *Occupational health guidelines for the management of low back pain at work*, (Faculty of Occupational Medicine, London).

Davis, K.G. and Heaney, C.A. 2000, The relationship between psychosocial work characteristics and low back pain: underlying methodological issues, *Clinical Biomechanics*, **15**, 389-406.

Dickinson, C.E., Campion, K., Foster, A.F., Newman, S.J., O'Rourke, A.M. and Thomas, P.G. 1992, Questionnaire development: an examination of the Nordic Musculoskeletal *Questionnaire, Applied Ergonomics*, **23**, 197-201.

Griffiths, A. 1998, The psychosocial work environment. In R. McCaig and M. Harrington (eds.) *The changing nature of occupational health*, (The Stationery Office, Norwich), 213-232.

Kendall, N.A.S., Linton, S.J. and Main, C.J. 1997, *Guide to assessing psychosocial yellow flags in acute low back pain: risk factors for long-term disability and workloss*, (Accident Rehabilitation & Compensation Insurance Corporation of New Zealand and the National Health Committee, Wellington, NZ).

Linton, S.J. 2001, Occupational psychological factors increase the risk for back pain: a systematic review, *Journal of Occupational Rehabilitation*, **11**, 53-66.

Merllie, D. and Paoli, P. 2001, *Ten years of working conditions in the European Union (Summary)*, (European Foundation for the Improvement of Living and Working Conditions, Dublin).

North, F., Syme, S.L., Feeney, A., Head, J., Shipley, M.J. and Marmot, M.G. 1993, Explaining socio-economic differences in sickness absence: the Whitehall II study, *BMJ*, **306**, 361-366.

Pincus, T., Burton, A.K., Vogel, S. and Field, A.P. 2002, A systematic review of psychological factors as predictors of chronicity/disability in prospective cohorts of low back pain, *Spine*, **27**, E109-E120.

Westgaard, R.H. and Winkel, J. 1997, Ergonomic intervention research for improved musculoskeletal health: a critical review, *International Journal of Industrial Ergonomics*, **20**, 463-500.

A TRIAL OF BACK BELTS FOR RIGID INFLATABLE BOAT CREWS IN THE ROYAL NAVY

RS Bridger and GS Paddan

Institute of Naval Medicine, Crescent Rd, Alverstoke, Hants, PO12 2DL
hhfd@inm.mod.uk, 02392 768220, fax 02392 504823

Crews from 3 Fisheries Protection Squadron vessels wore back belts during RIB journeys and boardings of fishing vessels. Data were captured for 252 man-boardings while wearing no belt and 156 man-boardings with the belt. There was significantly less low back pain when the belts were worn. The frequency of back pain dropped from about one half to about one quarter of RIB journeys ($p<0.02$). There was no evidence that wearing the belts caused increased risk taking by either coxswains or boat crews. Guidelines for a policy on belt use were developed. Ergonomic assessment of RIB design identified many risk factors for low back pain. Better design of seating on RIBs was identified as the best solution in the long-term. There would seem to be value in carrying out further research to determine, directly, the effect of back belts on the trunks of vehicle occupants. Very high vibration magnitudes were measured in the rigid inflatable boats.

Introduction

There is a high prevalence of low back pain amongst rigid inflatable boat (RIB) crews in the RN fisheries protection squadron. Boat crews are subjected to a variety of low back stressors including vibration and shock, constrained, twisted sitting postures and manual handling in confined spaces on board fishing vessels (Bridger and Powell, 1999). In the short term, no changes to RIB design were possible.

There is some evidence that tight abdominal belts stiffen the trunk of the wearer even during sudden, unexpected loading (Lavender et al., 2000). On an unstable boat, where crews lack any kind of back support and there are no firm handholds, belts might be beneficial in reducing the load on the back muscles. The increased intra-abdominal pressure might act as a "hydraulic splint" that helps the wearer to maintain an upright sitting posture.

Most of the scientific evidence on the use of back support belts for the prevention of low back injury in industrial manual handling operations suggests that these belts do not reduce injury rates. Anecdotally, it is reported that back belts are issued to RIB crews in the police force and are used by cross-country motorcycle riders, but no scientific evaluations appear to have been done.

Some industrial studies have shown decreased injury rates when belts are used, but the injuries that do occur have been more expensive to treat (Mitchell et al., 1994).

Although there is scope for improvements to RIB design and for changes in work practices, the availability of short-term palliative measures was limited. A requirement for a trial of back belts was identified on the grounds that:

- It is not known whether the wearing of back belts reduces back pain in those exposed to vibration and shock in moving vehicles.
- There is evidence that these belts do stiffen the trunk when worn tightly around the waist. This might be beneficial in practice.

The purpose of the trial was to test the hypothesis that belt wearing would reduce the frequency of occurrence of low back pain associated with boarding operations. Data on any other advantages or side effects associated with belt use were also captured. A secondary aim of the trial was to assess crew exposure to vibration during travel in RIBs.

Method

All subjects were volunteers and gave their informed consent to participate in this ethically approved trial. A crossover design was used based on the method of Snook et al. (1998). Baseline data on back pain and RIB use were collected. Subjects were then allocated to an experimental group for a patrol of approximately 6 weeks or to a control group. Experimental subjects received a back support belt ("Sallis White Belt"). Control subjects received instruction in six exercises of the kind normally recommended to back pain sufferers and were given routine advice on back care in everyday life (Faas et al., 1993). The groups were allocated on the basis of ship to prevent sharing of belts (all crews on a particular ship were issued with belts or participated in the exercise condition). After 6 weeks, the conditions were reversed.

All participants completed a short checklist at the beginning of the trial to provide information on all known predictors, confounders and effect modifiers for low back pain (e.g. previous back problem, taking pain killers for back pain, smoking, family history of back pain). Participants kept a personal back pain diary that they completed at the end of each RIB journey. Any back pain was rated on a scale of 1, "mild discomfort" to 10 "severe pain".

Bridge logs were kept to record the details of RIB journeys including those on board, journey length and duration, the sea conditions and how the boat was driven.

At the end of the trial, participants completed questionnaires on the belts and on the exercises. In order not to pre-judge any of the issues surrounding the use of the belts, the questionnaires were limited to open-ended questions designed to elicit spontaneous responses on matters of importance to the participants. Open-ended questions are held to be free from bias. Participants commented on the back belts at the end of the trial.

Trial Procedure

The investigator provided brief presentations to crews at the start of the trial (to explain the procedure and the purpose and to demonstrate correct donning of the belt, the exercises and give advice on routine back care).

Back Support Belts

Elastic back support belts with velcro fasteners ("Sallis white belts") were obtained in 4 sizes (small, medium, large and extra large). The belts were fastened at the front with velcro and had two additional velcro fasteners at either side of the main one.

All participants were given the opportunity to select a belt of the appropriate size before participating in the experimental session. Participants were instructed to breath-in and extend the trunk slightly before donning the belt and to adjust the fasteners to achieve a fit that was tight but not uncomfortable.

The author spent 5 days aboard HMS LINDISFARNE observing and assessing RIB operations and undertook several RIB journeys both with and without a back support belt.

Vibration Measurements

Four channels of acceleration (x-, y-, z-axis on the seat and x-axis on the backrest) were measured on four saddle-type seats (including the drivers' seat) in a rigid inflatable boat during travel at about 45 knots over 'rough' water. These axes of vibration are defined in British Standard BS 6841 (1987). Accelerations were measured using piezoresistive accelerometers mounted inside flexible plastic seat pads. The data were low-pass filtered at 100 Hz and digitised into a computer system at a sample rate of 526 samples per second. Each measurement was of 60-s duration.

The data were frequency-weighted and assessed in accord with BS 6841 (1987) to calculate r.m.s. accelerations and vibration dose values, VDVs. Vibration magnitudes greater than 1.25 ms^{-2} r.m.s. are likely to be categorised as being "very uncomfortable", and VDVs of 15 $ms^{-1.75}$ will "... *usually cause severe discomfort*" and increase the "... *risk of injury*".

Results

A combination of technical problems and bad weather caused serious disruption with the result that only 17 volunteers completed the trial procedure, returning questionnaires on the effectiveness of the back support belts. Seventeen subjects had the opportunity to try the back belts and complete the diaries.

RIB Journeys

Average journey distance and duration of 22 RIB journeys in which the belt was worn were 2.9 nautical miles and 23.4 minutes. For 107 journeys in which the belt was not worn, average duration and distance were 3.9 nautical miles and 25.4 minutes.

The number of journeys undertaken with and without belts in different sea conditions is presented below (Table 1).

Table 1. RIB Journeys with and without Belts in Different Sea Conditions

	Calm	Moderate	Rough	High
No Belt	24	66	15	2
Belt	3	13	6	0

Sometimes, RIB crews boarded more than 1 fishing vessel in a single day. Table 2 gives the frequency of boardings per day, with and without the use of back belts, as recorded by participants in their personal logs. On about fifty per cent of days, crews boarded only 1 fishing vessel. On about one third of the recorded days, 2 vessels were boarded. Only rarely, were there more than 4 boardings per day by the same individuals.

A total of 11 accidents were reported during the trial period. Most of these involved minor injuries. Two involved damage to RIBs and one resulted in the crew being flung from the boat. Only one incident (a minor injury to the shoulder following a slip) occurred when the belts were being worn.

Table 2. Percentage of Days in Which One or More Vessels were Boarded

	No. Boardings/Day (% of Total Boardings)			
	1	2	3	4 +
No belt (203 man days)	47	35	12	6
Belt (106 man days)	54	30	15	1

Low Back Pain

The data used in the analysis below are restricted to the 10 subjects who used a back belt for at least 1 RIB journey/day on at least 5 occasions (Table 3). The first analysis deals with the occurrence of any low back pain (LBP) during or after a journey and the second to low back pain rated at 3 or more (LBP3). These occurrences of low back pain are compared to pain experienced on days when the belt was not worn (at least 1 RIB journey/day on at least 5 occasions).

One subject experienced no LBP with or without a back belt. The remaining 9 subjects had LBP on one or more journeys. Eight of the nine subjects had a higher frequency of occurrence of LBP when wearing no back belt. The binomial test indicates a significant ($p<0.02$; 1-sided test) difference in occurrence of LBP with and without back belts.

One subject experienced no LBP3 with or without a back belt. The remaining 9 subjects had LBP3 on one or more journeys. All nine subjects had a higher frequency of occurrence of LBP3 when wearing no back belt. The binomial test indicates a significant ($p<0.01$; 1-sided test) difference in occurrence of LBP3 with and without back belts.

In response to the question, "Did the belts have any unwanted side effects?" 7 of 10 who used them on >5 journeys reported that there were none. Two reported discomfort after wearing the belts for >6 hours and 1 reported that the belt limited twisting movements. Thirteen out of 17 who used the belts at least once, reported no side effects.

Only 9 participants returned questionnaires concerning the exercises. Of these, 3 did not carry out the exercises (two because they exercised anyway and a third who already practised exercises to relieve a back condition). Of the remainder, 2 commented that the exercises could not be done in the boat, 3 commented that the exercises were easy to do and helped relieve back pain and 2 that sleeping in the "Fowler position" (with pillows under the neck and the knees) alleviated back pain resulting from the day's activities. Based on these comments and on personal observation, it seems unlikely that the exercises could be used to relieve back pain during boarding operations. The number responding was too small to enable any conclusions to be drawn about the benefits of back exercises at other times.

Table 3. Frequency of LBP (any level) During Journeys With and Without Back Belts*

	With Back Belt	No Back Belt
Average %journeys with LBP	24	47
Minimum %journeys with LBP	0	0
Maximum %journeys with LBP	57	88
Number of subjects with no LBP	4	1

* Based on totals of 252 man-boardings without the belt and 156 man-boardings with the belt.

Users' Comments

The majority of the subjective comments about the belts were that they supported the back during the RIB journeys, lessened back pain and were easy to use. There were fewer negative comments and these were mainly concerned with chafing due to the velcro fasteners and restriction of breathing.

Vibration Measures
Very high vibration magnitudes (of up to 2.5 ms^{-2} r.m.s.) were measured for the vertical axis on the seat in the RIB during travel at full speed over rough water; the vibration is mainly associated with the vertical slamming of the boat. Exposure to such vibration could be subjectively categorised as being "very uncomfortable" to "extremely uncomfortable". The highest VDV measured for a 60-s duration was 11.5 $ms^{-1.75}$ for the vertical axis (in the drivers' seat); exposure to such vibrations could reach a VDV of 15 $ms^{-1.75}$ after about 3 minutes. Measures in other seats showed a maximum duration of 22 minutes before reaching a VDV of 15 $ms^{-1.75}$. Lower vibrations were measured in the horizontal axes on the seat and on the backrest.

Acceleration measurements showed that the time histories for the vertical axis contained shocks (mainly due to the slamming motions of the RIB). High crest factors (up to 13.7) occurred for vertical motion. Where crest factors exceed a value of about 6, the VDV is considered to be the preferred indicator of vibration exposure severity compared to the RMS vibration magnitude (Griffin, 1990).

Discussion

Analysis of the back pain diaries indicated that, once issued, the belts were worn on all RIB journeys. Back pain was less severe when the belts were worn than when the belts were not worn, and the proportion of journeys with back pain dropped from about a half to about a quarter. This reduction in pain is consistent with the operation of a mechanism of support in which the belts stiffen the trunk. There is evidence from other studies that the belts increase the passive stability of the trunk by 34% (Ivanic et al., 2002) and are efficacious in stiffening the trunk (Lavender et al., 2000). Evidence from the present study indicates that belts can be effective in reducing low back pain when used on rigid inflatable boats.

There was concern that the increased feelings of support provided by the belts might lead to behavioural compensation in the form of more risk taking by wearers. A total of 11 accidents (including slips, trips and falls) were reported on the diaries and the bridge logs. Only one of these occurred when the belts were worn. Only two incidents resulting in damage to RIBs were recorded throughout the trial and, in neither of these, were belts worn.

Few unwanted side effects were reported. Discomfort after long hours of use (3-6) is not unexpected. Given that RIB journeys tend to last for less than 30 minutes and, on most days, there are only one or two boardings, this should not be a problem under the expected conditions of use. One "unwanted" side effect reported was that the belts limited twisting movements. Although restriction of twisting motions might be inconvenient, it is almost certainly advantageous, particularly in heavy seas.

Ergonomic problems caused by design deficiencies of the current generation of searider RIBs were readily apparent.

The very high vibration magnitudes in the RIBs show that the 'action level' suggested in BS 6841 for whole-body vibration exposure over a 24-hour period could be exceeded in a few minutes. Continued exposure to such vibrations could pose a health risk.

Conclusions

There is evidence that wearing of back belts reduces back pain in RIB crews.

Recommendations

It was recommended that the belts be available, on request, to crewmembers experiencing low back pain during FPS RIB duties but should not be issued as personal protective equipment, nor issued automatically or to those who do not suffer from low back pain. Prospective users should be advised that there is no evidence that the belts offer any protection from vibration, shock or any other stressors that may cause back injury and that there is no evidence that the belts prevent the development of chronic medical conditions involving the back. Users to be advised that the *only known benefit* of wearing the belts is reduced back pain. Known *side effects* are a feeling of support and increased confidence. These are to be regarded as illusory and on no account are belt wearers to attempt to lift heavier loads or lengthen RIB journeys just because they are wearing a belt. Belts are not to be held in common. Each person is to be issued with and responsible for their own belt at all times.

References

Bridger, R.S., Powell, W. 1999. Musculoskeletal pain in fisheries protection squadron RIB users. Unpublished MoD Report.

British Standards Institution 1987. Measurement and evaluation of human exposure to whole-body mechanical vibration and repeated shock. BS 6841. London.

Faas, A., Chavennes, A.W., van Eijk, JThM., Gubbels, J.W. 1993. A randomised placebo controlled trial of exercise therapy in patients with acute low back pain. *Spine, 18: 1388-1395.*

Griffin, M.J. 1990 Handbook of human vibration. Academic Press Limited, London. ISBN 0-12-303040-4.

Ivanic, P.C., Cholewicki, J., Radebold, A. 2002 Effects of the abdominal belt on muscle generated spinal stability and L4/L5 joint compression force. Ergonomics, 45:501-503.

Lavender, S.A., Shakeel, K., Andersson, G.B.J., Thomas, S.J. 2000. Effects of a lifting belt on spine moments and muscle recruitment after unexpected sudden loading. *Spine,* 25:1569-1577.

Mitchell, R.V., Lawler, F.H., Bowen, D., Mote, W., Asundi, P., Purswell, J. 1994. Effectiveness and cost-effectiveness of employer-issued back belts in areas of high risk for back injury. *Journal of occupational Medicine, 36:90-94.*

Rabinowitz, D., Bridger, R.S., Lambert, M. Lifting technique and abdominal belt usage: a biomechanical, physiological and psychological investigation. *Safety Science, 28:155 164, 1998.*

Rafacz, W., McGill, S.M. 1996. Wearing an abdominal belt increases diastolic blood pressure. Journal of Occupational and Environmental Medicine, 38, 925-927.

Shah, R.K. 1993. A pilot study of the traditional use of the patuka around the waist for the prevention of back pain in Nepal. *Applied Ergonomics, 24:337-344.*

Sheard, S.C., Pethybridge, R.J., Wright, J.M., and McMillan, G.H.G. Back pain in aircrew - an initial survey. *Aviat Space Environ Med, 1996. 67(5), 474-477.*

Snook, S.S., Webster, W.S., McGorry, R.W., Fogleman, M.T., McCann, K.B. 1998. The reduction of chronic specific low back pain through the control of early morning lumbar flexion. *Spine 23: 2601-2607.*

WORK-RELATED STRESS AS A RISK FACTOR FOR WMSDs: IMPLICATIONS FOR ERGONOMICS INTERVENTIONS

Jason Devereux

Robens Centre for Health Ergonomics
EIHMS, University of Surrey
Guildford, Surrey, GU2 7TE

There is the potential for work-related mental stress (psychological stressors and strains) to reduce the effectiveness of an ergonomics intervention for preventing WMSDs. There is epidemiological and psychophysiological evidence implicating work-related mental stress in the development of WMSDs. Ergonomic interventions in the workplace are needed to reduce the risks of physical and psychosocial work risk factors for musculoskeletal disorders via organisation design changes. In addition, individual susceptibility should become an increasing concern for Ergonomists. Methods for identification and strategies for solutions need to be considered in the future.

Introduction

This paper considers the latest scientific evidence concerning the potential relationship between work-related mental stress and work-related musculoskeletal disorders (WMSDs). The implications of such a relationship for Ergonomists making workplace interventions is also considered.

There is a major initiative across the European Union to provide good practice for preventing the leading occupational health problem in Europe – WMSDs. There is European consensus that musculoskeletal disorders can be work related and that ergonomics interventions in the workplace can reduce risks and the incidence of WMSDs (Buckle & Devereux, 1999; Op De Beeck & Hermans, 2000). Good practice is centred on reducing physical and psychosocial risk factors in the workplace (Buckle & Devereux, 2002; Cox et al., 2002; European Agency for Safety and Health at Work, 2002).

The latest on physical factors in the workplace that increase WMSD risk

Systematic critical literature reviews regarding workplace risk factors for WMSDs have been consistent in their findings (Ariëns et al., 2000; Hoogendoorn et al., 1999; Hoozemans et al., 1998; NIOSH, 1997). For musculoskeletal disorders affecting the neck region, high postural load has been shown consistently to be a risk factor (duration of sitting, twisting and bending of the trunk). For the upper limbs, there is strong evidence that the biomechanical load from a combination of repetition, force and posture increases the risk multiplicatively for musculoskeletal disorders affecting the elbow. The combination effects have also been shown to increase the risk of specific hand disorders, i.e. carpal tunnel syndrome and tendinitis .

In a systematic critical review of Display Screen Equipment (DSE) users, there were consistent study findings regarding increasing duration of DSE use and increasing risk of neck/shoulder and hand/wrist musculoskeletal disorders (Punnett & Bergqvist, 1997). The relationship was mainly dependent on the degree of repetitive finger motion and sustained muscle loading across the forearm and wrist. At least 4 hours of keyboard work per day appears to increase risk about two-fold compared to little or no keyboard work.

Some Hand-Arm Vibration Syndromes HAVS (for example, vibration-induced white finger) have clearer cause-effect relationships compared to other WMSDs. It is widely accepted that vibration is the main causal agent, however, the relationship between vibration and HAVS may also be modified by various environmental and individual variables (Bovenzi, 1998).

For the lower back, there has been consistency among critical reviews that manual handling and whole body vibration are risk factors.

Psychosocial factors in the workplace can increase WMSD risk

Psychosocial factors in the workplace are now widely accepted as contributing to the development of clinical signs and symptoms of WMSDs. In addition, they have also been included in HSG60(rev), the revised guidance on upper limb disorders in the workplace by the HSE, and are classified as psychosocial factors in the risk assessment (Health and Safety Executive, 2002).

Plausible models to explain the relationships between psychosocial work factors and WMSDs, as well as recent laboratory experimentation supporting the models, have provided support for an interactive relationship between physical and psychosocial risk factors in the workplace (Lundberg, 2002).

To date, only one epidemiological study has been conducted within a UK organisation to explore potential interaction effects between physical and psychosocial work factors. The study showed that high exposure to a combination of recognised psychosocial risk factors, high mental demands, low job control and poor social support, not only has an independent risk effect on WMSDs but also has an interactive effect on risk (Devereux et al., 1999; Devereux et al., 2002).

Of all the workers experiencing recurrent back problems in this study, about 15% of all cases were due to the relatively high psychosocial load, 50% of cases were due to the relatively high physical load and 15% of cases were because of the interaction effects between physical and psychosocial work risk factors.

The potential impact of individual psychological reactions on ergonomics interventions to reduce WMSDs

The extent of the interaction between physical and psychosocial work risk factors and WMSDs may potentially be modified by individual psychological reactions. There is evidence to support a relationship between individual psychological reactions (eg anxiety, depression and psychosomatic symptoms) and WMSDs. A review of the epidemiological literature indicates that studies have mainly been cross-sectional and subject to confounding, so it is unclear whether these so-called stress reactions are more likely to lead to the development of WMSDs or vice versa (Devereux & Buckle, 2000). There are plausible reasons why both relationships could be observed. The measures of stress reactions used in the available literature have not been similar with respect to duration, frequency or constructs used. Many of the measures used have also not been validated.

The existence of such a relationship could have a serious negative impact for ergonomics interventions that focus on physical and/or psychosocial workplace risk factors.

Interventions to reduce lifting or hand repetition rates, for example, may result in a 2 to 3-fold reduction in risk. However, despite the reduction in risk, workers may still continue to experience WMSDs because of other pathological pathways causing musculoskeletal damage, such as individual psychological reactions. These reactions may be due to prolonged exposure to psychosocial work risk factors.

Likewise, interventions that focus on reducing psychosocial work risk factors or both physical and psychosocial work risk factors may also result in about a 2-4 fold reduction in risk, but sustained anxiety and depression may mask the true impact of the intervention.

In addition, individual psychological attributes such as beliefs about the causes of job stress may also have an impact on the perceived health and health-related behaviour of workers, and may also have an impact on work organisations because of sickness absence, staff turnover etc. A recent qualitative/quantitative study identified that people possess elaborate beliefs about the causes and consequences of psychosocial work stressors, which subsequently predict psychological well-being and performance (Daniels et al., 2002).

The latest epidemiological evidence

An ongoing prospective epidemiological study (STRESSMSD Study led by Dr. J. Devereux), involving 8000 workers in 20 companies across 11 industrial sectors in the UK is designed to investigate the impact of mental work-related stress and lay beliefs of work stress on the development of WMSDs.

Preliminary results from the data support the following relationship. Individual lay beliefs concerning causes of work stress affect the risk of perceived job stress for workers associated with exposure to psychosocial work risk factors.

For example, workers with high exposure to perceived job demands (sometimes or often having to work very fast or intensively, with constant time pressure or pressured to work overtime) showed a 2 to 3-fold increase in risk of high perceived job stress. However, there was a 4 to 5-fold increase in risk for those workers who held a strong belief that job stress results from having to work too fast and in limited amounts of time, compared to workers who did not possess this belief.

A similar modification effect in the exposure-response relationship was observed for workers with strong beliefs concerning the following possible causes of stress:

- low managerial support
- low job control
- low job satisfaction

The data support the view that lay beliefs concerning the causes of perceived job stress can form an individual susceptibility to psychosocial work risk factors, such that there is an increased risk of perceiving job stress if exposed to the risk factors for which a strong causal belief is held.

The study also showed another exposure-response relationship. The higher the level of job stress, the greater was the risk of experiencing recurrent musculoskeletal problems in the previous year affecting the lower back, neck and hand/wrists (odds ratios 1.30-1.61 for moderate levels of job stress, 1.85-2.37 for very high or extreme levels of job stress, 95% confidence intervals greater than one). These results are from the cross-sectional base-line, however results from the prospective study will available next year.

The STRESSMSD study is now the only prospective multi-company study in the UK. The study is due to be completed in October 2003 and is funded by the Health and Safety Executive. A cohort such as this is needed to evaluate the effect of interventions that address physical and psychosocial work risk factors and individual psychological reactions for reducing WMSDs.

Implications for Ergonomic Interventions

Research which attempts to reduce risk to both work-related mental stress and WMSDs is still in its infancy (Pransky et al., 2002). The available literature has not addressed changes at organisational level to reduce risk to both work stress and WMSDs, but has primarily focused on specific job design changes, for example training and rest breaks. Other studies have focused on individual interventions, for example stress reduction and cognitive-behavioural techniques.

A case study is summarised in order to exemplify the meaning about changes at organisational level. A participatory ergonomics study in a delivery driver work system showed that interventions at organisational level were needed to reduce exposure to both physical and psychosocial work risk factors where the exposure levels were high (Devereux & Buckle, 1999). An intervention was designed around a model in a previous paper by the same authors (Devereux & Buckle, 1998). Modifications were made to work system goals by changing the "Delivery to the point of use" customer service package. The organisation made it clear to customers that a delivery driver would only provide this service if safe to do so. The decision was made by the delivery drivers. The intention was to reduce psychosocial work risk factors by giving delivery drivers greater control over manual handling behaviour for each delivery, greater decision-making authority concerning exposure to risk and greater managerial social support to eliminate hazards.

In addition, poor quality communication between call centre staff and delivery drivers could increase the daily load lifted, the exposure time for whole body vibration and also put the delivery driver behind schedule. The latter increased perceived time pressure and also increased the biomechanical load via increased lifting velocity and asymmetrical working postures. Therefore, recommendations were made to retrain call centre staff and for delivery drivers to meet call centre staff and discuss issues in communication. These interventions contributed to manual handling injuries being significantly reduced over the following two years.

The results provided in this paper indicate that there are two levels of primary intervention. Firstly, organisational level interventions are needed to influence work system design with respect to the work environment, specific job design, the tools and technology. Secondly, individual susceptibility is important and there is a need to identify individuals who are more likely to develop ergonomic injuries such as work stress and WMSDs. Improving self-recognition of exposure and effects and methods of health surveillance, particularly for those with high risk beliefs about work stress, may be useful in a work culture that is constructive and geared towards humanistic and encouraging norms.

The mismatch between the demands of the work system and individual capacities is more likely to be minimised if both sets of factors can be assessed and included in the work system design.

References

Ariëns, G. A. M., Van Mechelen, W., Bongers, P. M., Bouter, L. M., & van der Wal, G. (2000), Physical risk factors for neck pain, *Scandinavian Journal of Work Environment and Health*, **26**, 7-19

Bovenzi, M. (1998), Exposure-response relationship in the hand-arm vibration syndrome: an overview of current epidemiology research, *International Archives of Occupational and Environmental Health*, **71**, 509-519

Buckle, P. & Devereux, J. (1999), *Work-related neck and upper limb musculoskeletal disorders*. (European Agency for Safety and Health at Work, Bilbao, Spain)

Buckle, P. W. & Devereux, J. J. (2002), Work-related neck and upper limb musculoskeletal disorders: reaching a consensus view across the European Union, *Applied Ergonomics*, **33**, 207-217

Cox T., Randall, R., and Griffiths, A. (2002), *Interventions to control stress at work in hospital staff*. HSE Books, Sudbury

Daniels K., Harris, C., and Briner, R. B. (2002), *Understanding the risks of stress:A cognitive approach*. HSE Books, Sudbury

Devereux, J. and Buckle, P. A participative strategy to reduce the risks of musculoskeletal disorders. Hanson, M. A., Lovesey, E. J., and Robertson, S. *A. Contemporary Ergonomics* , 286-290. 1999. London, Taylor & Francis.

Devereux, J. J. & Buckle, P. W. (1998). The impact of work organisation design and management practices upon work related musculoskeletal disorder symptomology. In P. Vink, E.A.P. Koningsveld, & S. Dhondt (Eds.), *Human Factors in*

Organizational Design and Management – VI (pp. 275-279), Amsterdam: North-Holland.

Devereux, J. J., Buckle, P. W., & Vlachonikolis, I. G. (1999), Interactions between physical and psychosocial work risk factors increase the risk of back disorders: An epidemiological study, *Occupational and Environmental Medicine*, **56**, 343-353

Devereux, J.J. and Buckle, P.W. (2000), Adverse work stress reactions-a review of the potential influence on work related musculoskeletal disorders. *Proceedings of the IEA 2000/HFES 2000 Congress*, July 29-Aug 4, San Diego, U.S.A. 457-460

Devereux, J. J., Vlachonikolis, I. G., & Buckle, P. W. (2002), Epidemiological study to investigate potential interaction between physical and psychosocial factors at work that may increase the risk of symptoms of musculoskeletal disorder of the neck and upper limb, *Occupational and Environmental Medicine*, **59**, 269-277

European Agency for Safety and Health at Work (2002), *How to tackle psychosocial issues and reduce work-related stress* (Office for Official Publications of the European Communities, Luxembourg)

Health and Safety Executive (2002), Upper limb disorders in the workplace. HSE Books, Sudbury

Hoogendoorn, W. E., van Poppel, M. , Bongers, P. M., Koes, B. W., & Bouter, L. M. (1999), Physical workload during work and leisure time as risk factors for back pain: a systematic review, *Scandinavian Journal of Work Environment and Health*, **25**, 387-403

Hoozemans, M. J. M., Van der Beek, A. J., Frings-Dresen, M. H. W., Van Dijk, F. J. H., & van der Woude, L. (1998), Pushing and pulling in relation to musculoskeletal disorders: a review, *Ergonomics*, **41**, 757-781

Lundberg, U. (2002), Psychophysiology of work:stress, gender endocrine response and work-related upper extremity disorders, *American Journal of Industrial Medicine*, **41**, 383-392

NIOSH (1997), *Musculoskeletal disorders and workplace factors: a critical review of epidemiologic evidence for work-related musculoskeletal disorders of the neck, upper extremity, and low back* (DHHS (NIOSH) Publication No. 97-141, Cincinnati)

Op De Beeck, R. & Hermans, V. (2000), *Research on work-related low back disorders* (European Agency for Safety and Health at Work, Bilbao, Spain)

Pransky, G., Robertson, M. M., & Moon, S. D. (2002), Stress and work-related upper extremity disorders: Implications for prevention and management, *American Journal of Industrial Medicine*, **41**, 443-455

Punnett, L. & Bergqvist, U. (1997), *Visual Display Unit Work and Upper extremity musculoskeletal disorders. a review of epidemiological findings* (Swedish Institute of Working Life, Stockholm)

THE IMPACT OF DESIGN ON MSD AND PSYCHOLOGICAL ASPECTS IN ASSEMBLY LINES

Ma Teresa Del Carmen Ibarra Santa Ana

Institute of Occupational Health,
The University of Birmingham,
Birmingham, B15 2TT

All manufacturing systems involve human interaction into the processes in one way or in another. Still now days, when technology has facilitated processes, and in some cases extremely reduced the human intervention, the risk of work-related injuries or illness has not been eliminated or minimised enough from the manufacturing environments.
This paper presents the results of a study conducted in an automotive industry with two assembly lines. Both lines differ from each other mainly by their design. The research involved assessments of different areas within the company from quality and human resources to the application of two questionnaires to evaluate musculoskeletal disorders and stress among the shop floor employees.

Introduction

When analysing systems that are managed, operated and controlled by humans, which aim to produce goods in order to resolve problems or satisfied human necessities, it results impossible to dismiss or overlook the human factor involved in them. All manufacturing systems involve human interaction into the processes in one way or in another. Still now days, when technology has facilitated processes, and in some cases extremely reduced the human intervention, the risk of work-related injuries or illness has not been eliminated or minimised enough from the manufacturing environments. Even more, misuse of technology and resources could increase hazards in the workplace. Some times the inclusion of inadequate equipment for an operation and others the incorrect use of tools, equipment or the bad design of the workplace could be the source of musculoskeletal injures.
According to the World Health Organisation (WHO), in 1995, between the 30 and 50 percent of the employees are exposed to occupational physical hazards; and the same number report stress symptoms because of the psychological load, Vallejo (2000).
Vallejo (2000) has pointed out that it has been estimated approximately 120 millions of occupational accidents and 200 000 deaths occur every year; and even more, between 68 and 157 million of occupational injuries due to diverse causes worldwide. Since work is essential to our society and the nature of the task is predetermined, it could seem that there is little to do to change the situation. However, understanding the accidents and

occupational injuries mechanisms would situate us in a better position to design effective control and preventive strategies. Every job has been design to be productive with efficiency, the accidents and injuries are non-planned events and they can be prevented through an appropriate design and job planning (Vallejo, 2000).
For the purpose of this study it can be assume that the systems studied assemble the same product; the main difference is the design of the assembly lines. System A consists of an assembly line at floor level. The employees have to adopt awkward postures in order to reach every assembly point, handling heavy loads in some of the operations. System B on the other hand, has been designed considering ergonomic principles, the assembly line has been placed to different heights accordingly to the task in order to facilitate the operation, in many of the operations lifting devices have been placed to reduce the handling of heavy loads.

Methodology

The study within the company was divided in three steps:

1. Interview with different departments to obtain general information about: training schemes, quality programs, health and safety policy, shift work, rotation, flexibility within jobs/tasks and the level of ergonomic awareness
2. An assessment to the manual assembly operations with more musculoskeletal problems reported. This assessment was done applying direct observation methods using video. The postures and movements were later assessed using different tools and methods.
3. Application of two questionnaires to employees working in manual assembly operations in order to assess MSD and their association with work as well as occupational stress.

Results

The design characteristics of both systems were well reflected when analysing MSD and psychological disorders incidence. It was found that System A had more attendances due to musculoskeletal pain and discomfort as well as for psychological disorders such as anxiety and depression than System B. The number of days lost due to these two main reasons, MSD and psychological problems, was bigger for System A as well. However, both Systems reported high incidence in neck/back discomfort. 75% of the sample reported MSD through the questionnaires, being the most common lower back and knees discomfort. System A also reported pain on hands and shoulders while system B reported it on neck.
Significant difference ($p=.029$) was found for musculoskeletal discomfort reported between systems. The results are shown in Table 1; System A employees reported more pain and discomfort than employees on system B. The difference can be linked to the system design itself.

Table 1 Significant difference in MSD between systems

	System A *(n= 37)*	System B *(n= 35)*	F	p
MSD	2.35 *(±1.89)*	1.46 *(±1.48)*	4.956	.029*

After analysing the whole sample, it was decided to analyse each of the systems separately in order to assess them independently. Significant difference was found between employees with a height of less than 1.73m and those who measure 1.73m or more, as shown in Table 2. People shorter than 1.73m reported significantly less musculoskeletal disorders than those above the percentile. The difference could be related to the poor design of the system A where people have to lean or get inside of the car in order to perform their tasks. Shorter people has certain advantages since they have to bend less their backs and they find easier to get inside of reduce spaces than taller employees.

Table 2 Significant difference in MSD between height percentiles

	1st 50th percentile *(n= 17)*	2nd 50th percentile *(n= 16)*	F	p
MSD	1.71 *(±1.31)*	3.50 *(±1.97)*	9.616	.004**

System A reported more MSD in left shoulder and right hand than system B accordingly with a Chi-square analysis (Table 3 and 4) which could be associated with the assembly line design as well.

Table 3 Correlation between Left Shoulder and System

	No MSD in left shoulder	MSD in left shoulder	
System A	28	9	37
System B	34	1	35
	62	10	

*Yates Continuity Correction = 5.252 p=.022**

Table 4 Correlation between Right Hand and System

	No MSD in right hand	MSD in right hand	
System A	26	11	37
System B	32	3	35
	58	14	

*Yates Continuity Correction = 3.878 p=.049**

Significant correlation also was found for number of months that an employee has worked in the company and discomfort in the lower back. Those employees who have been more time in the company reported more discomfort in the lower back region, Table 5.

Table 5 Correlation between Lower Back and Months worked at the company

	No MSD in lower back	MSD in lower back	
1st percentile	22	10	32
2nd percentile	15	25	40
	37	35	

*Yates Continuity Correction = 5.755 p= .016***

An important result appeared when performing ANOVA analysis to the psychological questionnaire. Significant difference was found regarding satisfaction, sources of pressure and influence and control between systems, Table 6. Employees of System B reported more satisfaction and having more influence and control in their jobs than employees in System B. However, System B reported significantly higher stress than System A. The high satisfaction of employees in System B is linked with the fact that they have more influence and control over their work, since these two factors are correlated significantly. It could also be linked to the more ergonomically designed environment. The fact that system A reported less stress is surprisingly, since system A has a shorter production cycle and there is not as many equipment to help production as in System B.

Significant difference also was found between employees who had had major events as the death of a relative, etc.; those reporting being under pressure and physical well being.

Table 6 Significant Differences in PMI between Shifts

	System A *(n= 37)*	System B *(n= 35)*	F	p
Satisfation	38.78 *(±9.86)*	44.46 *(±7.22)*	7.691	.007
Sources of Pressure	101.24 *(±28.35)*	114.26 *(±26.38)*	4.180	.045
Influence and Control	25.73 *(±3.99)*	27.89 *(±4.90)*	4.214	.044

Discussion

The inappropriate design of the assembly lines and workplace resulting in employees having to adopt awkward postures and perform repetitive movements, are the main reasons of the high incidence of musculoskeletal discomfort reported. Regarding psychological factors, it was found a high number of absenteeism due to depression, nervous debility and other psychological problems. Knave (1991) mentions the use of absenteeism, sickness absence and personnel turnover statistics as predictors of ergonomics and psychosocial factors in the shop floor.

However, not only the physical characteristics within the work environment as tools, assembly lines and machinery design have an impact on the development of musculoskeletal injuries, it was also found that the way in which work has been organised plays a key role. Work design and organisation involve many important issues as training, environment, health, safety, welfare, communications, etc. The correct integration of all these factors affects the way in which employees feel at work. Luczak et al (1998) and Eklund (1999) mention the relation between the individual and the work environment emphasising the need of a holistic ergonomics approach in the workplace.

Regulations and risk assessments are key factors in the prevention of these hazards when assessing not only the physical side but also the psychological, sociological as well as the human interaction with the work environment.

As a result of this research, it has been found a link between stress and other variables as support, coping, satisfaction, type A behaviour, as well as mental and physical well being, and workplace design. Carayon et al., (1999) discusses the link between work organisation, job stress and musculoskeletal disorders. Even if no correlation between MSD and psychological assessment was found, it is clear that system A has not only more MSD but also more psychological problems than system B. It can be concluded that the better workplace design in system B has been reflected on the physical and the psychological sides, among employees.

However surprisingly, in the psychological questionnaire results, System B employees reported more stress than employees of system A. System B also reported more satisfaction and influence and control in their jobs. This could be the reason of why even if experimenting more sources of pressure, there are fewer reports of psychological problems in the medical service. Being satisfied with one's job and feeling that enough influence and control is exercised, seems to moderate the effects of stress.

Conclusion

1. Musculoskeletal disorders are still a significant problem among manufacturing industries.
2. A close relation between workplace design, involving more than just the physical aspects, and musculoskeletal disorders exists. Satisfaction, organisation and stress are all linked to physical and mental well being including musculoskeletal disorders.
3. Since stress and musculoskeletal problems have been found correlated with each other, it is necessary to introduce in the industry methodologies and procedures to assess the welfare of the employees in a holistic way, integrating the psychological side as well as the physical one.
4. Companies must try to identify not only those tasks or operations with high risk of musculoskeletal disorders but also those that could have a bad psychological effect among their employees in order to prevent the outcome by changing the working conditions, improving the organisational climate and enriching the jobs in order to reduce monotony and repetitiveness.
5. Companies should give more importance to the human factors within them and treat them as the most precious resource to fulfil their vision. More investment has to be made in systems that allow the early detection of task and operations with high risk of injury.
6. Ergonomics holistically seemed plays a key role in quality of working life and total quality management philosophies.

References

Carayon, P. et. Al, *Work Organisation, Job Stress, and Work-Related Musculoskeletal Disorders,* Human Factors, Vol 42, Number 4, December 1999, Human Factors Engineering Society.

Eklund, J.*Ergonomics and Quality Management –Humans in Interaction with Technology, Work Environment, and Organisation.* International Journal of Occupational Safety and Ergonomics, 1999, Vol.5, Number 2, pages 143-160.

Knave, B et al. *Incidence of work-related disorders and absenteeism as tools in the Implementation of Work Environment Improvements: The Sweden Post Strategy,* Ergonomics, 1991, Volume 34, Issue 6, Pages 841-848.

Luczak, H, et al. *Ergonomics and TQM,* in Karwowski, W and Salvendy G. 1998, Ergonomics in Manufacturing: Raising Productivity through Workplace Improvement, (Society of Manufacturing Engineers) pp 505-531.

Vallejo, J, *Taller para la detección de factores de riesgo ergonómico, Centro de Calidad Ambiental*, ITESM, Campus Monterrey, Noviembre 2000, México.

OCCUPATIONAL HEALTH AND SAFETY

STRESS IN THE WORKPLACE: IMPLICATIONS OF RECENT DECISIONS BY THE COURT OF APPEAL

Neil Morris[1] and Bianca Raabe[2]

[1]*University of Wolverhampton, Department of Psychology, MC Block, Wolverhampton WV1 1SB, UK*
[2]*The Open University in the West Midlands, 66–68 High Street, Harborne, Birmingham B17 9NB, UK*

In 1995 a precedent was set in Walker v Northumberland County Council when substantial damages for work related stress were awarded. In 2000 in Howell v Newport County Borough Council more than £250 000 damages were awarded. Several further cases were also awarded large sums but most of these decisions were overturned on appeal in February 2002. The Court of Appeal outlined the conditions under which a claim for damages for workplace stress arising from alleged employer negligence is likely to succeed. This paper disseminates and discusses the implication of the recent changes. These changes are radical and of considerable import to those advising stressed workers.

Introduction

When the workplace is believed to be the cause of emotional damage the victim usually claims that they are suffering from stress related problems. If the problems arising from this cannot be settled without recourse to the legal system then a civil action ensues and the claim is that the employer has been negligent. This is a 'tort' or civil wrong and the usual remedy is damages. Most cases for the tort of negligence, arising from injury in the workplace, are the result of some physical injury that, it is claimed, resulted from an employers failure to provide a safe environment. Only in the last 50 years has English law extended the tort of negligence to encompass 'psychiatric injury' (Barrett and Howells, 2000). Prior to this successful actions involving emotional distress were confined to cases involving 'nervous shock' ensuing from involvement in, or close observation of, events

causing or likely to cause severe physical injury or death. Thus the law recognised conditions such as post-traumatic stress disorder. However, in recent years, i.e., the last decade, some civil actions have succeeded when 'mental breakdown' has resulted from persistent work overload.
In December 2000 Janice Howell brought a successful action against her former employers, Newport County Borough Council. She received a settlement worth £254, 362 after the Borough Council admitted liability for intolerable working conditions at Maindee Junior School, Newport, Gwent. Mrs. Howell argued that she suffered unacceptable levels of stress while teaching at the school. Returning from extended sick leave with medical evidence that her previous working conditions were injurious to her health, she faced similar stressful conditions and took early retirement when she was unable to cope (reported in *The Times (London)* December 5th 2000. - www.thetimes.co.uk).

The Tort of Negligence

As noted above, in cases of alleged workplace induced stress actions are brought for the tort of negligence. For an action for negligence to succeed the injured party must show that the defendant owes them a duty of care and that this duty of care has been breached. This involves invoking the *neighbour* principle, that is, showing that the injured party could be considered to be the *neighbour* of the defendant. In Donoghue v Stevenson (1932 – cited in Owen, 2000) the neighbour principle was established and Lord Aitken explained that a neighbour was anyone so closely affected by an act that it is reasonable to expect that one could contemplate that one's acts or omissions might cause them injury. This establishes the important principle that, unless there is a specific statutory duty, a duty of care is *only* present if the court accepts that there is a reasonable expectation of possible injury. There is no liability if the employer could not have foreseen that such injury might occur. However, under common law, an employer has a duty to provide 'competent' fellow employees and a safe work environment. Employers, provided they are not using self-employed workers or those employed by an independent sub-contractor, are *vicariously* liable for the negligence of their employees but only *in the course of employment.* Thus a company has a duty to provide a safe environment for its staff and it is corporately liable for the negligence of its employees but not for their actions outside of work duties (Owen, 2000). Hence, a local authority, for example, whose management team fail to observe a *reasonable* duty of care to an employee is liable for the negligence of the management team. The crucial factor here is *what constitutes reasonable?* As noted above, there is no absolute duty of care so, for example, an action for negligence may fail if it would not be reasonably foreseeable that injury would occur. Injury *per se* does not constitute negligence. In the context of workplace stress there is case law that clarifies what may be considered reasonably foreseeable.

Walker v Northumberland County Council (1995) and it's implications

Walker v Northumberland County Council (1995) established that

'it may be negligence for an employer to expose employees to conditions where they are so stressed as to be caused psychiatric injury' (Barrett and Howells, 2000, p. 433).

The plaintiff, Walker, had been a social services officer in Blythe, in the north-east of England, for 17 years when he requested extra staff and management guidance as the caseload in the area increased. He was provided with no further staff or guidance and he had a 'nervous breakdown' and took three months sick leave following medical advice. On returning to work his workload was not reduced in line with the expectation he had of a phased re-entry. Subsequently, he had a second breakdown and was dismissed from his post on grounds of ill health. The local authority, while admitting that they owed Walker a duty of care to provide a reasonably safe working environment, argued that they had not breached that duty. They also argued that if the risk to Walker's health was reasonably foreseeable they still had not acted unreasonably in not reducing his workload because they faced budgetary constraints that made this difficult or impossible. Mr. Justice Colman found for Walker. In discussing the difficulties of establishing foreseeability and causation when the 'damage' was 'psychiatric' rather than physical he based his argued on the fact that Walker's first breakdown was so 'substantial and damaging' that there was a foreseeable risk of this re-occurring if Walker returned to the same working conditions. The crux of the case is that the first breakdown was not predictable *but* the second breakdown was foreseeable so the County Council was negligent in its treatment of Walker. Secondly, there was no precedent in English law that a statutory body could use matters of policy (for example, fiscal considerations) to *negate their obligation in law* to provide a reasonably safe work environment. This case is directly applicable to Janice Howell's case. Taken together these cases suggest that employers, especially statutory bodies, will need to respond in a much more effective manner when employees suffer 'stress injury' unless they are prepared to face very punitive damages.

When negligence has been established damages are usually awarded but these can, under the Law Reform (Contributory Negligence) Act 1945, be very substantially reduced if the plaintiff can be shown to have contributed to the damage inflicted. Thus it seemed likely that employers would have little option but to attempt to limit their liability by claiming contributory negligence. We have discussed how the claimant's physical and emotional constitution may be examined to reduce damages elsewhere (Morris and Raabe, 2001; 2002).

Assuming that a plaintiff feels that he or she has not contributed to their stress response and can defend this position in court then they may chose to bring an action for the tort of negligence against their employer. It would be necessary to show negligence in terms of the neighbour principle. Following the Walker and Howell cases, a strong case would be made if

a) substantial injury had occurred, for example, the plaintiff became unable to continue working in the foreseeable future.

b) following a previous severe stress related illness, with medical documentation, the employer made no attempt to modify work conditions when made aware of this. This is crucial because what may have been unforeseeable at first occurrence now provides the information necessary to predict a repeat of the stress reaction. The claimant is a *neighbour*, i.e., someone that one could contemplate would be affected by one's acts or omissions in a way that might cause them injury.

The Court of Appeal decisions.

Since the Walker and Howell cases there have been several large sums of money awarded as a result of actions for negligence resulting in stress related illness. However employers have, in a number of cases, appealed against these awards. In February 2002 most of these decisions were over-turned by a ruling from the Court of Appeal. The main points in the decision are shown in Table 1. These rulings set precedent for courts below the Appeal Court and thus, at the time of writing, represent the legal position with respect to acts for negligence leading to stress in the workplace.

Table 1: Summary of salient points of the Court of Appeals decision on 5th February 2002.

. For an action to succeed, the "signs of stress in a worker must be plain enough for any reasonable employer to realise something should be done". If it is not obvious then the employee must warn the employer of the problem. Employers are entitled to take what they are told by employees at face value.

. The onus is on the worker to decide whether to leave the job or carry on working and accept the risks.

. An employer will not be in breach of his or her duty of care in letting a willing employee continue in a stressful job if the alternative is dismissal or demotion. The employer is in breach of duty of care only if he/she fails to take reasonable steps bearing in mind the level of, and justification for, the risk, the potential consequences and the cost of prevention.

. There are **NO** occupations that should be regarded as intrinsically dangerous to employees' mental health.

. Employers who offer confidential counselling advice services with access to treatment are unlikely to be found to be in breach of their duty of care.

These major changes raise a number of points.

a) An employer is not expected to be unusually vigilant in detecting employee stress. The onus is on the employee to bring to the attention of the employer that he/she is stressed. Only in circumstances when the employee is so stressed that he/she has lost the capacity to report stress and the signs are overt does a duty to act upon this fall upon the employer. This is very much in line with the neighbour principle requirement for forseeability of injury. The more sinister implication is that employers are entitled to take what they are told at face value. An employee may well deny feeling stressed because he/she is afraid of the consequences for their future work prospects if they admit to feeling stressed or unable to cope with work. Such an admission may be construed as weakness or incompetence or as an open criticism of management practice. All of these outcomes are clearly inhibitory suggesting that this last point is an unwelcome precedent.
b) A worker who chooses to stay in their present employment despite feeling stressed is assumed to accept the risks associated with doing so. This is not a straightforward point in law. It appears to be a recommendation that the employee can choose to accept the risks of continuing in a stressful job and loose their right to a remedy in law as a result of accepting the risk. Clearly this is problematic because financial constraints might well compel an employee to follow this route but such a recommendation is also at odds with employment law. There are limits to the extent to which one can sign away basic rights in the workplace. This should prove to be an interesting precedent that will, we think, infringe the Human Rights Act (1998).
c) The employer does not breach the duty of care if dismissal or demotion is the only alternative and if the employee is willing to continue working under stress. Now we see aspects of Walker v Northumberland County Council (1995) being overturned. Mr. Justice Colman's argument that fiscal considerations cannot remove an obligation in law is diluted and the return to work by an employee who has no guarantee that work conditions will change may constitute consent to accept these practices. This has similar implications to b) above. In addition, it is not clear what effect accepting seriously health-damaging work will have on insurance premiums.
d) The view that there are no occupations *intrinsically* dangerous to mental health may be a fallacy in terms of psychological science but may be useful in law in as much as one will still have to show injury even if an occupation is potentially psychologically hazardous, there can be no strict liability.
e) We have dealt with the role of counselling at length in Morris and Raabe (2002) and this recent change in law increases the concern expressed in that paper that counselling provisions of too Spartan a nature to be useful may be introduced to protect employers from liability for stress in the workplace.

These legal changes should be of concern to both employers and employees. It is not useful to employers to simply be protected from paying damages if the workplace is stressful. Workplace stress is expensive in terms of lost revenue and it is in every ones interest to reduce it. Cases in which damages of around £250 000 have been awarded are not typical and have all involved an extraordinary level of workplace stress and unresponsiveness from management. The changes wrought by the Court of Appeal not only protect bad management but also show a lack of concern for employees that few employers wishing to have a productive workforce would welcome.

References

Barret, B. and Howells, R. (2000) *Occupational Health and Safety Law: Text and Materials* (2nd edition). London: Cavendish.

Morris, N. and Raabe, B. (2001) Psychological stress in the workplace: cognitive ergonomics or cognitive therapy? In M. Hanson (Ed.) *Contemporary Ergonomics 2001.* London: Taylor and Francis. 257-262.

Morris, N. and Raabe, B. (2002) Some legal implications of CBT stress counselling in the workplace. *British Journal of Counselling and Guidance,* **30,** 55-62.

Owen, R. (2000) *Essential Tort Law* (3rd edition). London: Cavendish.

Walker v Northumberland County Council (1995) **IRLR 35** (reprinted in Barret, B. and Howells, R. (2000) *Occupational Health and Safety Law: Text and Materials* (2nd edition). London: Cavendish. P. 122-125).

THE PRINCIPLE OF ERGONOMICS WITHIN EUROPEAN UNION DIRECTIVES

Vincent Kelly and Jason Devereux

Robens Centre for Health Ergonomics
University of Surrey
Guildford, Surrey, GU2 7TE

The principle of ergonomics underpins the general principles of prevention in the European Framework Directive. These principles can be classified into managing risk, adopting the work to the individual, policy, personal protection and training. These general principles of prevention apply to any occupational health and safety problem in the European Union. Despite ergonomics being central to the directive, there is no specific mention of psychosocial factors in the workplace. In view of advancing knowledge, a revision is recommended. In addition, the principle of ergonomics should be central to any new directive.

Introduction

The European Union (EU) legislation on health and safety is primarily contained within the Framework Directive (89/391/EEC) and a series of daughter Directives (The Council of the European Communities, 1989). The framework directive sets the overall framework for health and safety and lists the general principles of prevention. The daughter Directives deal with individual aspects of the work system, for example the Manual Handling Directive (89/269/EEC) and the Display Screen Equipment Directive (89/270/EEC) (The Council of the European Communities, 1990a; The Council of the European Communities, 1990b).

These directives have a scientific basis. In this paper the ergonomic principle underlying the general principles of prevention will be examined and the relevance to practice considered.

In view of the extent of work-related stress in the EU and its potential impact on work-related musculoskeletal disorders, as well as other health conditions, a new directive is likely. This paper considers whether the principle of ergonomics should underpin such a future directive within the general principles of prevention.

The general principles of prevention in the Framework Directive

The general principles of prevention are set out in Article 6(2) of the Framework Directive as:
(a) avoiding risks;
(b) evaluating the risks which cannot be avoided
(c) combating the risks at source
(d) adapting the work to the individual, especially as regards the design of workplaces, the choice of work equipment and the choice of working and production methods, with a view in particular to alleviating monotonous work and work at a predetermined work rate, and to reducing their effect on health.
(e) adapting to technical progress
(f) replacing the dangerous by the non-dangerous and less dangerous

(g) developing a coherent overall prevention policy which covers technology, organisation of work, working conditions, social relationships and the influence of factors relating to the working environment
(h) giving collective protective measures priority over individual protective measures; and
(i) giving appropriate instructions to the workers

These principles can be classified into 5 categories: managing risks (a, b and c); adapting the work system, technology and equipment to the individual (d, e and f); policy (g); personal protection (h); and instruction (i).

Managing risk

The general principles of prevention are exemplified in the requirements of the Manual Handling and Display Screen Equipment Directives. Taking the requirements of the Manual Handling Directive as an example, the principles are as follows: avoidance of manual handling where possible, and where manual handling cannot be avoided, assessment of the risks to reduce these to a minimum.

In the Manual Handling Directive, the definition of the manual handling of loads includes the expression "unfavourable ergonomic conditions", putting ergonomics at the centre of the management of manual handing risks. Unfavourable ergonomic conditions occur when there is an imbalance between the demands of the working task and the capacity of the individual. Contemporary ergonomics knowledge shows the demands of work to be of both a physical and a psychosocial nature (Devereux et al., 1999; Devereux et al., 2002). An overview of the scientific literature is provided in this issue of Contemporary Ergonomics by Devereux (2003).

Adapting the work system, technology and equipment to the individual

The basic tenet of adapting the work, technology and equipment to the individual is the application of good job design. Good job design involves identifying the important job characteristics from the worker's perspective and then designing accordingly.

The information required includes the nature of the organisation, for example its work system goals, future developments, company history, and business environment; job content, for example important physical and mental work activities, use of job aids and managerial tasks; job context, for example presence of stressors, performance conditions, work atmosphere, leadership climate and organisational culture; critical characteristics, for example skills, cognitive capacities and traits (Devereux & Buckle, 1998; Huang et al., 2002).

This information can be obtained from mission statements, company reports, written job descriptions, job holders' reports, colleagues reports and qualitative methods.

Once a job design has been operationalised, the characteristics of individual applicants can be assessed through a selection procedure. The closest match is identified between the job design and individual applicants. After this, the work must be adapted to the individual to eliminate any remaining mismatches. This process is important because the variation in individual characteristics makes it difficult, if not impossible, to specify and implement without testing. This is why ergonomics is an applied science rather than an engineering discipline where one can specify and implement with a guaranteed result.

Policy

Organisational policy on health and safety represents a genuine commitment to action. The general intentions, approach, objectives, criteria and principles upon which the policy bases its action should be stipulated (Health and Safety Executive, 1997). The three core issues that policy must take account are the organisation, personal factors and the job.

The general principles of prevention stipulate that a coherent overall prevention policy should cover technology, organisation of work, working conditions, social relationships and work environment factors. Balancing these essential elements is the requirement for an ergonomics model (Carayon et al., 1999).

Personal protection

Even with a system for managing risks, implementation of good job design and a clear organisational policy, an ongoing system of protection is required to identify any imbalance between the work and the individual, which may result in adverse health reactions.

The first requirement is supervision, to ensure that policy and work organisational practices are adhered to. A system must be in place to allow complaints about policy, work organisation practices etc to be formally acknowledged, investigated and, where necessary, acted upon. Appropriate health surveillance should be provided to ensure that possible damage from health and safety risks that occur at work are identified as early as possible.

Workers should have information concerning risks to health and safety within an organisation. Work systems are constantly evolving and therefore the balance between organisational factors, technology, job/task design, the work environment and the individual may be affected. A job design should involve balanced participation by workers in the development of a proposal for work changes that affect their health and safety. This is commonly referred to as a participatory ergonomics process. There are different definitions of participatory ergonomics but each highlights the involvement of those who perform the work tasks (Devereux & Buckle, 1999; Devereux et al., 1998).

Training

Managing risks, adoption of work to the individual and personal protection are functions of the employer. The objective is to ensure a balance between work and individual capacity.

Individual capacity can be adversely affected by individual behaviour, for example, an office worker not using a back rest while using a computer for 6 hours during the day. Therefore, to minimise any imbalance between work and the individual, behaviour must be consistent with work system goals, practices and policy.

This is a requirement to provide adequate instructions in the general principles of prevention but this has been expanded in article 12 of the Framework Directive to the training of workers. According to Pheasant and Stubbs (1994) the provision of training should involve knowledge, procedures, practices and skills.

However, in order for the training to be usable, it must be effective, efficient and satisfactory to the trainees. Therefore, the knowledge should address the needs of the job and the capacity of the worker to understand the knowledge given. The practices, procedures and skills should be developed taking account of the individual worker's capacity and implementation of the training. The training should be evaluated and the reactions, outcomes and impact of the training assessed on an ongoing basis to ensure satisfaction amongst workers.

Current Position

The general principles of prevention in the Framework Directive apply to any occupational health and safety problem in the EU. Back pain and work-related stress are the two leading occupational health problems in Europe according to the Third European Survey on Working Conditions (Paoli & Merllié, 2001).

In an attempt to address back pain caused by manual handling, a specific directive has been issued. Despite ergonomics being central to the directive there is no specific mention of psychosocial risk factors in the workplace. There is available evidence now showing these factors to be important for back pain, neck and upper limb disorders in general, and this is a consensus view across the EU (Buckle & Devereux, 2002; Op De Beeck & Hermans, 2000). From a scientific and practical perspective, European Directives should take into account advancing knowledge.

As yet, a directive has not been issued in relation to work-related mental stress which affects 28% of workers in the European Union (Paoli & Merllié, 2001). However, the European Agency for Safety and Health at Work, Bilbao, has published a report on managing work stress (Cox et al., 2000) and on reducing psychosocial issues to reduce work-related stress (European Agency for Safety and Health at Work, 2002). These reports indicate good practice.

Ongoing pioneering research, led by Dr. J.J. Devereux, is being carried out at the Robens Centre for Health Ergonomics investigating the relationship of working conditions, work-related stress and the potential impact of work-related stress upon the development of musculoskeletal disorders. The results of this research, as well as results from other key epidemiological studies across the EU, can potentially contribute to the development of a European Directive on work-related mental stress. According to the European Union Community Strategy on Health and Safety at Work for 2002-2006, work-related mental stress is a top priority issue on the agenda.

A European Directive on work-related mental stress could also have implications for the prevention of other work-related disorders such as back pain and neck and upper limb musculoskeletal disorders due to psychosocial work factors and work stress being involved in their development (Devereux, 2003). The principle of ergonomics should also be inherent within the prevention principles of work-related mental stress since it arises from an imbalance between the worker and the working task.

Conclusion

The principle of ergonomics is inherent within every aspect of the general principles of prevention set out in the Framework Directive. In these circumstances it is reasonable to conclude that the principle of ergonomics should be central to any new directive on work-related mental stress.

References

Buckle, P. W. & Devereux, J. J. (2002), Work-related neck and upper limb musculoskeletal disorders: reaching a consensus view across the European Union, *Applied Ergonomics*, **33**, 207-217

Carayon, P., Smith, M. J., & Haims, M. C. (1999), Work organisation, job stress and work-related musculoskeletal disorders, *Human Factors*, **41**, 644-663

Cox T., Griffiths, A., and Rial-González, E. (2000), Research on work-related stress. European Agency for Safety and Health at Work, Bilbao

Devereux, J. and Buckle, P. A participative strategy to reduce the risks of musculoskeletal disorders. Hanson, M. A., Lovesey, E. J., and Robertson, S. A. *Contemporary Ergonomics*, 286-290. 1999. London, Taylor & Francis

Devereux, J.J. (2003), Work-related stress as a risk factor for WMSDs: Implications for ergonomics interventions. *Contemporary Ergonomics*. London, Taylor & Francis

Devereux, J. J. & Buckle, P. W. (1998). The impact of work organisation design and management practices upon work related musculoskeletal disorder symptomology. In P. Vink, E.A.P. Koningsveld, & S. Dhondt (Eds.), *Human Factors in Organizational Design and Management – VI* (pp. 275-279), Amsterdam: North-Holland

Devereux, J. J., Buckle, P. W., & Vlachonikolis, I. G. (1999), Interactions between physical and psychosocial work risk factors increase the risk of back disorders: An epidemiological study, *Occupational and Environmental Medicine*, **56**, 343-353

Devereux, J. J., Buckle, P., & Haisman, M. F. (1998), The evaluation of a hand-handle interface tool (HHIT) for reducing musculoskeletal discomfort associated with the manual handling of gas cylinders, *International Journal of Industrial Ergonomics*, **21**, 23-34

Devereux, J. J., Vlachonikolis, I. G., & Buckle, P. W. (2002), Epidemiological study to investigate potential interaction between physical and psychosocial factors at work that may increase the risk of symptoms of musculoskeletal disorder of the neck and upper limb, *Occupational and Environmental Medicine*, **59**, 269-277

European Agency for Safety and Health at Work (2002), *How to tackle psychosocial issues and reduce work-related stress* (Office for Official Publications of the European Communities, Luxembourg)

Health and Safety Executive (1997), *Successful health and safety management*. HSE Books, Sudbury

Huang, G. D., Feuerstein, M., & Sauter, S. L. (2002), Occupational stress and work-related upper extremity disorders:concepts and models, *American Journal of Industrial Medicine*, **41**, 298-314

Op De Beeck, R. & Hermans, V. (2000), *Research on work-related low back disorders* (European Agency for Safety and Health at Work, Bilbao, Spain)

Paoli P. and Merllié, D. (2001), *Third European survey on working conditions*. Office for Official Publications of the European Communities, Luxembourg

Stubbs D. and Pheasant, S. (1994), *Manual Handling: an ergonomic approach*. The National Back Pain Association, Teddington

The Council of the European Communities (1989), Council Directive of 12 June 1989 on the introduction of measures to encourage improvements in the safety and health of workers at work (89/391/EEC), *Official Journal of the European Commission*, **L183**, 1-8

The Council of the European Communities (1990a), Council Directive of 29 May 1990 on the minimum health and safety requirements for the manual handling of loads where there is a risk particularly of back injury to workers 90/269/EEC (fourth individual Directive within the meaning of Article 16(1) of Directive 89/391/EEC), *Official Journal of the European Commission*, **L156**, 9-13

The Council of the European Communities (1990b), Council Directive of 29 May 1990 on the minimum safety and health requirements for work with display screen equipment 90/270/EEC (fifth individual Directive within the meaning of Article 16(1) of Directive 87/391/EEC), *Official Journal of the European Commission*, **L156**, 14-18

CONTEMPORARY ERGONOMICS PRACTICE – THE NEW ZEALAND SETTING

George S. Corbett & Dave Moore

George S. Corbett
UNITEC Applied Technology Institute
Private Bag 92025, Auckland
New Zealand

Dave Moore
Centre for Human Factors and Ergonomics
Massey University Campus at Albany
Auckland, New Zealand

New Zealand as a country is small but the ergonomics community is active, with ergonomics as a discipline comparing well internationally for social presence. Professional cohesion has been helped by a strong, dominant connection with Loughborough University in the United Kingdom. The largest concentration of ergonomists is in evidence- based injury prevention, design and evaluation in the primary industries where New Zealand already has a lead role is growing. International work experience and training is the norm for ergonomists - not the exception as it is in the UK. For ergonomists in New Zealand there are limitations however arising mostly from the geographic isolation. The political and regulatory climate is generally favourable for practice, but the state-level systems have weaknesses, including role-conflict between government bodies.

Ergonomics practice

This paper gives a brief outline of the state of Ergonomics practice in New Zealand and the political and regulatory climate that local ergonomists operate in.

A small set of islands in the South Pacific
New Zealand is a small country made up of the North and South Island's and numerous smaller islands. The North Island is volcanic and the South Island is mountainous and sparsely populated compared to the North.

The total population of New Zealand is approximately 3.8 million in comparison to the United Kingdom, which has 59 million. New Zealand has a landmass of 265,000 sq kilometres while the United Kingdom has only 245,000 sq kilometres. The capital city, Wellington, is located at the bottom of the North Island with the largest city Auckland also located in the north. The New Zealand climate is described as being maritime and can be very unpredictable because of its isolation and geography.

Table 1. International comparisons

	NZ	UK	Neth.	US	Canada	Nordic	Ireland
Land area in square km (000)	268	245	42	9343	9970	1153	70
Population (millions)	4	60	15	250	30	23	5
People per sq km	12	233	377	26	3	20	71
Members of IEA Federated Societies	120	1024	510	3655	547	1596	35
Members per million of total population	30	17	34	15	18	70	7
Certified Ergonomists [CREE equivalent]	12	60	35	827*	54	18	5
Certified Ergonomists per million total pop	3	0.8	2.3	3.3	1.8	1.3	1
1000s of Sq km per Certified Ergonomist	22	4	1	11	185	64	14

*Around 10% of these appear to be based overseas. This is also the case in New Zealand and possibly the same in other countries listed.

Where New Zealand Ergonomists work

There is not a formal survey available, but if we look at the professional roles of current key ergonomists in New Zealand (Certification Scheme members, New Zealand Ergonomics Society (NZES) post-holders, and any Past Presidents not in the former groups) as at December 2002 (n=22) we see the following.

Table 2. Positions

CoHFE* (Part of a Government-owned research institute, Forest Research, based in Rotorua in the central North Island)	4
University academic staff	4
Health providers / therapists	3
ACC (State owned Accident Compensation body)	3
Private ergonomics consulting	2
Working overseas	2
OSH (Government department: role comparable to HSE)	1
Civil Aviation Authority	1
Forestry industry	1
Dairy industry	1

* Centre for Human Factors and Ergonomics

Where New Zealand ergonomists were educated

For the same group of 22 people the following table shows where their major taught ergonomics qualification (Masters or equivalent) was gained.

Table 3. Education

Massey University NZ	11
Loughborough University UK	8
University College London	1
Surrey University	1
University of Wisconsin	1

The Massey University course was developed initially by Loughborough and UCL-trained ergonomists. Hence the structure, philosophy and perspective of ergonomics presented to students has remained close to that of these institutions.

This has led to the NZ ergonomics community having a less diverse set of perceptions about fundamental tenets-such as what ergonomics is and how it is practiced-than may have been the case with a mix of European and North American influences at inception.

More recently, the course has been re-structured to more closely meet the needs of the CREE based certification scheme, but again, those involved were predominantly Loughborough and/or Massey trained. Inevitably perhaps, these links have also resulted in continuing collaborative working relationships between the East Midlands and New Zealand.

This unusually singular approach to the discipline in New Zealand could be seen by some as professional impoverishment, but one real advantage has been the minimising of internal arguments within the domestic ergonomics community. New Zealand is also not divided into States or Provinces with there own local governance to negotiate. Professional cohesion and a single tier political system both greatly assist the process of getting reasonably swift beneficial changes at a national level.

Type of work undertaken by New Zealand Ergonomists.

Until it's entry into the EEC, Britain was by far the major destination for New Zealand's industrial output. Now it is fourth behind Japan, Australia and the United States. Primary industries (notably: dairy, meat forestry and wool) lead the NZ economy and therefore much of the work for research and intervention design work for ergonomists is linked to these rather than manufacturing, aerospace, or information technology industries as may be found elsewhere. There is increasing ergonomics activity in the dairy sector, New Zealand's biggest earner, both in processing and on the farm.

The only industry to have a dedicated team looking at ergonomics issues is forestry.

Over the last 15 years the Centre of Human Factors and Ergonomics has established a significant presence and has conducted international standard applied research in logging and silviculture. Funding comes predominantly from central government. This compares with models such as the Safety Associations in Ontario who have been able to maintain a presence in specific industries over several decades through industry levies.

Another area where ergonomists are working is as consultants. As in most countries consultants operate in most industries to some extent. A surge of contracted opportunities has followed the implementation of the Health and Safety in Employment Act (1992).

Emerging areas of work

The New Zealand Government is currently developing the New Zealand Injury Prevention Strategy (2002). This will inevitably influence the placing of funding for intervention-focussed research over the next few years.

Falls, both occupational and domestic are likely to be targeted. These remain the leading cause of hospitalisations in New Zealand. Drownings in New Zealand are a significant problem with so much water-based activity, but remain under-researched. Prevention strategies have been poorly resourced.

Traditionally difficult and expensive areas for research and/or interventions are increasingly being tackled by New Zealand ergonmists. Any industry or sector dominated by small contractors (80% of New Zealand companies employ six people or less) has been left somewhat behind in health and safety efforts by the bigger employers. These are more visible to the enforcement bodies, and have had to respond by putting in more resources as a result. Construction is therefore being targeted; in particular the residential sector where most work is done by hard to reach self-employed operators.

Agriculture and horticulture. A disparate industry of 80,000 plus farms plus suppliers and casual staff, represented by over 80 established consumer groupings. This is not an easy industry to research to the depth needed for detail intervention design, but a start has been made with significant funding awards finally being made to CoHFE in 2001 and 2002. Ongoing work on quad-bike (ATV) related injuries is attracting international interest. (Moore and Bentley, 2002)

In addition research related to tourism looks set to gain momentum. Injuries to tourists in New Zealand are costly to the country as well as the injured party as publicising of spectacular fatalities and so forth in the country of origin can result understandably in reduced bookings. New Zealand is highly regarded as an adventure tourism destination, but expectations regarding risk control vary between cultures and sub groups. Work by Bentley and others have highlighted high injury rates with activities of low perceived but high actual risk, such as horse riding and biking - as opposed to aviation sports which have high perceived risk but a low actual rate of compensatable injuries.

The working Environment

Occupations linked to Health Safety in New Zealand gained higher prominence with the introduction of the Health and Safety in Employment Act in 1992. The 1995 Regulations pertaining to this Act are also of relevance. There is the requirement included that 'all plant must be designed in accordance with relevant ergonomics principles'. This is clearly ambitious as no Regulation of this nature was acting on industry previously and much of the plant is imported. The Regulation has yet to 'bite', as OSH inspectors first themselves have to know what the 'relevant principles' are. The existence of a basic undergraduate Ergonomics paper within the Diploma of OSH (the entry-level qualification for such officers) at Massey University is helping this process however.

Accident Insurance

New Zealand was the first country in the world to introduce a system of 24-hour, comprehensive, compulsory, no-fault insurance cover for people with accident-related injuries and disabilities. It has a strong egalitarian tradition reflected in such legislation – it was also the first place where women enjoyed full voting rights.

This accident compensation scheme took effect in 1974. It replaced statutory workers' compensation schemes, compulsory third party motor vehicle insurance and criminal injuries compensation scheme. It also removed the common law right to sue for damages in return for support for injured people, regardless of who was at fault.

The Accident Compensation Corporation (ACC) is a Crown-owned entity with a board of directors appointed by the Minister for Accident Insurance.

ACC is active in all points of the injury prevention, treatment, and rehabilitation process. Personal support to make the results of the accident more comfortable. This can take many forms including independence allowance, care services, modification to homes and vehicles for those with a lasting disability.

Occupational Safety and Health

The Occupational Safety and Health Service (OSH) of the Department of Labour is responsible for the provision of occupational safety and health policy advise and services in New Zealand. It is comparable in function to the Health and Safety Executive in the UK. The predominant piece of occupational safety and health legislation is the Health and Safety in Employment Act 1992. Its principal objective is the prevention of accidents to employees at work. Under the act, employers have the responsibility for ensuring the health and safety of their workers by various means.

Employees are also responsible for ensuring their work does not endanger the health and safety of themselves or others. People who control a place of work, the self-employed and principals to contracts also have responsibilities to protect the health and safety of employees and others at work.

The uneasy relationship between OSH and ACC.

OSH is the service of the New Zealand Department of Labour with the responsibility of administrating and if necessary enforcing the Occupational Safety and Health Act 1992. As stated earlier the Accident Compensation Corporation is a State owned entity that administrates Accident compensation in New Zealand.

ACC collect data on workplace injuries and so do OSH and it would appear there is little sharing of data. Baker (2000) states, "Until very recently the majority of national exchanges were typically hostile and characterised by boundary disputes and overlaps, with frequent attempts at "turf" demarcation". ACC is considerably larger than the Occupational Safety and Health Service and due to its levy collection function enjoys influencing a larger flow of funds.

There is little doubt that the Accident Compensation Corporation has a more comprehensive source of data than the Occupational Safety and Health Service muster. Problematically, neither database can reveal the full extent of occupational injuries in New Zealand (Feyer, 2001 *et al.*, Cryer 1995) and the two rarely agree. In fact ACC's data indicates higher occupational fatalities than even those collated by the New Zealand Health Information Service mortality files. (Feyer, et.al. 2001).

Baker states that "one simple reason ACC has better data is because the payment of funeral expenses to the surviving family provides a greater incentive to report. OSH on the other hand has different functions- investigation and possible prosecution, which make fatal and non-fatal accidents less likely to be reported.

Conclusion

This paper gives a brief outline of the state of Ergonomics practice in New Zealand and the political and regulatory climate that ergonomists operate in.

The country is small but the ergonomics community is active and unusually cohesive.

The largest concentrations of ergonomists work in evidence-based injury prevention, design and evaluation, notably in the primary industries. Professional limitations arise mostly from the geographic isolation.

The political and regulatory climate is generally favourable for practice owing much perhaps to the egalitarian traditions of the country and the professional cohesion of the ergonomists. The state-level systems have weaknesses invariably though, including role-conflict between the main government bodies involved in health and safety.

References

Baker, V. (2002) Occupational Health and Safety in New Zealand; contemporary social research. In: Lloyd, M. Ed. Chapter 3. *Relationships and Dependencies between OSH, ACC, and employer Client.* The Dunmore Press Ltd, New Zealand.

Moore, D.J & Bentley, T. (2002). *All Terrain Vehicle Injuries on New Zealand Farms.* 6th World Conference on Injury Prevention and Control. Montreal.

Moore, D.J. & Legg, S.J. (1999) *Ergonomics Education and Professional Certification in a Small country: Developments in New Zealand.* In: Proceedings of the IEA-HES Symposium, "Strengths & Weaknesses, Threats & Opportunities of Ergonomics in front of 2000". Santorini, Greece.

New Zealand Department of Labour. (2002). *Fatal Accident Statistics notified to, and investigated by Occupational and Health Service. July 1990 to June 2002".*
Web site: http:// www.osh.dol.govt.nz/touch/press/1999/PR990902.shtml

Slappendel, C (1995). Ed. *Health and Safety in New Zealand Workplaces.* The Dunmore Press Ltd. Palmerston North New Zealand.

AN EMPLOYEE DRIVEN APPROACH IN TACKLING ERGONOMICS AT ETHICON SCOTLAND

Kevin Tesh

Safety and Ergonomics Manager
Ethicon Scotland
Edinburgh EH11 4HE

The recent Health and Safety Executive's (HSE) publication on Upper Limb Disorders in the Workplace HSG60 (rev) (HSE 2002) promotes a holistic approach to better management of acute work related musculoskeletal disorders (MSDs). The publication, based on evidence from a National Audit study (GAO 1997), reported that effective ergonomic interventions can drastically reduce MSDs if all of the following are included; senior management commitment, worker involvement, risk assessment, control measures, instruction and training and proper management of cases. ERGO, a J&J worldwide imitative, has been introduced at Ethicon Scotland and complements the HSG60 strategy. This paper discusses the mechanisms in place to establish ergonomics as a critical success factor in the business and details the steps involved in the ERGO process. Other in-house ergonomic champions allow the ergonomic culture to be maintained.

Introduction

Ethicon Scotland, part of the J&J organisation, has implemented a J&J ergonomic initiative called ERGO which uses a culture-driven approach to gain control and manage ergonomic risks. Driven by employees and overseen by the company ergonomist, ERGO seeks to build the culture through the use of processes, tools and cascade training provided within a systematic framework. Guidelines contained within ERGO are aligned with existing European standards. Key to the success of this system approach is ergonomics being mainstream and one of the 'critical success factors' for the Ethicon Scotland business, which is aligned with the Ethicon franchise. The strategic plan allows all parts of the business to see how they are organised and where each department wants to get to achieve a common goal. Ergonomics, part of the health, safety and the environmental department sets goals and how these will be achieved which is reviewed either monthly or quarterly.

The ERGO programme

The foundation for ERGO is the Ergonomic Maturity Ladder (EML), which contains six steps towards achieving an ergonomic culture (Figure 1). Each step involves a more thorough and detailed ergonomic analysis. Prior to taking the first step on the EML, Ethicon Scotland needed to establish an organisational structure with Ergo teams and an Operational Ergonomic Committee (OEC) as the organisation have a number of facilities within Scotland. The ergonomic structure was designed and aligned to fit in with the company's' health, safety and environmental organisational structure. A critical part of getting ERGO started is senior management commitment which was obtained from the Operations Director to fully support and implement the ERGO process.

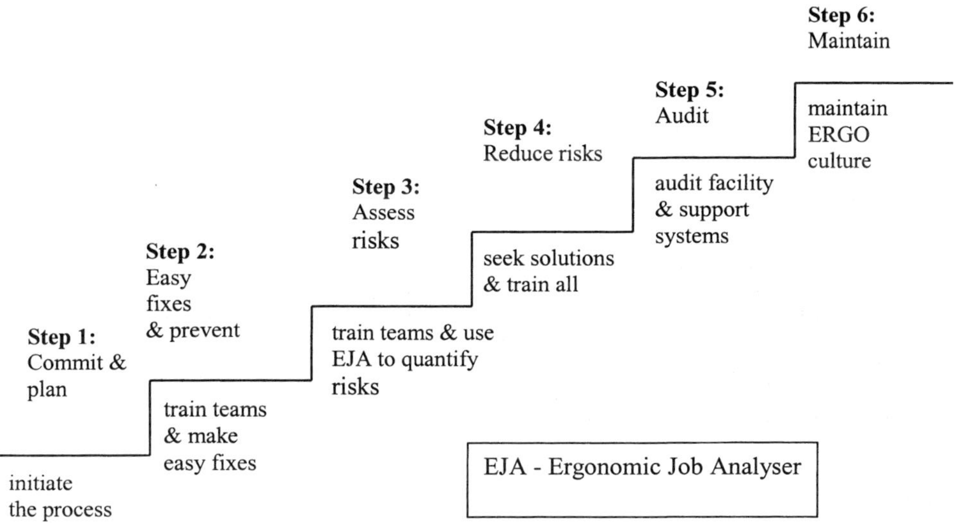

Figure 1 The ERGO EML

As part of the getting started stage, ERGO team members were introduced to the ERGO process. ERGO team members comprised of employees with different skills depending on the location of the team. Team sizes varied but on average contained 4-8 members with skills as production operators, team leaders, training and engineering. At Ethicon Scotland there are currently four teams, one at each of the facilities and a team from engineering department. Teams meet every 4-6 weeks to discuss progress on easy fixes (step 2) and other issues such as training requirements, ergonomic features and prizes for employees putting forward ergonomic suggestions.

Employee participation involved ERGO teams designing an ergonomic display board to act as a focal point for communicating ergonomic information to our employees. Located at high profile sites, the display boards featured a cartoon character ERGO Eric, developed by an employee, which is used to encourage employees to put forward easy fixes for ergonomic problems and solutions. One of the roles for the ERGO teams is to update and manage their respective ergonomic display boards.

Step 1 on the EML involves learning basic principles about ergonomics and the need to make improvements. ERGO team members are tested using a simple 21-question exercise to ensure that the broad principles are understood. Step 2 provides easy fix examples and accompanying suggestion sheet as well as a training video on how to identify easy fixes. As the name suggests easy fixes should be simple, cheap and quick modifications that can be introduced into the workplace to demonstrate the benefits of ergonomics. Distributed around each of the facilities are easy fix sheets housed in ERGO Eric holders with Eric dressed in the appropriate attire for that area. These holders are again positioned in high profile areas e.g. next to clocking in areas, health and safety boards to encourage employees to fill in suggestion sheets. Completed sheets are then posted in a suggestion box situated at the main ergonomic display boards.

It is important to recognise individual efforts in putting forward easy fixes and therefore an ERGO award programme has been set-up which is managed by the ERGO teams. Any employee putting forward a suggestion is rewarded with a small gift (bronze) which has the ERGO Eric logo printed on the prize. Each site awards both quarterly (silver) and annual (gold) prizes to the value of £50 and £200 respectively for the best-implemented easy fix suggestion over that period. Since the ERGO initiative has been running approximately 90 easy fixes have been suggested per year, which is equivalent to a suggestion from one in every fifteen employees.

Step 3 involves using a standard ergonomic assessment tool called the Ergonomic Job Analyser (EJA) to quantify job risks. ERGO team members are trained on how to use the assessment kits and provided with their own kits including video cameras to conduct their own ergonomic surveys. As the level of analysis is more detailed and thorough the training accompanying step 3 is more extensive lasting about two days, one day of theory and the other carrying out case study examples from the factory floor. The EJA assessment tool identifies combination of risk factors such as posture and repetition or force and repetition to establish task risks categories; okay, low moderate and high. Environmental, injury / illness data and employee responses risk factors are also considered as part of a comprehensive assessment tool. A similar assessment tool has also been developed for manual handling tasks called the Manual Handling EJA (MHEJA), which incorporates all risk elements contained within the Manual Handling Operations Regulations 1992 (HSE 1998).

Step 4 provides further problem-solving training for ERGO teams and develops plans to reduce the risks identified in the previous step. As part of the cascade training, ERGO skilled work teams are trained to help identify and solve ergonomic risks in specific areas known to give rise to ergonomic risk. Step 5 further develops an ergonomic culture by rolling out teaching programme to all employees on basic ergonomic concepts. Audits are also conducted to ensure risk reduction solutions are having the desired effect and support systems are reviewed and modified to include ergonomic best practice. Step 6 develops a plan to maintain the ERGO culture.

In addition and running parallel to the EML is the Medical Maturity Ladder (MML), which is run by our in-house occupational health team. For example at step 1 occupational health (OH) staff learn about ergonomics and the consequences of poor workplace design. On step 3 of the MML, the occupational health department provide additional support by proactively conducting health surveillance programmes among groups of employees to establish potential areas of operator ill health and discomfort at an early stage so that appropriate and early ergonomic interventions can be implemented. The occupational health department can also provide proper treatment and rehabilitation to employees with problems.

As well as the ERGO process which allows employees to be empowered with an ergonomic skills and tools to carry out the management of easy fixes, assessments and risk reduction measures other ergonomic champions assist to enhance the ergonomic culture within the organisation. Each champion group is trained to an appropriate level of competence to recognise ergonomic issues and either solve the problem or know who to turn to for assistance.

For example, OH staff have had training in basic ergonomic principles and use a checklist to identify potential risk areas. Safety representatives, team leaders and safe behaviour teams have received general awareness training and are knowledgeable and closely involved with the work processes to identify low risk ergonomic problems so as to prevent higher risk problems developing in the future. Engineers have received ergonomic design principle training to ensure that modifications to existing equipment or new processes are considered at the design stage. Every two years all employees receive ergonomic training as part of the Good Manufacturing Practice (GMP) initiative. Last year the topic was on seating and the importance of setting up the seat correctly to achieve the correct sitting height and comfortable working postures. The company ergonomist manages the ERGO programme but also raises the profile of ergonomics across all departments as they can all influence ergonomic risk. For example, product design will influence the manufacturing process and the type of packaging required to present the product.

Conclusion

Ethicon Scotland have employed a full time ergonomist since 1993 as they recognised the importance this discipline can make to achieving a risk-free workplace, productivity gains and employee job satisfaction. The ERGO programme has been successful in other Johnson & Johnson facilities in reducing the incidence and severity of ergonomics-related injuries and illness. The success of the Ethicon Scotland ERGO approach has recently been recognised as the programme was selected as an excellent example of creating the right organisational environment as part of the HSE launch of HSG60 (rev). Benchmarking of injury / illness data, alternate duties and therapy treatments is currently being compiled to measure the effectiveness of this long-term ergonomic strategy. Innovative programmes such as ERGO can loss their impact unless different initiatives are continually developed to raise its awareness and benefits. Updating the ERGO boards, changing the award prizes, running ergonomic competitions and running ERGO awareness days have or are being used to continually drive the ergonomic culture forwards.

References

Manual Handling. Manual Handling Operations Regulations 1992. Guidance on Regulations L23 (Second Edition) HSE Books 1998 ISBN 0 7176 2415 3

Upper limb disorders in the workplace HS (G) 60 (rev) ISBN 9 780 717 619788. HSE Books.

US General Accounting Office study: HEHS-97-163 August 1997 Worker Protection: Private Sector Ergonomics Programs Yield Positive Results.

IMPLEMENTATION OF THE PHYSICAL AGENTS (VIBRATION) DIRECTIVE IN THE UK

CM Nelson and PF Brereton

Health and Safety Executive, Magdalen House, Bootle L20 3QZ, UK
chris.nelson@hse.gsi.gov.uk

Abstract

The UK is required to introduce Regulations on Vibration at Work by 6 July 2005 to implement the Physical Agents (Vibration) Directive (2002/44/EC). This Directive builds on existing general duties on employers to manage health and safety, and introduces exposure action and limit values for both hand-arm vibration and whole-body vibration, thus setting minimum standards for the control of vibration risks in Europe.

The new Regulations should serve to strengthen the continuing work of the Health and Safety Executive (HSE) to reduce exposures to hand-arm vibration in British industry.

Implementation of the Directive for whole-body vibration presents a different challenge and the HSE is currently considering the form of appropriate guidance to accompany the Regulations. A holistic approach is proposed for drivers, involving a broad assessment of the ergonomic risks of the job and prioritising controls, setting vibration in context with other risk factors for back pain, particularly postural concerns and manual handling operations.

The HSE is currently engaged in informal consultation with its stakeholders, and intends to publish draft Regulations and guidance during 2003. The HSE welcomes any evidence-based opinions regarding its proposals for published guidance for employers and implementation of Regulations on vibration at work.

1. Introduction

Occupational exposure to vibration is known to be related to various risks to health, and is usually considered in two distinct categories: *hand-arm* vibration (transmitted from hand-held, hand-fed or hand-guided machinery) and *whole-body* vibration (transmitted to the body through the supporting surface (seat or floor), generally in a vehicle or mobile machine).

The effects of vibration on health are many and varied (Griffin 1990). Exposure to hand-transmitted vibration can result in hand-arm vibration syndrome (HAVS). This includes vascular, neurological and musculoskeletal components (such as episodic blanching of the fingers, numbness, tingling, reduced sensitivity to touch and temperature, weakened grip and pain). Whole-body vibration and shock have been associated with back pain and spinal injury and, occasionally, with a range of other health effects.

Since vibration exposures are associated with risks to health, employers have a duty, under general health and safety legislation, to manage and minimise these risks. The European Union adopted Directive 2002/44/EC (known as the Physical Agents (Vibration) Directive or PA(V)D) in 2002 (European Parliament and the Council of the European Union 2002). This will, for the first time, require UK legislation on vibration exposure at work. This paper explores the requirements of the new Directive and the current work programme of the Health and Safety Executive (HSE) for the implementation of national Regulations and guidance.

2. The development of the Directive

A proposal for a Directive on 'Physical Agents' was first produced by the European Commission in the early 1990s (European Commission 1994). This proposal applied the principles of the existing 'Framework' Directive on the health and safety of workers (Council of the European Communities 1989) in a consistent manner to a range of different 'physical agents' (noise, hand-arm and whole-body vibration, optical radiation and electromagnetic fields and waves). Daily exposure criteria (the 'threshold level', 'action level' and 'exposure limit value') were given in separate annexes for each 'agent' as triggers for control action by employers (minimising exposure and risk, providing suitable equipment and working methods, providing information and training, health surveillance, etc.).

This proposal made little progress, largely because scientific understanding of the effects on health was not sufficiently advanced for agreement on a common approach to all the physical agents. In early 1999, however, a new draft was offered by the German Presidency, which took forward the proposals for hand-arm vibration (HAV) and whole-body vibration (WBV) only, leaving the other physical agents to be dealt with later. After two years of negotiation and compromise by Member States, the French Presidency referred the outstanding issues to the Social Affairs Council in November 2000 where political agreement was reached in June 2001. The proposed Directive was then submitted for consideration by the European Parliament. Negotiations resulted in agreement of the Directive in early 2002 following a process of conciliation between the Parliament and the Council. Directive 2002/44/EC was published on 6 July 2002. Member States of the European Union now have until 6 July 2005 to implement the Directive in national legislation. In the United Kingdom this will require Regulations made under the enabling powers of the Health and Safety at Work, etc. Act 1974.

3. Requirements of the Directive

3.1 General duties

Most of the content of the PA(V)D re-iterates general duties of employers to assess and manage risks to health and safety, already in force under the Framework Directive (implemented in the UK by the Management of Health and Safety at Work Regulations 1999). However, the agreement between Member States on levels of exposure above which employers must take action is new, as is the introduction of an exposure limit.

3.2 Exposure action and limit values

The daily exposure criteria are all expressed as 8-hour (energy-equivalent) frequency-weighted accelerations, although vibration dose value (VDV) alternatives are given for WBV. The WBV limit value is greater than that in the original proposal and the 'short term' criteria in earlier drafts, which effectively set maximum vibration magnitudes notwithstanding the exposure times, have been removed.

- HAV daily exposure action value (EAV): 2.5 ms^{-2} A(8)
- HAV daily exposure limit value (ELV): 5 ms^{-2} A(8)
- WBV daily exposure action value (EAV): 0.5 ms^{-2} A(8) (or VDV 9.1 $ms^{-1.75}$)
- WBV daily exposure limit value (ELV): 1.15 ms^{-2} A(8) (or VDV = 21 $ms^{-1.75}$)

No technical definitions are given for the quantities used in the definitions of the EAVs and ELVs. The PA(V)D relies on references to standards: it calls on ISO 5349-1:2001 (BS EN ISO 5349-1:2001 (British Standards Institution 2001)) for definitions for HAV and on ISO 2631-1:1997 (International Organization for Standardization 1997) for WBV. Thus, the values for HAV are derived from the triaxial vibration magnitude (root-sum-of-squares of the three single axis values) and the WBV values are considered separately for each axis of vibration.

3.3 Determination and assessment of risks

The PA(V)D refers to the Framework Directive duty to carry out a competent risk assessment. In this context, employers are required to 'assess and, if necessary, measure' the vibration exposure of workers. It is made clear that vibration measurement is not expected in all cases; establishing probable vibration magnitudes, through the use of vibration information from equipment manufacturers, is specifically mentioned. This provides a link with the Machinery Directive (Council of the European Union 1998) which requires suppliers of machinery to provide users with information on residual risks (having designed and constructed the products to be safe) and to declare vibration emission values.

The risk assessment required by the PA(V)D is not simply the evaluation of daily exposure (indeed, a precise exposure value will not always be necessary). Employers are required to consider intermittent exposure, repeated shocks, workers 'at particularly sensitive risk', other factors such as workplace temperature and information from health surveillance. The risk assessment must also identify the measures required for appropriate avoidance or reduction of vibration exposure and provision of worker information and training.

3.4 Exposure control

The Directive includes a general duty requiring *risks* arising from vibration exposure to be eliminated or reduced to a minimum, 'taking account of technical progress and the availability of measures to control the risk at source'. This is based on the principles of prevention given in the Framework Directive (broadly, the well-known 'hierarchy of control') and is not dependent on the level of exposure.

If the EAV is exceeded, an appropriate formalised preventative programme must be set up by the employer to reduce to a minimum *exposure* to vibration and the *attendant risks*. Particular attention is drawn to:

- alternative working methods to reduce exposure;
- selection of appropriate work equipment;
- use of auxiliary equipment to reduce vibration risks;
- maintenance programmes;
- workplace design and layout;
- information, instruction and training;
- limiting exposure duration and magnitude;
- work schedules and rest periods; and
- clothing to protect against cold and damp.

Workers must not be exposed above the ELV and the Directive includes a requirement for immediate corrective action by the employer if the ELV is exceeded.

3.5 Information, training and consultation with workers

There are requirements to inform workers of the outcome of the risk assessment and provide them with information and training to enable them to understand the risks, the EAV and ELV, the measures taken to control risks and the actions they need to take to minimise risk, report symptoms, etc. Workers or their representatives must also be consulted and allowed to participate in all matters covered by the Directive.

3.6 Health surveillance

The text expands on the Framework Directive's requirements for appropriate health surveillance where the risk assessment has indicated a risk to health and/or when the EAV is exceeded. Health surveillance is intended to give an early indication of vibration-related ill health in individuals and also to provide information to the employer regarding the effectiveness of the control measures. Where ill-health is found, the employer has specific duties to review the risk assessment and control measures, consider redeployment of affected workers and arrange continued health surveillance.

3.7 Transitional periods and derogations

Member States have three years from the date of adoption of the Directive (6 July 2002) to implement the requirements in national legislation. Regulations in the UK should therefore be in force on 6 July 2005.

Following this date, Member States may apply a transitional period for the requirement to keep exposures below the ELV. These would only apply where the work equipment producing the vibration was put into use before 6 July 2007 and where it is not currently possible, with that equipment, to keep exposures below the ELV. The transitional periods would be for a maximum of five years (to 5 July 2010) generally or nine years (to 5 July 2014) regardnig equipment used in the agriculture and forestry sectors. The duties to minimise risk from vibration and to implement a programme of measures above the EAV are not affected.

Member States are permitted to grant derogations from the WBV exposure limit value in cases of air and sea transport where it is not possible to reduce the exposure below this level.

For WBV and HAV, derogation powers are also available to allow occasional exposures above the exposure limit value where the exposure is normally below the exposure action value. However, the exposure 'averaged over 40 hours' (i.e. a weekly average) must be less than the exposure limit value and there must be evidence that risks from this pattern of exposure are lower than those at the exposure limit value. Increased health surveillance is also a requirement for derogation.

4. Implementation – hand-arm vibration

4.1 Background

The HSE published comprehensive guidance on HAV in 1994 (Health and Safety Executive 1994). By following this and subsequent guidance (e.g. Health and Safety Executive 1994, 1997a, 1998a, 1998b, 1999, 2000, 2001a), employers should be able to satisfy the requirements of the existing law, including the Management of Health and Safety at Work Regulations 1999, with respect to HAV.

Employers are required to manage risks arising from exposure to vibration. When daily exposure regularly exceeds an action level of 2.8 ms^{-2} *A*(8), enforcing authorities expect reasonably practicable management actions (vibration-free work processes, vibration-reduced equipment, provision of information and training, warm clothing, health surveillance, etc.).

Although there are an estimated 1.2 million people at risk from HAV in the UK (Palmer *et al.* 1999a), British industry has made much progress during the last decade in reducing exposures. Drivers for this have included the publicity generated by compensation against employers in the civil courts, information campaigns and publications by the HSE and others, innovations in industrial processes and the increasing availability of power tools and other work equipment with reduced vibration emissions.

4.2 Changes to the action level

The HSE's existing recommended action level of 2.8 ms^{-2} *A*(8) was agreed through consultation with its stake-holders during the 1980s. This action level is a 'dominant axis' exposure value, obtained in accordance with the old BS 6842:1987 (British Standards Institution 1987). (This standard recommended measurement in three axes and the use of the greatest value for evaluating exposure.) The action and limit values given in the PA(V)D are three-axis (root-sum-of-squares) values obtained according to BS EN ISO 5349-1:2001 which has now replaced BS 6842:1987. The 2.8 ms^{-2} *A*(8) action level can be approximated by a three-axis value of 4 ms^{-2} *A*(8) (Nelson 1997) which can be compared with the EAV and ELV in the PA(V)D. Thus, it can be seen that the current action level lies between the future action and limit values of 2.5 ms^{-2} *A*(8) and 5 ms^{-2} *A*(8) respectively.

4.3 New guidance

The HSE is preparing revised guidance on HAV to be published in 2005. This will replace much of the existing portfolio, consolidating and updating the material, possibly in a single volume. It will include the new Vibration at Work Regulations and guidance for employers on compliance (through assessment and control of HAV and management of residual risks) together with more detailed material, in annexes, for the 'competent person' and for occupational health professionals.

Much experience has been gained since the present guidance on health surveillance programmes for HAVS was written. The new guidance will contain updated recommendations on health surveillance, recognising that, with the reduced action value, more people will be need to be included in health surveillance schemes provided by employers. The HSE is currently consulting on a proposal for a tiered, risk-based health surveillance protocol, which should provide a simple cost-effective means of identifying any new cases of vibration injury, while targeting specialist resources at the monitoring and management of existing affected individuals.

5.0 Implementation – whole-body vibration

5.1 Background

Numerous epidemiological studies have demonstrated a link between those occupations where exposures to WBV occur and an increased prevalence of low back pain (LBP). The problem with most studies has been isolating the WBV exposure from other contributory risk factors for back pain. There is also a high prevalence of back pain in the general population. However, Bovenzi and Betta (1994), investigating a population of tractor drivers, were able to control for several confounders by logistic modelling and concluded that WBV and postural load were independent contributors to the excess risk for LBP. Although a link between WBV exposure and low back disorders is widely accepted, no clear quantitative dose response relationship has yet been established and understanding in this area is far from mature. In a review of the literature on WBV (Bovenzi and Hulshof 1998), it was concluded that there is an association between occupations involving exposure to whole-body vibration and the risk of LBP and other spinal disorders; however, predictive relationships could not be produced from the existing data. Another review (Lings and Leboeuf-Yde 2000) concluded that evidence for a causal link between WBV and LBP is weak and that, although it was sensible to reduce WBV exposures so far as possible, there was no quantitative dose-response relationship and a safe exposure limit could not be established. Palmer *et al.* (1999b) studied occupational exposures to WBV. They found high prevalences (40-60%) of LBP and sciatica in the general population, but observed that any increases in the prevalence ratios with increased vibration exposure were small and that vibration exposure was no more significant than, for example, lifting heavy weights and working with the hands above head height. This, and other studies (see Stayner 2001) provide evidence to support the view that WBV is one of several stressors which may contribute to occupational LBP (others include poor posture, prolonged sitting and manual handling tasks) but that it can rarely be identified as the sole cause.

This was recognised by the HSE in its guidance on WBV for employers, 'In the driving seat' (HSE 1997b) which includes working posture and manual handling in its advice on risk assessment This leaflet identifies people who may be at risk (particularly drivers of off-road mobile machines in the agriculture, forestry, quarries and construction sectors) and recommends sensible precautions, such as ensuring that the vehicle or machine is suited to the task, that it is properly maintained, that appropriate seating is fitted and used correctly (particularly in the case of suspension seats) and that operators have training in the risks and in correct adjustment and operation of the machine and its seat. This allows the assessment of risks and the development of appropriate action plans to achieve compliance with the general duties in the Management of Health and Safety at Work Regulations 1999 (i.e. the Framework Directive).

5.2 Exposure action and limit values

The PA(V)D introduces numerical exposure criteria for WBV exposures for the first time. The option to select the EAV and/or the ELV from an *A*(8) value (an 8-hour 'time-weighted' r.m.s. acceleration value) or a *VDV* ('fourth-power' vibration dose value) reflects different opinions amongst Member States regarding the 'correct' method for evaluating exposures. The *VDV* method is more responsive to shocks and impacts, and has arguably a more appropriate time-dependency than the r.m.s. method. Research (e.g. Sandover 1997) has shown that repeated shocks, which occur in industrial operation of mobile machines, can result in fatigue stress fractures of the vertebrae.

While the *A*(8) method seems likely to be favoured by most national authorities in Europe, the HSE favours the use of *VDV* for the EAV, since it is the better method for reflecting risk to health from whole-body vibration and shock. (The HSE is, however, proposing that the *A*(8) option should be chosen for the ELV in the interests of a common minimum standard in Europe.) The implications of the choice between the two methods for defining the EAV are discussed fully elsewhere (Nelson and Brereton 2002)

5.3 New guidance

Musculoskeletal problems affecting the back are the most reported work-related illnesses in manual workers in the UK, with an estimated prevalence of 642,000 in 1995 (HSE 2001b). A number of laboratory-based human volunteer studies, and experiments on cadaver materials, have demonstrated the feasibility of vibration-induced low back injuries. However, as discussed above, it is rarely possible to identify whole-body vibration as the sole cause of back pain in an individual, or even a group of workers. Other ergonomic hazards are likely to be significant contributors to the risks and may interact with WBV. Poor posture and badly managed manual handling operations can also cause chronic or acute injury. Vibration can also lead to muscle fatigue, increasing susceptibility to injuries from manual handling or whole-body shocks. Sitting posture affects the transmission of vibration and shock through the body, and this also seems likely to affect susceptibility to vibration injury.

With the introduction of Regulations on WBV, which include numerical exposure action and limit values, there is potential for inappropriate allocation of resources to reduce vibration, while the principal risk factor for back pain remains unidentified. The HSE is therefore preparing new guidance, to coincide with the introduction of the new Vibration at Work Regulations. This is likely to include:

- simple WBV guidance for 'low risk' activities – mainly road vehicle operators; and
- back pain guidance for 'higher risk' activities – such as drivers of off-road machines and industrial trucks.

The former should be sufficient for the estimated seven million drivers who are exposed to WBV but where risk management will be straightforward. The latter will encourage employers to assess risk of back pain holistically, considering three main risk factors for back problems in drivers of mobile machinery:

- poor or constrained posture in the driving position;
- manual handling operations; and
- exposure to whole-body vibration and shock.

This should allow the identification of any significant risk presented by WBV or the other causes of back pain, and appropriate targeting for control.

The vibration assessment itself will need to establish whether the EAV (and the ELV) is likely to be exceeded. For vehicles most likely to present a risk of injury from whole-body vibration and shock (usually off-road machinery or machinery with minimal suspension) the manufacturers of the machinery will often be the best source of information on vibration risk and its control.

The guidance will encourage employers to assess WBV exposure with a view to reducing it, by means such as:

- maintaining roadways and surfaces;
- enforcing speed limits when the roughness of the surface leads to excessive vibration in a machine or vehicle;
- confirming that the vibration of the machine or vehicle in use is amongst the lowest achievable for the application;
- checking that the design performance of seat (and other) suspensions is matched to the vehicle vibration characteristics;
- checking that suspensions are performing as designed;
- ensuring that employees have been given information and training to enable them to protect themselves from vibration risk and that they comply with company policy.

While the Directive contains a requirement for *appropriate* health surveillance, there are considerable difficulties in identifying specific disorders which are clearly attributable to WBV. Low back pain and other lumbar disorders are the main health effect of concern. At present it is not considered that any valid diagnostic technique exists for the detection of changes which can reliably indicate the early onset of low back pain, as this is a symptom and not necessarily a specific disease/condition. It therefore seems likely, at the time of writing, that the guidance will advise that there is no *appropriate* health surveillance for WBV-related health risks at present, but that a less formal programme of health monitoring will be good practice.

6.0 Consultation

Since the adoption, in July 2002, of the PA(V)D, the HSE has been conducting an informal consultation process with employers' organisations, manufacturers' organisations, trade unions, consultants, etc. The Health and Safety Commission will publish two Consultative Documents (for HAV and WBV respectively) during 2003, containing draft Regulations and the draft guidance, and inviting public comment. Following the final Government decision on the details of the Regulations, the final version of the guidance will be prepared for timely publication prior to the introduction of the new Regulations in July 2005.

The HSE wishes to know the views (backed by data or other evidence) of industrialists, trade unions, occupational health professionals and other interested parties on the issues raised here and in the Consultative Documents mentioned above. In particular, the HSE would welcome information on industrial WBV exposures, and associated ergonomic issues, particularly those determined for full shifts or 24-hour periods, that will enlighten decisions on the choice of *VDV* or *A*(8) for the EAV and/or ELV and inform the development of appropriate guidance which targets industries of concern.

The HSE website now has a page dedicated to vibration (www.hse.gov.uk/vibration), and this will contain the relevant documents for the consultation process with regular updates.

7.0 Conclusions

Hand-arm vibration is already well-known in the UK; the HSE's expectations of industry in this area have been widely promulgated during the last decade and are consistent with the requirements of the PA(V)D. The introduction of the Regulations on HAV should not bring any surprises to industry. The reduced action level and the anticipation of the ELV will, in some cases, require increased efforts to eliminate or reduce exposures and a wider availability of health surveillance. However, the approaches to the successful management of HAV are well established. (These can include process re-engineering, selection of lower vibration equipment, restriction of tool use time, information and training, etc.) It is expected that progress made since the introduction in 1994 of the existing guidance and action level will continue. The likely transition periods provide several years for industry to eliminate those vibration exposures remaining above 5 ms^{-2} $A(8)$.

For whole-body vibration, some excessive exposures which are under the control of the employer will be identified and controlled as a direct result of the introduction of Regulations. For example, industrial trucks designed for use on smooth factory and warehouse floors can produce high WBV exposures when operated on outdoor surfaces; this is an avoidable risk which can be eliminated. However, the most severe industrial exposures will be in the use of off-road vehicles (agricultural tractors, forestry, construction and quarrying machinery, etc.). There is potential for some reduction in levels of WBV and shock from interventions on operating techniques (chosen terrain, driving speed, etc.). However, it is often not reasonable to limit daily exposure times to achieve reduced daily exposures. The main opportunities for exposure control lie in the selection of machinery suitable for the work. The proposed long transitional periods for introduction of enforcement of the ELV reflect an understanding of this. Furthermore, significant exposure to WBV rarely occurs without other ergonomic stressors (such as posture problems) and any occupational back pain may not necessarily be attributable to vibration. It is therefore important to take a holistic approach to the assessment and control of risk, concentrating on vibration only where it is the dominant contributor to risk.

The Physical Agents (Vibration) Directive places duties on employers regarding the management of risks from mechanical vibration. It clarifies the existing general duties, already defined in the Framework Directive, by defining exposure action values above which certain prescribed management actions must be in place and by setting limits for daily exposure. It also complements the duties of manufacturers of work equipment (set out in the Machinery Directive) to control vibration risks to which the users of their products are exposed.

8. References

BOVENZI, M. and BETTA, A. 1994, Low back disorders in agricultural tractor drivers exposed to whole-body vibration and postural stress, *Applied Ergonomics,* **25, 231-241**.

BOVENZI, M and HULSHOF, C.T.J. 1998, An updated review of epidemiological studies on the relationship between exposure to whole-body vibration and low back pain, *Journal of Sound and Vibration,* **215 (4), 595-611**.

British Standards Institution 1987, BS 6842, *Measurement and evaluation of human exposure to vibration transmitted to the hand.*

British Standards Institution 2001, BS EN ISO 5349-1, *Mechanical vibration and shock – Measurement and evaluation of human exposure to hand-transmitted vibration – Part 1: General requirements.*

Council of the European Communities 1989, Council Directive 89/391/EEC on the introduction of measures to encourage improvements in the safety and health of workers at work.

Council of the European Union 1998, Council Directive 98/37/EC on the approximation of the laws of the Member States relating to machinery.

European Commission 1994, Amended proposal for a Council Directive on the minimum Health and Safety requirements regarding the exposure of workers to the risks arising from physical agents, *Official Journal of the European Communities,* **OJ 37 (C230), 3-29**.

European Parliament and the Council of the European Union 2002, Directive 2002/44/EC on the minimum health and safety requirements regarding the exposure of workers to the risks arising from physical agents (vibration). *Official Journal of the European Communities*, **OJ L177, 6.7.2002, p13**.

GRIFFIN, M.J. 1990, *Handbook of Human Vibration.* Academic Press, ISBN 0-12-303040-4.

Health and Safety Executive 1994, *Hand-arm vibration*, HS(G)88, HSE Books ISBN 0-7176-0743-7.

Health and Safety Executive 1997a, *Vibration Solutions. Practical ways to reduce the risk of hand-arm vibration injury*, HS(G)170, HSE Books ISBN 0-7176-0954-5.

Health and Safety Executive 1997b, *In the driving seat,* Leaflet IND(G)242L, HSE Books.

Health and Safety Executive 1998a, *Health risks from hand-arm vibration: advice for employees and the self-employed*, Leaflet IND(G)126(rev1), HSE Books.

Health and Safety Executive 1998b, *Health risks from hand-arm vibration: advice for employers*, Leaflet IND(G)175(rev1), HSE Books.

Health and Safety Executive 1999, *Hand-arm vibration syndrome*, Pocket card for employees, INDG296P, HSE Books.

Health and Safety Executive 2000, *The Successful management of hand-arm vibration*, CD-ROM, HSE Books ISBN 0-7176-1713-0.

Health and Safety Executive 2001a, *Power tools: how to reduce vibration health risks,* Leaflet INDG338, HSE Books.

Health and Safety Executive 2001b, *Health and safety statistics 2000/01*, ISBN 0 7176 2110 3, HSE Books.

International Organization for Standardization 1997, ISO 2631-1, *Mechanical vibration and shock – Evaluation of human exposure to whole-body vibration – Part 1: General requirements.*

LINGS, S. and LEBOEUF-YDE, C. 2000, Whole-body vibration and low back pain: a systematic, critical review of the epidemiological literature 1992-1999, *Int Arch Occup Environ Health* **(2000) 73: 290-297**.

NELSON, C.M. 1997, Hand-transmitted vibration assessment – a comparison of results using single axis and triaxial methods, *Presented at the United Kingdom Group Meeting on Human Response to Vibration held at the ISVR, University of Southampton, Southampton, SO17 1BJ, England, 17-19 September 1997.*

NELSON, C.M. and BRERETON, P.F. 2002, Exposure action and limit values for whole-body vibration: implementing the new directive, *Presented at the 37th United Kingdom Conference on Human Responses to Vibration, held at Department of Human Sciences, Loughborough University, UK, 18 – 20 September 2002.*

PALMER, K.T., COGGON, D., BENDALL, H.E., PANNETT, B., GRIFFIN, M.J. and HAWARD, B.M. 1999a, Hand-transmitted vibration: Occupational exposures and their health effects in Great Britain, *Contract Research Report 232/1999,* HSE Books, ISBN 0-7176-2476-5.

PALMER, K.T., COGGON, D., BENDALL, H.E., PANNETT, B., GRIFFIN, M.J. and HAWARD, B.M. 1999b, Whole-body vibration: Occupational exposures and their health effects in Great Britain, *Contract Research Report 233/1999*, HSE Books, ISBN 0-7176-2477-3.

SANDOVER, J. 1997, High acceleration events in industrial exposure to whole-body vibration, Summary and concluding report, *Contract research report 134/1997,* HSE Books, ISBN 0-7176-1361-5.

STAYNER, R.M. 2001, Whole-body vibration and shock: a literature review. Extension of a study of overtravel of seat suspensions, *Contract Research Report 333/2001*, HSE Books, ISBN 0-7176-2004-2.

The views expressed in this paper are those of the authors and do not necessarily represent the opinion or policy of the Health and Safety Executive or any other government body.

SLIPS, TRIPS AND FALLS

THE ROLE OF SURFACE WAVINESS IN FRICTION AT SHOE AND FLOOR INTERFACE

Wen-Ruey Chang[1], Raoul Grönqvist[2] and Mikko Hirvonen[2]

[1]*Liberty Mutual Research Institute for Safety, Hopkinton, MA 01748, USA*
[2]*Finnish Institute of Occupational Health, FIN-00250 Helsinki, Finland*

Friction is widely used as an indicator of surface slipperiness in slip and fall incidents. Surface texture affects friction, but it is unclear which surface characteristics are better correlated with friction and are preferred as potential interventions. Transition friction under 3 different mixtures of glycerol and water as contaminants was correlated with the surface parameters generated from the tile surfaces. The surface texture was quantified with various surface roughness and waviness parameters using 3 cut-off lengths to filter the measured profiles for obtaining the surface roughness or waviness profiles. Surface waviness parameters calculated from the filtered profiles were shown to be better indicators of friction than commonly used surface roughness parameters, especially when a short cut-off length was used or when the viscosity of the contaminant was high.

Introduction

Slips and falls are a serious problem. The annual direct cost of occupational injuries due to slips, trips and falls in the U.S.A. is estimated to exceed 6 billion U.S. dollars (Courtney *et al.*, 2001). In the total direct workers' compensation for occupational injuries due to slips and falls, falls on the same level accounted for 65% of claim cases and, consequently, 53% of claim costs (Leamon and Murphy, 1995).

Friction is widely used as an indicator of surface slipperiness. As summarized previously (Chang, 2001), higher friction could potentially improve slip resistance, and dynamic friction is more applicable to human walking than static friction. Surface texture plays a crucial role in friction (Chang *et al.*, 2001). However, it is not clear which surface characteristics are better correlated with friction and, therefore, are preferred as potential interventions for slip and fall accidents.

A majority of the published studies on the effects of surface textures on friction related to slip and fall accidents focused on surface roughness. Among common surface roughness parameters, several were reported to correlate strongly with friction related to slip and fall accidents. The surface roughness parameters R_a, R_{pm}, Δ_a and R_k are defined in Table 1. As summarized previously (Chang, 2001), these parameters were shown to have strong correlations with friction. In addition, several floor materials were used to cover the desired range of surface roughness in some of the previous studies and these friction measurements reflected the combined effect of floor materials and surface roughness as indicated previously (Chang, 2001).

Table 1. The Definitions of Surface Roughness and Waviness Parameters

Surface Roughness Parameters:

R_a - arithmetical average of surface heights or the center line average of surface heights

R_k - kernel roughness depth

R_{pm} - average of the maximum height above the mean line in each cut- off length

Δ_a - arithmetical mean of surface slope

Surface Waviness Parameters:

W_a - arithmetical average of surface heights or the center line average of surface heights

W_p - maximum height of the profile above the mean line within the assessed length

W_{pm} - average of the maximum height above the mean line in each cut- off length

W_q - root mean square of surface heights

W_t - maximum peak to valley height in the assessed length

W_{tm} - average of peak to valley height in each cut-off length

W_y - maximum of peak to valley in all cut-off lengths

$W\lambda_q$ - root mean square measure of spatial wavelength

$W\Delta_a$ - arithmetical mean of surface slope

$W\Delta_q$ - root mean square of surface slope

Surface profiles contains both roughness and waviness components. The surface profile measured with a profilometer is filtered with a proper selection of a filtering length, also known as the cut-off length, to obtain the surface profiles of either surface roughness or surface waviness. The roughness profile contains the surface components with short wave lengths compared with the selected cut-off length, while the waviness profile contains the surface components with long wave lengths. The surface parameters are calculated from the filtered profiles. The importance of surface roughness or surface waviness in determining interface friction might depend on the size of the contact area and the magnitude of deformation at the contact interface. Surface waviness could be more important than surface roughness when the contact area is large compared with the cut-off length used. Pocket size profilometers with a built-in cut-off length of 0.8 mm are widely used by safety professionals due to their portability and low cost. A dangerous forward slip that could lead to a fall usually occurs shortly after heel strike (Strandberg and Lanshammar, 1981). The size of the contact area between shoe and floor at this critical moment is much larger than 0.8 mm in length (Harper *et al.*, 1961). Therefore, surface waviness measured with this cut-off length could be more important than surface roughness. However, most of the publications in the literature only investigated the role of surface roughness, not surface waviness. The objective of this study was to investigate whether surface waviness parameters are better representations of friction than the surface roughness parameters at 3 different cut-off lengths. The surface parameters that had the strongest linear correlation with the transition friction between neolite and quarry tiles measured under 3 different contaminants were identified.

Test apparatus and design of experiment

Sand blasting, controlled by air pressure at the inlet and exposure time, was used to systematically alter surface texture on unglazed quarry tiles. The dimension of each tile surface was approximately 15.2 by 15.2 cm. The exposure time was 2, 4, 6 and 7 minutes. The air pressure at the inlet was 2.76×10^5 and 5.52×10^5 Pa (40 and 80 psi). The distance between the tile and the exit nozzle of the sand particles was 7.6 cm. The nozzle was kept normal to the tile surface. Sand particles of "black beauty," which consisted of 99 to 100% coal slag, were used. Five different processes, consisting of different combinations of exposure times and air pressures, were selected for the sand blasting. Eight tiles were generated under each sand blasting process for a total of 40 tiles.

A dynamic apparatus (Grönqvist *et al.,* 1989) to simulate a slip was used to measure transition friction in this experiment. An artificial foot was used for mounting the shoe to the system. A piece of neolite, a standard test material, was attached to the heel of the shoe. The tile was attached to a force plate for normal and shear force measurements at the shoe and tile interface. The horizontal velocity of the shoe was maintained at a constant speed of 0.40 (± 0.02) m/s. The shoe was positioned slightly above the floor surface initially and was lowered onto the surface at a speed of 0.1 m/s. The time interval of 50 ms, for averaging the normal and friction forces and friction coefficient, began 100 ms after heel touch-down, which was the instant the normal force exceeded 100 N. The applied normal force and a shoe-tile contact angle of 700 ± 20 N and 5 degrees, respectively, were used. The measured transition friction represents the available friction during one of the most critical moments in a slip.

A profilometer was used to make 6 parallel surface profile measurements, 10 mm apart in the direction of friction measurements, at the center of each tile. Three cut-off lengths, 0.8, 2.5 and 8 mm, and the 2CR PC filter, a recursive filter with a phase correction, were used. The measured profiles were stored in a computer. Numerical filtering was performed on these stored profiles to obtain the filtered profiles of surface roughness and surface waviness. The assessed length for the surface parameter calculations consisted of 4 cut-off lengths. Since the main focus of this paper was on the surface waviness parameters, only the few surface roughness parameters reported to have a strong correlation with friction were included in this comparison. The definitions of all the surface parameters used, 4 roughness parameters and 10 waviness parameters, are shown in Table 1. Graphic illustrations of most of the surface roughness parameters can be found in the literature (Chang *et al.*, 2001).

The contaminants in this experiment were glycerol mixed with water at 85%, 70% and 50% weight ratios to cover a range of viscosity. The neolite surface was sanded to resemble the shape of a worn shoe. This surface was sanded by 180 grit abrasive paper with an orbital sander, whenever there was a change in tile or glycerol content, to maintain a consistent surface condition. The test sequence was randomized with glycerol contents and sand blasting processes. All of the 40 tiles were used repeatedly with different glycerol contents. Right before and after the friction measurements, the tiles were sequentially cleaned in an ultrasonic bath with water. The tile surfaces were then wiped with 50% ethanol between the ultrasonic bath and friction measurements. Five repeated friction measurements were taken for each tile and glycerol content.

Results and discussions

The results of an analysis of variance (ANOVA) indicated that different sand blasting processes and glycerol contents resulted in statistically significant differences in friction coefficients ($p<0.0001$). Also, different sand blasting processes and cut-off lengths resulted in statistically significant differences in all surface parameters ($p<0.05$).

The means and standard deviations (STD) of the transition friction coefficients for the tiles generated from each sand blasting process under different glycerol contents are shown in Table 2. They were calculated from the average friction coefficients of the 8 tiles from a single sand blasting process.

Table 2. The Mean and Standard Deviation of Friction Coefficients

Sand Blasting Process		Glycerol Content		
		50%	70%	85%
1	Mean	0.480	0.170	0.014
	STD	0.027	0.020	0.005
2	Mean	0.674	0.283	0.038
	STD	0.054	0.041	0.007
3	Mean	0.719	0.366	0.078
	STD	0.019	0.021	0.010
4	Mean	0.715	0.392	0.106
	STD	0.034	0.018	0.020
5	Mean	0.703	0.429	0.137
	STD	0.010	0.016	0.063

The linear correlation coefficient between each average surface parameter and average measured friction among all 40 tiles was calculated for each cut-off length and glycerol content. The top 8 surface parameters that had strong correlations with friction are shown in Table 3. As shown in Table 3, the linear correlation coefficients for most of the surface parameters with the glycerol content of 70% are higher than those with 50% and 85%. The results indicated that a longer cut-off length resulted in a higher linear correlation coefficient for 70% glycerol content and that the cut-off length of 2.5 mm resulted in a slightly higher correlation coefficient than the other two cut-off lengths used in this experiment for glycerol contents of 50 and 85%. However, the correlation coefficients of the parameters for the cut-off length of 8.0 mm are higher on average than the other two cut-off lengths. The surface parameters R_{pm}, R_a and $W\Delta_a$ had the highest correlation coefficients averaged across all three cut-off lengths with average correlation coefficients of 0.839, 0.829 and 0.828, respectively. The results indicated that the waviness parameter $W\Delta_a$ appeared to have a stronger correlation with friction than the roughness parameters under a short cut-off length or when the glycerol content was 85%.

The surface parameter $W\Delta_a$ had the highest average correlation coefficient across 3 glycerol contents at the 0.8 mm cut-off length. $W\Delta_a$ versus the measured friction coefficient for all three glycerol contents and the cut-off length of 0.8 mm is shown in Figure 1. $W\Delta_a$ had correlation coefficients of 0.791, 0.862 and 0.697 for glycerol contents of 50%, 70% and 85%, respectively, for the 0.8 mm cut-off length.

The 0.8 mm cut-off length is widely used in surface roughness research related to slip and fall accidents (Chang *et al.*, 2001). Results from the current study indicated that surface waviness ought to be measured when using this cut-off length rather than surface roughness. Also when glycerol content was increased, the surface waviness parameters appeared to correlate stronger with the measured friction than the surface roughness parameters. The surface roughness parameters appeared to be better predictors of friction than the surface waviness parameters when the cut-off length of 8 mm was used.

Table 3. Surface Parameters with Top Linear Correlation Coefficients for 3 Cut-off Lengths

Cut-off Length	50% Glycerol		70% Glycerol		85% Glycerol	
	SP	r Value	SP	r Value	SP	r Value
0.8	R_{pm}	0.872	$W\Delta_a$	0.862	W_a	0.753
mm	Δ_a	0.813	W_{tm}	0.860	W_q	0.724
	R_k	0.811	R_{pm}	0.858	W_p	0.714
	R_a	0.809	R_a	0.850	W_{tm}	0.708
	$W\Delta_a$	0.791	W_p	0.844	$W\lambda_q$	0.704
	$W\Delta_q$	0.779	W_a	0.838	$W\Delta_a$	0.697
	W_{tm}	0.750	W_q	0.834	W_{pm}	0.672
	W_{pm}	0.724	W_{pm}	0.832	W_t	0.666
2.5	Δ_a	0.907	R_{pm}	0.942	W_a	0.858
mm	R_{pm}	0.826	R_k	0.935	W_p	0.846
	R_a	0.786	R_a	0.924	W_q	0.838
	R_k	0.785	$W\Delta_a$	0.916	W_{tm}	0.835
	$W\Delta_q$	0.776	$W\Delta_q$	0.910	R_k	0.831
	$W\Delta_a$	0.764	W_a	0.910	$W\Delta_a$	0.826
	W_{pm}	0.744	W_{pm}	0.902	R_{pm}	0.817
	W_t	0.744	W_q	0.901	W_{pm}	0.814
8	Δ_a	0.868	R_a	0.975	R_k	0.854
mm	R_a	0.831	R_k	0.970	R_{pm}	0.847
	R_{pm}	0.812	R_{pm}	0.966	$W\Delta_a$	0.840
	R_k	0.811	$W\Delta_a$	0.964	$W\Delta_q$	0.834
	$W\Delta_q$	0.801	$W\Delta_q$	0.963	R_a	0.828
	$W\Delta_a$	0.796	W_{tm}	0.912	W_{tm}	0.786
	W_y	0.717	W_y	0.891	W_y	0.745
	W_{tm}	0.715	Δ_a	0.857	W_t	0.727

SP: Surface Parameter

The results indicated that surface waviness, in addition to the surface roughness, is also a critical factor in determining the interface friction. When the glycerol content was 50%, surface roughness parameters were more important than surface waviness parameters. As the glycerol content was increased, the surface waviness parameters became more important than the surface roughness parameters. This implies that the shoe surfaces can penetrate contaminants when the viscosity of the contaminant is low and establish direct contact in most of the areas within the contact area. However, when the viscosity of the contaminant is increased, it becomes difficult to penetrate the contaminant, especially in the trough areas, and contact is made only near the crests of the surfaces. When a short cut-off length is used, the surface roughness measurement reflects the surface features at the microscopic level. As the cut-off length is increased, more surface waviness features are included in the surface roughness measurements. This helps explain why the surface roughness parameters had a stronger correlation with the measured friction when a longer cut-off length was used.

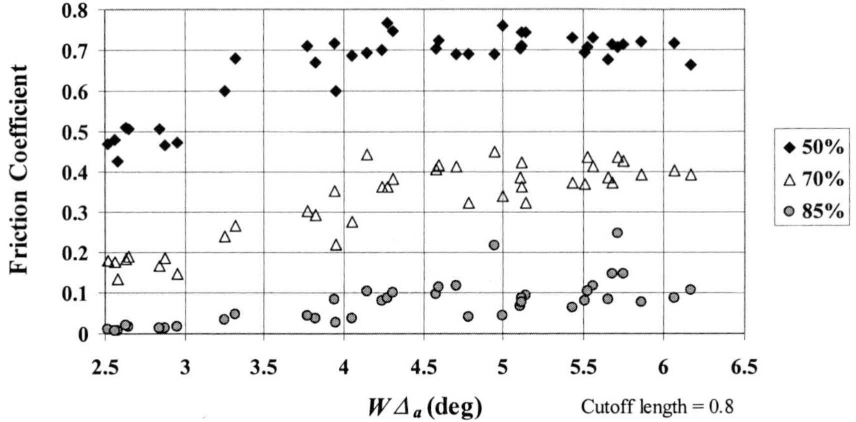

Figure 1. Friction Coefficient vs. $W\Delta_a$ with 0.8 mm Cut-Off Length with 3 Different Glycerol Contents

Conclusions

The results presented in this paper indicate that, in contrast to the current common practice of measuring surface roughness, surface waviness should be measured when a short cut-off length is used or when the viscosity of the contaminant is high. This implies that the shoe surfaces can penetrate the contaminants when the viscosity of the contaminant is low and establish direct contact in most of the areas within the contact area. However, when the viscosity of the contaminant is increased, it becomes difficult to penetrate the contaminant, especially in the trough areas, and contact is made only near the crests of the surfaces.

References

Chang, W.R. 2001, The effect of surface roughness and contaminant on the dynamic friction of porcelain tile, *Applied Ergonomics*, 32, 549–558

Chang, W.R., Kim, I.J., Manning D. and Bunterngchit, Y. 2001, The role of surface roughness in the measurement of slipperiness, *Ergonomics*, 44, 1200–1216

Courtney, T.K., Sorock, G.S., Manning, D.P., Collins, J.W. and Holbein-Jenny, M.A. 2001, Occupational slip, trip, and fall-related injuries–can the contribution of slipperiness be isolated? *Ergonomics*, 44, 118–1137

Grönqvist, R., Roine, J., Järvinen, E. and Korhonen, E. 1989, An apparatus and a method for determining the slip resistance of shoes and floors by simulation of human foot motions, *Ergonomics*, 32, 979–995

Harper, F. C., Warlow, W. J., and Clarke, B. L. 1961, *The forces applied to the floor by the foot in walking*, National Building Research Paper 32, DSIR, Building Research Station, (London: Her Majesty's Stationary Office)

Leamon, T. B., and Murphy, P. L. 1995, Occupational slips and falls: more than a trivial problem, *Ergonomics*, 38, 487–498

Strandberg, L. and Lanshammar, H. 1981, The dynamics of slipping accidents, *Journal of Occupational Accidents*, 3, 153-162

BEING LED UP THE GARDEN PATH: WHY OLDER PEOPLE ARE STILL FALLING AT HOME

C L Brace, R A Haslam, K Brooke-Wavell, P A Howarth

Health and Safety Ergonomics Unit,
Department of Human Sciences,
Loughborough University,
Leicestershire,
LE11 3TU
UK

Falls in the home are a major problem for older people of which there are many recognised risk factors. A group of older people were interviewed about fall safety and then monitored over the subsequent year. The fall rate amongst the sample was 29%. There were no significant differences amongst the fallers and non-fallers in terms of physical abilities or reported behaviour. However, it was concluded that the design of the environment played a role in the reported falls. Opportunities exist to reduce the risk of older people falling in and around the home, with respect to improved design. If the incidence of falls can be reduced, older people can stay in their homes for longer, whilst enjoying a higher quality of life.

Introduction

Falls among older people are a major health concern. It has been well documented over the years that a third of individuals over 65, and nearly half of those over 80, fall each year (Prudham and Evans, 1981). Approximately half of all recorded fall episodes that occur among independent community dwelling older people happen in their homes and immediate home environments (Lord *et al*, 1993). Fall related incidences are influencing factors in nearly half of the events leading to long-term institutional care in older people (Kennedy and Coppard, 1987). Clearly, if the incidence of falls can be reduced, people can live longer, more healthily and more independently in their own homes, with a better quality of life.

Falls pose a threat to older persons due to the combination of high incidence with high susceptibility to injury. The tendency for injury because of a high prevalence of clinical diseases (e.g. osteoporosis) and age-related physiological changes (e.g. slowed protective reflexes) makes even a relatively mild fall dangerous.

Over 400 potential risk factors for falling have been identified, which are commonly split into categories of intrinsic and extrinsic risk. However, individual fall incidents are generally multifactorial. Intrinsic factors are age and disease related changes within the individual that increase the propensity for falls, e.g. decreased balance ability, disturbed gait, cognitive impairment, reduced strength, impaired vision, illness, and side effects from use of medication. Research has estimated that intrinsic risk factors play a role in approximately 50% of falls amongst a combined group of institutionalised and community dwelling older people (Rubenstein and Josephson, 1996). Extrinsic factors are environmental hazards that present an opportunity for a fall to occur, including floor surfaces (textures and levels), loose rugs, objects on the floor (e.g. toys, pets), poor lighting, problems with walking aids and equipment, ill-fitting footwear, sensory surround and feedback (audio and visual), placement of furniture. Although environment-related risk factors are reported to be causal in around 33% of falls, it has only been recently that detailed work has been done to look at the design of some areas of the home environment in relation to older people, falls and independent living (Brace *et al.*, 2002). Previous research (Hill *et al*, 2000; Brace *et al.*, In Press) has identified important behavioural factors which affect the risk of older people falling in the home, e.g. rushing, carrying objects; these findings indicate that behaviour contributes to approximately 35% of falls. It has also been found that behaviour patterns change after a fall episode; general psychological state and experience can have an effect on the individual, affecting confidence and fear of falling.

This paper describes how a group of older people, after being interviewed about fall safety, participated in a fall notification scheme over the course of the subsequent year. The relationship between physical abilities, behaviour, and design of the home environment, have been investigated. The paper discusses the falls in terms of type, severity and cause of fall, and how behaviour has changed as a result of these incidences and since the initial interview.

Method

Interviews were conducted with 177 older people, aged 65 – 99, in their own homes. Issues explored by the interviews were factors affecting risk of falling in the home including age-related aspects and self-perceived safety. The immediate and longer term consequences of having a fall and the value and acceptability of preventative measures were also discussed. The interviews involved detailed discussion of different areas of the home, with regard to specific risk factors, and the interviewee's fall history and origin. In addition, standard anthropometric dimensions of interviewees were recorded, along with other measurements of strength, balance and coordination, and vision. The interviewees were monitored over the subsequent year, and all reported falls were recorded using a notification system, quantifying event, behaviour, and location.

Participation was invited for the Fall Notification Scheme immediately after each home interview. Of the original interviewed participants, 156 felt able to participate in the follow up study (86% response rate). Consenting participants were briefed and freepost envelopes and fall notification postcards were distributed. Participants were strongly encouraged to alert the researcher as soon as possible after a fall within the home (or that of a friend or relative), using the postcard or by a free telephone call. A very positive emphasis was given to participants that the researcher was interested in all falls regardless of whether or not an injury was sustained. When alerted, the researcher used an assessment during a telephone interview to illicit all possible information about the fall/s. At six months, personalised reminder letters (including further notification postcards, freepost envelopes and reminder details of the study) were distributed to all participants. At twelve months a semi-structured telephone interview was conducted with each individual to gather any further/missing fall data and to further investigate changes in fall related behaviour and perceptions of fall risk.

Results

Sample population

At the end of the year, 69 falls (range 1 – 12, mean 1.57, SD 1.78) were reported from a total of 44 individuals, within a final sample of 150 participants (after 2 deaths, 4 unobtainable), a fall rate of 29%. The majority (88%) of falls reported were recorded within 2 months of the incident occurring. The mean age of participants was 76 years (range 66-91, SD 6.7); an unexpectedly high proportion of reported falls were from males (41%, n=28). Of the sample, 36% of participants aged over 75 had fallen and 36% of participants aged over 80 had fallen.

Location

The majority of falls occurred within a person's own home (86%). The remainder occurred in the homes of friends or family (14%). As may be expected, most falls occurred between the hours of 6am and 6pm, as the majority of activities and movements are undertaken during these hours. As found in previous research (Brace *et al*, In Press), the majority of reported falls occurred in the garden (41%). The second most common location for falling was the living area of the home (25%), followed by the kitchen (9%) and bedroom (9%). Types of falls were varied but not dissimilar from Home Accident Surveillance System (HASS) data, Table 1.

Table 1. Types of falls sustained amongst the sample

Fall Type	% falls in this study		% falls in older population *
Slip	24	} 63	} 48
Trip/stumble	39		
Fall on/from stairs/steps	8		12
Fall on/from ladder/stepladder	3		1
Fall from building/structure	1		<1
Other fall from one level to another	13		10
Slip/trip no fall	0		<1
Body part gave way	9		1
Unspecified	3		27

*1999 HASS data, DTI 2001

Injury and recovery

A large proportion (84%) of fall episodes resulted in some level of injury, with varying levels of severity (from bruising to fractures). 14% of falls resulted in hospital treatment and 11% in visits to the GP. The remaining 75% of fall-incurred injuries were self treated by the individuals concerned. Recovery time also varied according to severity of injury. Just under half of fall episodes resulted in an immediate recovery (22%) or a complete recovery within a number of days (23%). However, over half of reported incidents required further time for a complete recovery which involved several weeks (39%) or months (14%). One person died as a result of a fall.

Factors causing falls

The majority (66%) of falls were multifactorial in their causation (see quote below). In just over a third of cases (34%), individuals reported that only one factor had played a role in the incident. Table 2 gives a breakdown of the perceived causal factors in falls among the sample.

> *"I was putting bread out for the birds in the garden and I tripped on a raised paving slab on the patio. I was hurrying as it was cold out... I just flew forward onto my hands. It was first thing in the morning after breakfast and I was feeling a bit dopey as I was taking some new medication that was making me feel a bit peculiar. I think that certainly had something to do with it."*

Table 2. Causal factors in falls (multiple responses)

Causal Factor	% of resulting falls
Behavioural	
Rushing	39
Trip over (unfixed) object, e.g. clutter	27
Fall from height	14
Inappropriate use of footwear	13
Carrying (reported to affect balance)	9
Carrying (reported to obscure vision)	7
Overstretching/bending	5
Pet related activity, e.g. trip	5
Light level	4
Extrinsic	
Slippery surface (outdoors)	23
Trip over (fixed) object, e.g. furnishing	23
Uneven surface	16
Walking aid error	11
Slippery surface (indoors)	7
Intrinsic	
Balance	16
Body part gave way/collapsed	9
Poor vision	6
Dizziness	5

Falls, fall history and physical abilities

In this study, fallers were generally slightly older than non-fallers (mean 76 years compared to 74 years of age), and they had had more falls previously than non-fallers (mean of 1.07 compared to 0.77 previous falls per person). However, these differences were not significant. Although a proportion of falls were reported to be due to intrinsic risk factors (Table 2), no differences were found for general fall related health problems between fallers and non-fallers (93% and 91% respectively), mobility problems (55% and 59%), scores for activities of daily living (7.3 and 7.6), mean number of prescribed medications being taken (2.1 and 2.3 medications per day), type of accommodation lived in, or cohabitation status (43% of both groups lived alone).

Falls and reported behaviour in the home

There were no significant differences between fallers and non-fallers for reported changes in behaviour; 61% of fallers and 63% of non-fallers reported taking more care in their behaviour since the initial home interview. Similar numbers of fallers and non-fallers reported a fear of falling at the final interview (41% and 42% respectively). Furthermore, there where no significant differences in behaviour (reported in the previous interview), e.g. rushing, leaving items on the floor, carrying, undertaking DIY activities, between fallers and non-fallers. However, it is apparent that hurrying and keeping clutter on the floor are major contributors to the falls among this sample population, Table 2.

Falls and design of the home environment

The most common fall types of slipping and tripping (Table 1) were to a large extent caused by objects on the ground, and slippery and uneven surfaces (Table 2). These factors lead to implications for design with respect to: aesthetic, non-slip floor surfaces (indoor and outdoor); redesign/removal of (fixed) trip hazards, e.g. steps, raised thresholds; and, improvements in texture, type and level of surface. Of the 41% of falls that occurred in the garden, all were due to at least one of these factors.

Discussion

After more than two decades of research into the area, a sufficient quantity of convincing studies have been completed to give reliable estimates of fall incidence among community dwelling older people. Approximately 30% in the >=65 year age group , 40% in the >=75 year age group and 50% in the >=80 year age group, fall each year (Prudham and Evans, 1981; Tinetti et al., 1988). The figures found in this study are slightly lower than these estimates. It is unknown whether this is a reflection of uncaptured fall data, behavioural changes due to improved awareness, or simply owed to chance.

Cummings *et al.* (1988) found that in retrospective studies, recall of falls among older people is underestimated by 13-22% compared to prospective studies, depending on the time period of recall. Although most falls reported in this study were recorded within a short time period, it is possible that incidences may have been forgotten or overlooked.

It can be expected that older people who have a history of falling are more likely to fall again in future, due to frailty and loss of confidence. A slightly increased fall rate was found in this study although the finding was not statistically significant.

It is apparent that behavioural risk factors are major contributors to the falls amongst this sample population (Table 2). It could therefore be suggested that behavioural factors are strong elements in many falls. Little information exists due to the fact that behavioural risk factors are often ignored and unrecorded components due to a strong medical research perspective, concentrating on intrinsic risk factors (Brace *et al.*, 2002).

The strong relationship between falls in the garden and extrinsic risk factors demonstrates a need for improved design and a 'design for all' ethos. As well as reducing the risk of falling, physical activity has been shown to modify a significant number of the risk factors associated with falling (Skelton and McLaughlin, 1996) so it seems inappropriate that some older people are trapped inside their homes, lacking the confidence and the ability to manage the environmental challenges of their gardens, surroundings that could well be used to improve health, quality of life and independence.

Conclusions

If falls are to be reduced amongst this population, it is imperative that all aspects of fall causation are well considered and appropriately addressed. Although not shown to be a statistically significant factor in this study, behaviour is still an area that requires further investigation. Opportunities exist to reduce the risk of older people falling in and around the home, with respect to the design of products and buildings. Clearly, if the incidence of falls can be reduced, people can live for longer, both more healthily and more independently, in their own homes.

Acknowledgements

The authors wish to acknowledge the support of the Department of Trade and Industry (DTI) who sponsored part of this research. The views expressed, however, are those of the authors and do not necessarily represent those of the DTI.

References

Brace, C.L., Haslam, R.A., Brooke-Wavell, K., Howarth, P. IN PRESS, *Behavioural Factors Resulting in Falls Among Community Dwelling Older People.* (Department of Trade and Industry: London)

Brace, C.L., Haslam, R.A., Brooke-Wavell, K., Howarth, P. 2002, Reducing Falls in the Home Among Older People - Behavioural and Design Factors. In P.T.McCabe (ed.) *Contemporary Ergonomics 2002,* (Taylor and Francis, London), 471-476.

Cummings, Nevitt, M.C., Kidd, S. 1988, Forgetting falls. The limited accuracy of recall of falls in older community-dwelling women. *J Am Geriatr Soc* 1994; **42**: 1110-7.

Hill, L.D., Haslam, R.A., Howarth, P.A., Brooke-Wavell, K., and Sloane, J.E., 2000, *Safety of Older People on Stairs: Behavioural Factors*. (Department of Trade and Industry: London) DTI ref: 00/788.

Department of Trade and Industry (DTI). 2001, *HASS listings for 1999, for males and females aged 65 and above*. (Department of Trade and Industry, 2001)

Kennedy, T.E. and Coppard, L.C. 1987, The prevention of falls in later life. *Dan Med Bull* 1987; **34**: 1-24.

Lord, S.R., Ward, J.A., Williams, P., and Anstey, K.J. 1993, Physiological factors associated with falls in older community-dwelling women. *Australian Journal of Public Health*; **17** (3): 240-5.

Prudham, D., and Grimley-Evans, J. 1981, Factors associated with falls in the elderly. *Age and Ageing* **10**:101.

Rubenstein, L.Z. and Josephson, K.R. 1996, Falls. In R.A.Kenny (ed.) *Syncope in the Older Patient*, (Chapman and Hall, London).

Skelton, D.A. and McLaughlin, A.W. 1996, Training functional ability in old age. *Physiotherapy*; 82 (**3**): 159-67.

Tinetti, M.E., Speechley, M., Ginter, S.F. 1988, Risk factors for falls among elderly persons living in the community. *N Eng J Med* 1988; **319**: 1710-7.

FATIGUE

AN OVERVIEW OF RESEARCH ON FATIGUE IN SUPPORT SHIPPING IN THE OFFSHORE OIL INDUSTRY

Andy Smith[1] , Tony Lane[2] and Mick Bloor[2]

[1]Centre for Occupational and Health Psychology and [2] Seafarers International Research Centre, Cardiff University, 63 Park Place, Cardiff, CF10 3AS

The present paper reviews findings from a project on fatigue in ships working in the offshore oil industry. The results show that long working hours, varying shift patterns, reduced manning and problems with motion and noise are often present. These factors are often associated with perceptions of reduced safety. Combinations of these factors are also associated with impaired health. A review of the literature and analyses of accident data provided little information on the impact of seafarers' fatigue. Analyses of onboard measures of performance, alertness and sleep suggest that fatigue offshore is not a general problem but that certain job characteristics, such as working at night, are associated with reduced alertness and impaired performance. Further research is now required to determine whether this view holds up across a range of ship types.

Introduction

Mounting concern with seafarer fatigue is widely evident among maritime regulators, insurers, ship owners, trade unions and welfare agencies. We are carrying out a research programme to investigate this topic and the first phase of the research was concerned with specific comparisons between offshore oil support shipping and the offshore oil industry. The overall objectives of the research are: to predict worst case scenarios for fatigue, health and injury; develop best practice recommendations appropriate to ship type and trade; and produce advice packages for seafarers, regulators and policy makers. This topic has been investigated using a variety of techniques to explore variations in fatigue and health as a function of the voyage cycle, crew composition, watch keeping patterns and the working environment. The methods involve:

- A review of the literature
- A questionnaire survey of working and rest hours, physical and mental health
- Physiological assays assessing fatigue, rhythm adjustment and cardiovascular risk
- Instrument recordings of sleep quality, ship motion, and noise
- Self-report diaries recording sleep quality and work patterns
- Objective assessments and subjective ratings of mental functioning

- Analysis of accident and injury data

A detailed account of the project is given in Smith, Lane and Bloor (2001).

The literature review

The purpose of this literature review was to identify existing research into fatigue, health and injury rates among employees engaged in exploration, production and support roles in the offshore oil industry. Although the contribution of fatigue to occupational injuries has been extensively researched in other industries, findings cannot automatically be applied to the offshore industry. This is because of the unique combination of conditions offshore workers have to cope with, including motion, extreme weather, noisy working conditions and demanding work and rest patterns. The offshore oil industry has gone through major economic, structural and technological changes in recent years, leading to reduced manning of installations and ships, increased automation, increased workload and decreased job security. This has put increased pressure on both seafarers and installation workers who often feel they have to work extra hours and shifts in order to keep their jobs.

Seafaring research indicates that:

- Collision risk is greatest during the early hours of the morning, suggesting a circadian influence. Fatigue has been proposed as a contributory factor.
- Seafarers report elevated levels of anxiety, perceived work load, dissatisfaction with shift schedules and sleep problems.
- Environmental factors are related to sleep disturbances, fatigue and stress.
- Motion adversely affects cognitive and psychomotor performance.
- Circadian adaptation can only be partially achieved at best on 4-on/8-off shift systems.

Furthermore, injury studies of seafarers have paid little attention to factors such as hours-into-shift, days-into-tour, job area, risk perception, personality characteristics, time of day of the incident. In addition, reporting of accidents needs to be more formal and systematic with enough detail to enable investigation of the underlying causes of accidents and make comparisons between the different aspects of the offshore oil industry. One of the major problems is the lack of exposure data, which makes it impossible to calculate accurate accident rates. Sample sizes also tend to be very small which makes it hard to draw firm conclusions. Future research needs to encompass a range of issues that affect offshore workers because it is the combination of factors that make the offshore oil industry unique.

The Survey

The current survey was therefore designed to identify all aspects of the working environment that may lead to fatigue, and affect the health and general well-being of personnel employed in all sectors of the offshore oil industry. The aims of the survey were:

- To assess the work and rest patterns of seafarers and offshore installation workers

- To assess the extent to which working hours, shift patterns and time spent offshore are associated with fatigue, injuries, and poor physical and mental health of crewmembers
- To assess the impact of time spent offshore on leave time

The questionnaire (33 pages) was designed to encompass all aspects of an offshore worker's life, and assessed the nature of tours of duty, work and rest patterns, fatigue and sleep, health-related behaviours and general health and well-being. It was divided into the following three sections:

1. Offshore: included questions relating specifically to work patterns, and subjective measures of attitudes towards work.
2. On leave: included subjective measures of health and well-being, and health related behaviours such as eating, drinking, smoking and exercise.
3. Life in general: included a number of standardised scales of well-being, such as the General Health Questionnaire (GHQ), the Profile of Fatigue-Related Symptoms (PFRS), the Cognitive Failures Questionnaire (CFQ) and the MOS Short Form Health Questionnaire (SF-36).

The questionnaire was distributed to the home addresses of members of the seafaring officer's union, NUMAST and the installation worker's union, MSF. Secondly, questionnaires were distributed to seafarers onboard offshore oil support vessels operating in the UK sector, by visiting researchers. A short version of the questionnaire was sent to a group of onshore workers, as a control for items specifically relating to fatigue. Results were also compared with normative data from three other sources: a sample of onshore workers currently participating in a study of workplace stressors, a random sample of the working population taken from the Welsh Health Survey (1998), and data from the Bristol Health and Safety at Work Study (1999).

The main findings were:

Although exposed to similar working environments, the results demonstrate a number of important differences in work patterns between the two groups of offshore workers. Seafarers for example, are more likely to work fixed shifts and longer tours of duty than installation personnel. However, both groups work considerably longer hours per week than a comparison group of onshore workers.

Both samples were of similar mean age, although socio-economic differences did exist between the populations of seafarers and installation personnel studied. Largely as a result of targeting methods, the majority of the seafarers surveyed were officers, whereas the majority of installation respondents were classified as manual workers. (The effects of these differences in socio-economic status were statistically controlled).

A number of significant findings emerged from the analyses, which highlight the potential impact of demanding work schedules and lack of sleep. These can be summarised as follows:

- Installation workers reported that they were generally worse off in terms of psychological well-being compared to either seafarers, or a comparison group of onshore workers (when age, gender and socio-economic status were controlled for)
- Rotating shift patterns (e.g. changing from day to night work, or vice versa) impacted negatively on mental health
- Long work hours were associated with increased reports of emotional distress

- Perceptions of fatigue tended to be highest amongst personnel who also reported psychological, somatic and cognitive difficulties
- Cognition and physical and mental well-being were negatively influenced by insufficient or poor quality sleep
- There was evidence to suggest that working practices offshore also have a negative effect on well-being during the early stages of leave

It is evident therefore, that the majority of offshore personnel feel that current working patterns and practices are in some ways detrimental to their health and safety. However, a number of issues require further clarification: firstly, to what extent current shift systems are responsible for the apparently poorer health of installation personnel, and secondly, to what extent self-selection and medical screening buffer the influence of difficult working conditions.

Combined effects of work characteristics on health

Recent analyses (McNamara & Smith, in press) have examined associations between combinations of hazards and health. The results show that individuals exposed to the greatest combination of negative factors report the most health problems.

Analysis of injury databases

Two datasets, containing injury-related information from support vessels were analysed in terms of temporal and environmental factors. Dataset 1 was obtained from a large multinational oil company, and dataset 2 from the MCA. Unfortunately, there were several problems inherent within both datasets, namely lack of exposure information and large amounts of missing data within significant temporal and environmental variables.

Nonetheless, incidents were examined in terms of time of day, hours into shift, days into tour and motion (inferred from sea state and wind force). Further analyses were conducted specifically examining injury severity (i.e. fatalities, serious injuries) and injury type (i.e. burns/scalds, strains/sprains) in terms of the above temporal and environmental factors, in order to see if this would add to information gained from total incidence analyses.

Although the two datasets deal in the main with different injury types (the majority of incidents detailed in dataset 1 are relatively minor, whereas in dataset 2 they are, for the most part, major) they generate a very similar pattern of results. In both cases, injury frequency demonstrates a time of day effect (incidence is higher between 0900-1600 hours) although there is no evidence that fatigue is a causal factor. Injuries do not peak significantly during any time periods associated with natural circadian troughs (i.e. post lunch or in the early hours of the morning). However, some evidence is presented to suggest that sleep inertia (or unfamiliarity with the work; type of work being carried out at this time) impacts upon injury frequency, as injury rates are highest in the first few hours of a shift. With regards days into tour, (it was only possible to analyse information from dataset 1 in this instance) it seems that the occurrence of injury is greatest in the first week. However, it should be noted that database 1 contains a limited amount of data regarding fatalities and major injuries.

With regards environmental factors, accidents seem more likely to occur in calm conditions. However, this somewhat peculiar finding may simply be a reflection of work patterns: in other words, it is more than likely that a greater proportion of personnel are exposed to potential incidents in calm conditions, as they are more likely to be working.

Analyses of temporal factors in terms of injury severity and accident and injury type, failed to yield additional information to that provided by overall incidence rates. In conclusion, the present analyses provide little evidence for a major role of fatigue in offshore accidents. This does not mean that fatigue plays no part, rather it shows that it is impossible to determine the impact of fatigue from data of this type, largely due to the inadequacies inherent within current reporting systems.

Logbooks

Logbooks were completed by 58 onshore day workers and 42 offshore workers doing day shifts. The results showed that the two groups differed significantly on a number of sleep variables. Offshore workers slept for a shorter time, woke up more often during the night, had greater difficulty falling asleep, and were less likely to consider that they had a deep sleep or enough sleep. Although these differences are statistically significant the magnitude of the effects are small. Offshore workers also considered their jobs to be less physically demanding than the onshore sample.

Comparisons were then made between different offshore groups. These analyses compared 31 installation workers and 29 seafarers. 42 were day workers and 18 night workers. 25 were in the first week of their tour of duty and 35 were in either their second or third week offshore. The results showed that installation workers felt less alert at the start of the day. Those working nights reported lower alertness at the end of the working day even though they perceived their job to be less physically demanding. Day workers starting their tour of duty awoke more frequently than those in their second or third week of the tour. The reverse was true for night workers. Sleep duration was reduced for the first night offshore, especially for installation workers doing nightshifts. The alertness levels at the end of the first day were lower for the seafarers than installation workers, with the reverse pattern being present on days 6 and 7. Physical effort was perceived by the day workers to decrease over the week whereas night workers perceived it to increase.

43 volunteers completed the weekly log while they were on leave. 22 were installation workers and 21 seafarers. 34 had worked day shifts before leave and 9 had worked nights. Of these 43 participants 22 had just started their leave and 21 were on their second week of leave. The results showed clear evidence that sleep duration and alertness were abnormal at the start of leave.

Assessments aboard ship

Initial analyses of onboard measures of performance, alertness and sleep suggested that fatigue offshore is not a general problem present at all times in all personnel. However, certain job characteristics, such as working at night, were associated with reduced alertness and impaired performance. Noise exposure was also found to be associated with sleep problems and impaired performance. Analyses of combinations of factors (e.g. noise and night work) suggest that the effects of different job characteristics are largely independent (Wellens et al., in press). Further research is now required to determine whether this view holds up across a range of ship types. In addition, it is essential to examine situations where combinations of potentially harmful factors are present for at the moment we may have information about "best practice" rather than the "worse case" scenario.

Recommendations

The literature review has confirmed the absence of information on seafarers' fatigue. It is essential, therefore, that further research is conducted on this topic and the present project shows that this is feasible. The analyses of injury data show that better reporting systems are necessary and collaboration is now needed to develop and validate new systems. Our survey questionnaire is a useful tool for providing initial descriptions of potential problems and can now be used as a screening tool providing justification for more detailed investigations. The onboard measurements of performance show that features of work offshore (e.g. shift system; noise) may have an impact on safety. These methods can now be applied to the assessment of other ships and to make recommendations about suitable working practices. The research has shown that the effects of offshore work on quality of life while on leave requires further investigation.

Acknowledgement

The research described in this article was supported by the Maritime and Coastguard Agency, the Health and Safety Executive, NUMAST and MSF. We would also like to acknowledge the contribution made by the ship and installation owners and their employees who volunteered to participate in the research. The data collection and analyses have been carried out by Rachel McNamara, Alison Collins, Victoria Cole-Davies, Neil Ellis, Geoff Boerne and Jo Beale.

References

McNamara, R.L. and Smith, A.P. The combined effects of fatigue indicators on health and well-being in the offshore oil industry. *Archives of Complex Environmental Studies, in press*

Smith, A., Lane, A. and Bloor, M. 2001, Fatigue Offshore: A comparison of offshore oil support shipping and the offshore oil industry. Seafarers International Research Centre (SIRC). ISBN: 1-900174-14-6.

Wellens, B.T., McNamara, R.L., Ellis, N. and Smith, A.P. Combined effects of shift work and occupational noise exposure on performance tasks in a seafaring population. *Archives of Complex Environmental Studies, in press.*

A CROSS-VESSEL SURVEY OF SEAFARERS EXAMINING FACTORS ASSOCIATED WITH FATIGUE

Paul Allen, Ailbhe Burke and Neil Ellis

Centre for Occupational and Health Psychology
Cardiff University
63 Park Place,
Cardiff, CF10 3AS

As part of the second phase of the seafarers fatigue project a questionnaire based survey was administered through onboard field research and a mail shot, replicating phase one. The short sea and coastal shipping industry was investigated in contrast with the offshore oil industry examined in the first phase. Higher fatigue and lower health were found in the phase two sample, suggesting the importance of vessel type in determining levels of fatigue. Ship based fatigue differences were detected in the stage two sample, prompting further investigation of factors which define ships at a practical level. Tour length and watch / duty schedules may be particularly useful in terms of characterising vessels and therefore by extension these factors may be of use in terms of ultimately accounting for seafarers fatigue.

Introduction

In phase one of the seafarers fatigue project the offshore oil industry was examined (Smith, Lane and Bloor, 2001). This initial study represented the first comprehensive investigation of fatigue at sea, adopting a series of subjective and objective measures and involving onboard field research. A questionnaire based survey was also administered to both onboard participants and to a larger sample via a mail shot. In the second phase of the project the short sea and coastal shipping industry was examined in order to widen the scope and generalised relevance of results from phase one. Whilst slight adjustments were made to the onboard measures between phases one and two (Ellis *et al*, 2003), the survey format was identical to that previously adopted. Results from the phase two survey will form the basis of this paper.

Survey content

The questionnaire used in the survey was identical to that used in phase one of the project and consists of three sections:

1. *Offshore:* Questions in this section refer specifically to time spent offshore, encompassing measures of work and rest patterns, and subjective measures of attitudes towards work.
2. *On leave:* Questions in this section relate to time spent on leave/at home and include subjective measures of health and well-being, fatigue, sleeping patterns, and health-related behaviours such as eating, drinking, smoking and exercise.
3. *Life in general:* Questions in this section are designed to measure incidence of accidents and injuries, and general health and well-being, using a number of standardised scales such as the General Health Questionnaire (GHQ), the Profile of Fatigue Related Symptoms (PFRS), the Cognitive Failures Questionnaire (CFQ), and the MOS Short Form Health Questionnaire (SF-36).

Sample

In parallel with phase one of the project, two techniques were used to administer the survey to seafarers. Firstly, the questionnaire was sent out in a mail shot to the officers union NUMAST and four individual seafaring companies. Secondly, the questionnaire was completed by participants recruited as part of the onboard study. The ships involved with onboard testing represented the same four companies targeted with the mail shot. Characteristics of the onboard sample are described in the accompanying paper by Burke *et al* (2003). The mail shot sample is described below.

Mail shot sample

The survey was sent to two ferry/freight companies and two tanker companies. The number of questionnaires sent out and response rates are shown in Table 1 below. As can be seen from Table 1, response rates were relatively low, however these rates were approximately comparable across the different sub-samples which reduces the possibility of any selective group biases. The mail shot sample had a mean age of 45.0 years (range 17- 66, SD=9.72). This was high due to the large proportion of NUMAST respondents in this sample. NUMAST respondents represent officers and those in senior positions. In terms of education, 54.7% of respondents reported completing GCSE's / 'O' levels, which again may have been skewed by the large number of officers in the sample.

Table 1. Mail shot response rates

Company	Number of Questionnaires Sent	Number of Questionnaires Received	Response Rate (%)
Numast	2740	539	19.7
Ferry Co.1	650	137	21.1
Ferry Co.2	110	35	31.8
Tanker Co.1	110 Polish / 90 English	30	15.0
Tanker Co.2	250	48	19.2

Phase 1 vs. Phase 2 comparison

A key concern within the second phase of the project was to determine the extent to which the findings from the offshore oil industry could be generalised to short sea and coastal shipping. Table 2 below shows the type of vessel respondents reported working on in both stages (mail shot and onboard sample combined). As the table shows, phase one focused upon industrial type vessels, whilst in phase two passenger carrying vessels were most highly represented.

Table 2. Phase 1 and Phase 2 vessel types

	Vessel type	% Respondents
Phase 1 (seafarers only)	Offshore support	26.3 (n=147)
	Supply vessel	29.0 (n=162)
	Standby vessel	13.4 (n=75)
	Tanker	4.2 (n=23)
	Other	27.2 (152)
Phase 2	Passenger ferry	40.3 (n=372)
	High speed passenger ferry	9.0 (n=83)
	Freight	21.1 (n=195)
	Tanker	16.2 (n=150)
	Dredger	6.6 (n=61)
	Other	6.8 (n=63)

The next stage was to assess relative levels of fatigue within the two phases.

Levels of fatigue

In order to assess relative levels of phase one and phase two fatigue, four measures were adopted from the survey. In addition to the Profile of Fatigue Related Symptoms (PFRS) fatigue subscale, three separate factors were derived as detailed below. Constituent factor items were kept consistent across the two phases.

General Fatigue Symptoms. Factor analysis was conducted on seven items addressing symptoms of fatigue. On each item respondents had to rate the extent to which they had experienced the specific symptom whilst at sea, from 1 (very) to 5 (not at all). The symptoms were; 1. Confusion, 2. Lethargy, 3. Poor quality sleep, 4. Depression, 5. Tension, 6. Loss of concentration and 7. Increased use of caffeine. In phase one these items clustered on two factors, with only items 3 and 7 loading most highly on the second factor. Whilst only one factor was extracted in the phase 2 analysis, the two least loading items were again 3 and 7. A combined factor was therefore calculated as the average of scores across items 1, 2, 4, 5 and 6.

Fatigue at work. Factor analysis was conducted on all six items within the 'feelings at work' section of the questionnaire. The first three items were found to load on a single factor relating to fatigue at work. These items assessed the respondents' typical state during work, their level of tiredness at work and how often they felt sleepy at work. This finding was consistent across both phases, and therefore a combined factor was calculated as the average of items 1 to 3.

Fatigue after work. The last three items in the 'feelings at work' section were found to load on a separate factor addressing fatigue after work. These items assessed levels of physical and mental tiredness at the end of the working day, and how tense the respondent felt after the working day. Again this finding was consistent across both phases, and therefore a combined factor was calculated as the average of items 4 to 6.

In Table 3 below phase one and phase two fatigue scores are shown across the four fatigue scales detailed above. On all scales a higher score indicates higher levels of fatigue (score ranges: General fatigue = 1 to 5; Fatigue at work = 1 to 6: Fatigue after work = 1 to 4: PFRS fatigue scale = 12 to 84).

Table 3. Mean levels of fatigue in phases one and two

	General Fatigue symptoms	Fatigue at work*	Fatigue after work*	PFRS Fatigue scale**
Phase 1	1.39 (n=494, S.E = 0.04)	3.63 (n=550, S.E = 0.04)	2.35 (n=559, S.E = 0.03)	25.4 (n=553, S.E = 0.53)
Phase 2	1.43 (n=793, S.E = 0.03)	3.74 (n=905, S.E = 0.03)	2.46 (n=911, S.E = 0.02)	29.3 (n=907, S.E = 0.47)

*= differ significantly at the p<0.05 level. ** = differ significantly at the p<0.01 level

As Table 3 above shows, phase two consistently reported higher levels of fatigue. The next question to address was whether these fatigue based differences had any direct relation to reported health status.

Health status

Table 4 compares the two phases using the Short Form Health Survey (SF-36) section of the questionnaire which assesses eight health dimensions. A low score on all subscales indicates poorer functioning, except the 'bodily pain' scale where a low score indicates an absence of significant pain. As Table 4 shows, phase two respondents consistently reported poorer health except on the bodily pain subscale, thus supporting an intuitive link between fatigue and health scores. The next question to address concerned what actually causes the between stage differences that have been identified.

Table 4. Comparison of health status in phases one and two

	phase 1 (mean)	phase 2 (mean)
Physical functioning	**90.0** (n=556 , S.E = 0.67)	**89.6** (n=901 , S.E = 0.49)
Role-physical **	**87.9** (n=536 , S.E = 1.14)	**83.1** (n=909, S.E = 0.95)
Bodily pain *	**77.9** (n=559 , S.E = 0.87)	**75.1** (n=930 , S.E = 0.74)
General health **	**70.4** (n=557 , S.E = 0.73)	**67.0** (n=907 , S.E = 0.63)
Vitality **	**64.0** (n=558 , S.E = 0.76)	**59.7** (n=904 , S.E = 0.64)
Social functioning **	**84.9** (n=554 , S.E = 0.87)	**81.2** (n=917 , S.E = 0.76)
Role-emotional **	**88.3** (n=527 , S.E = 1.17)	**81.5** (n=914 , S.E = 1.06)
Mental health *	**74.8** (n=555 , S.E = 0.70)	**72.5** (n=904 , S.E = 0.58)

*= differ significantly at the p<0.05 level. ** = differ significantly at the p<0.01 level

Accounting for the between phase differences in fatigue and health status

Whilst the between phase differences that have been found are interesting in isolation, these differences prove most useful in terms of highlighting factors directly associated with seafarers fatigue. Ship type in particular was identified as potentially useful in accounting for between phase differences, therefore this factor was examined independently within the phase two sample. Clearly any ship based fatigue differences would have considerable applied relevance.

Ship type differences

Five ship groupings were made within the phase two survey sample into which most respondents naturally fell (Passenger, High speed passenger, Freight, Tanker and Dredger). Any unclear cases were put into a sixth 'other' category.

Levels of fatigue

In Table 5 fatigue after work scores are shown for the six ship categories.

Table 5. Ship comparison of fatigue after work

Vessel	After work fatigue (Mean)
Passenger	**2.57** (n=363 , S.E = 0.03)
High speed Passenger	**2.57** (n=79 , S.E = 0.07)
Freight	**2.37** (n=189 , S.E = 0.05)
Tanker	**2.32** (n=147 , S.E = 0.05)
Dredger	**2.39** (n=61 , S.E = 0.06)
Other	**2.36** (n=62 , S.E = 0.07)

As shown in Table 5, there were found to be differences between the vessels in terms of self-reported levels of fatigue. A series of analyses of co-variance (ANCOVA) co-varying for age, education and socio-economic status (SES) found significant between-vessel differences on three of the four fatigue scales (fatigue at work ($p<0.01$), fatigue after work ($p<0.01$) and PFRS fatigue scale ($p<0.01$)). Differences on the general fatigue symptoms scale were also approaching significance ($p=0.063$). The most marked difference across the four fatigue scales was that the passenger and high speed passenger respondents in particular reported noticeably higher levels of fatigue than the tanker respondents. Evidence was also shown for a broader industrial / passenger split with the passenger vessels (passenger ferry, High speed passenger ferry, freight) generally reporting higher levels of fatigue than the more industrial type vessels (tanker, dredger, other). Therefore it appears that to a large extent phase one versus phase two fatigue differences can be accounted for by the effect of ship type. The next area to consider is whether the concept of ship type can account for health status differences that were found between the two phases.

Health status

In Table 6 the different vessel types are compared in terms of health status. As the table illustrates, significant differences were found between the ships in terms of specific health scores (e.g. vitality). On other scales, however, differences were less pronounced (e.g. mental health) or negligible (e.g. general health). Conclusions are therefore more difficult to draw concerning the relative health status of different ship types. The only justifiable conclusion is that there is marginal evidence on a number of the health scores to support a broad industrial / passenger division as previously identified. The fact that a ship based comparison did not reveal clear health differences raises questions in terms of the use of this variable in accounting for the unequivocal phase based differences which were found. In terms of understanding this result two possibilities therefore emerge; 1. Ship type is not the most crucial variable distinguishing phase one from phase two, or more plausibly, 2. Vessel differences within phase two do not reflect the magnitude of vessel differences between the phases. Assuming the second possibility to be true, the next stage is to identify which factors are crucial within the concept of ship type in terms of understanding fatigue. Therefore whilst phase differences arguably pointed towards ship type, ship type must also point towards a number of factors, such as tour length or shift type, which actually characterise ships on a practical level.

Table 6. Ship comparison of SF-36 Health scores

Vessel	Vitality** (Mean)	Mental Health (Mean)	General Health (Mean)
Passenger	56.7 (n=362, S.E = 0.95)	72.0 (n=359, S.E = 0.85)	66.7 (n=362, S.E = 0.90)
H.speed passenger	56.4 (n=78, S.E = 2.24)	72.1 (n=78, S.E = 2.14)	68.5 (n=81, S.E = 2.46)
Freight	60.6 (n=187, S.E = 1.52)	70.0 (n=186, S.E = 1.39)	67.2 (n=188, S.E = 1.49)
Tanker	64.7 (n=143, S.E = 1.75)	74.5 (n=148, S.E = 1.50)	66.6 (n=145, S.E = 1.62)
Dredger	60.7 (n=61 , S.E = 2.12)	76.3 (n=61, S.E = 1.78)	63.8 (n=59 , S.E = 2.36)
Other	65.3 (n=62 , S.E = 1.90)	74.8 (n=61 , S.E = 2.07)	69.5 (n=60 , S.E = 2.20)

**= differ significantly at the $p<0.05$ level.

Conclusion

Through examination of phases one and two of the Seafarers Fatigue Project differences were found in terms of levels of fatigue and also health status. These differences were identified as primarily reflective of the different vessel types studied within the two phases. Whilst vessel differences are interesting, however, in terms of identifying the causes of fatigue the concept of vessel type itself needs to be disassembled into key defining factors.

Acknowledgments

The research described in this article is supported by the Maritime and Coastguard Agency, the Health and Safety Executive, NUMAST and the Seafarers International Research Centre. We would also like to acknowledge the contribution made by the ship owners and seafarers who have participated in the research.

References

Burke, A.M., Ellis, N. and Allen, P.H 2003, The impact of work patterns on stress and fatigue among offshore worker populations. *Contemporary Ergonomics 2003.*

Ellis, N., Allen, P.H. and Burke, A.M. 2003, The influence of noise and motion on sleep, mood, and performance of seafarers in the coastal and short sea shipping industry. *Contemporary Ergonomics 2003.*

Smith, A.P., Lane, A.D. and Bloor, M., (2001). *Fatigue Offshore: A Comparison of Offshore Oil Support Shipping and the Offshore Oil Industry.* Seafarers International Research Centre. ISBN 1-900174-14-6.

THE IMPACT OF WORK PATTERNS ON STRESS AND FATIGUE AMONG OFFSHORE WORKER POPULATIONS

Ailbhe Burke, Neil Ellis and Paul Allen

Centre for Occupational and Health Psychology,
Cardiff University,
63 Park Place,
Cardiff, CF10 3AS

This study examined the effects of tour length on stress and fatigue in seafarers in the coastal and short sea shipping industry, in terms of both self report and objective measures. Firstly, a brief outline of the sample and measures used will be given. Then, some background on the issue of tour length is provided. This will be followed by analysis of length of tour for this study in terms of its impact on various measures used in testing. These included self-reports of sleep quality, fatigue, stress levels and mood and performance on reaction time and attention tasks and objectively measured sleep quality. These findings are then outlined and discussed, and the role of tour length in seafarers stress and fatigue is evaluated.

Assessment of Fatigue Onboard Ship

The unique combination of stressors present in the offshore environment - e.g. extreme weather conditions, noise, motion and demanding work schedules - mean that research findings from other transport industries and onshore populations cannot automatically be applied to seafarers. As well as collecting survey data, it was felt important to actually go onboard ships to gather more detailed information. In this part of the research, a variety of objective indicators and subjective reports were used in assessing seafarers' fatigue.

Sample

177 participants were recruited in total by researchers who visited seven ships, operating in the UK sector. These consisted of 3 small oil tankers, 2 passenger ferries, a freight ferry, and a fast ferry. This sample was compared with the survey sample and the two were found to be generally similar, although the onboard sample were younger on average, which may be attributable to the higher proportion of officers in the survey sample, or to the comparatively young crew of the fast ferry. This is compared with the

phase one onboard sample of 144 workers from the offshore oil industry in order to assess generalisability of findings between phases.

Age
Participants were generally older in the phase one sample. This may again be partially accounted for by the relative youth of the fast ferry crew, and also the relative youth of those working on ferries compared to those working on the offshore oil support ships studied in phase one (see Table 1)

Table 1: Mean ages of subjects by vessel type

Group	N	Mean	SD
Phase 1	144	41.31	9.82
Pipe Layer	18	40.78	10.14
Dive support vessel	81	42.04	8.63
Shuttle tanker	19	38.84	12.85
Supply Vessel	12	44.00	8.16
Standby/supply Vessel	14	38.86	12.51
Phase 2	177	36.07	11.40
Freight	27	39.11	10.45
Tankers	24	41.83	12.67
Passenger Ferries	71	37.28	9.90
Fast ferries	55	30.49	11.07

There were more mixed nationality crews in phase 2, with only 63.8% (n=113) of crews being from the British Isles, in comparison to 91.2% (n=134) in phase 1. Other nationalities in phase 2 included Spanish (20.3%, n=36), Polish (13.0%, n=23), and Canadian (2.8%, n=5).

Length of tour
The typical tour length was shown to differ between the two phases, with the majority of participants in phase 1 (68.3%, n=99) working 4 weeks on/4 weeks off tours, in comparison to phase 2 in which the majority worked 1 week tours (34.4%, n=61) (see Table 2). However, again this was skewed by tour length on the fast ferry, which never exceeded seven days.

Table 2: Tour length

Tour length	Phase 1	Phase 2
1 week	----	34.4% (n=61)
2 weeks	2.8% (n=4)	15.3 (n=27)
3 weeks	4.1% (n=6)	6.2% (n=11)
4 weeks	68.3% (n=99)	3.4% (n=6)
5 weeks	6.9% (n=10)	0.6% (n=1)
6 weeks	2.1% (n=3)	1.1% (n=2)
7 weeks	9.0 (n=13)	----
8 weeks	6.9 (n=10)	10.2% (n=18)
8+ weeks	----	29.0% (n=51)

Phase 1 and phase 2 participants were tested at a similar stage into the tour, with the highest proportion of subjects being tested in week 1 (43.7% in phase 1, and 48.0% in phase 2) (see Table 3).

Table 3 . Weeks into tour at testing

Weeks in tour	Phase 1	Phase 2
week 1	43.7% (n=62)	48.0% (n=85)
week 2	26.1% (n=37)	13.0% (n=23)
week 3	16.2% (n=23)	6.8% (n=12)
week 4	4.2% (n=6)	5.6% (n=10)
week 5	2.1% (n=3)	4.0% (n=7)
week 6	1.4% (n=2)	4.5% (n=8)
week 7	5.6% (n=8)	3.4% (n=6)
week 8	0.7% (n=1)	5.1% (n=9)
week 8+	----	9.6% (n=17)

Procedure

Volunteers participated in four sessions overall, which were scheduled before and after work on the first and final day of their testing, typically an interval of 5-7 days. During these sessions, before and after work questionnaires (henceforth 'logbooks') recording food intake, medication, breaks, caffeine consumption, smoking, sleep, symptoms of fatigue and perception of work related issues were completed. Performance tasks measuring reaction times, errors and lapses of attention as well as subjective reports of alertness, hedonic tone (happiness, sociability) and anxiety were also administered during these sessions.

Further objective measures involved sleep recording to assess sleep quality, noise measurements from different areas of the ship, and measurements of the pitch, roll, and heave dimensions of motion. The survey was completed by participants onboard ship in their own time during this testing interval.

Tour Length - Background

Seafaring may be regarded as a very important occupational area for study, having high accident rates and more deaths per capita than any other industry in Britain. It is therefore necessary to examine the issues that make this so. However, studies of seafarers have, until now, paid little attention to potentially important factors such as days-into-tour. Intuitively the expectation is that longer tours of continuous duty would be more detrimental in terms of cumulative effects leading to more fatigue and poorer health. It is indeed the case that some research on installation workers (Collinson, 1998; NUMAST, 1992) has indicated that tours exceeding 2 weeks show increased injury rates, and that adverse physiological changes may be related to tour lengths exceeding one week.

However Forbes' (1997) study found that accident frequency among installation workers was greatest at the beginning of a tour, specifically during the first tour week, and then declined steadily over the course of a tour. Thus clearly, these mixed findings indicate that this intuitive sense that longer tours are more detrimental in health and fatigue terms requires re-examination.

Generally indications from phase one (Smith *et al*, 2001) also affirmed that shorter tours are not necessarily better in terms of seafarers' well-being. Although the accident data is limited, most accidents were found to occur in the first week of tour regardless of actual tour length. The logbook data confirmed that the impact of their work and work environment on seafarers may change over the course of the tour. For example, sleep duration was reduced for the first night offshore but improved with days into tour. Also, seafarers showed improvements in alertness across the week, and analyses of effects of days into tour on those working night shifts showed that those more than 5 days into tour (average 18 days) made fewer errors on performance tasks than those less than 5 days into tour (average 3 days).

It was found that many phase one respondents reported feeling 'below par' on returning to their vessel/installation after a period of leave. Approximately half of all respondents felt that adjusting to life offshore took at least 2-3 days and felt their performance to be affected during this period of adjustment. This may partially account for these findings, since the period of adjustment takes up a smaller proportion of longer tours than of shorter tours.

Tour Length – Phase Two

This analysis was conducted to yield information about changes in seafarers' stress, fatigue and performance levels over a discrete period of time, in order to assess the impact of offshore work on seafarers as a function of time into tour. This was done using a mixture of data from both the logbooks and the objective measures. Survey data were not relevant to this analysis since the measures within the survey were only completed once, not at the specific onboard testing intervals.

The period of analysis matched the period of the performance testing onboard, i.e. typically 5-7 days, and the analysis itself had two layers. Firstly, the aim was to reveal any fatigue and performance related differences there may be as a result of a specific period – that is to uncover what, if any, effect working over approximately a week long period has on self-reports of fatigue, work-related variables, and objectively measured performance and sleep. Secondly, these effects were analysed as a function of time into tour, to assess whether longer tours mitigated or exacerbated these effects. This analysis was restricted to the logbooks only.

Approximately forty participants (the fast ferry sample) were excluded from this analysis since their entire tour was only 6-7 days long, and it was felt that including them would bias the data set, by confounding ship type with tour length.

Key findings:

- Levels of job stress, job effort, alertness and outcomes on some of the performance measures were more negative further into tour
- Habituation to noise levels aboard ship seems to occur fairly consistently as a function of time into tour.
- Sleep appears to improve further into tour, which may help account for the relatively low levels of fatigue in this sample.

Both the logbook and the objective measures provide evidence that the cumulative effect of work, both across the day and across the working week, may influence both levels of fatigue and performance. Across the working week, job stress was found to increase and this was mirrored by slower task reaction times and lower levels of alertness. This may indicate that over longer periods, seafaring work has a detrimental effect on individual fatigue and performance.

There is also some evidence from the logbooks that seafarers' sleep improves as a function of time into tour. Although an actual improvement in sleep was not recorded by the objective measures, no impairment in sleep was identified either so at the very least these do not contradict the logbook findings. Also, generally habituation to noise levels onboard was observed as a function of days into tour.

Key differences on the first fortnight vs. after two weeks analysis:

First fortnight of tour:

- Physical effort significantly lower after seven days
- General health significantly worse after seven days

After first two weeks of tour:

- Almost no change in sleep across the testing interval
- Physical effort significantly higher after seven days
- Weather becomes more of an issue across the testing interval, despite no change in actual weather conditions
- Support from fellow workers, self-regulation of own work, and work satisfaction are all affirmed more highly after seven days for this sub-sample

This further analysis of the logbook data indicates that any cumulative effects over the testing interval vary as a function of weeks into tour. There is some evidence of habituation, and some evidence of cumulative negative effects of time at sea, e.g. fewer effects of noise are observed further into tour, whereas the subjective impact of motion increases. This first fortnight/after second week of tour split is supported by the manifest differences between the two sub-samples. For example, from day one to day seven of the first fortnight of tour, there is a significant increase in self reported work stress and lack of sleep, and even though physical effort decreases over the seven day period, general health is reported as worse. After the first fortnight of the tour there are fewer negative day one day seven differences. Again, general health was found to be worse on day 7. Stress is mostly the same, sleep appears stable over a week long period after the first fortnight of the tour. There are some indications that longer tours may be better in that there is higher affirmation of receiving support from fellow workers, self-regulation of work and work satisfaction further into tour.

Conclusion

Thus tour length seems to be an important factor in stress and fatigue at sea, and also to affect sleep and other exposure variables. Furthermore, the effects of work and work related issues on seafarers over the work period vary as a function of time into tour. These time into tour differences may even indicate that in some ways, longer tours are actually less detrimental in terms of fatigue and work related exposure variables than shorter ones. This finding is supported by analyses of the survey data, indicating that the longer the tour, the lower the fatigue as measured by the PRFS scale and the fatigue at/after work factors.

Thus it seems that tour length may in fact prove quite an important factor in addressing issues of fatigue and health, and indeed accident and injury in seafaring. Further analysis is currently being undertaken to extend our understanding of the importance of this issue.

Acknowledgements

The research described in this article is supported by the Maritime and Coastguard Agency, the Health and Safety Executive, NUMAST and the Seafarers' International Research Centre. We would also like to acknowledge the contribution made by the ship owners and seafarers who have participated in the research.

References

Collinson, D.L. 1998, "Shift-ing lives": Work-home pressures in the North Sea oil industry. *Canadian Review of Sociology and Anthropology*. **35** (3), 301-324

Forbes, M.J. 1997, *A study of accident patterns in offshore drillers in the North Sea.* Dissertation prepared for the Diploma of Membership of the Faculty of Occupational Medicine of the Royal College of Physicians.

NUMAST 1992, *Conditions for Change.* London, NUMAST.

Smith, A.P., Lane, A.D. and Bloor, M. 2001, *Fatigue Offshore: A Comparison of Offshore Oil Support Shipping and the Offshore Oil Industry.* Seafarers International Research Centre. ISBN 1-900174-14-6.

THE INFLUENCE OF NOISE AND MOTION ON SLEEP, MOOD, AND PERFORMANCE OF SEAFARERS

Neil Ellis, Paul Allen, and Ailbhe Burke

Centre for Occupational and Health Psychology,
Cardiff University,
63 Park Place,
Cardiff, CF10 3AS.

The effects of noise and motion have been widely studied in transport industries. However little research has examined these combined factors, especially in the short sea shipping industry. This paper describes the second phase of a project looking at seafarers fatigue, in which objective measures of noise and motion onboard short sea shipping vessels were examined in order to assess their influence on performance, mood and sleep. Data were collected from 177 participants on 7 vessels. As in previous offshore research, noise and motion levels were shown to affect both performance and mood. However no effect of noise or motion was seen on measures of sleep. Multiple regression analyses further indicated that noise predicted a number of the performance variables. However it also indicated that other factors, such as tour length, influenced performance and mood.

Introduction

Although a number of studies have looked at the effect of noise and motion in the work environment (Smith and Ellis, 2000; Hygge *et al,* 1998; Powel and Crossland, 1998; Lewis and Griffiths, 1997), few have examined both variables together, especially in the offshore shipping industry. Such factors may be especially relevant offshore, as seafarers are continuously exposed to these conditions, which may affect not only performance at work, but also health and well being (Collins *et al*, 2000). A recent study has investigated effects of motion and noise on seafarer fatigue in the offshore oil industry (Smith *et al*, 2001). The results showed that high levels of exposure to noise were associated with increased alertness, and greater focusing of attention, but also more lapses of attention, sleep disturbance, and poorer sleep efficiency. Fewer associations were found between motion and performance measures, although motion was found to be positively related to the number of errors on a categoric search task. However this study had a number of methodological shortcomings. Motion was measured in degrees, ignoring acceleration within these dimensions, and the sampling time frame was relatively slow (every 3 seconds). There were also a number of problems with the measurement of noise, as this was measured in 24-hour blocks making analysis of the influence of noise at specific times, such as during the work period, impossible. These methodological shortcomings were addressed in the second stage of the project, which examined fatigue in the coastal and short sea shipping industry.

The specific aims of this paper are to examine the influence of noise and motion on performance, mood, and sleep, and to draw comparisons with the findings from the study of the support ships in the offshore oil industry.

Methods

This study was a direct continuation of that conducted by Smith *et al*, (2001), and the methods used were generally the same. Objective measures and subjective reports were used to assess levels of fatigue, performance, sleep efficiency, and mood. A brief overview of the measures used is outlined below, with any changes from the earlier project described.

Sleep measures
Sleep efficiency was measured using the Actiwatch system (For more details see Smith *et al*, 2001). Participants wore the sleep watch on two occasions, during the sleep period before the testing session on day 1, and during the sleep period prior to testing on the seventh day. Subjects from the fast ferry (n=55) were not included in the analysis of sleep measures as they did not sleep on the vessel.

Performance measures
Mood, subjective alertness, and performance were assessed using a battery of computerised tasks (see Smith *et al*, 2001, for details). Participants completed these tasks both before and after their shift, on day 1 and 7 days later. Mood was assessed using 18 bi-polar visual analogue scales (after Herbert *et al*, 1976). Three main factors were derived from the scale: alertness, hedonic tone, and anxiety. The first performance task was a variable fore-period simple reaction time task. In this task a box appeared in the centre of the computer screen, and at varying intervals (from 1-8 seconds) a target square appeared in the centre of the box, which the participant responded to by pressing a key on an external response box as quickly as possibly. Mean reaction times were recorded to the nearest millisecond for each minute of the task. The second task was a choice reaction time task involving focused attention (Broadbent *et al*, 1986). In this task three warning crosses appeared on the screen followed by target letters (either A or B) in the centre of the screen. Participants were required to respond by pressing the left key of the response box if the target was an A and the right key if it was a B. On certain trials distracters were presented either side of the letter, for example A's, B's, or asterisks. Within the test a number of factors were measured including reaction times to the target alone and with distracters, the percentage correct, occasional long reaction times, speed of encoding of targets which were the same or different from the previous trial, and a measure of focusing of attention (the Eriksen effect). The third task, a categoric search task, was similar to the focused attention task but the volunteer did not know in which of two locations the target would appear. Again participants were required to respond to target letters (either A or B), as quickly and accurately as possible. These letters were presented either in the centre, or the far left, or the far right of the screen. Distracters (in this case digits) were also presented on some trials. This task measured a number of factors including reaction times and percentage correct for targets presented alone and with distracters, targets in near and far locations, targets that were the same or different from

the previous trial, and targets presented in the same or a different location to the previous trial.

Vessel Motion

Motion of the vessel was measured continuously throughout the testing period using the Seatex MRU H.2 as in phase 1 (see Smith *et al*, 2001, for details). However a number of adjustments were made. As in phase 1 pitch, roll and heave, were recorded (in degrees), however accelerations within these dimensions (metres/second) were also additionally recorded. The sampling rate was also increased, and data were logged for these dimensions every third of a second. From this data root mean squared (RMS) displacement scores (the standard deviation of the raw values) were calculated for acute time periods, and for motion of the vessel overall. The acute time periods included an hour period around the time of testing, the time during the work shift on day 1 and day 7, and the sleep period on day 1 and 7. The overall motion of the vessel during the visit was summarised using the time period between the end of shift on day 1 and the start of shift on day 7. Motion data were only recorded onboard the tankers (n=24), as the motion on the other vessel types was negligible.

Noise levels

Noise was measured using CEL-460 Dosimeters as in phase 1 (see Smith *et al*, 2001, for details). However in this second phase, the dosimeters were set to record consecutive 1-hour blocks data continuously throughout the visit, not single 24-hour blocks as in phase 1. Two dosimeters were used; one was located on the vessel's bridge, and the other was located in a cabin. From this data average equivalent sound levels (Leq) were calculated again for the testing and sleep periods, and for noise levels over the 7 day period. No noise data were available for two of the ships studied (n=34) due to problems with the dosimeters.

Procedure

Participants initially completed a familiarisation session in which they ran through a shortened version of the performance tasks. On completion of the tasks they were given an actiwatch and asked to wear this during the sleep period prior to the shift they were to be tested on. They were then tested immediately before, or as close to the beginning of their shift as possible. At the end of their shift they returned and repeated the tasks. Seven days later (or as close to the seventh day as possible) participants were again given a sleep watch for the sleep period prior to testing, and asked to wear it during their sleep. Participants then completed performance tasks before and after their shifts, using the same procedure as day 1.

Participants

Data were collected from 177 participants from seven vessels within the short sea shipping industry. These ships included 3 small oil tankers, 2 passenger ferries, a freight ferry, and a fast ferry. The mean age of participants was 36.07 years (s.d.=11.40), with crew generally consisting of mixed nationality crews. 63.8% of the volunteers were from the British Isles, 20.3% (n=36) were Spanish, 13.0%, (n=23) were Polish, and 2.8%, (n=5) were Canadian. 72.3% (n=128) were marine crew, with the highest proportion of participants (34.4%, n=61) being on a 1 week on/ 1 week off shift pattern. Nearly half (43.78%, n=85) of participants were test in the first week of their tour. However, this bias towards 1 week tours, and being tested in the first week of tour was due to the large number of participants from the fast ferry (see Table 1), who worked predominantly one week on/ one week off tours.

Table 1. The percentage of subjects from each vessel type

Vessel Type	Frequency
Freight ferries	15.3% (n=27)
Tankers	13.6% (n=24)
Passenger Ferries	40.1% (n=71)
Fast ferries	31.1% (n=55)

Results

Of those recruited into the study, 80.2% (n=142) completed testing on both day 1 and 7, the remainder being tested only once (due to practical considerations).

Noise - Performance tasks

A significant correlation was found between noise levels at the time of testing and the speed of encoding of new information on the focused attention task both for day 1 before and after shift ($r=-.182$, $n=137$, $p=.033$, and $r=-.197$, $n=134$, $p=.022$ respectively), suggesting that as noise levels increase the ability to encode new information in a known location improves. No significant differences were found for the other performance measures, and given the number of analyses conducted the above result may be a chance effect.

Noise - Subjective mood

No correlations were found between alertness and noise measures at acute periods. However significant difference were found between low and high noise vessels (as defined by a median split) for levels of anxiety, for the before shift test on day 1, and for the before and after shift test on day 7, with those on noisier vessels reporting lower levels of anxiety.

Noise - Sleep

Sleep quality was not found to be influenced by noise, either during the sleep period, or as a function of motion in general.

Motion - Performance measures

Performance measures were not found to relate to any of the measures of motion, either at test periods, or over the course of the visit.

Motion - Subjective mood

There was a significant difference between alertness for the high and low motion vessels (see table 2), with those on low motion vessels generally appearing more alert.

Table 2. Mean alertness scores (SD in parentheses) for high and low motion ships

	High Motion	Low Motion
Alertness		
Day 1, before work	239.28 (24.94)	287.35 (78.83)
Day 1, after work	225.00 (28.36)	286.12 (73.78)
Day 7, before work	227.57 (28.08)	277.36 (76.41)
Day 7, after work	217.86 (29.01)	285.67 (65.64)

Motion - Sleep
Measures of sleep efficiency were not found to be influenced by motion during the sleep period, or by the motion of the vessel over the seven day period.

Multivariate analysis
Although the univariate analysis seems to indicate that noise and motion may influence performance and mood (although not sleep measures), such analysis ignores the impact of other factors. This may have two possible effects. Firstly, variations in performance and mood that were attributed to noise and motion may actually be due to other related factors (i.e. ship type). Secondly, variations in these measures due to noise and motion may be masked by the influence of other factors. Therefore in order to disentangle this effect multiple regression analyses were conducted, allowing assessment of effects of noise when the influence of other factors are taken in to account. Due to the limited amount of motion data available (i.e. tankers only, n=24) multiple regression analyses were not carried out on the motion data.

As in the univariate, analysis the multiple regressions indicated that speed of encoding on the focused attention task was predicted by noise on the bridge at the time of testing (r^2 =.068). However multiple regressions also identified a number of the outcome variables which were influenced by measures of noise not shown in the initial unvariate analysis. Speed of encoding on the categoric search task was found to be predicted by noise on the bridge (r^2 =.151), and mean reaction time on the simple reaction time task was also found to be predicted my overall noise levels on the bridge (r^2 =.058). In addition, average overall levels of cabin noise were found to be predictive of changes in both alertness and hedonic tone between the day 1 and 7 tests (r^2 =.043, and r^2 =.091 respectively), with high noise exposure leading to reduced alertness and hedonic tone. Although anxiety was found to be related to noise levels in the initial analysis, the multiple regression suggested that variation in levels of anxiety are actually predicted by tour length (r^2 =.032), and not overall levels of noise. Again, there no effects of noise on sleep were found.

Discussion

This study offers some support for the view that performance and mood are influenced by noise levels on offshore vessels. As in our earlier research, the effect of motion on performance and mood was less pronounced. However, the specific measures that were affected in phase 2 were different from those in phase 1. This may reflect a number of factors. Firstly these differences may relate to the variation in the ship types studied in the two phases, and the characteristics of these vessel types. For example, in phase 1 intermittent noise such as alarms and doors closing was common. However, in the present study noise was more continuous, and produced by the engine or discharging pumps. The lack of effect of motion on performance and mood may also reflect the vessels studied, as the ships in the present study worked in coastal areas, and were generally exposed to less severe weather condition, and lower levels of motion than ships studied earlier. The differences between the findings of the two phases may also reflect the increased sensitivity of the measurement used both for noise and motion, which allowed a more accurate assessment of the effect of these environmental factors at the time of test. In contrast to phase 1, motion and noise did not affect sleep measures. Again, this may relate to the less severe environment in phase 2.

Although motion and noise influenced both performance and mood, the study also indicated that there are a number of other factors such as age and tour length that affect performance and mood offshore. It may be more appropriate to consider noise and motion as part of a multitude of factors which influence performance and mood offshore, in order to obtain a clearer and more accurate picture.

Acknowledgements

The research described in this article is supported by the Maritime and Coastguard Agency, the Health and Safety Executive, NUMAST and the Seafarers International Research Centre. We would also like to acknowledge the contribution made by the ship owners and seafarers who have participated in the research.

References

Broadbent D.E., Broadbent M.H.P. and Jones J.L. 1986, Performance correlates of self-reported cognitive failure and obsessionality, *British Journal of Clinical Psychology*, **25**, 285-299

Collins A., Matthews V. and McNamara R. 2000, *Fatigue, Health and Injury Among Seafarers and Workers on Offshore Installations: A Review*. Seafarers International Research Centre Technical Report Series, ISBN No: 1-900174-12-X

Herbert M., Johns M.W. and Dore C. 1976, Factor analysis of analogue scales measuring subjective feelings before sleep and after sleep. *British Journal of Medical Psychology*, **49**, 373-379.

Hygge S., Jones D.M. and Smith A.P. 1998, Recent developments in noise and performance. In N.L. Carter & R.F.S. Job (ed.) *Noise as a Public Health Problem*, Volume 1 (PTY Ltd, Sydney), 321-328.

Lewis C.H. and Griffiths M.J. 1997, Evaluating the motion of a semi-submersible platform with respect to human response. *Applied Ergonomics*, **28 (3)**, 193-201.

Powel W.R. and Crossland P. 1998, *A literature review of the effects of vessel motion on human performance*. INM Technical Report No 98027.

Smith A.P. and Ellis N. 2002, *Objective measurement of the effects of noise aboard ships on sleep and mental performance*. The 2002 International Congress and Exposition on Noise Control Engineering, Dearborn, MI, USA. August 19-21, 2002.

Smith A P, Lane T, and Bloor M 2001 *Fatigue Offshore: A Comparison of Offshore Oil Support Shipping and the Offshore Oil Industry*. Seafarers International Research Centre, ISBN No: 1-900174-14-6

DISPLAY SCREEN EQUIPMENT

LARGE SCALE DSE RISK ASSESSMENT PROGRAMMES: A CASE STUDY

Terry Cloke

Human Applications
139 Ashby Road
Loughborough
Leicestershire
LE11 3AD

In a landmark court case in 1993, brought by the National Union of Journalists against Reuters, Mr Justice Prosser famously quoted as stating that the term RSI had "…no place in the medical dictionary". The NUJ lost on that occasion, but more recent successful compensation claims have cited the absence of risk assessments. "If it's not written down, it didn't happen" seems to be the attitude of the courts. This alarmed one national newspaper group who realised they were vulnerable in this area. They started a programme of risk assessments covering everyone from the editor down to the newest recruit on the fashion desk. This paper describes the practicalities of carrying out a large-scale programme of display screen equipment (DSE) risk assessments on 500 editorial staff who were well informed, stressed and often highly cynical about health and safety.

Introduction

In a landmark court case in 1993, brought by the National Union of Journalists (NUJ) against Reuters, Mr Justice Prosser famously quoted as saying that the term repetitive strain injury (RSI) had "…no place in the medical dictionary". At the time he was widely mis-reported as saying that RSI did not exist! The NUJ lost their case on that occasion, but in recent years there have been many successful RSI compensation claims and some of them have advantageously cited the absence of risk assessments as a contributory factor to the injury. "If it's not written down, it didn't happen" seems to be the attitude of the courts. This alarmed one national newspaper group who realised they were vulnerable in this area. The decision was taken to train internal risk assessors and at the same time involve an external contractor to make a start on the backlog of approximately 2000 assessments. The use of an external contractor gave credibility to the assessment process and supported the internal assessors while they gained sufficient experience to take over the running, monitoring and reviewing of the assessment programme.

This paper describes the practicalities of carrying out a large-scale programme of display screen equipment (DSE) risk assessments of approximately 500 editorial staff. Everyone was assessed from the editor right down to the newest recruit on the fashion desk. The assessments had to be completed within a strict timescale and on budget. Unlike traditional office workers, journalists have highly irregular work patterns, daily copy deadlines and high levels of stress. In addition to this, they were highly sceptical about the benefits of risk assessments. All these factors presented particular challenges for the team of assessors. This paper reports on how these logistical problems were overcome along with practical advice on carrying out risk assessments on a well informed, stressed and often cynical group of DSE users.

Legal Background

It's now been ten years since the Health and Safety (Display Screen Equipment) Regulations 1992 became UK law, but many employers seem to have done little or nothing to comply with the requirements of the Regulations. During this period there has been a dramatic change in employees' willingness to take their employers to court to seek compensation for work-related injuries. Computer users are bombarded with adverts in the press and on television from legal firms offering a "no win, no fee" claims service. Within the last year, the Health and Safety Executive (HSE) have issued an Improvement Notice under the DSE Regulations to a large Police Force. (Crown immunity from health and safety legislation for the police ended in 1997.) And yet. complying with the Regulations is not difficult. Put briefly, the Regulations emphasise to need to manage risk, which means risk assessment to identify the problems, risk reduction to do something about them and risk monitoring to check for any new risks that are subsequently introduced into the job. It is worth noting that the HSE stress the need for risk assessments to be carried out by competent people and recommend that companies should develop in-house assessment expertise whenever possible.

The Regulations also require the employer to make sure that jobs are a mixture of screen-based and non screen-based work or to provide formal rest breaks. Most jobs have these changes of activity built-in, but environments such as call centres, where computer use is intensive, may require supervised, formal breaks. As far as health surveillance is concerned the Regulations require the employer to provide, at the users' request, eye and eyesight testing. The computer workstation itself must meet a minimum standard referred to in the Regulations as The Schedule. This covers not only the hardware (computer, desk, chair, etc.) but also the immediate environment around the workstation (lighting, heating, noise, etc.) and the user computer interface (software).

One of the most neglected requirements of the Regulations is that users of the workstation should be given training in its safe use. Many companies provide their computer users with good quality adjustable chairs and desks but fail to give any guidance on good posture or in how to use these adjustments effectively. Too often employers and users of the equipment assume that being comfortable is the same as being safe, which is far from reality. Training is needed to raise awareness of the long-term, cumulative effect of sustained bad posture.

What Are The Risks?

Musculo-skeletal disorders

In a recent special issue of 'The Ergonomist' published in 2002 Trevor Shaw of the HSE presented some startling statistics on computer usage and the incidence of musculo-skeletal disorders (MSDs). In 1998/99 there were around 223,000 office-based premises employing about 3.8 million people. Statistics collected for the period 1998-2000 showed that computer operators and data processing operators had the highest average annual incidence rate of MSDs, with an estimated 72 per 100,000 workers. Respondents to an HSE survey were asked their opinion of what caused conditions affecting their upper limbs and neck. Many people (38%) thought that repetitive work or typing was the cause while 28% believed that the posture they adopted at work caused the problem. The Health and Safety Commission's MSD 'Priority Programme' has set targets to reduce the incidence of MSD illness by 20%, and the number of working days lost due to MSD by 30%, by 2010. The HSE have identified two other main hazards associated with computer use. These are visual fatigue and stress.

Visual fatigue

Medical evidence shows that using computers does not cause damage to the eyes, but it may make some people with pre-existing vision defects more aware of them. Prolonged use of the computer screen can also lead to temporary visual fatigue with symptoms such as blurred vision, sore eyes, itchy eyes and headaches. Other factors associated with the workstation can contribute to visual fatigue, such as glare from badly designed lighting, reflections on the screen making text difficult to read, reflection of windows that are not fitted with blinds and instability of the screen image (e.g. flicker, jitter and swim).

Stress

A survey carried out by the HSE in 1999 reported that as many as 1 person in 5 believe that they have suffered from work related stress. Amongst the most likely causes are poor job design, high repetition, poorly designed software and social isolation. These organisational factors are in addition to the personal stressors such as marriage, divorce, bereavement, financial problems, moving house, etc. While it is often difficult to link particular symptoms to a specific cause, it is generally agreed that the risk of stress can be significantly reduced by following good ergonomic principles in the design of equipment, the workstation, the physical environment, the job and the software.

The Risk Assessment Programme

It was against this background of the legal requirements of the Regulations and the possible effects on health associated with computer use that the programme of risk assessments was planned.

Planning the assessments

The assessment programme was being managed by the newspaper's Health and Safety Department with a strict limit on the number of assessment days available and a target date for completion. Every member of the editorial staff who used a computer, however briefly, was designated as a DSE 'user' and was to be given an assessment. The newspaper group publishes four titles and the plan was to take each title in turn and complete the assessments of all editorial staff before moving on to the next title. The number of days allocated to the assessment of the first title meant that an average of ten assessments per assessor per day had to be completed if the project was to finish within budget and on time.

A short pilot assessment programme was run and this established that the target number of assessments could only be achieved if a strict appointment system was used. Journalists' work patterns are highly variable and driven by breaking news stories and daily copy deadlines. Different sections of the paper, (e.g. news, foreign news, business, sport, weekend, etc.), had different peak periods, so these had to be taken into account to make sure the target quota of assessments was met. To spread the burden of such an intensive programme of assessments, a team of four assessors was assembled. Up to three assessors were on site at any given time, but each assessor would only carry out assessments on two consecutive days before spending time back at the office writing up the assessment reports.

In addition to the individual workstation assessment records the Health and Safety Department was given weekly progress reports showing the number of assessments completed along with a summary of the follow-up actions. In order to assist with the purchase of accessories, and to spot trends, this summary was compiled as an Excel spreadsheet broken down into categories showing the numbers of wrist rests, footrests, document holders, monitor stackers, CPU relocations, chair repairs and replacement chairs required.

The assessment process

The assessment process was designed to meet the requirement in the Regulations to carry out a "suitable and sufficient analysis of the workstation" as well as providing training in the adjustment and correct use of the workstation. Before being assessed each user was asked to fill in a pre-assessment questionnaire covering all aspects of the Schedule (i.e. equipment, environment and software) as well as questions about IT training, posture training, eyesight tests and any work related health concerns they might have. The questions were all worded in such a way that answering "no" to a question indicated that the user had a potential problem. This questionnaire did not constitute an assessment but provided evidence of the user's perception of their problems and a formal record of their input into the process.

The assessment started with some general questions about the person's use of the computer to establish how likely they were to be at risk from the three main hazards discussed above. Then the user was taken through a step-by-step guide in how to adjust the chair and set up the workstation for safe and comfortable use. As problems were identified these were recorded, along with any remedial actions, on the assessment record. Finally the questionnaire was checked to make sure that all the problems identified by the user that been dealt with and recorded on the assessment record.

The final stage of the assessment was to type up the assessment record and document the actions on the summary spreadsheet. The purchase and distribution of the various accessories (footrests, document holders, wrist rests, etc.), as well as the repair or replacement of chairs, was managed by the Health and Safety Department.

Assessment Findings

A total of 476 DSE users were assessed of which 190 (40%) were female and 286 (60%) were male. By far the most frequently recommended item was the footrest. The majority of users set the seat height too low for comfortable keyboard use in order to place their feet flat on the floor. This resulted in static muscle load in the shoulders and upper arms. When the seat was set to the correct height for comfortable keyboard use, a total of 117 footrests were required (almost 25% of all users).

Lack of training in the correct adjustment of the chair was a factor in almost every case. Users were provided with an excellent ergonomic chair that provided good lumbar support and fully met the Schedule (i.e. seat height adjustable, backrest adjustable in height and tilt). In addition, the chair had height-adjustable, reversible arms that could also be easily removed if they prevented the user getting sufficiently close to the desk. The seat pan was adjustable in depth to accommodate a wider range of leg length. Despite all these adjustments the only setting that was used on a regular basis was the seat height. Training in how to adjust the chair was simple to carry out and resulted in greater awareness of the benefits of carrying out risk assessments. Many users had been openly cynical about the need for assessments, but were willing to admit that correctly adjusting the chair produced a noticeable improvement in comfort. A total of 32 (7%) chairs were found to be unserviceable and were replaced. A further 17 (3.5%) chairs had minor faults that were repaired.

Surprisingly, given that the majority of users were journalists and presumably well informed about health issues, many users failed to make the connection between poor posture and physical aches and pains. Working from documents laid flat on the desk next to the keyboard and repeated twisting of the head and neck to view first the screen, and then the document was widespread. This frequently correlated with complaints of neck discomfort. When asked about the use of a document holder some users replied that they thought they were only for copy typists. A total of 59 document holders (12%) were issued as a result of the assessments.

Another example concerned a laptop user complaining of frequent neck and shoulder pains. The laptop was used in a docking station, the chair was correctly adjusted and the user sat square on to both keyboard and screen. Eventually the user volunteered that they used the laptop at home at the weekends for up to 4 hours at a time, but didn't think that this could be the cause of the problem. They were given a laptop stand and separate keyboard for home use and the symptoms disappeared within 2 weeks.

The height and position of the display screen featured in many assessments. Users assumed that the position of the screen on the desk, the viewing height and viewing distance had all been correctly set by the IT department when the equipment was installed. Consequently users were unwilling to change the position of the screen and would often sit twisted to view a screen that was offset to the left or right. Those bold enough to try to move screens often found the cables were trapped in the cable management system of the desk. As a result of the assessments 114 screens had to be repositioned, raised or lowered. A significant amount of this work would have been unnecessary if the screens had been properly positioned at the time of the initial installation. This highlighted to need to raise awareness of workstation ergonomics for IT staff.

Conclusions

The initial 'hit' of approximately 500 users allowed the newspaper group to kick-start the assessment programme. Few of the risks identified lay outside the skills of an assessor trained in basic DSE assessment. Furthermore, the health and safety team learned how to manage a large scale assessment programme and were able to make estimates on the resource demands from the rest of the organisation.

Additional training of DSE assessors allowed the organisation to build up a team of trained personnel who were able to take the programme forward with support from the health and safety team.

References

The Ergonomist, Special Issue: Office Ergonomics, November 2002, No. 389

DEALING WITH AN IMPROVEMENT NOTICE FOR DISPLAY SCREEN EQUIPMENT: A CASE STUDY

Vicky Ball & Nigel Heaton

Human Applications
139 Ashby Road
Loughborough
Leicestershire
LE11 3AD

You might expect police forces to receive an improvement notice about 'scary' stuff like arresting criminals, putting surveillance officers in 'risky' situations or the use of personal protective equipment. However, in 2001 the HSE issued an Improvement Notice on a Police Force for contravening the Health and Safety (Display Screen Equipment) Regulations 2, 3, 6 and 7. The Improvement Notice focused the minds of the police force on occupational health issues. They realised that whilst they were dealing with the 'big scary stuff' the 'smaller, less scary' issues had gone largely unnoticed. This paper looks at the complexities involved in helping an organisation to deal with an Improvement Notice and identifies how it could have been prevented in the first place.

Introduction

The Health and Safety (Display Screen Equipment) Regulations 1992 (DSE Regs) have been around for 10 years and although many organisations are aware of their existence they have managed to 'get away' with doing very little about them. The Regulations require employers to conduct risk assessments of all users and all workstations, to provide information and training in the risks associated with the use of their workstation and to ensure that all workstations meet the Schedule of minimum requirement to the DSE Regs, among other things.

This implies that employers have to take a proactive approach and go out and look for problems. This in turn means that once you identify problems you have to do something about them and it would seem that some would rather not. This could be due to a lack of awareness and understanding of the requirements of the DSE Regs or a lack of resources or purely a matter of prioritisation.

Although it is legitimate to prioritise when managing risk it is also important that your priorities are based on sound knowledge and understanding of the risks faced. Unfortunately, many employers still fail to realise the potentially damaging nature of musculo-skeletal disorders and the effect they can have on people's daily life and happiness at work and ultimately their productivity.

Police Forces only became the subject of the Health and Safety at Work Act 1974 (and all the other health and safety Regulations) in 1997 as a result of the Police (Health and Safety) Act 1997. The requirement for Police Forces to manage proactively all aspects of health and safety is a relatively new requirement.

This scenario was the basis of an inspection by the Health and Safety Executive (HSE) after complaints from a member of a Police Force suffering from a musculo-skeletal disorder (MSD). As a result of the investigation the HSE served an Improvement Notice on the Police Force for contravening section 2, 3, 6 and 7 of the DSE Regs. The investigation showed that they had failed to:

- conduct DSE risk assessments
- meet the schedule of minimum requirements for all workstations
- provide information and training to users on the risks associated with using their workstations

This prompted the Occupational Health Department (OHD) to take action. Initially, they looked for a quick fix answer and bought a computer based training (CBT) package to be put on to the Intranet to administer training and information to all users. The CBT package procured included a self-assessment questionnaire which was designed to be completed by the individual user and the feedback from which could be used to identify users with problems who may require further individual assessment.

The implementation of the package took a few months, there were many problems with compatibility. The Health and Safety Advisor for the Force also left during this time and the OHD realised that they required outside, professional help to manage the training and assessment process.

A meeting was held with the appointed consultants, Head of Human Resources, Head of Training and the Head of Occupational Health in order to identify the state of play and the way forward. It was identified that there was a lack of infrastructure to support the assessment process. Line managers were not aware of their responsibilities or the importance of the risks posed by the use of DSE in light of the 'scarier' tasks performed by uniformed police officers.

It was apparent that although the CBT package was a great start it did not solve the problem of having to conduct DSE risk assessments on all users and all workstations. Many organisations fall into the trap of thinking that CBT packages will conduct DSE assessments and fulfil their legal obligation but in fact they should only be used as a tool to identify, filter and ultimately prioritise the need for full assessments by a competent risk assessor.

With this in mind it was decided that a number of initiatives would be undertaken to ensure that the DSE assessment process was understood and supported by all in an attempt to secure its success. It was at this time that a new Health and Safety Advisor was appointed to manage the process.

The DSE Audit

Before deciding how to meet the Improvement Notice, an audit was undertaken looking at the scale of the problem and identifying a suitable and sufficient response. Applying the principles contained in HS(G)65, auditing was undertaken at both the policy and practice levels. Benchmarking different types of users, enabled the Police Force to understand how to progress.

Managers and workers were interviewed, results from the use of the software package were reviewed and a small number of representatives were assessed.

A simple signal detection model was applied to assess the relationship between users of the software and their real or perceived problems. In essence, an 'expert' assessor determined whether the user had correctly identified problems, missed problems, identified problems that were not there or correctly identified safe working.

36% of software users reported 'high priority' problems. In all cases, the expert assessment backed this up. 54% of software users reported 'low priority' problems. The expert assessment indicated that 71% of those users did actually have 'high priority' problems.

The good news from the audit, was an indication that over 50% of all of the problems identified could be eliminated by very simple actions. A further 40% could be eliminated by a simple risk assessment conducted by a competent assessor.

DSE Risk Assessment Training Courses

It was decided that there was a need for the conduct of more competent DSE assessments and therefore line managers were be asked to nominate suitable, reliable and committed staff to be trained as DSE risk assessors. The nominated staff then attended a 2 day training course after which they had to sit a 30 minute exam and submit a sample assessment for coursework in order to demonstrate their understanding. Feedback was provided on the coursework submitted to ensure that mistakes were not repeated and on the successful completion of both tasks a certificate was issued.

The number of staff trained was aimed at a ratio of 1 assessor to 40 staff. Nomination was difficult because the implications of putting assessors forward meant time and resources would need to be allocated both for the course and for subsequent assessing.

Line Manager Training

It soon became apparent that line managers were not aware of their legal duties under the DSE Regs and wider health and safety legislation and many were simply not taking the process seriously. The changes in legal status of Police Forces had caused some confusion about roles and responsibilities and the accountability for health and safety. This was indicated by the audit. In order to focus their minds and gain their support a series of 2 hour line manager briefings were held for more senior managers. The content of the briefing included their legal responsibilities, a brief explanation of the risks their staff face when using DSE and an explanation of the purpose of the CBT and the forms used in DSE assessments and the requirements of them once received.

Many of the line managers left the briefings with a determined attitude to ensure that their staff had completed the CBT, that assessors were nominated for training and staff that were reporting problems were assessed.

Getting people trained and using the CBT to prioritise and then getting assessors to conduct assessments took time. Due to the geographic spread of the Force and the need to have assessors spread throughout the area, arranging convenient locations and dates was a problem.

Use of the CBT package to identify potential problems and thus priorities was difficult because the way that the questions had been phrased and the way that the filters had been designed identified a high percentage of people as priorities. It was also apparent that due to the high number of people reporting potential problems the Police Force was creating a loaded gun, i.e. they had been informed of problems but did not have the resources to deal with everyone at the same time which ultimately meant that people with problems could have been left to wait.

During this time the consultants conducted assessments that were identified as high priority by Occupational Health and helped the H&S Advisor to manipulate the CBT package filters to provide more meaningful priorities.

It also became apparent at this time that the increase in awareness of the risks associated with DSE use and the potential development of MSDs caused others to think about the risks they faced and the problems they were experiencing. As a result, the consultants carried out a number of ergonomics assessments on people that did not use DSE but felt the tasks that they performing were causing or exacerbating MSDs.

Policy

During the training and briefing period the consultants liased with OHD and Human Resources to create a DSE Policy. This included responsibilities, a statement of commitment, arrangement for review and monitoring, eye testing arrangements and arrangements for homeworking. It was decided that the policy would be reviewed after 6 months.

Conclusion

After two visits and discussions with the HSE, the Improvement Notice was lifted on the proviso that the Notice had been satisfied but they would be back to inspect within the next 12 months to ensure they were satisfied with progress.

Many lessons have been learned from the experience. It is apparent that the HSE are actively interested in the enforcement of non compliance with the DSE Regs. Raising the awareness of the benefits of ergonomics interventions has a knock on effect, staff felt that they had been considered and problems were reduced and in some cases eliminated. A top down and bottom approach is essential, although it has recently become apparent that middle management need training to understand their responsibilities. It is important to provide channels of communication for feedback about the process. Many of the problems were identified through conversations with staff who were not actively involved in the process.

The project has been extended to wider, risk management issues aimed at establishing the Force as a leader in health and safety. By the end of 2002 a comprehensive structure should be in place to ensure that not only DSE issues but all occupational health and safety is effectively managed. The Health and Safety Advisor has also recruited an assistant as the Force realised that health and safety is a bigger job than first thought!

References

Health and Safety at Work etc. Act 1974
Health and Safety (Display Screen Equipment) Regulations 1992
Successful Health and Safety Management - HS(G)65, HSE Books
Management of Health and Safety at Work Regulations 1999
Police (Health and Safety) Act 1997

WORKING FROM HOME

Vicky Ball & Nigel Heaton

Human Applications
139 Ashby Road
Loughborough
Leicestershire
LE11 3AD

Every day thousands of people all over the country settle down to a day's work in the comfort of their own home. Homeworkers are inadequately represented in official statistics but figures from the Labour Force Survey show that there are two quite different trends underlying the recent rapid growth of homeworking. Whilst the traditional forms of manual, assembly work persists, there is also the emergence of other forms of non-manual work, generally well paid, making considerable use of information technology (IT) and only partially carried out at home. Although appealing to some not everybody is suited to the lifestyle it requires and not all employers are aware of the health and safety implications of working from home. This paper will examine the issues associated with homeworking for both employee and employer.

Introduction

Homeworkers are inadequately represented in official government statistics. However, figures from the Labour Force Survey (LFS) show that underlying the recent rapid growth of homeworking are two quite different trends. Whilst the traditional forms of (sometimes) exploitative manual work carried out primarily in the home persists, there is also the emergence of other forms of non-manual work, generally well paid, making considerable use of information technology and only partially carried out at home. In recent years, as office space becomes a premium for organisations, employers are more willing to create opportunities to work from home. This paper will look at issues associated with working from home for both employer and employee.

Legal Background

Although there are no specific laws or regulations relating to working from home, the Health and Safety at Work Etc Act 1974 (HSWA) and other regulations still apply. The HSWA places a general duty on employers to secure the health, safety and welfare of people at work and of others who might be affected by the work activity. Employees are required to take reasonable care of their own health and safety and that of others. This duty also extends to work undertaken in the home.

It is worth discussing here the notion of "choice". An employee who "chooses" to work at home has exactly the same legal rights and protection as an employee who is required to work at home. It is explicitly not a defence to claim that an employee was not adequately protected at home because they "chose" to work at home. Nor can an employer claim that they did not know an employee was working at home.

The Management of Health and Safety at Work Regulations 1999 also apply, they emphasise the requirement for risk management, which includes risk assessment, reduction and monitoring. These regulations also require the provision of information, training, instruction, supervision and health surveillance from employers. They require employees to take responsibility for their own health and safety and to report any problems.

Associated Risks

Due to the variety of tasks and activities carried out by homeworkers it would be impossible to go into specific detail about all of the risks involved. However, it is possible to focus on the general risks faced by this group, and divided them into physical and social / psychosocial risks.

Physical Risks

As most homes have not been designed with work in mind, the environment is unlikely to incorporate features that may minimise health hazards. Working conditions can be cramped and ergonomically unsuitable. Homeworkers are less likely to have undergone risk assessments and due to the nature of the work, less likely to receive information, advice, equipment and training about health and safety.

It is often the case that homeworkers will adapt an area of their house, commonly a bedroom or downstairs room to accommodate the task they are required to perform. However, in the majority of cases this space is also used for its originally intended domestic purpose and often the designated space is small to minimise the disruption to other household members. As a result of working in cramped or unsuitable conditions the homeworker may be at risk from an increased likelihood of developing a musculoskeletal disorder.

For example, if a homeworker is required to use a computer at home they could use a 'space saving' computer workstation. Whilst saving space, this kind of arrangement, used for prolonged periods could be expected to contribute to musculoskeletal problems. Such workstations often fail to meet the mandatory Schedule of minimum standards contained in the Health and Safety (Display Screen Equipment) Regulations 1992.

Although many homeworkers tend to adapt current space some choose to create extra space for their homeworking task. If the establishment of the home working space involves any building work, such as conversion of a loft space, there are strict building regulations, which must be adhered to. It is worth contacting an architect and approaching building control and planning departments for advice before starting work to ensure that the building adheres to relevant regulations.

The risk of musculoskeletal injuries exists in many different jobs. There are a number of contributory factors within many homeworking tasks that should be addressed to reduce the occurrence of such problems, these include excess force, repetition and static or twisted posture.

Typical tasks that lead to the development of musculo-skeletal disorders (MSDs) include excessive use of display screen equipment (DSE), manual handling operations and tasks involving static muscle load. For example, many homeworkers receive their supplies in bulk. The handling of these supplies may well constitute a risk, especially as it is likely that the homeworker is alone. If this is the case, homeworkers should receive training in the basic principles of good handling (eg how to avoid, split loads, move and handle safely, etc). Inadequate tools for the job often exacerbate problems when combined with repetitious work over long periods of time.

The layout of the workspace can also contribute to the development of MSDs. If the homeworker is required to reach excessively or twist they will be at an increased risk of MSDs. The principles of ergonomics, "fitting the task to the person" should be applied so that the workspace is arranged in the best way for the task performed by the homeworker.

The visual aspect of many tasks carried out by homeworkers can be very demanding. Those engaged in tasks such as assembly work and computer use are often required to use a focal range within their near vision in order to perform tasks. If a focal distance is held for prolonged periods the homeworker will experience visual fatigue. Therefore, it is important that breaks are taken where the eyes can adopt a different focal length and the muscles are replenished with oxygen. There is no evidence of permanent damage to the eyes associated with this type of work.

The majority of homeworkers work alone, that is they have no colleagues within their workspace and often no other house members who are in at the time they are working. This fact creates great difficulties in "raising the alarm." Just like many other lone workers, homeworkers are somewhat invisible and therefore neither their employers nor anybody else would know if they were in danger. Our experience with organisations who utilise home workers is that unless explicit plans are made, it is often assumed that lone workers will 'look after themselves.' One of the most important provisions for this group is regular contact and communication with others, not only to ensure that there are no problems but also to provide support.

Psychosocial Risks

There are a number of factors associated with homeworking that can contribute to mental stress. These include isolation, over work, boredom and too much personal control.

Numerous studies have been conducted on the degree of 'social support" which employees encounter and its effect on psychological strain. There is consistent evidence to show that employees with greater access to support from others (their boss, colleagues or subordinates) experience lower levels of stress. Homeworkers often do not receive any social support. They spend the majority of their day alone, without social interaction. People need social stimulation from others, whether this is a telephone call from a colleague or a chat by the photocopier, it is vital for mental health.

It takes a certain type of person to get up and start work in their own home when there are domestic tasks to be completed. Some people are able to balance their work and job stressors with the demands of domestic life, whilst others find it difficult. Homeworkers often know the amount of work that has to be done and are left to get on with it. Management of their workload is down to them. Some employers may require the homeworker to work normal office hours so that they can be contacted, other employers simply require the workload to be completed by a certain date. There is, therefore, a lot of emphasis on the discipline and motivation of the homeworker. Our experience is that the partitioning of work and non-work life can be a significant stressor. Partitioning, particularly when others are involved (e.g. partners, children, etc.), will need careful management. One organisation organises seminars for homeworkers and their partners aimed at explaining exactly what homeworking is and how it works.

Identifying Problems

Work related mental stress should be included in a risk assessment, as it is a workplace hazard and should be managed and controlled in the same way as other hazards. However, identifying homeworker's problems can be very difficult due to the fact they work remotely. Lines of communication should be made available to ensure that homeworkers can contact someone, both inside and outside of the organisation, especially in emergency.

If it is discovered that a home worker is suffering from work related stress it is imperative that action is taken to ensure that the 'stressors" are eliminated or reduced and that the employee receives adequate support. The organisation should have a policy for dealing with stress and a procedure to follow if homeworkers feel that they are suffering from work related mental stress.

Risk Assessment

Employers are required to conduct suitable and sufficient risk assessments of significant risk. Due to the variety of activities undertaken at home it is impossible to detail a specific risk assessment, therefore generic risk assessments should be conducted which should include consideration of:

- Manual Handling
- Work Equipment and PPE
- Electrical Equipment
- COSHH
- FIRE
- DSE

These generic assessments can then be tailored to specific homes. This requires some form if home based assessment.

The problem with assessing the risks to which a homeworker is exposed, is that employers do not have a right of entry into the homes of their employees. A homeworker can refuse an assessor entry to their home, which would make undertaking risk assessments difficult. Therefore, it is acceptable to go through the assessment procedure with the homeworker and ask them to provide reliable and valid information about their workspace and working practices. Remember that employees have a duty of care towards their own health and safety.

Where a homeworker refuses entry to their property the assessor could ask for photographic or video evidence of their workspace. Alternatively, they could request that the assessment be conducted via the telephone so that the homeworker is in their usual surroundings and is not reliant on memory.

Due to the possible difficulties when assessing homeworkers and the problems of competency in self-assessment, some employers choose to make it part of the homeworker's contract that they will be allowed entry for assessment and inspection. This also has implications for the assessors themselves as they are lone workers going into others' homes and are therefore potentially exposed to risk.

Once control measures are in place, managers and homeworkers have a responsibility to ensure that they are implemented effectively. Typically, this will mean ensuring that homeworkers are encouraged to follow the safe systems of work, that the equipment provided is used, etc. The manager must ensure that the homeworker is supported and that the training provided is followed. Our experience is that safety issues should form part of the quarterly or six-monthly review meetings that routinely occur between homeworkers and managers. Standing items should include discussions on the effectiveness of controls and equipment and any problems that are occurring.

The Provision of Training and Information

The content of the training will largely depend on the work required to be carried out at home and will mainly focus around learning to perform the task. However, time must also be spent on the issues associated with working from home. This will include the social and psychological implications of working from home, such as how to balance work time with domestic and social time.

The content must also focus on the risks to the health and safety of the homeworker identified by risk assessment and the subsequent control measures to eliminate or reduce the risks. Information and instruction will be closely linked to the training. Most organisations use a combination of printed literature (eg homeworker booklets) combined with information held electronically (eg available from the organisation's Intranet). Magazines such as Which? have published information for homeworkers (see Which? January 2002).

Monitoring

It is probably worth holding a formal review meeting for homeworkers on a regular (eg annual) basis. This can be in the form of a workshop facilitated by a competent health and safety person, and should focus on the issues associated with homeworking. In our experience, it is worth having a "theme" to the workshop, such as 'safe working with computers at home" or "hosting visitors".

It is important that homeworkers feel integrated into the organisation. The organisation needs to demonstrate a commitment to the homeworker's health, safety and welfare, so that homeworkers continue to feel a part of the organisation.

Homeworking Policy

The organisation should have a homeworking policy which forms the framework for procedures and practices. It provides clarity, demonstrates commitment and develops confidence in the organisation's approach to working from home. Whilst the normal health and safety policy for the organisation should lay out the broad framework for the management of health and safety, organisations need to have specific guidelines for homeworkers and their managers to demonstrate how the homeworker's risks are managed.

Associated Costs

There are potential costs associated with homeworking that must be considered by both the homeworker and their employer. There are travelling expenses, capital gains tax, business rate and council tax implications associated with working from home and homeworkers are advised to contact the Inland Revenue.

Negotiations must also be held about the utility cost associated with working from home and contributions from the employers. Many power utility companies have a policy that to qualify for the minimum tariff premises must be used wholly for domestic purposes. The homeworker must also ensure that they are insured for working from home. A standard home and contents policy is unlikely to cover the home office equipment but specific policies targeted at home offices have been produced to replace the plethora of computer, office and home policies previously designed to confuse the homeworker. Generally, the employer must meet these additional costs.

Future of Homeworking

Homeworkers are owed a duty of care by their employer. Any breach of this duty may leave the employer liable for a negligence claim. Similarly, the employer's duty extends to those affected by the behaviour and work of the homeworker (a vicarious liability).

The production of competent, detailed risk assessments are vital. Ultimately, organisations should assume that they will need to visit at least a sample of homeworkers, if not all. Employers will also need to provide homeworkers with enough information and training to enable them to contribute to the risk assessment process, possibly by conducting self-assessments or walkthroughs.

Realistically, with suitable terms and conditions and depending on the type of work undertaken, the advantages of homeworking will almost inevitably outweigh the risks. Just removing the homeworker from driving to and from work is probably a major contribution to the homeworker's safety. However, employers must understand their duties and be clear that ignorance of what happens in the homeworker's home is absolutely no excuse if homeworkers are put at risk.

References

"A Statistical Portrait of Working at Home in the UK: Evidence from the Labour Force Survey", ESRC Future of Work Working Paper no.4 by Alan Felstead, Nick Jewson, Annie Phizacklea and Sally Walters/
Croner's *Managing Off Site Safety: Homeworking and Mobile Working*
Health and Safety at Work etc Act 1974
Health and Safety (Display Screen Equipment Regulations) 1992
Management of Health and Safety at Work Regulations 1999
Which? January 2002. P52-53

VISION

TELLING TIME: THE ERGONOMICS OF MOUNTING A LARGE PUBLIC EXHIBITION

David Wooding, Kevin Purdy, Alastair Gale and Mark Mugglestone

Applied Vision Research Unit
University of Derby
Kingsway House
Kingsway
Derby DE22 3HL

The world's largest eye movement experiment was undertaken in early 2001 as part of the UK National Gallery exhibition on 'Telling Time'. The experiment entailed Gallery visitors being able to examine several paintings, individually displayed on a computer monitor, whilst their visual search behaviour was discretely monitored. As someone examined the paintings then others were able to watch the individual's visual search patterns on a large plasma computer display. The system was designed to be fully autonomous with the computer program automatically starting when an individual's eye was located in an appropriate spatial position in front of the monitor. The system was configured to be accessible by as wide a range of individuals as possible and in the event nearly 10,000 people took part. This paper describes the ergonomic considerations involved in mounting such a large autonomous public exhibition, together with the compromises and limitations found to be necessary in the final implementation.

Introduction

As part of the National Gallery exhibition on 'Telling Time', which ran from September 2000 to January 2001, an experiment was planned which would allow visitors to see just how they visually examined paintings over a short time period. A selection of paintings was to be individually displayed on a computer monitor. This would be examined by an individual and his/her visual search of the various paintings recorded by some discreet means. Other members of the public would be able to watch the visual search patterns, in real time, on a large plasma computer display. Subsequently, the individual themselves could then watch a rerun of how they had examined each painting.

The idea behind the experiment was to develop an interactive exhibit which would run in the Gallery without any intervention from an experimenter. The experiment was to be: accessible to as many people as possible; require virtually no instructions for usage; be extremely usable, and provide rich pictorial and informative feedback. Primarily mounting the experiment involved developing a suitable system and relevant software which would present an individual with a selection of paintings to view and both record and display the visual search behaviour of that person as they examined these paintings. This paper considers the main ergonomic issues regarding mounting such a complex interactive exhibit.

The principal aspect in the design of the experiment was that a high throughput of members of the public was required. Thus the overall time that it took each person to take part in the experiment was paramount. For an individual to participate s/he had to: firstly sit in the correct location in front of the computer monitor (more precisely one of their eyes was required to be spatially located at the same height as the centre of the monitor and the recording system had to detect this eye and monitor its movements); self-calibrate their point of gaze with the recording system; enter some demographic information about themselves, and then visually examine a series of digitised images of paintings. It was estimated from the outset that the overall time for an individual to participate should be less than a target of about three minutes.

In a typical laboratory study, using eye movement recording, then some 10-20 participants may take part under the precise and continuous control of an experimenter who would: carefully situate them with regard to the equipment; calibrate them; then run the experiment, and all the time monitor the resultant data stream to ensure that no problems occurred.

In marked contrast the design here required the equipment and software to achieve all these goals autonomously without any experimenter intervention – the experimenters being over 200 km away.

Eye movement recording technique

The first ergonomic requirement was the choice of an appropriate eye movement recording system. There are a number of potential techniques (see Young and Sheena, 1975, for a review) and within an applied environment infra red (IR) video-based systems are often used. Previous attempts using eye-trackers in public displays have raised issues concerning system reliability and usability (e.g. Buquet et al., 1988; Glenstrup and Engell-Nielsen, 1995). To record visual search behaviour in this exhibition there were broadly three main potential recording approaches which were considered; head mounted, head free and head fixed.

Head Mounted

Head mounted eye movement recording systems require careful fitting to each individual's head by a skilled researcher and so were discounted for use in this autonomous system.

Head Free

This approach uses a remote eye tracker (c.f. Merchant, 1974) and allows the participant some small degree of 3D free head movement, typically circa some 30cms^3. The chief advantage of this method is that there are no attachments to the participant and so it is very user friendly. However, the disadvantage in an exhibition setting, is that for each new participant one of their eyes has first to be automatically found in 3D space by the system and this eye then needs to be tracked accurately as the person's head moves.

Whilst this can be done very well by current commercially available systems, it is not ideal for an autonomous approach as finding and tracking the eye automatically can take a finite amount of time. With no experimenter present it would then be difficult to limit in size any potential head movement of the participant. Additionally, as the system hunts for an eye, there is the possibility of false detections of an apparent eye caused by other 'non-eye' IR reflective objects.

In a laboratory situation these spurious detections can be discounted easily by the experimenter. Therefore the approach has advantages but also some disadvantages in the situation planned here which could slow down the throughput of observers.

Head Fixed

A fixed head approach requires that the participant is restrained somewhat, for instance using a chin rest. Thus the X (horizontal) and Z (distal) spatial location of the to-be-monitored eye is known and the system then only needs to find this eye in the Y (vertical) plane. Consequently, this is both simpler and faster than a full head free approach and less likely to give rise to rogue detections of pseudo-eye artefacts.

The Selected System

After detailed consideration a head free system (Applied Science Laboratories remote oculometer; model ASL 504) was utilised. With this recording system it would be perfectly possible to have a person, of any height, seated in front of the display monitor and then speedily move the equipment (i.e. both the display monitor and the eye movement recording camera) vertically until the system detected one of their eyes appropriately. This approach was originally planned as being the most acceptable to Gallery visitors but was then eventually rejected.

Consequently it was decided to opt for a hybrid approach which utilised the remote oculometer in conjunction with an arrangement of a fixed pair of 'goggles' to provide some elements of a head fixed situation, thereby incorporating the advantages of both approaches. The remote system itself required no attachments to the individual but the goggles located their eyes spatially in approximately the correct 3D position with respect to the display monitor, therefore making it fast and easy for the system to actually locate the appropriate eye. Additionally, sensors were built into the goggles to detect when someone's head was in place; this information then acted as a switch to initiate the whole system.

Usability issues

System

The overall system was designed to be fully autonomous. To achieve this the software was written so as simply to require the power to be turned on, with no other keyboard or mouse operation (these devices not being available to the participant). If the software crashed then simply turning the power off and then on again would reboot it. The experimental program began automatically when someone sat in front of the monitor. No Applied Vision Research Unit staff were present during the exhibition, except for routine maintenance operations and to download the amassed recorded data from the computer.

Participant

To increase participant throughput, minimal instructions for using the system were displayed alongside the exhibit; although it was not necessary to read these in order to operate the exhibit successfully. One of the main functions of the instructions was to alert spectacle wearers to potential difficulties due to their spectacles producing additional IR reflections, which could be mistaken by the eye movement system for eye features leading to faulty data. Such individuals were requested to attempt the experiment without use of spectacles, if their visual acuity allowed it. Expert optometric advice was used to determine the best viewing distance between the participant position and the monitor so as to enable as many of these people, who normally used reading glasses, as possible to view the images without using their glasses.

The system was designed to operate with a participant's eye spatially located at a particular height. This fixed eye height would be, by its nature, a careful compromise between lowering tall people and raising small people. A height of 1225mm from the floor was chosen, as this was felt to be around the lowest at which a 95th percentile male adult could reasonably be seated for a short period of time, with their legs stretched out before them. This 'minimum' height would therefore also form the lowest achievable height to which smaller people would need to be raised. It was decided that this distance would be acceptable in combination with a step or footrest to aid mounting and dismounting. A stool whose height was adjustable over a very wide range (375mm – 690mm) was constructed specially to enable people from age 8 years upwards (this lower age being based on the Gallery's known demographic profile of visitors) to use the exhibit. The stool was removable to allow the exhibit to be accessible to wheelchair users.

Operation

When a participant's head was located in the goggles then the system began automatically. The goggle sensors were designed with some tolerance in order to allow a participant to be able to move their head slightly without the system thinking that they had left and a new participant had arrived. This tolerance was again a careful compromise as the same sensors were used to define when participants were adequately positioned in a small space, as well as otherwise merely present. The system initiated by displaying some on-screen instructions and information which served to force the participant look at the screen during which time the eye movement recording system first detected one of their eyes and then performed some initial calculations to ensure correct eye capture by the software.

The participant then had to undergo a calibration phase where they looked at known spatially separated points on the monitor. To increase ease of use this was accomplished by the system instructing the participant simply to 'follow the bee'. An animated 'bee' was then displayed which 'flew' around the monitor and settled at certain (calibration) points for short periods of time. The first part of the bee's flight formed the calibration routine, and the second part was used to check (and log) the accuracy of the calibration. If the system was unable to collect usable data of sufficiently high quality from this calibration phase then the system thanked the individual for participating and requested them to let someone else take part.

For participants who calibrated successfully then specific participant information (e.g. gender, age, artistic training, etc.) was recorded. This was done using their eye fixation locations as 'eye buttons' (Jacob, 1991; Sibert and Jacob, 2000). To do this, questions were presented at the top of the monitor screen with alternative answers displayed below, each one within a large coloured 'box'. All the participant had to do to select an answer was to stare at the appropriate box briefly until the computer detected this decision and provided feedback to acknowledge the decision.

Participants were then shown some three images of paintings from the Gallery, each image being displayed for 20 seconds – this time being arrived at through various pilot studies. A large number of paintings had previously been selected for potential display from which some 140 were actually used. Further details of the experimental design are described by Wooding *et al.* (2002).

Whilst an individual was actually examining some images of paintings then their visual search behaviour was relayed live to the large plasma screen to enable others to see it. Immediately after someone had taken part then these recordings were again replayed which enabled the participant to see what they had actually looked at. At all other times (e.g. when no-one was taking part, when someone was undergoing calibration or entering demographic information) this monitor replayed selections from previous recordings of individuals looking at different paintings so as to encourage people to participate.

Results

The experiment ran successfully for 89 days. In this time 110,000 people visited the exhibition and 9,884 of them actually took part in the experiment. Of these, some 5,446 successfully completed viewing all three of the images of paintings which they were shown, making it the world's largest eye movement experiment.

Where participants were unsuccessful in taking part then they mainly failed to achieve a useful calibration (3,890 people). In eye movement recording it is essential to obtain an adequate calibration, as otherwise it is difficult, if not impossible, to accurately reflect which parts of an image have been looked at by a participant. Obtaining a good calibration for individuals autonomously is clearly very difficult and the current approach of following a moving target ('bee') worked well.

Our recorded experimental information demonstrated that the exhibit was almost constantly in use whenever the National Gallery was open. The total median participant time was 232s, which was well within our initial estimated target time for participant usage.

Conclusions

Previous attempts to mount large public exhibitions of eye movement recording have been variable in their success. This particular study was not only very successful but was also specifically designed as an experiment to gather scientific data on how a large number of individuals studied a number of different paintings.

The hybrid eye movement recording approach worked very well and was a sensible compromise that allowed as many individuals to take part as possible. It would have been even more successful if it had been possible to calibrate additional individuals. Calibration is inherently difficult as it partly depends upon the adequate reflection of incident IR from the individual's eye (which can vary with the age of the individual). Additionally, the system needs to collect suitable data from the various calibration points, which requires some co-operation from the individual in appropriately tracking the moving target.

Fixing the spatial positioning of the participant's eye worked well as a technique, although it was informally noted that individuals tended not to adjust the height of the stool, preferring to lean down or stand as appropriate. This suggests that the stool was not sufficiently useable in this particular application, with participants trying to accommodate the eye height through their own means. Had the method used been more successful it is probable that even more participants would have been calibrated acceptably. Where participants did not adjust the stool height appropriately then they may have presented their face to the eye movement system at a less than optimal angle, through stretching or bending, and also to have moved position more during the experiment, through discomfort and lack of appropriate physical support. These factors would have led to failed calibration attempts.

Data analyses emanating from the experiment are complex and will be ongoing for several years. Wooding (2002) has detailed some of the initial data analyses techniques and information is routinely updated on our web site (http://ibs.derby.ac.uk/gallery).

References

Buquet, C., Charlier, J.R. and Paris, V. 1988, Museum application of an eye-tracker. *Medical and Biological Engineering and Computing*, 26, 277-281.

Glenstrup, A. and Engell-Nielsen, T. 1995, *Eye Controlled Media: Present and Future State.* Minor Subject Thesis, Laboratory of Psychology, University of Copenhagen (available at http://www.diku.dk/users/panic/eyegaze/)

Jacob, R.J.K. 1991, The Use of Eye Movements in Human-Computer Interaction Techniques: What You Look At is What You Get. *ACM Transactions on Information Systems*, 9(3), 152-169.

Merchant, J. 1974, Remote measurement of eye direction allowing subject motion over one cubic foot of space. *IEEE Transactions on Biomedical Engineering*, 21, 309-317

Sibert, L.E. and Jacob, R.J.K. (2000) Evaluation of Eye Gaze Interaction. *Proceedings of ACM CHI 2000 Human Factors in Computing Systems Conference*, (pp. 281-288), Addison-Wesley/ACM Press.

Young, L. and Sheena, D. 1975, Survey of eye movement recording methods, *Behavior Research Methods & Instrumentation*, Vol. 7(5), 397-429.

Wooding D.S. 2002, Eye movements of large populations: II: deriving regions of interest, coverage and similarity. *Behaviour Research Methods, Instruments & Computers*. In press.

Wooding, D.S., Mugglestone, M.D., Purdy K.J. & Gale A.G. 2002 Eye movements of large populations: I. Implementation and performance of an autonomous public eye tracker. *Behaviour Research Methods, Instruments & Computers*. In press.

Acknowledgements

The exhibit was staged in co-operation with various staff at the National Gallery, London, where Alexander Sturgis was the curator of the "Telling Time" exhibition.

Applied Science Laboratories (ASL) provided the ASL 504 eye tracker and other technical help during development.

The exhibit was developed with support from Derby University Enterprises Limited.

VARIATIONS IN THE DISCOMFORT AND VISUAL FATIGUE OF CYTOLOGY SCREENERS

J May, H Cowley and AG Gale

*Applied Vision Research Unit,
University of Derby
Kingsway House,
Kingsway,
Derby; DE22 3HL
Tel/Fax: 01332 593131*

The National Health Service Cervical Screening Programme (NHSCSP) screens four million women a year to detect pre-cancerous changes of the cervix. The aim of this study was to investigate the visual and muscular fatigue experienced by cytology screeners and how these varied in a typical screening laboratory. Vision, mood and subjective discomfort were measured periodically throughout the day. The number of slides each screener examined was also recorded. Some visual qualities and feelings of pain were found to worsen across the shift, and the breaks taken were insufficient to fully alleviate this. Participants were less likely to be relaxed and calm at the end of the shift. Alertness decreased significantly throughout the day and tiredness increased.

Introduction

It is estimated that the NHS Cervical Screening Programme (NHSCSP) prevents between 1,100 and 3,900 cases of cervical cancer each year (Sasieni et al. 1996). The microscopy work of cytology screeners requires them to adopt fixed postures, perform repetitive movements with their hands, and to perform close intensive visual work. They have therefore been found to suffer from job related discomfort, 84% in Sweden (Karlquist and Tapio, 1980) and 78% in the UK (Hopper et al., 1997). Discomfort is most prevalent in the neck, shoulders, wrists and back (Hopper et al., 1997, Rizzo et al., 1992). Kalavar and Hunting (1996) found that screeners working for over 6 hours per day had a greater prevalence of musculo-skeletal symptoms than those working for shorter time periods. Cytology screeners also report visual problems (Karlqvist and Tapio, 1980, Hopper et al.1997). Soderberg et al (1983) found that 80% of microscope operators experienced visual strain and this increased with the time spent using a microscope. It is important to understand exactly when and how cytology screeners become fatigued throughout the day to ensure that appropriate breaks are taken. This is important for the health of the cytology screener and also for the women being screened

Method

Six cytology screeners from a Hospital Cytology Department participated in the study. They were monitored for four days while undertaking their normal screening duties. All the participants worked for two and a half hours, had a 15 minute break and then worked another two hours and twenty minutes. They then had a half hour lunch break and worked for one hour before finishing their shift.

At specified times, (before shift, after 1st hour, after 2nd hour, after 3rd hour, after 4th hour, before lunch, after lunch, after shift) participants were asked to give self-reported ratings of mood and fatigue. This questionnaire used the UWIST Mood Adjective Checklist (UMACL; Matthews, Jones and Chamberlain, 1990) and a Symptomatic Questionnaire developed by Lie and Watten (1994). Any change in visual performance were objectively measured (using the Titmus Vision Screener) by testing the participants' vision (before shift, before lunch, after lunch and after shift). Visual acuity, accommodation, convergence and phorias were tested. Additionally, the screeners recorded the number of slides they examined and this was confirmed by video data.

Results

Quantity of Work

There was no significant difference in the screening rate (number of slides per hour) (p= 0.806) and the average time to screen each slide (p= 0.481) throughout the day.

Visual Performance

Table 1 indicates significant differences (ANOVA) in visual performance.

Table 1. Significant differences in vision screening results

Visual Test	Distance	Subject	Time of day	Subject * Time of day
Accommodation		0.001		
Lateral Phoria	Far	0.001		
Vertical Phoria	Far	0.009		
Acuity Both	Far	0.002		
Acuity Right	Far	0.004		
Acuity Left	Far	0.001	0.009	
Lateral Phoria	Near	0.001		0.017
Vertical Phoria	Near	0.002		
Acuity Both	Near	0.001	0.003	0.014
Acuity Right	Near	0.001		
Acuity Left	Near	0.001	0.037	0.043

Figures 1 and 2 indicate some of the significant differences found in the participants' visual acuity across the four sessions. The y-axis represents the line of the eye chart that they were able to read accurately. Participants' mean visual acuity (far) for the left eye decreased before lunch and increased again afterwards, but then deteriorated at the end of the shift. The mean visual acuity (near) for both eyes decreased before lunch and did not improve but continued to decrease slightly. While there was an interaction between the time of day and subjects' visual performance on some measures, there was not always a clear pattern of deterioration.

Figure 1. Differences in visual acuity for the left eye at far distance

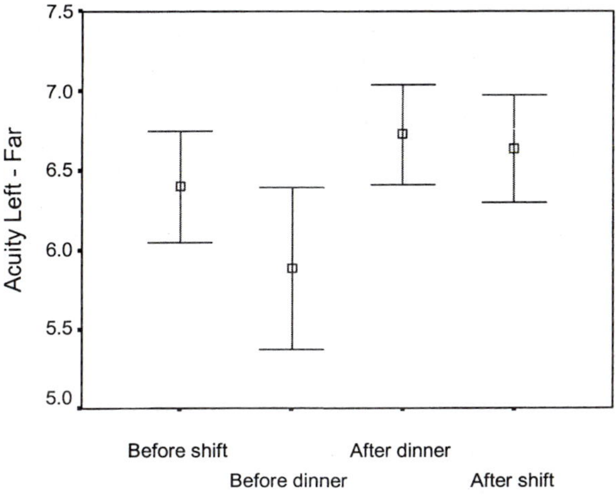

Figure 2. Differences in visual acuity for both eyes at near distance

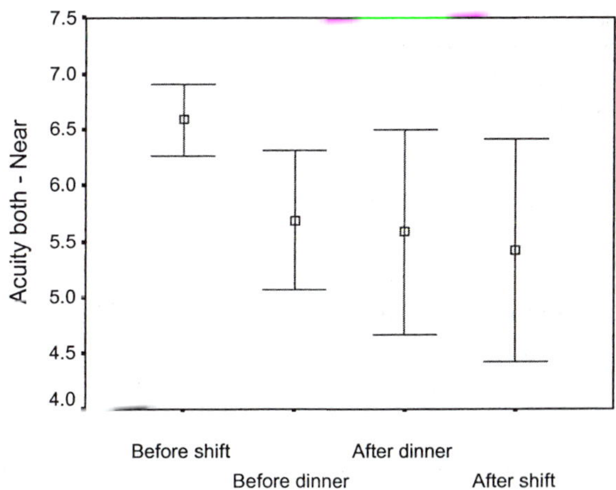

Questionnaire

Significant differences were found for the different sessions, these are shown in Table 2.

Table 2. Significant differences over time for self-reported fatigue

Currently feeling pain, tension or aching:	**df**	**F**	**Sig.**
Behind or around the eyes	7,143	2.582	0.015
In the neck	7,143	2.391	0.024
Headache	7,143	2.175	0.040
Eyeball pain when screening	7,143	2.383	0.025
Concentration problems	7,143	2.340	0.027
Indicate how your mood is at the moment	**df**	**F**	**Sig**
Energetic	7,143	2.384	0.025
Relaxed	7,143	2.887	0.007
Tired	7,143	5.626	0.000
Calm	7,143	2.200	0.038
Restful	7,143	2.837	0.008
Composed	7,143	2.677	0.012
Alert	7,143	5.002	0.000

Figures 3 and 4 for example, indicate some of the significant changes in response to the questions concerning pain and tension throughout the day. The values on the y-axis indicate the participants mean responses at various times of the day. A response of 1 indicates no pain, aching or tension and a response of 7 indicates severe pain, aching or tension. Basically self-reported pain, tension or aching behind or around the eyes and neck, headache, eye ball pain and concentration problems increased on average until the participants' had a break for lunch. These factors then decreased after the half hour lunch break and then increased again at the end of the shift.

Figure 3. Currently feeling pain, tension or aching behind or around the eyes

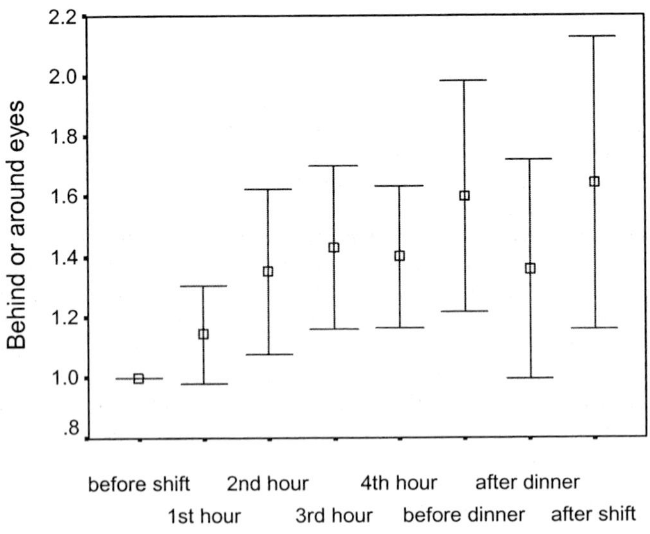

Figure 4. Currently feeling pain, tension or aching in the neck

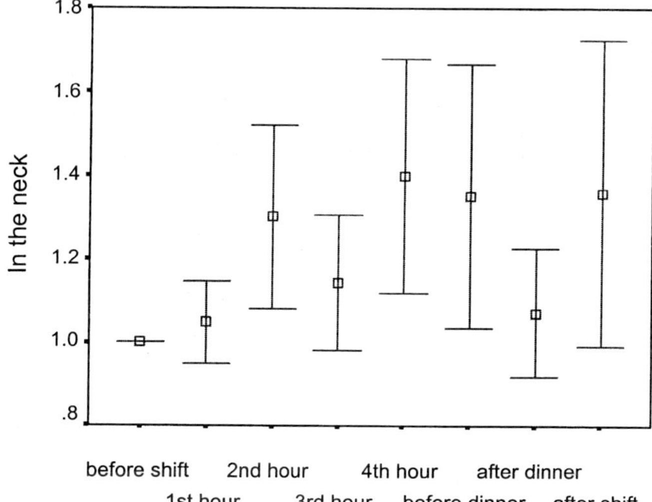

Regarding self reports of mood, participants were significantly less relaxed, calm, restful and composed at the end of the day. The half hour lunch break only resulted in an increase in feelings of being relaxed and restful. Participants' alertness decreased significantly throughout the day and tiredness increased, but with the half hour lunch break having a positive effect on both these variables.

Discussion

No significant differences were found regarding the quantity of work performed throughout the day. The measurement of slide screening time or quantity of slides screened however, does not make any allowances for slide difficulty and thus is not a particularly good measure of work performed. Slide screening performance may therefore be a better measure to use.

Screeners' visual acuity varied significantly throughout the day with a negative effect being found for undertaking five or more hours of screening. The half hour lunch break was shown to be important in improving visual acuity (far) or at least slowing down the decline in visual acuity (near). Half an hour was not sufficient time for near visual acuity to recover, but far visual acuity appeared to recover in this time. It is not known if screeners compensate for such visual changes by altering the way in which they use the microscope. There was also found to be wide variation in visual performance measures between individual participants over the four sessions and more subjects may need to be measured to validate these findings.

Participant's self-reported feelings of pain, were lower than expected, although for questions concerning headache and concentration problems participants use a rating of four and five. Several self-reported feelings of pain (visual and musculoskeletal), headache and concentration problems increased across the hours of the shift, with all factors showing a significant improvement over the half hour lunch break. This break was insufficient however, to reduce the pain to the same levels as at the start of the shift. Some factors (e.g. neck pain,) were shown to slightly improve after the 15 minute rest break, but others were not (e.g. eye pain). Participants were less likely to be relaxed, calm, restful and composed at the end of the shift than at the start. The half hour lunch break improved restfulness, but did not affect how calm and composed the participants were. Alertness was shown to decrease significantly throughout the day and tiredness increase, the half hour lunch break had a positive affect on these variables as expected.

Conclusions

The quantity of work did not deteriorate throughout the day, although quality was not measured. Screeners' visual acuity worsened with the time on the task but having a break to some extent relieved this effect. Self-reported pain and fatigue (visual and muscular) was low, but was shown to increase throughout the day. Breaks also had a positive effect on the subjective discomfort recorded. A half hour lunch break was insufficient for some aspects of visual performance and for feelings of pain to fully recover. A fifteen minute mid-morning break did have a slight positive effect on some factors but not on others and may therefore be insufficient Feelings of alertness decreased over the day and tiredness increased. Breaks were shown to be very important and may need to be lengthened.

References

Hopper, J., May, J. and Gale, A., 1997, Screening for Cervical Cancer. In: S.A. Robertson (ed.) *Contemporary Ergonomics 1997* (Taylor and Francis, London), 38-39.

Kalavar, S.S. and Hunting, K.L. 1996, Musculoskeletal symptoms among Cytotechnolgists, *Laboratory Medicine*, **27**, (11), 765-769.

Karlquist, L and Tapio, S,. 1980, The working environment of cytology laboratory technicians. In M. Ostberg and E. Moss (ed.) 1984 Microscope work. *Proceedings of the International Conference on Occupational Ergonomics, Rexdale, Ontario, Canada.* Human Factors Association of Canada 402-406.

Lie, I. and Watten, R.G. 1994, VDT work, oculomotor strain, and participative complaints: an experimental and clinical study, Ergonomics, **37**, (8), 1419-1433.

Mathews, G., Jones, D. and Chamberlin, A.G., 1990, Refining the measurement of mood: The UMIST Mood Adjective Checklist, *British Journal of Psychology*, **81,** 17-42.

Rizzo, P., Rossignol, M. and Gauvin, J.P., 1992, A One Week Incidence Study of the Symptoms Experienced by Hospital Microscopists, *The Economics of Ergonomics. Proceedings of the 25th Annual Conference on the Human Factors Association of Canada, Hamilton, Ontario, Canada, October 25-28.*

Sasieni, P.D., Cuzick J. and Lynch-Farmery, E., 1996, Estimating the efficacy of screening by auditing smear histories of women with and without cervical cancer. The National Co-ordinating Network for Cervical Screening Working Group. *British Journal of Cancer*. **73,** (8), 1001-5.

Soderberg, I., Calissendorff, B., Elofsson, S., Knave, B. and Nyman, K., 1983, Investigation of visual strain experienced by microscope operators at an electronics plant, *Applied Ergonomics*, **14,** (4), 297-305.

THE FRAMING EFFECT DIMINISHES PICTORIAL DEPTH PERCEPTION: IMPLICATIONS FOR INDIRECT MANIPULATION VIA VIDEO LINKS

Anthony H. Reinhardt-Rutland

School of Psychology,
University of Ulster at Jordanstown,
Newtownabbey,
Co. Antrim,
Northern Ireland,
BT37 0QB

Indirect visual-motor manipulation via video links - for example, in minimally-invasive ("keyhole") surgery and the handling of dangerous materials - reduces the information available for depth perception: depth information is mainly *pictorial*. Degradation of depth perception may also arise from the *framing effect*: depth is underestimated when viewing through a frame between the observer and the scene. In video links, the monitor's case acts as a frame for the VDU screen. To determine if the framing effect may apply to depth conveyed pictorially, depth was judged in surfaces with true and pictorial slants-in-depth relative to the frontal plane; surfaces were rectangular and trapezoidal in shape. Consistent with previous studies, the framing effect reduced the perceived slant-in-depth of a surface conveying true and pictorial slant by about 20%. A substantial reduction (70%) affected a truly frontal surface that was pictorially slanted. The latter result suggests that the framing effect may be of major importance in judging depth conveyed by video links.

Introduction

An increasing number of visual-motor manipulations can be performed indirectly via a video link. This is particularly valuable in cases where directly performed manipulations are costly, dangerous, or otherwise undesirable. Keyhole surgery and the handling of hazardous materials are good examples. In keyhole surgery, a miniature camera records the site of operation, which is transmitted for viewing at a video monitor. However, the information-processing issues with indirect operation via video links can be formidable.

One important problem concerns the loss of information for depth perception (Perkins *et al*, 2002; Reinhardt-Rutland *et al*, 1999). Depth information can be classified as *monocular* (eg, accommodation), *binocular* (convergence and retinal disparity), *motion* (eg, motion parallax) and *pictorial*. Being the consequence of inborn design-features, the first three types have often been regarded as of primary importance for depth perception. Since all three are attenuated or convey the flatness of the video screen, this might suggest that viewing via a video link is not feasible. While such a position is clearly no longer tenable, it does suggest the need for caution.

The final type of information - pictorial information - refers to depth conveyed in 2D representations of depth. Examples of pictorial information include occlusion - a closer object obscures a more distant object - and a variety of cues based on our knowledge of structured space: for example, many surfaces in the environment are rectangular in shape, so a trapezoidal image is likely to arise from a rectangular surface slanted-in-depth relative to the frontal plane. Underlying knowledge can be extremely subtle: for example, triangular surfaces do not in principle provide strong pictorial information for their orientation-in-depth, yet participants seem able to employ internalized models of "typical" triangles in making slant judgments (Reinhardt-Rutland, 1996a,b). Pictorial information can be powerful in the laboratory, often overriding other types of information (Reinhardt-Rutland, 1995a; Stevens and Brookes, 1988). However, pictorial information is easily manipulated to be misleading: even though trapezoidal surfaces are rare in most people's experience, a trapezoidal image may as easily arise from a trapezoidal surface in the frontal plane as from a slanted rectangular surface.

In normal viewing, the various types of depth information are in accord, so there is redundancy amongst them. It is this redundancy that permits a degree of depth perception during attenuated conditions of viewing, including those applying to VDUs: in the latter case, there is conflict between types of depth information - because of the flatness of the VDU screen - so depth perception of the viewed scene relies mainly on pictorial information, perhaps enhanced by any motion in the image.

Improvements in depth information may arise with stereoscopic viewing (Wenzl *et al*, 1994) - but not without introducing other problems. Stereoscopic depth perception can take a while to achieve, which is important if critical stages of operation require switching between normal and stereoscopic viewing. Also, prolonged stereoscopic viewing is tiring: the duration of operations may need to be short. Finally, stereoscopic systems add considerable complexity and expense, because of the need for a second camera and the hardware required for sequencing the stereoscopic images (Reinhardt-Rutland and Ehrenstein, 1996).

To the potential problems with depth perception during viewing via video links, the present paper adds the *framing effect*. This is the underestimation of depth differences between points in space when viewing through a frame placed between the participant and the viewed scene (Eby and Braunstein, 1995; Reinhardt-Rutland, 1999). The phenomenon has clear implications for indirectly-viewed manipulations, since the VDU monitor's case acts as a frame for the screen. Investigation of the framing effect has concerned three-dimensional scenes: true depth matches pictorial information for depth. To decide if the framing effect might apply to VDU monitors, one needs to ascertain whether the framing effect influences two-dimensional representations of depth - scenes that are truly in the frontal plane, but which convey pictorial depth information.

Method

Stimuli

Stimuli were rectangular and trapezoidal surfaces having real and/or simulated pictorial slants-in-depth relative to the frontal plane. Viewing was stationary-monocular. Surfaces were:

(a) a 10 cm by 15 cm rectangular surface in the frontal plane. This surface is labelled T0P0, where T indicates the true orientation and P indicates the pictorial orientation: in both cases this was 0 deg relative to frontal. The frame should not affect perceived slant-in-depth of this surface.

(b) a trapezoidal surface actually slanted at 30 deg to frontal, but matching pictorially the rectangular surface with frontal orientation, that is pictorially 0 deg to frontal: this surface is labelled T30P0. Under the viewing conditions without the frame, such a surface should be perceived as frontal, or close to frontal (Reinhardt-Rutland, 1995a). Hence, there should be little if any framing effect with this surface.

(c) a 10 cm by 15 cm rectangular surface slanted at 30 deg to frontal. This surface is labelled T30P30, where T indicates the true orientation and P indicates the pictorial orientation relative to the frontal plane. Eby and Braunstein's (1995) evidence suggests that a framing effect should affect this surface.

(d) a trapezoidal surface actually frontal but pictorially matching the slanted rectangular surface, labelled T0P30. This last surface corresponds to viewing via a video link and is therefore of principal interest.

The left vertical edge of each surface was always 1 m from the participant and subtended 5.9 deg of visual angle. The right vertical edge receded from the participant in the T30 stimuli.

In half the conditions, the surface was viewed through a rectangular frame 32 cm long by 20 cm high. The frame was in the frontal plane, located between the experimental surface and the participant; it was 0.2 m in front of the experimental surface and 0.8 m from the participant.

The surfaces and frame were made of textureless matt-white card. A table supporting the surfaces and frame, and other areas around the surfaces and frame, were painted matt black. Lighting was ultraviolet (Philips TLGW08 8 watt UV tube) so that only the surfaces and frame were visible. The luminance of experimental surfaces and frame approximated 0.2 cd m^{-2} (Hagner S1 photometer), while the luminance of other surfaces was below measurable levels.

Participants

Forty-eight male and female undergraduates reporting uncorrected, normal vision acted as participants.

Design and procedure

Each experimental surface was viewed monocularly one at a time from one end of the table supporting the surface and frame. Participants estimated the orientation-in-depth of each surface employing a verbal procedure, in which the participant envisaged the surface as the "minute-hand" of a clock; the table was the "face" of the clock along which the participant viewed the experimental surface / "minute-hand". As examples, 12 min would be the response to an experimental surface perceived as being at 18 deg to frontal, while 10 min would be the response to a stimulus perceived as being at 30 deg to frontal. This procedure therefore involves a familiar, everyday metric which participants find straight-forward to use (Reinhardt-Rutland, 1995a, 1996b). Responses in minutes of time are readily converted into degrees of angle relative to the frontal plane by the equation $y = 6(15 - x)$, where y is degrees of angle and x is minutes of time.

To prevent order effects that might be problematic in a repeated-measures design, half the participants viewed the surfaces in the absence of the frame and half viewed in the presence of the frame. Viewing order of the surfaces was random across participants.

Results and discussion

Results are shown in Figure 1. Without the frame, judgments mainly depended on pictorial information: the pictorial slants of T30P30 and T0P30 were slightly overestimated, while T0P0 and T30P0 elicited near-frontal judgments. This is consistent with previous evidence (Reinhardt-Rutland, 1995a).

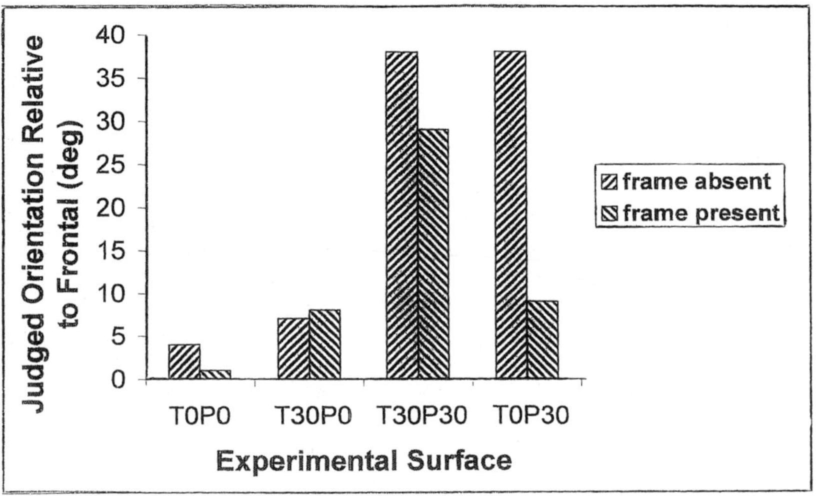

Figure 1. Means of participants' judgments of surface orientation-in-depth in degrees relative to the frontal plane.

With the frame, judgments of T0P0 and T30P0 were little altered, which is expected given the near-frontal responses without the frame. However, perceived slant was reduced for T30P30 by about 20%; this is consistent with Eby and Braunstein's (1995) results with stimuli with both true and pictorial slant. For T0P30 - which is closest to viewing via a video link in that slant is only conveyed pictorially - the reduction in perceived slant was much greater at about 70%.

These impressions are confirmed in a two-way mixed-model analysis-of variance. Differences in judged orientations were all significant at the $p = 0.05$ level or better: between surfaces ($F = 17.3$;

df 3, 138; $p < 0.005$), between presence and absence of frame ($F = 4.9$; df 1,46; $p = 0.031$), and regarding the interaction between the two factors ($F = 4.0$; df 3,138; $p = 0.010$).

The framing effect therefore affected those surfaces perceived as slanted. Since TOP30, the surface whose slant was conveyed pictorially alone, elicited the greatest effect and corresponds to viewing via a video link, it can be concluded that the framing effect may be a serious problem in judging depth via a video link. The nature of this problem concerns absolute judgments of depth. In the present case, surface orientation-in-depth may be difficult to judge. In the case of discrete objects in the viewed image, their separation-in-depth may be difficult to judge. One limitation in the applicability of the present experiment is that it entailed monocular viewing: this was partly to allow comparison with the previous studies of the framing effect. However, it should be noted that a considerable proportion of the population - about 10% - lacks stereoscopic function (Sachsenweger and Sachsenweger, 1991). Furthermore, informal comments suggest that some practitioners of keyhole surgery lack stereoscopic function.

The practical importance of the framing effect also depends on the quality of depth perception required from the transmitted VDU image. If a purely *ordinal* judgment is required of discrete objects represented in the image, then often the pictorial cue of occlusion is available, or can be achieved by movement of the camera recording the site of manipulation: the more distant object is, or becomes, partly hidden by the nearer object. Such limited depth information may well be satisfactory in many cases. For example, among surgical procedures performed indirectly, cholecystectomy is popular: this is removal of the gall-bladder, which under the conditions of operation can be separated well from other tissue and structures around it (Treacy and Johnson, 1995), so enhancing occlusion and ordinal depth perception.

However, there is an ever-increasing desire for more sophisticated operations to be performed by keyhole surgery. Also, indirect manipulation has attractions in repairing complex and safety-critical structures such as aircraft wings. The problem in keyhole surgery is exacerbated by the poor transmitted image because of lens contamination: this affects keyhole surgery generally, but is likely to be more of a problem when the target tissue or structure is not well-defined (Grimbergen, 1996; Tendick *et al*, 1993). In such cases, the necessary levels of skill - including depth perception - need to be appreciated before possible training procedures can be determined and these operations become routine practice.

One issue that remains is the question of the degree to which the framing effect can be overcome by appropriate experience. Evidence suggests that training needs to be adequately devised: there is not good transfer from conventional surgery to keyhole surgery (Perkins *et al*, 2002). In part, this could be due to the presence of the framing effect only in keyhole surgery. The framing effect appears to be part of a general tendency in depth perception for participants to be affected by the *equidistance tendency* - that is, the viewed scene appears flat - in conditions not just of reduced information, but also when the task is an unfamiliar one. For example, Reinhardt-Rutland (1995b) reported that the use of an unfamiliar metric entails an increase in reported equidistance. This suggests that the framing effect may be amenable to reduction, if the appropriate experience is given.

References

Eby, D. W. and Braunstein, M. L. 1995, The perceptual flattening of three-dimensional scenes enclosed by a frame, *Perception,* **24,** 981-993

Grimbergen, C. A. 1996, Man-machine aspects of minimally invasive surgery. In A. M. L. Kappers, C. J. Overbeeke, G. J. G. Smets and P. J. Stappers (Eds.), *Studies in Ecological Psychology,* (Delft University Press, Delft), 7-10

Perkins, N., Starkes, J. L., Lee, T. D. and Hutchison, C. 2002, Learning to use minimal access surgical instruments and 2-dimensional remote visual feedback: How difficult is the task for novices? *Advances in Health Sciences Education*, **7**, 117-131

Reinhardt-Rutland, A. H. 1990, Detecting orientation of a surface: The rectangularity postulate and primary depth cues, *Journal of General Psychology*, **117**, 391-401

Reinhardt-Rutland, A. H. 1995a, Perceiving the orientation in depth of real surfaces: Background pattern affects motion and pictorial information, *Perception,* **24**, 405-414

Reinhardt-Rutland, A. H. 1995b, Verbal judgments of a surface's orientation-in-depth in degrees of angle: equidistance tendency, motion ineffectiveness, and automaticity, *Journal of General Psychology,* **122**, 305-316

Reinhardt-Rutland, A. H. 1996a, Depth judgments of triangular surfaces during moving monocular viewing, *Perception*, **25**, 27-35

Reinhardt-Rutland, A. H. 1996b, Perceiving the orientation-in-depth of triangular surfaces: Static-Monocular, moving-monocular, and static-binocular viewing, *Journal of General Psychology,* **123**, 19-28

Reinhardt-Rutland, A. H. 1999, The framing effect with rectangular and trapezoidal surfaces: Actual and pictorial slant, frame orientation, and viewing condition, *Perception,* **28**, 1361-1371

Reinhardt-Rutland, A. H. and Ehrenstein, W. H. 1996, Depth perception and stereoscopic systems for minimally invasive surgery. In S. A. Robertson (ed.), *Contemporary ergonomics 1996,* (Taylor & Francis, London), 391-396

Reinhardt-Rutland, A. H., Annett, J. M.and Gifford, M. 1999, Depth perception and indirect viewing: Reflections on minimally invasive surgery, *International Journal of Cognitive Ergonomics,* **3**, 77-90

Sachsenweger, M. and Sachsenweger, U. 1991, Stereoscopic acuity in ocular pursuit of moving objects, *Documenta Opthalmologica,* **78(1-2)**, 1-133

Stevens, K. A. and Brookes, A. 1988, Integrating stereopsis with monocular interpretations of planar surfaces, *Vision Research,* **28**, 371-386

Tendick, F., Jennings, R. W., Tharp, G. and Stark, L. 1993, Sensing and manipulation problems in endoscopic surgery: Experiment, analysis and observation, *Presence,* **2**, 111-113

Treacy, P. J. and Johnson, A. G. 1995, Is the laparoscopic bubble bursting? *Lancet,* **346** *(suppl.)*, 23

Wenzl, R., Lehner, R., Vry, U., Pateisky, N., Sevelda, P. and Hussein, P. 1994, Three-dimensional video-endoscopy: clinical use in gynaecological laparoscopy, *Lancet,* **344**, 1621-1622

VISUAL ACUITY IN NEGOTIATING THE BUILT ENVIRONMENT: EYE MOTION DATA AND SPEED

Pirashanthie Vivekananda-Schmidt[1], Anthony H. Reinhardt-Rutland[2], T. Jim Shields[3] and Roger S. Anderson[4]

[1]Department of Primary Care Sciences,
Keele University,
Staffordshire,
England ST5 5BG, UK

[2]School of Psychology,
University of Ulster, Jordanstown,
Co. Antrim,
Northern Ireland BT37 0QB, UK

[3]School of the Built Environment (FireSERT),
University of Ulster, Jordanstown,
Co. Antrim,
Northern Ireland BT37 0QB, UK

[4]School of Biomedical Science (Vision Science),
University of Ulster, Coleraine,
Co. Londonderry,
Northern Ireland BT52 1SA, UK

In a study relating to emergency evacuation of buildings, participants with normal or impaired visual acuity walked routes differing in complexity in a library, while wearing portable eye-mark apparatus. Dependent variables were speed of locomotion and number of eye motions between classes of object along the route. Speed did not differ significantly between normal and impaired acuity, but number of eye motions did in the simple route. We conclude that, while poor acuity does not necessarily have an adverse effect on locomotion, this may be at the expense of cognitive load: perhaps those with poor acuity may miss evacuation information in the stress of an emergency.

Introduction

The history of fire safety and emergency evacuation of buildings has tended to be governed by engineering considerations. However, there is increasing acceptance - not least as a result of major tragedies such as Kings Cross - that human performance must be taken into account. One issue concerns individuals with visual impairment and their ability to negotiate an evacuation route.

We are conducting a series of studies directed to simulated visual impairments of acuity, contrast sensitivity and extent of visual field, all of which properties are implicated in many forms of low vision, such as macular degeneration and cataract.

The concern of the present report is impaired visual acuity. Previous evidence suggests that impaired visual acuity has rather little effect on locomotion, compared with defects in contrast sensitivity and size of visual field (Beggs, 1991; Marron and Bailey, 1982), a result consistent with evidence regarding the visual system (Anderson, 1986). Nonetheless, it is worth re-examining the conclusion regarding visual acuity in more detail: it is important to know if there are likely to be other issues that arise under the stress of a real emergency evacuation.

To this end, we used light-weight and portable eye-mark apparatus which is now available to assess patterns of eye motions. In particular, we wanted to determine whether the seeming lack of effect on locomotion speed might be at the expense of the processing of other information in acuity-impaired individuals. A procedure employing eye-mark apparatus in assessing locomotion in low vision has been followed previously in at least one other study: Turano *et al* (2001) compared a small sample of normal-sighted participants with two participants suffering retinitis pigmentosa. For the latter, more viewing was reported as "goal directed" and "straight ahead", than was reported as "object recognition" and "layout determination". The implication was that the participants with retinitis pigmentosa devoted increased information-processing to the immediate problems of locomotion.

Methodology

Location of study

The study took place in the library at the University of Ulster, Jordanstown during a time of limited use of the library; at no time did any library user obstruct or affect the locomotion of a participant. Two routes were identified that differed in complexity as judged from the number of changes in direction required. For the chosen routes, illumination at eye level was 150 lux or above. This compares with CIBSE (1994) recommendations of 150 lux around bookshelves in libraries and 100 lux around corridors, passageways and stairs.

One route - termed *simple* - involved no change in direction except at the end of the route: the participant followed a route between two long rows of tables, until reaching a prominent supporting beam; the target was behind this supporting beam. The length of this route was 15.5 m.

The other route - termed *complex* - involved five 90° changes in direction, because of the positioning of obstacles (tables and bookshelves); again, the final part of the route involved a target hidden from view until the end of the route. The length of this route was 20.2 m.

Participants

Participants were students and staff familiar with the library. Nine had normal acuity (mean age 28 years), defined as ≥ 0.04 LogMAR units. Eleven had reduced acuity (mean age 24 years); to achieve as near uniform reduction of acuity as possible, the latter wore spectacles of between 1 and 4 D, if necessary. So adjusted, their mean acuity was 0.91 logMAR units (SD 0.13).

Apparatus

Eye motion data were obtained from an ASL eye tracking system with head mounted optics (Model 501: Applied Scientific Laboratories, Bedford MA): one camera recorded eye position from the pupillary outline and corneal reflex, while the other recorded the scene ahead of the participant. The superimposed output of the two cameras was recorded; to allow free locomotion of participants, the optics were connected via a 30m long shielded cable to the processor supplied as part of the tracking system. The latter was mounted on a movable trolley: one of the experimenters was able to pursue the locomoting participant, while pushing the trolley.

Analysis of eye motion data was aided by Observer software (Noldus, Wageningen, Netherlands), which allows frame-by-frame recording of rapidly-changing events, such as saccades.

Procedure and design

The participant was calibrated next to the starting point of the route and given directions. He/she was instructed to walk normally, look for a fire exit sign at the end of the route and stop at the sign. As intimated earlier, the fire exit sign was not visible from the start of the route, only coming into the participant's field of view at the end of the route. Completion of the route was timed by one experimenter for determination of speed of locomotion. The order of locomoting the routes was randomized over participants. A second experimenter ensured that the shielded cable did not become trapped during any trial.

The design involved two independent variables: route (repeated-measures) and acuity (independent-participants). Dependent variables included speed of locomotion and eye movements. Specifically, the latter refers to movement - particularly saccades - between the predominant obstacles: *tables*, which extend relatively low in the visual field and *bookshelves* which extend to relatively high in the visual field. Evidence suggests that relatively low obstacles can cause more information-processing problems than do relatively high obstacles (Patla *et al*, 1991), particularly for individuals with low vision (Adkin *et al.* 1996; Long *et al,* 1990). This might be evinced in increased eye motions on the tables in the present study. A third category of object - *elsewhere* - refers to objects not on the routes, such as walls and windows of the library.

Results and discussion

Speed of locomotion

Data for speed on the two routes are summarized in Table 1. Analysis-of-variance showed that the route variable was significant ($F = 19.15$, $p < 0.005$): this is consistent with the differing complexities of the routes. However, the difference between normal and reduced acuity was not significant ($p>0.05$); this is consistent with previous evidence (Beggs, 1991;

Marron and Bailey, 1982). The interaction between route and acuity was also not significant.

Table 1. Speed of locomotion through the routes (metres per second)

Route	Normal Acuity	Reduced Acuity
Simple	0.96 (0.32)	1.03 (0.29)
Complex	0.56 (0.18)	0.56 (0.16)

Eye movements: simple route
Results for numbers of eye motions between the three categories of object in the simple route are shown in Table 2.

In the simple route, bookshelves were located at the end of the route. Therefore, participants with normal acuity were making many more eye movements to the end of the route than were participants with reduced acuity. Tables were close to participants throughout the route; there was little difference between the groups regarding eye movements on to tables. Number of eye-motions elsewhere was greater for normal acuity. These findings are reflected in a significant interaction between acuity and object-type ($F=3.58$, $p<0.05$). The two main effects were not significant.

Table 2. Mean number of eye-movements between object categories on simple route

Category of objects	Normal Acuity	Reduced Acuity
Bookshelves	32.56	16.56
Tables	12.33	13.11
Elsewhere	21.89	15.00

Patla and Vickers (1997) have shown that during locomotion in normal vision people tend to plan their locomotion well before reaching obstacles. In showing that participants in the reduced-acuity group gaze less at the end of the route, the present finding indicates that their route planning is performed later than is the case of participants in the normal-acuity group. An implication is that participants in the reduced-acuity group have a heavier cognitive load than participants in the normal-acuity group.

Eye movements: complex route
Results for numbers of eye motions between the three categories of object in the complex route are shown in Table 3.

The complex route entails tables and bookshelves throughout the route, so the relationship between distance of viewing point and obstacle is not readily determined. The results suggest that the reduced acuity group make more eye-motions overall in this more complex environment, suggesting that they need to sample the route more often.

Table 3. Mean number of eye-movements between object categories on complex route

Category of objects	Normal Acuity	Reduced Acuity
Bookshelves	20.20	23.36
Tables	5.30	9.82
Elsewhere	25.1	30.82

However, neither the main effects nor the interaction regarding the complex route achieve statistical significance. Despite this, the data are consistent with those for the simple route in indicating a heavier cognitive load for the reduced acuity group.

The difference between the simple and complex routes regarding eye movement data might be explained by the nature of the decision-making processes required for the two routes. In the simple route, it is possible that a long-range ballistic response can be performed on the basis of information available at the beginning of the route: unimpaired participants would follow this strategy. However, impaired participants are required to make several short-range responses at various points along the route. In the complex route, ballistic responses are ruled out because of the changes in direction. The unimpaired participants have relatively little advantage over the impaired participants: both groups follow the complex route with short-range responses based on more-or-less continuous monitoring of the visual input from the scene.

Conclusions

We conclude that poor acuity does not adversely affect locomotion in our chosen built environment, since speeds for normal and poor acuity groups are similar. However, the changes in the number of saccades between the different areas in the visual field suggest that there may be a greater cognitive load for the reduced acuity group. The impaired participant is required to sample the scene ahead more often and over a longer time to ensure adequate negotiation of the route. An implication is that - in a real evacuation - evacuees with poor acuity may have less opportunity to detect important detail in available evacuation information, for example regarding signage.

A broadly comparable conclusion might be drawn from the study of Turano *et al* (2001) with participants suffering retinitis pigmentosa, despite their rather different categorisation of eye movement data.

References

Adkin, A.L., Patla, AE. and Elliot, D.B. 1996, Effects of age related cataract in obstacle avoidance strategies. *Proceedings of Ninth Biennial Conference,* Canadian Society for Biomechanics, Vancouver.

Anderson, G.J. 1986, Perception of self-motion – Psychophysical and computational approaches, *Psychological Bulletin,* **99**, 52-65

Beggs, A. 1991, Psychological correlates of walking speed in the visually impaired, *Ergonomics,* **34**, 91-102

CIBSE (1994), *Code for Interior Lighting*. The Chartered Institute of Building Services Engineers, London

Long, R.G., Rieser, J.J. and Hill, E.W. 1990, Mobility in individuals with moderate visual impairments, *Journal of Visual Impairment and Blindness*, **84**, 111-118.

Marron, J.A., and Bailey, I.L. 1982, Visual factors and orientation-mobility performance. *American Journal of Optometry and Physiological Optics,* **59**, 413-426

Patla, A.E., Prentice, S.D., Robinson, C. and Neufeld, J. 1991, Visual control of locomotion: strtegies for changing direction and for going over obstacles, *Journal of Experimental Psychology, Human Perception and Performance*, **17**, 603-634

Patla, A.E. and Vickers, J.N. 1997, Where and when do we look as we approach and step over an obstacle in the travel path? *Neurophysiology,* **8**, 3661-3665

Turano, D. A., Geruschat, D. R., Baker, F. H., Stahl, J. W. and Shapiro M. D. 2001, Direction of gaze while walking a simple route: persons with normal vision and persons with retinitis pigmentosa, *Optometry and Vision Science*, **78**, 667-675

A FIELD STUDY INVESTIGATING THE RELATIONSHIP BETWEEN VISUAL DISCOMFORT AND THE MENSTRUAL CYCLE IN VDU WORKERS

Stacy A Clemes and Peter A Howarth

Visual Ergonomics Research Group
Department of Human Sciences
Loughborough University
Leicestershire
LE11 3TU

Reports of visual discomfort have been shown to vary over the menstrual cycle when women have worn an HMD to view a nauseogenic virtual environment. This paper reports the results of a field study investigating whether women performing intensive VDU work, at four telephone call centres, showed similar changes. Participants were each categorised into one of four groups: i) an experimental group (naturally cycling females, n = 59), ii) oral contraceptive users (n = 38), iii) postmenopausal women (n = 21), and iv) men (n = 26). Visual discomfort was measured using a visual symptoms questionnaire completed at the beginning and end of the shift, twice a week for five weeks. Ratings of visual discomfort did not change over the menstrual cycle for the experimental group (p = 0.73), nor were there any differences between the experimental group and any of the control groups (p = 0.56).

Introduction

We have previously found differences in the amount of visual discomfort reported over the menstrual cycle when women wore a head-mounted display (HMD) to view a nauseogenic virtual reality (VR) computer game (Clemes and Howarth, 2001). A greater change in visual discomfort whilst playing the game was reported during the second week of the menstrual cycle, which mirrored the changes in nausea symptoms associated with the use of this equipment. Such variations are also consistent with other changes in sensitivity, such as hearing thresholds (Elkind-Hirsch et al., 1992), which occur over the cycle.

The purpose of this study was to determine whether visual discomfort in other situations is influenced by the menstrual cycle, with the aim of relating the findings to the changes seen with VR immersion. Individuals performing intensive VDU work, at four telephone call centres in the UK, took part in this study. Call centre workers were chosen because previous field studies have shown that individuals whose employment involves concentrated visual work exhibit changes in reports of visual discomfort over the working day (e.g. Howarth and Istance, 1985).

Early findings of this research indicated that a relationship between the menstrual cycle and visual discomfort did appear to exist, an increase in visual discomfort was reported during the second week of the menstrual cycle in 20 naturally cycling pre-menopausal females (Clemes and Howarth, 2002). However, findings from a recent laboratory study, where visual discomfort was induced by volunteers wearing negative spherical lenses to complete a task on a VDU, have shown no relationship between visual discomfort and the cycle (to be published).

This paper reports the results of the completed field study investigating this issue.

Method

Overview

Four British Telecom Directory Assistance call centres were used in this research (Figure 1). Directory Assistance operators were chosen because their job involves intensive visual work using a VDU.

Figure 1. A Directory Assistance call centre

The study received ethical approval from the Loughborough University Ethical Advisory Committee.

Participants

Participation in the study was voluntary. All volunteers were given an initial health screening which enabled them to be classified into one of four groups: i) the experimental group, consisting of pre-menopausal women not taking any form of hormonal contraception (n = 59), ii) a control group, consisting of pre-menopausal women taking an oral contraceptive (n = 38), iii) a second control group consisting of postmenopausal women (n = 21), and iv) a third control group consisting of men (n = 26). The study was run over a five week period to ensure that those in the experimental group were tested during each week of a complete menstrual cycle.

During the initial screening, participants provided written informed consent and were assigned a number. The use of these numbers ensured that participants remained anonymous throughout the duration of the study. Participants were unaware of the true aim of the study, and of being categorised into one of the above four groups.

Participants in the experimental and oral contraceptive groups initially provided information about their menstrual cycle by writing down the start date of their last menstrual period on the health screen questionnaire. It was anticipated that all pre-menopausal female participants would have started a new cycle during the five weeks of the study, and at its completion these participants each received a letter explaining the true aim of the research and requesting that they once again provide the start date of their last menstrual period. For those volunteers who did not provide the second start date (28%), the information provided at the beginning of the study was used to determine their menstrual cycle stage.

Measurement of visual discomfort

Subjective ratings of visual discomfort were collected using a standard visual symptoms questionnaire (Howarth and Istance, 1985). This questionnaire is structured so that participants first rate symptoms of visual discomfort (tired eyes, sore eyes, irritated eyes, runny eyes, dry eyes, burning eyes, blurred vision and double vision) preparing them to make an overall rating of general visual discomfort on a 7-point scale, with '1' being no discomfort and '7' being severe discomfort.

Procedure

Participants received a pack of twenty questionnaires at the outset. One of these was completed at the beginning of the shift, and one at the end, on each designated study day (Mondays and Fridays). Following the completion of each questionnaire, participants detached it from the pack and placed it in a collection box. This procedure ensured that they were not influenced by their responses on previous questionnaires.

Data analysis

Data collected from both designated study days were used in the analysis. A 'cycle week' was assigned to each questionnaire completed by the pre-menopausal women. A four-week 'pseudo cycle' (corresponding to the first four available weeks of the study) was assigned to the postmenopausal female and the male groups. Unless participants had missed a day, the value was calculated by averaging the two sets of data collected during that week.

In the experimental group, comparisons were made between each week of the menstrual cycle for the pre shift ratings, for the post shift ratings, and for the change in rating of general visual discomfort. Similarly, for the control groups the pre shift ratings, the post shift ratings and the change in rating were compared for each week of the 'pill cycle' (for the oral contraceptive group) and over the four-week pseudo cycle applied to the postmenopausal female and the male groups.

In addition, comparisons were made between the four study groups, by comparing the mean pre shift ratings, post shift ratings and the change in rating of general visual discomfort reported by each group.

Results

Experimental group
Of the 59 participants in the experimental group, 80% reported having regular 28 day menstrual cycles on the health screen questionnaire completed during the initial screening. The average menstrual cycle length of this group was 28.2 days (SD = 2.4).

Pre shift ratings - comparisons between the cycle weeks
When comparing the pre shift ratings reported during each week of the menstrual cycle, no difference between the cycle weeks was evident in the experimental group ($p = 0.66$, Friedman test). Similarly, pre shift ratings did not vary over the pill cycle in the oral contraceptive group ($p = 0.61$), or over the pseudo cycles of the postmenopausal female ($p = 0.64$) and the male groups ($p = 0.10$).

Post shift ratings - comparisons between the cycle weeks
Likewise, when comparing the post shift ratings reported during each week of the menstrual cycle, no difference between cycle weeks was evident in the experimental group ($p = 0.94$, Friedman test) or in the oral contraceptive group ($p = 0.85$), or over the pseudo cycles of the postmenopausal female ($p = 0.59$) and male groups ($p = 0.25$).

Pre and post shift ratings - comparisons between the four study groups
The responses from the four study groups did not differ from each other in the mean ratings of visual discomfort reported at the beginning of the shift ($p = 0.66$, Kruskal-Wallis test), or in the mean ratings reported at the end of the shift ($p = 0.70$, Kruskal-Wallis test).

Change in ratings - comparisons between the cycle weeks
Figure 2 displays the mean *change* in rating reported during each week of the corresponding cycle for the four study groups, and Figure 3 shows the mean change in rating reported during each week of the menstrual cycle for the experimental group alone. While a consistent decline in rating is present over the menstrual cycle, the change in rating reported over the shift did not differ significantly between the cycle weeks for the experimental group, indicating that the decline was a chance occurrence ($p = 0.73$, Friedman test). Nor did the change in rating differ over the corresponding cycle weeks of the three control groups (oral contraceptive group, $p = 0.99$; postmenopausal female group, $p = 0.80$; male group, $p = 0.13$, Friedman test).

From the responses given by the postmenopausal female and male groups (Figure 2) it can be seen that there were no order effects present, since the four-week pseudo cycle assigned to these two groups was based on the calendar week of the study.

Change in ratings - comparisons between the four study groups
When comparing the responses from the four study groups, there were no differences between the groups in the mean change in rating reported over the shift ($p = 0.56$, Kruskal-Wallis test).

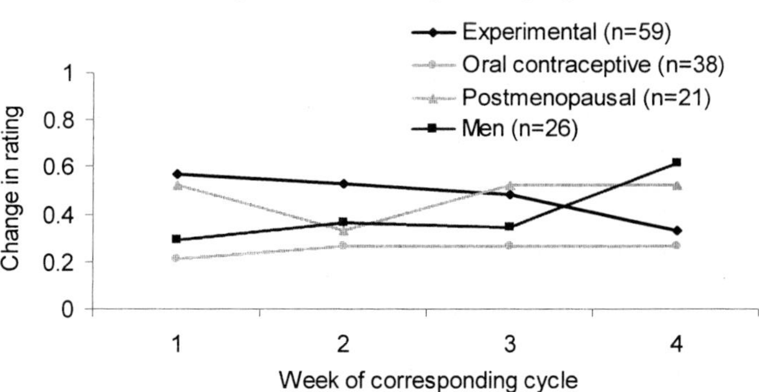

Figure 2. Mean change in general visual discomfort reported over the corresponding cycle for each group of volunteers

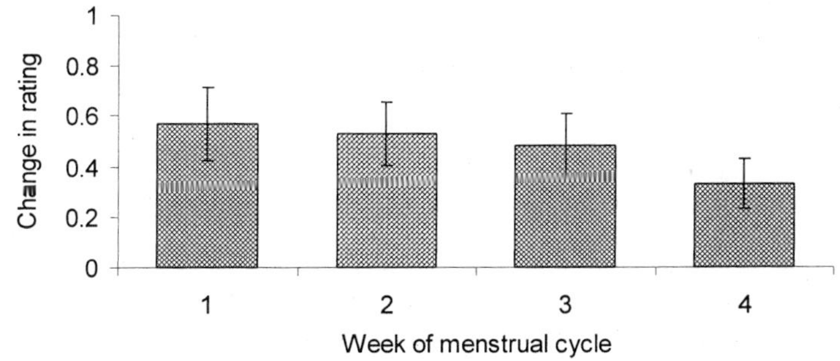

Figure 3. Mean change in general visual discomfort reported over the menstrual cycle for the experimental group

Discussion

Unlike the previously seen changes in visual discomfort reported over the menstrual cycle when women viewed a nauseogenic virtual environment through an HMD (Clemes and Howarth, 2001), the results of the current study have shown no such relationship between visual discomfort and the menstrual cycle in women performing intensive VDU work.

The change in rating of general visual discomfort reported over the shift did not vary between the weeks of the menstrual cycle ($p = 0.73$) in the 59 naturally cycling females used in this study. Similarly, when comparing the responses of the experimental group with those of the three control groups, no differences in the amount of visual discomfort reported were evident (p = 0.56).

Although the preliminary findings of this research suggested a possible relationship between the menstrual cycle and visual discomfort (Clemes and Howarth, 2002), the findings from the completed study (which includes this earlier data) indicate that no such relationship exists.

These findings support those of a recent (to be published) laboratory study where volunteers wore negative spherical lenses, to induce excess accommodation, whilst completing a simple task on a VDU. In that study the change in general visual discomfort reported over the task did not vary over the menstrual cycle in 16 naturally cycling volunteers.

In conclusion, the results of the current study indicate that the previously-seen differences in visual discomfort over the menstrual cycle with VR use were not a consequence of the performing of an intensive visual task. We suggest that the vection of the virtual reality game produced different amounts of nausea at different stages in the menstrual cycle, and this led to differences in general bodily discomfort. The change in visual discomfort is then a consequence of these general changes rather than having a specific ocular origin as might otherwise have been supposed.

References

Clemes, S.A. and Howarth, P. A. 2001, Changes in virtual simulator sickness susceptibility over the menstrual cycle, *Presented at the 36th United Kingdom Group Meeting on Human Responses to Vibration*, held at Centre for Human Sciences, QinetiQ, Farnborough, UK

Clemes, S.A. and Howarth, P. A. 2002, A field study investigating the relationship between visual discomfort and the menstrual cycle in VDU workers: Preliminary findings. In McCabe P T. (ed.) *Contemporary Ergonomics 2002*, (Taylor and Francis, London), 540-545.

Elkind-Hirsch, K.E., Stoner, W.R., Stach, B.A. and Jerger, J.F. 1992, Estrogen influences auditory brainstem responses during the normal menstrual cycle, *Hear.Res.* **60**, 143-8.

Howarth, P.A. and Istance, H.O. 1985, The association between visual discomfort and the use of visual display units, *Behaviour and Information Technology*, **4**, 131-49

Acknowledgments

We would like to thank Steve Longden and Tim Moody for giving us permission to undertake this research at the four British Telecom Directory Assistance Centres. We would also like to thank Sally Harris, Rita Ison Jacques, Margo May, Lynne Taylor, Jackie Ward and Phil Warren for their help in running the study across the four centres, and all of the employees who took part in the study.

This research was supported by the Vision Research Trust and the EPSRC.

APPLIED PHYSIOLOGY

PREDICTING THE RELATIVE METABOLIC AND CARDIOVASCULAR DEMANDS OF SIMULATED MILITARY LOAD-CARRIAGE TASKS

JA Lyons[1], AJ Allsopp[1] and JLJ Bilzon[1,2]

[1]*Environmental Medicine Unit, Institute of Naval Medicine, Alverstoke, Gosport, Hampshire PO12 2DL, UK*
[2] *Headquarters Army Training and Recruiting agency, Trenchard Lines, Upavon, Pewsey, Wiltshire, SN9 6BE*

The purpose of the experiment was to model the metabolic and cardiovascular demands of load-carriage tasks and their relationships to measures of aerobic fitness and indices of body composition. Thirty-six (*n* = 36) subjects (28 male and 8 female) walked on a treadmill at 4-kph for 60-min with 5-min on gradients of 0, 3, 6 and 9%, whilst carrying backpack loads of 0, 20 and 40-kg. During each 5-min stage, indirect respiratory calorimetry and heart rate data were collected. All male and 3 female subjects completed all phases of the experiment (*n* = 31). A model incorporating load (kg), gradient (%), gender, LBM (kg), FM (kg) and $\dot{V}O_2\,max$ (ml.min^{-1}) was produced. Indices of body composition as well as absolute aerobic power are important factors in predicting the relative metabolic demands of load-carriage.

Introduction

The ability to carry heavy loads over long distances for prolonged periods of time is a core functional competency for many military personnel, particularly the infantry soldier. Personnel must not only be capable of performing such arduous physical tasks, but must have the physical reserve to be combat fit on completion of the march (Scott and Ramabhai, 2000). The ability to accurately predict prolonged load-carriage capability using surrogate tests is essential, both for selection and the annual fitness assessment of military personnel. Many military organisations currently use non-load-carrying field tests of aerobic fitness (e.g. timed 2.4-km run, multistage fitness test) to predict fitness for task.

Anecdotal evidence suggests that such field tests (Cooper, 1968) do not accurately predict prolonged load-carriage capability of individuals. A number of studies have clearly demonstrated that such non-loading-carrying aerobic fitness tests impose a systematic bias against heavier personnel (Vanderburgh and Flanagan, 2000) who may outperform lighter subjects during prolonged load-carriage tasks (Bilzon *et al*, 2001).

Whilst body mass may be important in predicting prolonged load-carriage capability, it would seem logical, at least theoretically, that indices of body composition would be more sensitive predictors. Indeed, a high lean body mass is strongly correlated with absolute maximal oxygen uptake ($\dot{V}O_2\,max$) and has previously been shown to predict load-carriage performance (Buskirk and Taylor, 1957). In addition body fat is 'dead weight' when performing load-bearing activities and has previously been observed to impair performance of load-carriage tasks (Haisman, 1988).

The aim of this experiment was to model the metabolic and cardiovascular demands of load-carriage tasks and test the hypothesis that measures of aerobic fitness and indices of body composition (e.g. lean body mass, fat mass) will be related to these demands. Such a model, it was hypothesized, would be more sensitive than previous models of Pandolf *et al* (1977) and Duggan and Haisman (1988) in predicting the relative metabolic demands of load-carriage for an individual.

Methods

Subjects

Thirty-six (n = 36) healthy volunteers (28 male and 8 female) acted as subjects for this experiment, which was carried out with the approval of the Ministry of Defence (Navy) Personnel Research Ethics Committee. Written consent to participate was provided by all subjects after the nature of the experiment had been explained to them.

Preliminary measurements

Following a full medical examination, percentage body fat was estimated from skinfold thickness at four sites (Durnin and Womersley, 1974). Nude body mass was then determined (Sartorius ISI 20, Göttingen, Germany). The maximal oxygen uptake ($\dot{V}O_2\,max$) and heart rate (*HRmax*) of each subject was then determined during uphill treadmill running (Taylor *et al*, 1955). The preliminary tests and main experiments were performed in an environmental chamber controlled at a mean dry bulb temperature of 20°C and wet bulb temperature of 12°C, with an air velocity of 0.5 $m.s^{-1}$, giving a relative humidity of 36%.

Procedures

Each subject reported to the laboratory on a further occasion when they were required to walk at 4-kph for a total of 60-min (12 × 5-min) whilst wearing lightweight combat clothing. Prior to exercise each subject was instrumented with a heart rate monitor (Polar Vantage NV, Kempele, Finland). Each subject performed 5-min of exercise on a 0, 3, 6 and 9% gradient (20-min) whilst carrying no load. Following a 20-min recovery period subjects donned the 20-kg backpack and repeated the experimental procedure. Following a further 20-min recovery period, the experiment was repeated whilst carrying 40-kg.

During the final 3-min of each 5-min exercise period, expired gases were continously collected and recorded at 20-sec intervals (Sensormedics VMax29, Yorba Linda, CA) and heart rate at 5-sec intervals. Subjects were also asked to give their perceived rating of exertion at the end of each 5-min exercise period (Borg, 1982). From indirect respiratory calorimetry, the volume of expired gas ($\dot{V}E$), oxygen uptake ($\dot{V}O_2$) and carbon dioxide production ($\dot{V}CO_2$) were calculated. The mean $\dot{V}O_2$ and heart rate responses during each final 3-min period were used to represent the 'steady-state' metabolic and cardiovascular demand of each exercise period.

Calculations
Indices of body composition were derived, firstly by dividing LBM by FM (LBM/FM ratio) and secondly, by dividing LBM by 'dead mass' (DM), according to the following formula:

$$LBM\,/\,DM\,ratio = lean\ body\ mass\,/\,(fat\ mass + external\ load) \qquad (1)$$

Statistical analysis
Statistical analyses are based on the male and female subjects that completed the whole experiment ($n = 31$). Relationships between the subjects' physiological responses to the experimental conditions and their physical characteristics were assessed using the Pearson Product Moment correlation. Further analyses were performed using multiple linear regression techniques. Differences between group mean data were assessed by Analysis of Variance (ANOVA) for repeated measures. Differences and relationships were considered statistically significant if $P<0.05$.

Results

Metabolic demands
Relative $\dot{V}O_2\,max$ (ml.kg^{-1}.min^{-1}) produced the strongest correlation with the metabolic demand of exercise ($\%\dot{V}O_2\,max$) when the subjects were carrying no load on the 0% (r = -0.56, $P<0.01$) and 9% (r = -0.74, $P<0.01$) gradients. During the 40-kg load-carriage experiments, the correlation between the metabolic demand of exercise ($\%\dot{V}O_2\,max$) and relative $\dot{V}O_2\,max$ (ml.kg^{-1}.min^{-1}) was stronger on the 0% gradient (r = -0.64, $P<0.01$) and weaker on the 9% gradient (r = -0.67, $P<0.01$). The variables producing the strongest correlations with the metabolic demands of the 40-kg load-carriage experiment on the 0% gradient were absolute $\dot{V}O_2\,max$ (r = -0.86, $P<0.01$) and the LBM/DM ratio (r = -0.85, $P<0.01$). These variables also produced the strongest correlation when the gradient was increased to 9%.

Using multiple linear regression techniques the variables producing the strongest correlation with the metabolic demand ($\%\dot{V}O_2\,max$) of load-carriage tasks were external load, gradient, gender, $\dot{V}O_2\,max$, LBM and FM. The following equation produced a strong correlation (r = 0.94, $P<0.01$) with $\%VO_2\,max$ and a standard error of the estimate (SEE) of 3.6 $\%\dot{V}O_2\,max$, which is 10.7% of the mean metabolic demand of all of the load-carriage tasks (Fig. 1):

$$\begin{aligned}\%\dot{V}O_2\,max \;=\; & 23.62 + (0.34\times load) + (2.05\times gradient) + (4.89\times gender) \\ & - (0.0073\times VO_2\,max) + (0.26\times LBM) + (0.17\times FM)\end{aligned} \qquad (2)$$

where *load* is the external load carried (kg), *gradient* is the treadmill gradient (%), *gender* is the gender of the subject (1=male, 2=female), $\dot{V}O_2\,max$ is maximal oxygen uptake in $ml.min^{-1}$, *LBM* is lean body mass (kg) and *FM* is fat mass (kg).

Figure 1 Relationship between the actual 'steady-state' metabolic demand ($\%\dot{V}O_2\,max$) and the predicted metabolic demand ($\%\dot{V}O_2\,max$) of the load-carriage tasks using equation 2 (r = 0.94, *P*<0.01, SEE = 3.6 $\%\dot{V}O_2\,max$, *n* = 372).

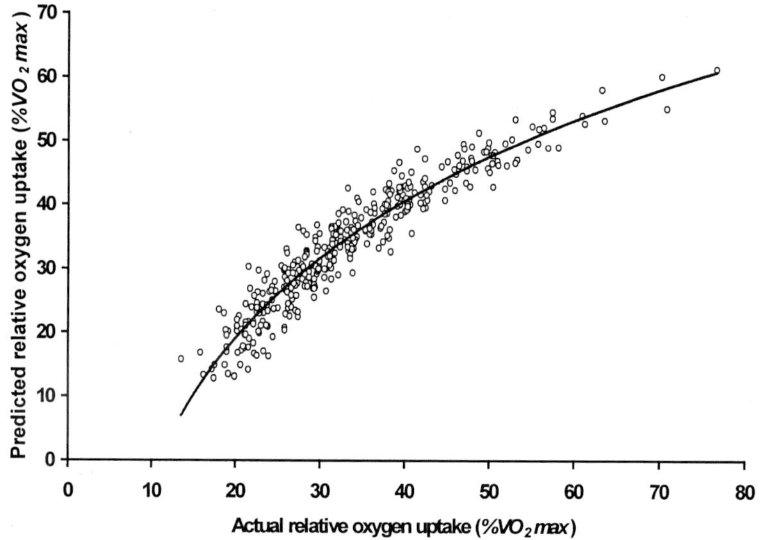

Cardiovascular demands

Relative $\dot{V}O_2\,max$ ($ml.kg^{-1}.min^{-1}$) produced the strongest correlation with the heart rate response to exercise when the subjects were carrying no load on the 0% (r = -0.40, *P*<0.05) and 9% gradients (r = -0.52, *P*<0.01). Both absolute $\dot{V}O_2\,max$ (r = -0.72, *P*<0.01) and the LBM/DM ratio (r = -0.75, *P*<0.01) produced strong correlations with 'steady-state' heart rate. The same 2 variables produced strong correlations when carrying 40-kg on the 9% gradient.

The variables producing the strongest correlation with the heart rate response to exercise were external load, gradient, gender, $\dot{V}O_2\,max$, LBM and FM. The following equation produced a strong correlation (r = 0.81, *P*<0.01) with heart rate and a standard error of the estimate (SEE) of 13.6 $b.min^{-1}$, which is 12.2% of the mean cardiovascular demand of all of the load-carriage tasks:

$$HR(b.min^{-1}) = 105.9 + (0.67 \times load) + (3.15 \times gradient) + (5.68 \times gender) - (0.0178 \times VO_2\,max) + (0.48 \times LBM) + (1.16 \times FM) \quad (3)$$

Discussion

The main finding of this experiment was that absolute maximal oxygen uptake ($\dot{V}O_2\,max$) and the LBM/DM ratio were the two variables most closely associated with the percentage metabolic ($\%\dot{V}O_2\,max$) and cardiovascular demands of heavy (40-kg) load-carriage tasks. Together these variables produced a strong correlation with the percentage metabolic demands ($\%\dot{V}O_2\,max$) of the 40-kg load carriage task (r = 0.92, $P<0.01$). In comparison, relative $\dot{V}O_2\,max$ (ml.kg^{-1}.min^{-1}), which is currently used by many military organisations to predict the fitness of personnel for load-carriage tasks, produced only a moderate correlation with the percentage metabolic demands ($\%\dot{V}O_2\,max$) of the 40-kg load-carriage task (r = -0.64, $P<0.01$). Furthermore, measures of absolute $\dot{V}O_2\,max$ (ml.min^{-1}), LBM and FM can be used in conjunction with external load, gradient and gender to produce a model which accurately predicts the percentage metabolic demands ($\%\dot{V}O_2\,max$) of load-carriage activities across a range of loads and gradients (r = 0.94, $P<0.01$). As the percentage metabolic demands ($\%\dot{V}O_2\,max$) of sustained tasks are closely associated with maximal work times, it is thought that such a model will be useful in planning critical occupational tasks which encompass sustained load-carriage (Bink, 1962).

It is well established that the energy expenditure per kg of load carried is equal to the energy expenditure per kg of body weight up to approximately 30-kg (Goldman and Iampietro, 1962). The present data are supportive of this notion, which explains why the percentage metabolic demands ($\%\dot{V}O_2\,max$) of the 0 and 20-kg load-carriage tasks were closely associated with relative $\dot{V}O_2\,max$ (ml.kg^{-1}.min^{-1}). However, it has also been suggested that when carrying increasingly heavier loads, humans become less efficient and the energy expenditure per kg of load increases sharply (Bobbert, 1960). Again, the present data are supportive of these earlier findings, showing that during the 40-kg load-carriage experiments the increase in energy expenditure per kg of load was significantly greater than when carrying 20-kg.

The inefficiency observed during heavy load carriage also explains why larger individuals, with greater total body mass and lean body mass (LBM), often outperform their lighter colleagues. Both total body mass (r = 0.76, $P<0.01$) and LBM (r = 0.91, $P<0.01$) produced strong correlations with absolute $\dot{V}O_2\,max$ (ml.min^{-1}) in the present experiment. As such, larger individuals will tend to have a greater absolute $\dot{V}O_2$ (ml.min^{-1}) reserve during load-carriage. By expressing the body composition data as a ratio of LBM to 'dead mass' (DM = fat mass + external load), strong correlations with the percentage metabolic demands ($\%\dot{V}O_2\,max$) of the 40-kg load-carriage tasks were observed. This novel body composition index has not, to our knowledge, previously been published.

Producing a multivariate model to predict the percentage metabolic demands ($\%\dot{V}O_2\,max$) of load-carriage tasks has several advantages. Firstly, such a model highlights the individual parameters which could be used to select and assess the capacity of individuals for specified occupational tasks which involve prolonged load-carriage. Secondly, it allows energy expenditure to be predicted where direct measurements are too difficult or intrusive. Finally, such a model may be useful in predicting maximal work times and devising optimal work/rest schedules for specific tasks.

From the results of this experiment it is concluded that measures of body composition and absolute $\dot{V}O_2 max$ (ml.min^{-1}) should be used, as opposed to measures of relative $\dot{V}O_2 max$ (ml.kg^{-1}.min^{-1}), when assessing the physical capacity of personnel for heavy load-carriage tasks. Such criteria will reduce the systematic bias against heavier personnel observed when using non load-carrying field tests of relative $\dot{V}O_2 max$ to predict load-carriage capability.

Acknowledgements

The authors are grateful to the Royal Marine and British Army personnel who volunteered as subjects for this experiment. This work was sponsored by the Ministry of Defence (Navy).

References

Bilzon, J. L. J. Allsopp, A. J. and Tipton, M. J. 2001., Assessment of physical fitness for occupations encompassing load carriage tasks. *J Occ Med*, **51**, 357-361.

Bink, B. 1962, The physical working capacity in relation to working time and age. *Ergonomic*, **5**, 25-28.

Bobbert, A. C. 1960, Energy expenditure in level and grade walking. *J Appl Physiol*, **15**, 1015-1021.

Borg, G. A. V. 1982, Psychophysical bases of perceived exertion. *Med Sci Sports Exerc*, **14**, 377-381.

Buskirk, E. and Taylor, H. L. 1957, Maximal oxygen intake and its relation to body composition, with special reference to chronic physical activity and obesity. *J Appl Physiol*, **11**, 72-78.

Cooper, K. H. 1968, A means of assessing maximum oxygen intake. *J Am Med Assoc*, **203**, 201-204.

Duggan, A. and Haisman, M. F. 1992, Prediction of the metabolic cost of walking with and without loads. *Ergonomics*, **35**, 417-426.

Durnin, V. G. J. A. and Womersley, J. 1974, Body fat assessment from total body density and its estimation from skinfold thickness: measurements on 481 men and women aged from 16 to 72 years. *Br J Nutr*, **32**, 77-97.

Goldman, R. F. and Iampietro, P. F. 1962, Energy cost of load-carriage. *J Appl Physiol* 17: 675-676.

Haisman, M. F. 1988, Determinants of load carrying ability. *Appl Ergonomics* , **19**, 111-121.

Pandolf, K. B., Givoni, B and Goldman, R. F. 1977, Predicting energy expenditure with loads while standing or walking very slowly. *J Appl Physiol* **43**: 577-581.

Scott, P. A. and Ramabhai, L. 2000, Load carrying: *in situ* Physiological responses of an infantry platoon. *Ergonomics* SA **1**: 18-24.

Taylor, H. L., Buskirk, E. and Henschel, A. 1955, Maximal O_2 intake as an objective measure of cardio-respiratory performance. *J Appl Physiol* **8**: 73-80.

Vanderburgh, P. M. and Flanagan, S. 2000, The backpack run test: A model for a fair and occupationally relevant military fitness test. *Mil Med* **165**: 418-421.

GENERIC TASK-RELATED OCCUPATIONAL REQUIREMENTS FOR ROYAL NAVAL (RN) PERSONNEL

James L. J. Bilzon[1,2], Emily G. Scarpello[1], Emma Bilzon[1], Adrian J. Allsopp[1]

[1] *Environmental Medicine Unit, Institute of Naval Medicine, Alverstoke, Gosport, Hampshire, PO12 2DL*
[2] *Headquarters Army Training and Recruiting Agency, Trenchard Lines, Upavon, Pewsey, Wiltshire, SN9 6BE*

The physical tests and standards used by many occupational groups have historically been arbitrary in nature, a practice that has come under increasing scrutiny in recent years. A series of experiments were therefore conducted to establish occupational task-based tests and standards (TBTs) for Royal Naval (RN) personnel. A total of 172 RN personnel volunteered for these experiments, which were designed to: identify the anthropometric requirements for operating various safety hatches and doors onboard a RN Frigate (TBT1); quantify the metabolic demands of shipboard fire-fighting tasks and establish an aerobic fitness standard (TBT2); identify a battery of tests to predict performance of shipboard casualty carrying tasks (TBT3). A battery of generic occupationally relevant tests and standards are therefore recommended to predict occupational task performance.

Introduction

Physical tests and standards have historically been used by many organisations to assess the capability of personnel for demanding occupations. This is particularly true in the military, where it is widely acknowledged that individual physical capability may directly influence the combat effectiveness of the organisation. Whilst the physical tests traditionally used by military organisations had some occupational relevance, the standards associated with them were in the main arbitrary. Arbitrary physical standards may unnecessarily discriminate against capable personnel and, more importantly, may lead the organisation to predispose individuals to tasks which they are not physically capable of performing. As such practices may result in injury to individuals and litigation against the organisation, the use of arbitrary standards has been the subject of increasing scrutiny. In determining appropriate physical tests and standards, objective criteria must be utilised in order that the 'occupational requirements' are fair, valid and justifiable.

Furthermore, the criteria should reflect the demands of the job and refer only to critical or essential components of the job (Fraser, 1992).

As a first step in identifying relevant occupational requirements (i.e. tests and standards) for the Royal Navy (RN), traditional physical test criteria were reviewed and a number of task analyses of critical job components were conducted. These critical and generic components of the job were identified as shipboard fire-fighting, casualty carrying and escaping through various hatches and safety doors onboard a typical RN vessel. Failure to perform such tasks to a minimum acceptable standard could put the individual, their colleagues and ultimately the vessel and the whole ships complement at severe risk in an operational scenario.

This paper describes a series of experiments, which were performed to develop generic task-related occupational requirements for RN personnel, otherwise known as task-based tests (TBTs). The task simulations for these experiments were endorsed by Naval 'subject matter experts'. The aim of the experiments was to develop a battery of tests and standards to be used to assess the ability of RN personnel to perform generic occupational tasks commensurate with survival at sea.

Methods

Three experiments were conducted, which aimed to identify: an anthropometric test to simulate escaping through a bulkhead door on a RN vessel (TBT1); an aerobic fitness standard commensurate with shipboard fire-fighting tasks (TBT2), and; a battery of strength and anaerobic fitness tests commensurate with shipboard casualty carrying (TBT3). All volunteers gave their informed written consent to act as subjects for these experiments, which were conducted following approval from the Ministry of Defence (Navy) Personnel Research Ethics Committee.

TBT1 (anthropometry)

Thirty healthy Royal Naval (RN) volunteers acted as subjects for this experiment. Subjects height and body mass were determined and percentage body fat was estimated from skinfold thickness at four sites. Grip strength was assessed in the dominant hand and various other body circumferences were measured. The mean (SD) physical characteristics of the male and female subjects are given in Table 1.

Table 1. Group mean (SD) physical characteristics of the male and female subjects from experiments TBT1, TBT2 and TBT3

	TBT1		TBT2		TBT3	
	Male (n = 20)	Female (n = 10)	Male (n = 34)	Female (n = 15)	Male (n = 52)	Female (n = 41)
Age (years)	29 (7)	27 (7)	26 (7)	26 (6)	28 (5)	29 (6)
Height (cm)	177 (79)	163 (53)	178 (7)	166 (5)	178 (6)	165 (6)
Body Mass (kg)	86 (13)	69 (11)	77 (11)	65 (12)	81 (9)	67 (12)
Body fat (%)	22 (5)	30 (5)	17 (4)	26 (5)	15 (5)	27 (7)
Vo_2max (ml/min/kg)			53 (5)	43 (8)	53 (8)	41 (6)

All of the experiments were performed onboard a RN Type 23 Frigate. Each subject was randomly assigned to complete three tasks as quickly as possible. The tasks were a vertical climb through an escape hatch (EH), open and secure a kidney hatch (KH) from above and below, and open and secure a bulkhead door (BD) from both sides. Each subject completed each task three times in different clothing ensembles: a fire-fighting suit and breathing apparatus (FF); a nuclear, biological and chemical defence ensemble (NBCD); and action working dress with anti-flash hood (AWD). Performance times were recorded (seconds) and the results of an observational analysis were used to determine which factors influenced performance times during each task. Individual performance times were subsequently correlated with anthropometric measurements.

TBT2 (aerobic fitness)

Forty-nine healthy RN volunteers acted as subjects for this experiment and each underwent a full medical examination. Subjects height and body mass were determined and percentage body fat was estimated from skinfold thickness at four sites. Lean body mass (LBM) and fat mass (FM) were estimated from these data. Grip strength was measured in the dominant hand and each subject completed a continuous incremental treadmill run to fatigue, for the direct assessment of maximal oxygen uptake (Vo_2max). The mean (SD) physical characteristics of the subjects are given in Table 1.

The main experiments were performed on a ship simulation unit at the RN Fire Training School. Each subject was randomly assigned to complete five 4-min simulated shipboard fire-fighting tasks, with at least 60-min recovery between each. Tasks were performed at a fixed rate commensurate with the minimum acceptable performance during operational fire-fighting. The 'steady-state' metabolic demand of each task was assessed via a portable indirect calorimetry system (Cortex Metamax, Germany). The tasks were liquid foam drum carrying (DC), extinguisher carrying (EC), boundary cooling (BC), hose running (HR) and ladder climbing (LC). The methods are described in greater detail elsewhere (Bilzon *et al*, 2001a). Statistical comparison of the metabolic demand the tasks was made using analysis of variance techniques with significant differences identified using Tukey's T-method of multiple comparisons.

TBT3 (strength and anaerobic fitness)

Ninety-three healthy RN volunteers acted as subjects for this experiment and each underwent a full medical examination. Subjects height and body mass were determined and percentage body fat estimated using four point bio-electrical impedance. Lean body mass (LBM) and fat mass (FM) were estimated from these data (see Table 1).

Subjects were randomly assigned to perform two casualty carrying simulation tasks around a predetermined circuit to simulate ladder climbing and walking the length of the deck of a RN Frigate. To simulate the free-carry (FC) a half manikin was used to simulate the head-end handler (37-kg). To simulate the stretcher carry (SC) a set of dumbbells at fixed width were used to simulate the head-end of the casualty and stretcher (41-kg). In addition to the carry simulation tasks, each subject was required to perform the following field tests of physical fitness: a timed 2.4-km run; a Multi-Stage Fitness Test (MSFT); grip strength; upright pull; standing broad jump; maximum number of pull-ups in 1-min; maximum number of press-ups in 1-min; maximum number of sit-ups in 1-min; and maximum number of 20-m shuttle sprints in 2-min. The relationship between the criterion casualty carrying rate and physical performance measures was assessed using the Pearson Product Moment correlation and multiple regression analysis. All data are presented as mean (SD) and were accepted as being statistically significant if $P<0.05$.

Results

TBT1 (anthropometry)

All 30 subjects completed all of the tasks satisfactorily and within the times given for minimum acceptable performance. The group mean (SD) performance times for each task in the different clothing ensembles are given in Table 2. It is evident from these data that the FF ensemble limited performance on the EH and KH tasks, but not the BD task. The BD task appeared to be limited by functional reach. Not surprisingly, therefore, performance time during the BD task was strongly associated with the anthropometric measurements of stature ($r = -0.62$, $P<0.05$) and reach ($r = -0.54$, $P<0.05$).

Table 2. Group mean (SD) escape times (seconds) in action working dress (AWD) fire-fighting suit (FF) and NBCD clothing (NBCD) during the escape hatch (EH), kidney hatch (KH) and bulkhead door (BD) tasks

	Escape Hatch (EH)	Kidney Hatch (KH)	Bulkhead Door (BD)
AWD	100 (28)	31 (6)	18 (5)
FF	155 (54)†	52 (10)†	20 (8)
NBCD	110 (35)	37 (8)	18 (7)

† Symbol denotes that value was significantly different from AWD and NBCD ($P<0.05$).

TBT2 (aerobic fitness)

Almost all of the subjects reported that: they had received adequate instruction (99.6%); the fire-fighting task reflected operational tasks (96.8%); and that the work rates (78.2%) were representative. With the exception of the drum carry (DC) task, where only 4 of the 15 female subjects were able to perform the task at the prescribed work rate, all subjects were able to complete the five fire-fighting tasks successfully. The task eliciting the highest *'steady-state'* metabolic demand was the DC task ($P<0.01$), which was equivalent to 82% $\dot{V}O_2$max (Table 3). The BC task elicited the lowest ($P<0.01$) metabolic demand and was equivalent to 47% $\dot{V}O_2$max. There were no differences in metabolic demand between the EC, HR and LC tasks (~78% $\dot{V}O_2$max). Further analysis revealed that the four female subjects who were able to complete the DC task successfully had a group mean $\dot{V}O_2$max of 54 ml/min/kg, which was greater than ($P<0.01$) the remaining females (39 ml/min/kg) and similar to the males (53 ml/min/kg). Furthermore, subjects with a $\dot{V}O_2$max greater than 43 ml/min/kg were able to perform each task with lower cardiovascular strain ($P<0.01$) and with more energy being derived from aerobic metabolism ($P<0.05$).

TBT3 (strength and anaerobic fitness)

Of the 93 volunteers, 52 male and 37 female volunteers completed all eleven components of the study. The relationships between FC and SC carry task performance and the physical performance and anthropometric measures are given in Table 4. Many of the physical performance and anthropometric measures produced a high correlation with FC and SC task performance. Of all the physical performance measures the standing broad jump (SBJ) produced the strongest independent correlation with FC ($r = 0.84$, $P<0.01$) and SC ($r = 0.81$, $P<0.01$) task performance. Furthermore, the derived anthropometric variable of lean body mass to dead mass (fat mass + casualty mass) ratio (LBM/DM) produced strong independent relationships with FC ($r = 0.87$, $P<0.01$) and SC ($r = 0.85$,

$P<0.01$) task performance. Using multiple linear regression techniques, the variables producing the strongest relationship with FC carry rate were, SBJ, LBM, DM, 20-m sprints, press-ups, sit-ups and grip strength ($r = 0.89$, $P<0.01$, SEE = 9).

Table 3. Group mean (SD) metabolic demand (ml/min/kg) and relative metabolic demand (% Vo_2max) of the five fire-fighting tasks

	Vo_2 (ml/min/kg)	Vo_2 (% Vo_2max)
Boundary Cooling (BC)	23 (6)‡	47 (10)‡
Drum Carrying (DC)	43 (6)†	82 (12)†
Extinguisher Carrying (EC)	39 (4)†‡	79 (9)†
Hose Running (HR)	38 (5)†‡	77 (10)†‡
Ladder Climbing (LC)	38 (5)†‡	77 (9)†‡

Symbols denote that values were significantly different from: † BC ($P<0.01$) and ‡ DC ($P<0.05$).

Table 4. Correlation coefficients based on the linear relationship between FC and SC task performance ($m.s^{-1}$) and physical performance measures (n = 89)

	Free Carry (FC)	Stretcher Carry (SC)
Body Mass (kg)	0.40‡	0.42‡
Body Fat (%)	-0.75‡	-0.73‡
Lean Body Mass (kg)	0.76‡	0.76‡
LBM/DM ratio	0.87‡	0.85‡
Height (cm)	0.64‡	0.65‡
SBJ (cm)	0.84‡	0.81‡
Grip Strength (N)	0.71‡	0.71‡
Upright Pull (N)	0.77‡	0.79‡
MSFT (Vo_2max)	0.67‡	0.67‡
2.4-km run (Vo_2max)	0.62‡	0.62‡
20-m shuttle sprints (n)	0.60‡	0.56‡
Sit-ups (n)	0.56‡	0.58‡
Press-ups (n)	0.69‡	0.70‡
Pull-ups (n)	0.72‡	0.72‡

Symbol denotes a significant relationship with carry task performance: ‡ $P<0.01$.

Discussion

The results of these experiments demonstrate that, where it would be dangerous, expensive or impractical to directly assess the occupational fitness of Royal Naval (RN) personnel, their effectiveness may be determined by the use of surrogate tests and task simulations. Taken together as true occupational requirements, the battery of physical tests and standards proposed would ensure that, in the main, successful employees would be capable of critical, generic sea survival duties. Whilst the use of such tests and standards would indirectly discriminate against a greater percentage of the female population, it should be remembered that the same people would not be capable of performing the 'true' occupational tasks, placing them at graver risk of injury. With the exception of a proposed reach test, there are very few males or females that could not achieve the required standards for TBT2 and TBT3 with appropriate physical training.

Whilst all of the subjects recruited for the first experiment (TBT1) were able to perform the tasks within the required time, it was apparent that shorter subjects with less reach struggled to perform the escape through a bulkhead door task. However, the ability to perform this task is likely to depend on several variables (e.g. height, arm length, flexibility, grip strength), and it is therefore impossible to suggest a height requirement at which employees will fail the task. It is therefore proposed that a direct simulation of the ability to reach and operate the top clip of the bulkhead door be used at the recruitment stage, to directly assess the ability to operate safely onboard a RN vessel (TBT1).

From the results of the experiment into the metabolic demand of simulated shipboard fire-fighting tasks, it has also been possible to propose an aerobic fitness standard, which is not substantially different from those proposed for civilian fire-fighting. RN Personnel are required to sustain a combination of the hose running, ladder climbing and boundary cooling tasks whilst wearing full fire-fighting clothing and breathing apparatus. The mean metabolic demand of these three tasks was 32.8 ml/min/kg Vo_2. Personnel must be able to sustain such levels of activity for 20-30 min. Fire-fighters wearing self-contained breathing apparatus is able to sustain a maximum work intensity equivalent to 80% Vo_2max for a 20-30 min duration (Louhevaara *et al*, 1986). Therefore, it is recommended that all RN personnel attain a Vo_2max of 41 ml/min/kg (TBT2).

The aim of the third experiment was to identify a battery of physical tests, which would accurately predict performance of casualty carrying tasks performed by seagoing RN personnel (TBT3). This general approach is supported by the fact that one of the subjects sustained an injury during the task simulations. The use of surrogate tests to predict performance is therefore more desirable than a true task simulation. Physiologically, the tests which produced the strong multivariate relationship with FC carry task performance, seem logical. A high lean body mass relative to the 'dead mass' being carried (fat mass + casualty mass), has previously been observed to influence carry task performance (Bilzon *et al*, 2001b). The minimum acceptable performance time for the casualty carrying tasks is 3-min, which is equivalent to an actual FC carry rate of 34 (regression analysis). A score that should be achieved by all RN personnel (TBT3).

In conclusion, the battery of task simulations (TBT1) and surrogate tests (TBT2 and TBT3) proposed would ensure that seagoing RN personnel have the physical capacity commensurate with the performance of critical and generic sea survival duties. RN training should be tailored toward achieving these occupational requirements.

References

Bilzon, J.L.J., Scarpello, E. G., Smith, C. V., Ravenhill, N. A. and Rayson, M. P. 2001a, Characterization of the metabolic demands of simulated shipboard Royal Navy fire-fighting tasks, *Ergonomics*, **44**, 766-780

Bilzon, J.L.J., Allsopp, A.J. and Tipton, M.J. 2001b, Assessment of physical fitness for occupations encompassing load-carriage tasks, *J Occ Med,* **51**, 357-361

Fraser, T. 1992, *Fitness for Work*, (Taylor and Francis, London)

Louhevaara, V., Smolander, J., Korhonen, O. and Tuomi, T. 1986, Maximal working times with a self-contained breathing apparatus, *Ergonomics*, **29**, 77-85

Acknowledgements

The authors are grateful to the RN personnel who volunteered as subjects for these experiments. This work was sponsored by the Ministry of Defence.

THE PHYSICAL DEMANDS OF ARMY BASIC TRAINING

Mark P. Rayson, David A. Wilkinson, Erik Valk & Alan M. Nevill

Optimal Performance Limited,
Old Chambers, 93-94 West Street
Farnham, Surrey GU9 7EB
United Kingdom

This study investigated the physical demands of basic training among male and female recruits in different Arms. Recruits were monitored throughout their waking day for three 3-day periods. The physical demands were found to be high. While the aerobically fittest recruits coped with moderate cardiovascular strain of 25-30%VO_{2max}, the least fit worked at a high level of 35-40%VO_{2max}. Physical Activity Levels were high, but not sufficient to challenge energy balance. Energy Expenditure was higher in the males than the females (16.1 v 12.1 $MJ.day^{-1}$), but physical activity levels did not differ. Female recruits experienced much greater cardiovascular strain than their male colleagues due to lower fitness levels. Body mass remained unchanged with a small 'exchange' of fat mass for fat free mass. The small energy deficit of 0.13 $MJ.day^{-1}$ was viewed as a positive adaptation to training.

Introduction

Over the last decade the British Army has made great progress in its policy and practice towards the wider integration of female soldiers. The implementation in 1998 of job-related gender-free physical selection standards (PSS(R)) for British Army applicants and the integration of men and women undergoing the Common Military Syllabus for Recruits (CMS(R)) imposed a common selection and training process on both genders. While this policy provides Equal Opportunities to both genders it has come at a cost of higher injury rates and medical discharges of female recruits.

There have been several studies in recent years investigating the effectiveness of the training programme and exploring the prevalence of injury and medical discharge during CMS(R). Rayson et al. (1996) showed that improvements in the various measures of fitness varied considerably, but averaged between 0 and 10% - fairly modest improvements given the potential that 10 weeks of training offers. Mason et al. (1996) confirmed suspicions that too much was being required of recruits. Typically recruits spent more than 3 hours per day at heart rates above 60% of their maximum heart rate (HR_{max}). The least fit recruits spent 6 hours of their 12-hour day above 60%HR_{max} placing them in the cardiovascular training zone for 6 hours (ACSM, 1998). Williams, Rayson & Jones (1999) investigated potential modifications to the Physical Training (PT) syllabus during CMS(R), the major new ingredient being formal resistance training. The modified PT programme brought about greater improvements in many of the fitness scores and military task performance than the control programme.

Gemmell (2000) conducted a retrospective study of medical discharges (MD) from CMS(R) over 2 periods of 9 months before and after the implementation of PSS(R). The change resulted in a small rise in MD during the first 6 weeks of CMS(R) from 4.4% to

5.8%, discharges that were attributed to Defect on Enlistment. Of those who remained and passed the 6-week point, MD rates rose from 3.9% to 6.1%. MD due to overuse injuries rose from 1.6% to 2.4%. In females MD due to overuse injury rose from 4.7% to 11.1%, compared to marginal increases in the males from 1.2% to 1.5%.

Casey & Owen (2001) measured energy expenditure (EE) and energy intake during Induction Week and the first 4 weeks of CMS(R). They reported mean EE to be 16.8 MJ (4000 kcal) per day and estimated that recruits were in chronic energy deficit amounting to about 25% of EE (4.2 MJ/day or 1000 kcal/day). However, it seems unlikely that such a huge energy deficit existed, as the resultant mass loss would have been around 1 kg per week or 12 kg over CMS(R). Further, this finding does not concur with most of the previous studies on CMS(R), most of which showed a gain in body mass.

Method

The objectives of this study were to assess the physical demands of CMS(R) and to comment on their acceptability, and to make comparisons between the demands on recruits of different gender, fitness and in different Arms. We monitored four platoons undergoing basic training at Army Training Regiment Winchester, comprising 1 Infantry (Inf) Light Division Platoon (all male) and 3 Adjutant General's Corps (AGC) Platoons. The 3 AGC Platoons comprised 1 all male, 1 all female, and 1 mixed gender Platoons. Recruits in all 4 platoons were briefed orally, provided with written information sheets and asked to sign consent forms. A sub sample of up to 20 recruits from each Platoon was then selected to participate. Permission to conduct the study was provided by the University of Birmingham's School of Sport and Exercise Science Ethics Sub-Committee and by the military authorities. The data collection phase spanned from 4 June to 12 October 2001. We collected data on all participants during waking hours (approximately 06:00 to 22:00) for 3-day periods on 3 separate occasions during weeks 1, 6 and 9 of CMS(R). In the mixed gender AGC Platoon we performed some more detailed measurements of energy expenditure for 10 consecutive days in weeks 1 & 2 and 8 & 9.

Height, body mass and body composition were measured on the day prior to the start of the measurement Period 1 in week 1. Body mass and body composition were measured again on the final measurement day in Period 3 in week 9. Body mass was measured on Seca digital scales (Cranlea Limited, Birmingham, UK) and body composition was estimated using the Bodystat 1500 electrical impedance system (Bodystat, Douglas, Isle of Man). We used Tracmor 3 accelerometers (Philips, the Netherlands) to monitor physical activity and to estimate EE in those not taking isotope, to provide an index of stress, or *absolute* load on individual recruits. Tracmor 3 - a prototype device - is reportedly "the most widely validated system for detecting body movement in free-living subjects" (Levine, Baukol & Westerterp, 2001) and is the smallest and lightest of all such devices (20 x 70 x 8 mm, 30 g). Validation studies using the predecessor Tracmor 1 and 2 have confirmed the validity of this piezo-electric technology in various populations (e.g. Meijer et al., 1989; Bouten et al., 1996; Pannemans et al., 1995; Westerterp & Bouten, 1997). The device has three uni-axial motion sensors, mounted orthogonally. The device was attached to the subject using an elasticated belt worn around the waist. Tracmor data were recorded continuously and downloaded to a PC at the end of the data collection periods. From these data, Physical Activity Counts (PAC) from the x, y and z-axes and Total Energy Expenditure (TEE) in mega Joules (MJ) were calculated. Basal Metabolic Rate (BMR) was estimated using the equation of Schofield (1985) incorporating body mass. Physical Activity Level (PAL) is a worldwide-accepted index of physical activity, expressed as a multiple of BMR (World Health Organization, 1985).

We used the Polar Team System (BodyCare, UK) Heart Rate (HR) Monitors to monitor cardiovascular response, providing an index of strain, or *relative* load on individual recruits. HR indices were calculated according to the recommendations of Howley (2001) and the ACSM (1998). Typically, we have cited the duration (minutes) and proportion of duration (%) spent above the %HRR thresholds during the whole day (approximately 06:00 to 22:00), and the working day (08:30-17:10).

Results and Discussion

Table 1 shows the physical and physiological characteristics of the recruits. There were no differences in any of the variables between platoons if gender is accounted for. However, the males were taller (1.75 ± 0.08 vs 1.66 ± 0.05 m, $p<0.001$), heavier (67.7± 8.6 vs 61.8 ± 7.1 kg, $p=0.004$), had a lower % body fat (10 ± 4 vs 23 ± 4 %, $p<0.001$) and ran 2.4 km faster (597 ± 44 vs 767 ± 65 s, $p<0.001$) than the females.

Table 1. Physical and physiological characteristics of the recruits (Mean ± SD).

Platoon	Mixed AGC		Male AGC	Male Inf	Female AGC
	Male	Female			
Number	9	9	17	18	19
Age (yr)	21.7 ± 4.1	19.2 ± 2.9	21.6 ± 3.9	18.9 ± 2.4	19.6 ± 3.4
Height (m)	1.74 ± 0.05	1.67 ± 0.04	1.77 ± 0.08	1.74 ± 0.09	1.66 ± 0.06
Mass (kg)	68.1 ± 10.5	60.4 ± 7.8	68.0 ± 7.9	67.1 ± 8.8	62.4 ± 6.9
Body fat (%)	12 ± 4	22 ± 3	11 ± 4	8 ± 4	24 ± 4
2.4 km run (s)	595 ± 45	746 ± 72	609 ± 40	588 ± 48	777 ± 61

Estimated percentage of maximum oxygen uptake sustained (%VO_2max)

Figure 1 shows the range of %VO_{2max} sustained by all recruits over each 3-day monitoring period during CMS(R). The mean %VO_{2max} over the *whole day* was 37 ± 4% during week 1 (N=66), 36 ± 5% during week 6 (N=50) and 33 ± 5% during week 9 (N=40). These mean values exceed the conservative limit of 33%VO_{2max} (Mital and Shell, 1984; Haisman, 1987) and are below the upper limit of 40%VO_{2max} (Astrand and Rodahl, 1986) for an 8-hour day that we have adopted in this report. While on first consideration the proportion of time spent above these thresholds may seem reassuring and reasonable, it should be born in mind both that these figures refer to a time period of approximately double the standard 8-hour working day and they refer to *mean* values for the group.

To compare %VO_{2max} across the three data collection periods, data from 39 recruits who completed the entire prescribed CMS(R) programme over the 9 days were used. There was a difference in %VO_{2max} over the three monitored periods, with weeks 1 and 6 being higher than week 9 (36 ± 4, 36 ± 5 and 33 ± 5%VO_{2max} respectively, $p<0.001$).

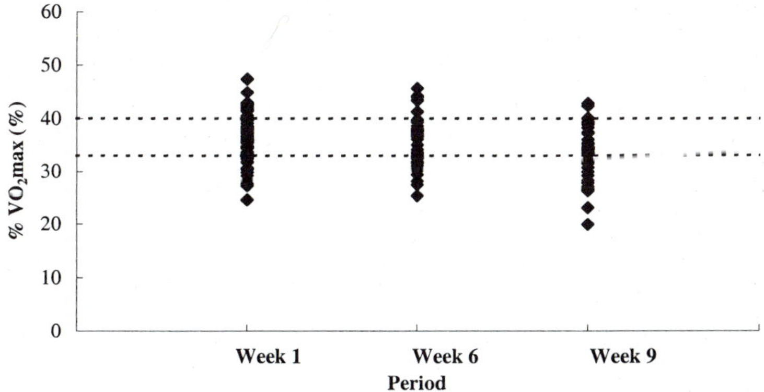

Figure 1. %VO_{2max} in all recruits over each 3 whole day period during CMS(R)

Figure 2 shows the range of $\%VO_{2max}$ sustained during the *working day* by all recruits over each 3-day monitoring period during CMS(R). The mean $\%VO_{2max}$ was 37 ± 4% during week 1 (N=66), 38 ± 6% during week 6 (N=50) and 35 ± 5% during week 9 (N=40). During week 1, 89% of all recruits spent their working day above the conservative limit of $33\%VO_2max$ and 27% spent their working day above the upper limit of $40\%VO_{2max}$. Even during week 9 – the least demanding of the 3 periods monitored - 27% were still above $33\%VO_{2max}$ and 7% were above $40\%VO_{2max}$. Sustaining these work intensities over prolonged working hours per day and over the extended period of 12 weeks of CMS(R) is likely to result in chronic fatigue, increasing the risk of injury.

To compare $\%VO_{2max}$ during the working day across the three monitoring periods during CMS(R), data from 39 recruits who successfully completed all the prescribed CMS(R) programme over the 9 days were used. There was a difference in $\%VO_{2max}$ over the three monitored periods, with weeks 1 and 6 being harder than week 9 (37 ± 4, 38 ± 6 and 35 ± 5 $\%VO_{2max}$ respectively, $p<0.03$).

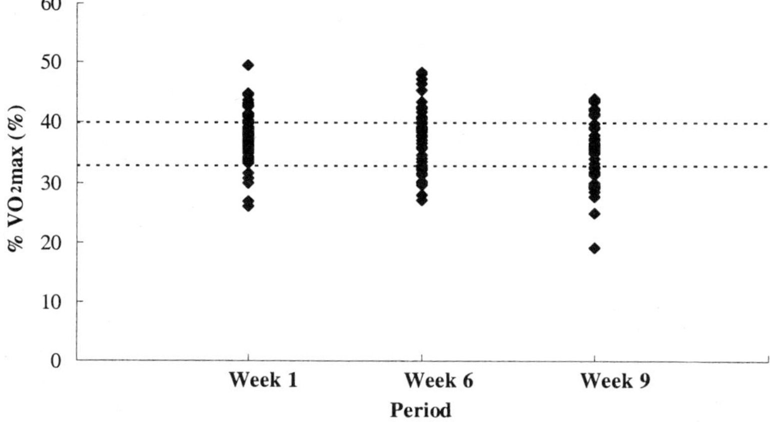

Figure 2. $\%VO_{2max}$ in all recruits in each 3 working day period during CMS(R)

Physical Activity Count, Energy Expenditure and Physical Activity Level

The PAC recorded during the whole day from the Tracmors (cumulative 3 day totals) from all recruits were $3.5x10^{10} \pm 7.7x10^9$ (week 1, N=69), $3.8x10^{10} \pm 7.7x10^9$ (week 6, N=40) and $3.8x10^{10} \pm 9.8x10^9$ (week 9, N=34). Using the equation developed by Rayson and Wilkinson (2001) to estimate EE, these values correspond to 14.3 ± 1.8, 14.8 ± 1.1 and 14.8 ± 2.1 $MJ.day^{-1}$ respectively, providing an overall mean daily EE of 14.6 $MJ.day^{-1}$ or 3480 $kcal.day^{-1}$. Estimated EE in individual recruits ranged from 10.7 to 19.6 $MJ.day^{-1}$.

When PAC values were converted to PAL, the respective mean PAL values were 2.1 ± 0.2, 2.1 ± 0.1 and 2.1 ± 0.2 for weeks 1, 6 and 9. In the general population PAL ranges between 1.2 and 2.2 to 2.5. Bouten et al. (1996) defined categories of activity, where a PAL of less than 1.60 equated to 'low', between 1.60 and 1.85 equated to 'moderate', and greater than 1.85, as 'high'. Westerterp has suggested that it is not until PAL values reach 2.5 that people have problems maintaining energy balance. The mean values of 2.1 that we found fall into Bouten's 'high' activity category, but were below the threshold value for maintaining energy balance suggested by Westerterp. PAL ranged in individual recruits from 1.80 to 2.56 (3-day period averages). Three recruits exceeded a PAL of 2.5 on the 3-day averages. Although they were over the theoretical limit for sustaining energy balance, they represent only 2% of the data, only exceeding the threshold value of 2.5 on a single 3-day period. Although PAL is high and EE is moderately high during CMS(R), there does not appear to be any widespread problem with unsustainably high physical activity levels for maintaining energy balance.

Body composition and aerobic fitness

Over the course of the first 9 weeks of CMS(R), body mass remained unchanged in the 50 recruits who remained in training (65.9 ± 8.9 vs 66.3 ± 8.1 kg, p=0.406). Estimated fat mass decreased by 1 kg on average from 9.0 ± 4.7 to 8.0 ± 3.5 kg (p=0.002) while estimated lean body mass increased by 1.3 kg on average (57.0 ± 9.3 to 58.3 ± 8.4 kg, p<0.001). Assuming an energy cost of protein synthesis of 21 $MJ.kg^{-1}$ and an energy expenditure of 35 $MJ.kg^{-1}$ for adipose tissue metabolism, a loss of 1.0 kg adipose tissue and a gain of 1.3 kg of lean muscle mass would equate to a net energy deficit of approximately 7.7 MJ. Over the 9-week monitoring period, this would equate to an energy deficit of approximately 0.13 $MJ.day^{-1}$ (30 $kcal.day^{-1}$). This finding conflicts with the data reported by Casey and Owen (2001). Our findings suggest there was a very small energy deficit over the course of CMS(R). The changes in body composition, with an 'exchange' of fat mass for fat free mass, are viewed as a positive adaptation to training, not a concern.

The 2.4 km run used to assess the change in aerobic fitness during CMS(R). Run time improved by 5.6% from 647 ± 82 to 611 ± 67 s (p<0.001) in the 46 recruits who performed both tests. There was no difference in the percentage improvement between the males in any of the platoons or between the females in either of the platoons. The females showed greater improvements (7%) than the males (4%) (p<0.05). 5.6% is a fairly modest improvement to show for 11 weeks of physical training, especially given that run time should reflect improvements in VO_{2max}, lactate threshold and pace judgment. Using these run times to estimate VO_{2max} we derived values of 46.8 and 48.5 $ml.kg^{-1}min^{-1}$, which equates to an improvement of only 3.6% in VO_{2max}. While this value is not dissimilar to the 2.7% increase in directly measured VO_{2max} in 131 male recruits at ATR Bassingbourn (Rayson et al, unpublished), it demonstrates the relative ineffectiveness of CMS(R) to improve aerobic fitness, perhaps the single most important aspect of fitness to military performance.

Physical demands between Arms (male AGC vs male Inf)

Surprisingly, the male AGC worked at a higher %HRR than the male Inf (32 ± 3 vs 28 ± 3%, p=0.013) during the whole day over the 9-day period. The mean %HRR was similar in week 1 (33 ± 3 vs 33 ± 4%HRR, p=0.760), but was higher for the male AGC recruits during week 6 (34 ± 5 vs 29 ± 4%HRR, p=0.006) and week 9 (29 ± 4 vs 25 ± 5%, p=0.010). The same findings were apparent during the working day. The total PAC, estimated EE and PAL for the 9 data collection days were not different between platoons. The difference in cardiovascular strain might be attributed to small differences in fitness, or a differing work ethic of the platoon staff.

Physical demands between genders

As anticipated, the male AGC had a lower mean %HRR (24 ± 2%HRR) for the whole day over the 9-day period compared to the female AGC (33 ± 2%HRR, p<0.001), equating to a 38% premium among the females. This difference was evident in all three periods of data collection. The same findings were apparent during the working day, though the difference was less, with the mixed male AGC working at a lower %HRR than the female AGC (26 ± 2% vs 31 ± 5%HRR, p=0.020) equating to a 19% premium for the females. These differences are primarily due to the lower level of aerobic fitness in female recruits. The total PAC and PAL were not different between genders, but the estimated EE was (16.1 ± 0.5 vs 12.1 ± 0.3 $MJ.day^{-1}$, p<0.001), primarily due to the greater body mass and resting metabolic rate of the men.

Physical demands in recruits of different fitness

Within the mixed AGC platoon there was a positive correlation between 2.4 km run time and %HRR during both the whole day and the working days (r=0.54 and 0.49, respectively, both p<0.05) during week 1. While the fastest recruits operated over 15 hour days at 25-30 $\%VO_{2max}$, the slowest recruits operated at 35-40 $\%VO_{2max}$. This level of cardiovascular strain in the least fit exceeds guidelines for an 8-hour day. There were no differences in PAC (p=0.507) or PAL (p=0.327) suggesting that all recruits, irrespective of fitness perform the same amount of physical activity in both absolute terms and relative to their resting metabolic rates.

This work was supported by the Ministry of Defence. We would like to convey our acknowledgements and thanks to our army client, the staff and recruits at ATR Winchester, Prof Klaas Westerterp and Guy Plasqui at the University of Maastricht, Lynsey Wyatt, and Philips Research in Eindhoven.

References

ACSM 1998. The recommended quantity & quality of exercise for developing & maintaining cardiorespiratory and muscular fitness, and flexibility in healthy adults. *Medicine & Science in Sports and Exercise*, **30** (6), 975-991

Astrand P.O. & Rodahl K. 1986, *Textbook of Work Physiology*, Third Edition, (McGraw-Hill International Editions)

Bouten C.V., Verboeket-van de Veene W.P., Westerterp K.R., Verduin M & Janssen J.D. 1996, Daily physical activity assessment: comparison between movement registration and doubly labeled water. *J. Appl. Physiol.* 81: 1019-1026

Casey A. & Owen G. 2001, *Nutrition, stress and injury during Army Phase 1 Training.* DERA/CHS/PPD/TR010145, DERA, Farnborough, UK

Gemmell I.MMcK. 2000, *Overuse injury patterns in British Army recruits following changes to selection and training techniques*. Thesis, The Faculty of Occupational Medicine of The Royal College of Physicians, London

Haisman M.F. 1988, Determinants of load carrying ability. *Applied Ergonomics*, 19, 2, 111-121

Howley E.T. 2001, Type of activity: resistance, aerobic and leisure versus occupational physical activity. *Med. Sci. Sports Exerc.* 33 (6), S364-S369

Levine J.A., Baukol P.A. & Westerterp K.R. 2001, Validation of the Tracmor Triaxial accelerometer system for walking. *Med. Sci. Sports Exerc.* 33 (9), 1593-7

Mason M.J., N'Jie M., Holliman D. & Rayson M.P. 1996, *The physical demands of basic training: a pilot study*. DRA/CHS(HS2)/CR96/019. DRA, Farnborough, UK

Meijer G.A., Westerterp K.A., Koper H., Ten Hoor F. 1989. Assessment of energy expenditure by recording heart rate and body acceleration. *Medicine & Science in Sports and Exercise*, **21**, 343-347

Mital A. & Shell R. 1984, Determination of rest allowances for repetitive physical activities that continue for extended hours. In, *Proceedings of the Annual International Conference, Industrial Engineers*, 637-645

Pannemans D.L., Bouten C.V., Westerterp K.R. 1995, 24h energy expenditure during a standardized activity in young & elderly men. *Eur. J. Clin Nutr.*49(1),49-56

Rayson M.P. & Wilkinson D.A. 2001, *Energy Expenditure during CMS(R).* Technical Report dated 20 November 2001, Optimal Performance Limited, Farnham, UK

Schofield W.N. 1985, Predicting basal metabolic rate, new standards and review of previous work. *Hum. Nutr. Clin. Nutr.* 39 (Suppl. 1), 5-41

Westerterp K.R. 1999, Physical activity assessment with accelerometers. *Int. J. Obesity* **23**, Suppl 3, S45-S49

Westerterp K.R. & Bouten C.V. 1997, Physical activity assessment: comparison between movement registration and doubly labelled water method. *Z. Ernahrungswiss* 36(4), 263-7

Williams A.G., Rayson M.P. & Jones D.A. 1999, The effects of basic training on material handling ability & physical fitness of BA recruits. *Ergonomics*, 42, 8, 1114-1124.

WHO 1985, *Energy and protein requirements.* Report of a joint FAO/WHO/UNU Expert Consultation. Geneva, WHO (Technical Report Series 724).

ESTIMATING ENERGY EXPENDITURE USING GLOBAL POSITIONING SYSTEM (GPS)

Mark P. Rayson

Optimal Performance Limited,
Old Chambers, 93-94 West Street,
Farnham, Surrey GU9 7EB
United Kingdom

This paper describes a pilot study to assess the feasibility of using an innovative technique involving GPS data collection and map overlays to monitor the movement and estimate the energy expenditure of people working in a field environment. Using data from a team of four soldiers carrying out patrolling tasks in a rural area we were able to demonstrate the viability of the technique. Data on distances, gradients and speed of the soldiers are presented, together with estimated power and oxygen uptake, enabling discussion about absolute and relative work intensities and the sustainability of the work rate. This method appears to offer a promising way ahead for monitoring work or exercise in a field environment, though at present it remains technically challenging and as yet, unvalidated.

Introduction

Measuring movement and energy expenditure (EE) during occupational tasks, especially in a field environment has always presented a challenge to ergonomists. A number of options exist but all have shortcomings. The method that is widely regarded as the gold standard of measuring EE is the doubly labelled water method, which measures the elimination rate of two isotopes from the body (Westerterp and Saris, 1992). However, the high cost of this method and the need for access to isotope ratio mass spectrometry makes it unsuitable for many ergonomists to use. Consequently, to date, physiologists have been limited to a number of indirect measures of EE to estimate physical demand. These have included heart rate monitoring (ideally calibrated for each individual against oxygen uptake during each specific activity), indirect calorimetry (oxygen uptake) measured in the field, and pedometers which measure the number of steps taken. However, there are shortcomings associated with these techniques also, including user-friendliness, robustness, and reliance upon expensive technology and expert staff etc.

Innovative motion sensing technology involving accelerometers using piezo-electric sensors offers the potential to overcome many of these limitations. Accelerometers have been shown to be useful and valid tools for monitoring activity in free-living adults (e.g. Meijer et al, 1989; Westerterp, 1999). The small devices are worn around the waist and measure accelerations and decelerations caused by movement of the body, providing both the occurrence and intensity of movement. The data are stored within the self-contained device and downloaded at a suitable occasion by a connecting cable to a PC where they are converted a Physical Activity Count and possibly to EE. We have just completed validating a European-manufactured device for just that purpose (Rayson and Wilkinson, 2001). However, there are shortcomings too with accelerometers – they are insensitive to static effort, upper body movement, to gradients, and to external loads that are carried.

This paper explores the use of Global Positioning System (GPS) technology to monitor movement and, we believe for the first time, to estimate EE. The objective of this pilot study was to explore the viability of using GPS in the field to estimate EE in soldiers performing patrolling in a rural environment.

Method

24 trained soldiers were monitored over a 6-day military exercise in Scotland. In this paper, detailed data are presented from 4 of the group only, during a single physically demanding task lasting approximately 100 minutes, to demonstrate the concept.

The battery powered data-logger which was attached to a GPS receiver engine was housed in an aluminium box and carried in the top of each soldier's rucksack such that the receiving aerial was facing upwards. The data-logger recorded binary information from the receiver to save memory space. At the end of each day the data were downloaded to a computer and processed from the binary to Geographicals to Orthogonals (GB Ordnance Survey grid). An AXIS2000 GIS Application System was designed to import these data into a data set, which could then be displayed overlaid on mapping (often ortho-rectified aerial survey photography) and could be analysed and edited. The AXIS system afforded considerable power of analysis using a special geographically-orientated scripting language. To obtain more accurate information about vertical movement than GPS alone provides (required for the estimation of EE) the GPS coordinates were used to interpolate the ground levels from a 10m-grid data set of the surface. Output was then generated which included coordinates, height, time, distance-from-last-point (2D and 3D), time-interval, speed (m/sec; km/hr), rise-in-ground-level, and percent gradient into a convenient output of comma-separated values (csv). These values were fed into an equation (Pandolf et al., 1977) requiring body mass (measured as nude body mass), load mass (all external load including clothing, weapon and equipment), terrain factor (assumed as 1.1 equating to 'light brush'), velocity (derived from the GPS data) and gradient (derived from the GPS overlay onto contoured maps), to estimate power and oxygen uptake.

$$\text{Metabolic cost (watts)} = 1.5*W + 2*(W+L) * (L/W)^2 + T*(W+L) * (1.5*V^2 + 0.35*V*G)$$

Where: W=body mass (kg); L=load mass (kg); T=terrain factor; V=velocity (m/s); G=gradient (%)

Results

The subject data are shown in Table 1 – age, height, body mass, percent body fat, fat free mass, VO_{2peak} in both $l.min^{-1}$ and $ml.kg^{-1}.min^{-1}$, and load both in absolute units (kg) and as a percentage of body mass are shown.

Table 1. Subject characteristics and load carried

subject	age	height	body mass	fat	ffm	VO_{2peak}	VO_{2peak}	load	load
#	yr	mm	kg	%	kg	$l.min^{-1}$	$ml.kg^{-1}min^{-1}$	kg	%BM
14	22	1680	68.2	14.7	58.2	2.72	45.8	31.0	45.5
18	19	1655	53.6	6.7	50	2.73	50.9	34.3	64.6
21	23	1810	76.6	10.1	68.9	3.68	48.1	35.2	46.0
23	17	1740	72.2	12.2	63.4	3.53	48.9	30.6	42.0
mean	20.3	1721	67.7	10.9	60.1	3.17	48.4	32.8	49.5
sd	2.8	69	10.0	3.4	8.0	0.51	2.1	2.3	10.2
min	17	1655	53.6	6.7	50.0	2.72	45.8	30.6	42.0
max	23	1810	76.6	14.7	68.9	3.68	50.9	35.2	64.6

Speed and Distance
The mean patrolling speed was calculated as 2.45 (sd 1.95) km.hr^{-1} and the total distance covered was 3929 m. Table 2 describes the data on speed and gradient.

Table 2. Speed and Gradient

	Speed m.s^{-1}	Speed km.hr^{-1}	Gradient %
mean	0.69	2.45	0.6
sd	0.65	1.95	16.7
min	0.02	0.08	
max	9.14	13.20	

Power and Oxygen Uptake

Table 3 shows the mean and standard deviation (sd) Power output (watts) and VO_2 (l.min^{-1}) as predicted by the Pandolf equation for each subject during the 100 minute patrolling task. The mean Power was 553 (sd 255) Watts and the mean VO_2 was 1.63 (sd 0.75) l.min^{-1}, equating to approximately 51% VO_{2peak}.

Table 3. Estimated Power and Oxygen Uptake

	W	l.min^{-1}	W	l.min^{-1}	W	l.min^{-1}	W	l.min^{-1}	W	l.min^{-1}
mean	535	1.58	528	1.56	605	1.78	544	1.60	553	1.63
sd	252	0.74	222	0.65	284	0.84	261	0.77	255	0.75
min	115	0.34	158	0.47	131	0.39	109	0.32	128	0.38
max	3122	9.21	2807	8.28	3520	10.39	3225	9.52	3168	9.35

Ascent and descent

The total ascent and descent in metres amounted to 100 m and 92 m, respectively during the 100 minutes.

Discussion

The objective of this pilot study was to explore the viability of using GPS to monitor the movement of people and to estimate their EE whilst working in the field. This objective was achieved and data from a single task have been presented in the previous section. From the 5 days of data that were collected during this pilot study (not presented here), one task performed by a team of 4 soldiers lasting 100 minutes was selected for detailed presentation in this paper. The overall mean oxygen uptake for the 5 days of rural patrols was found to be 1.10 l.min^{-1} (sd 0.46) or 15.0 ml.kg.$^{-1}$min.$^{-1}$ - substantially lower than the mean value of 1.63 (sd 0.75) or 24 ml.kg.$^{-1}$min.$^{-1}$ reported during this one task. This discrepancy was anticipated given that we concentrated on a particularly active task on a single day.

The data from this single task indicated that the physical demand was very variable during the 100 minutes. For example, the mean speed of 2.45 km.hr^{-1} sounds modest, but it masks considerable variation, with speeds varying from close to zero (when the patrol stopped for whatever reason) to 13.2 km.hr^{-1}, which equates to a moderate running speed. Considering that soldiers were at all times carrying loads that averaged 33 kg, amounting to approximately 50% of their body mass, and that the terrain was undulating, it is not surprising that the estimated work rate rose to over 3000 W and 9 l.min^{-1} for brief periods.

These are equivalent to estimated power outputs during a 100 m sprint (Astrand and Rodahl, 1986), the primary energy sources being the anaerobic breakdown of high-energy phosphate compounds.

However, it is not the actual data that are reported here that are of primary interest but rather the concept of collecting these types of data using GPS technology combined with map overlays and the possible interpretation that it allows. This method enables the ergonomist to estimate the power or EE associated with a given occupation or a given task in absolute terms. All of the variables measured can be presented in any meaningful unit of time (e.g. the peak VO_2 during any 1 minute period, the maximum speed covered in 10 seconds), providing versatility of analysis. The use of GPS to track people lends itself especially well to rural communities or to work that is truly 'in the field', where work inside buildings or in the close proximity to buildings is rare. Nearby buildings, vehicles, trees or people can interfere with GPS-based information as the signal from the satellites to the receiver may be blocked. Ideally the receiver should have a good clear view of the sky. By 'good clear view of the sky' we mean that the receiver can see lots of satellites and that the signals it receives are via pure line of sight paths.

The accuracy of positional information on GPS is generally considered to be better than 2 metres in the X and Y axes after the traces have been differentially corrected. Since the US government have turned off the intentional errors on the standard system there is little difference in the position obtained from standard and differential systems operation in the clear. However the standard traces are taken directly from the receiver's low power GPS engine that uses the four highest elevation satellites in its position calculation and the differential traces are calculated with commercial surveying software on a PC using all the available satellite data. There are other differences between the systems but in essence the surveying software is better at removing erroneous points than the standard set-up. We estimate that information on gradient, interpolated from the 10 m contour grid is also accurate to about 2 metres. The accuracy of our predicted power and oxygen uptake using the Pandolf equation remains unknown and requires validation before use in earnest.

This method, if found to be valid, can also be used to make some qualitative interpretation of work intensity. For example, Wilson and Corlett (1995) proposed a work intensity classification system based on VO_2. Tasks that require a VO_2 of over 2 $l.min^{-1}$ are classified as 'extremely heavy' and those using 1.5-2.0 $l.min^{-1}$ as 'very heavy'. Tasks using 1.0-1.5 $l.min^{-1}$ are considered 'heavy', 0.5-1.0 $l.min^{-1}$ as 'moderate' and less than 0.5 as 'light'. By this classification system our subjects would be working at a 'very heavy' work intensity, if this intensity was sustained throughout the working day. Other occupations that fall into this category of 1.5-2.0 $l.min^{-1}$ include heavy industrial work, heavy gardening and agriculture. The mean VO_2 for the 5-day period of 1.1 $l.min^{-1}$ equates to 'heavy' work.

All activities (leisure and work) and the intensity of the activities can also be classified according to their metabolic equivalent, where 1 MET equates to resting metabolic rate, or approximately a VO_2 of 3.5 $ml.^{-1}kg^{-1}.min$. In their textbook, Wilmore and Costill (1994) cite the MET values typically associated with a number of occupational tasks. Sitting at a desk is allocated 1.5, bricklaying and plastering 3.5, and digging 7.5, for example. The 100-minute task conducted in this study had a mean MET equivalent of 7.1, broadly equivalent to digging, though the mean MET for the 5-day period was 4.3 METs.

The American College of Sports Medicine's (ACSM) Position Stand (American College of Sports Medicine, 1998) extends these classification systems further by classifying physical activity intensity, based on physical activity lasting up to 60 minutes in METS by *age category*. For young males a work intensity of 7.1 METs falls at the upper limit of the moderate work intensity (4.8-7.1 METs) and just below the hard category (7.2-10.1 METs). A work intensity of 4.3 METs falls in the light work intensity. The different durations of the work tasks (100 minutes in our task vs. 60 minutes in the ACSM classification) should be born in mind when interpreting these data. Further, none of these classification systems takes into account the physical fitness of the employee. It is physical fitness that will largely determine the strain on the individual employee.

Over the 5 days the overall mean $\%VO_{2max}$ was 33% (SD 6), with mean values per day ranging from 26.4% (sd 6) to 37% (sd 7). On the most demanding day, 50% of soldiers performed at a mean work intensity above $40\%VO_{2max}$. Many studies (Mital and Ayoub, 1989; Astrand and Rodahl, 1986) have demonstrated that the demands during an 8-12 hour day should not exceed 30-40% of the individual maximal capacity, though higher intensities of work may be sustained for shorter periods.

In summary, GPS when combined with other tools such as contoured maps and Pandolf's equation was found to be a viable method for monitoring the movement, and estimating the EE of workers in a field setting. However, this innovative method is currently both technically challenging and time consuming and its validity for estimating EE remains unknown.

This work was supported by the Ministry of Defence. I would also like to acknowledge the contributions of Andy Baird who constructed the data-loggers attached to the GPS receiver engines, to Eric Malcomson who programmed the AXIS2000 GIS Application System to import the data into a dataset, and to Dr Steve Cole for enabling this work to be conducted.

References

Astrand P.O. and Rodahl K. 1986, *Textbook of Work Physiology*, Third Edition, (McGraw-Hill International Editions)

American College of Sports Medicine 1998. The recommended quantity and quality of exercise for developing and maintaining cardiorespiratory and muscular fitness, and flexibility in healthy adults. Position stand. *Medicine and Science in Sports and Exercise*, **30** (6), 975-991

Ayoub M.M. and Mital A.1989, *Manual Materials Handling*, (Taylor and Francis, London)

Meijer G.A., Westerterp K.A., Koper H., Ten Hoor F. 1989. Assessment of energy expenditure by recording heart rate and body acceleration. *Medicine and Science in Sports and Exercise*, **21**, 343-347

Pandolf K.B., Givoni B. and Goldman R.F. 1977. Predicting energy expenditure with loads while standing or walking very slowly. *J. Appl. Physiol.*, **43**, 4, 577-581

Rayson M.P. and Wilkinson D.A. 2001, *Energy Expenditure during CMS(R)*. Technical Report dated 20 November 2001, Optimal Performance Limited, Farnham, UK

Westerterp K.R. 1999, Physical activity assessment with accelerometers. *Int. J. Obesity* **23**, Suppl 3, S45-S49

Westerterp K.R. and Saris W.H.M. 1992, Limits of energy turnover in relation to physical performance, achievement of energy balance on a daily basis. In C Williams & JT Devlin (ed) *Food, Nutrition & Sports Performance* (E & FN Spon, London), 12-16

Wilmore J.H. and Costill D.L. 1994, *Physiology of Sport and Exercise*, (Human Kinetics)

Wilson J.R. and Corlett E.N. 1995, *Evaluation of Human Work*, Second Edition, (Taylor and Francis, London).

AN AUTOMATIC GEARSHIFT LOGIC FOR DISCRETE BICYCLE TRANSMISSIONS

Huai-Ching Chien and Ching-Huan Tseng

Department of Mechanical Engineering, National Chiao Tung University, Hsinchu, Taiwan 300, R.O.C. E-mail: chtseng@cc.nctu.edu.tw
Tel: +886-3-5726111 Ext. 55129 Fax: +886-3-5717243

Considering the characteristics of a discrete transmission on bicycles, this study proposes an automatic gearshift logic for discrete bicycle transmissions. This logic calculates shifting occasions of a transmission for the individual. Shifting gears at these shifting occasions can keep the individual pedaling in the suitable cadence region, namely satisfy his/her ergonomic necessity.

Introduction

In recent years, the multi-speed bicycles become the main stream on the market. As shown in Figure 1, the chainwheel, freewheel and the chain compose the freewheel-type transmission system of a multi-speed bicycle. The chainwheel and the freewheel respectively includes several chainrings and cogs that contain different numbers of teeth. The front and rear derailleurs guide the chain to engage larger chainrings or cogs, or guide the chain to drop to smaller chainrings or cogs, to produce different gear ratios for a cyclist.

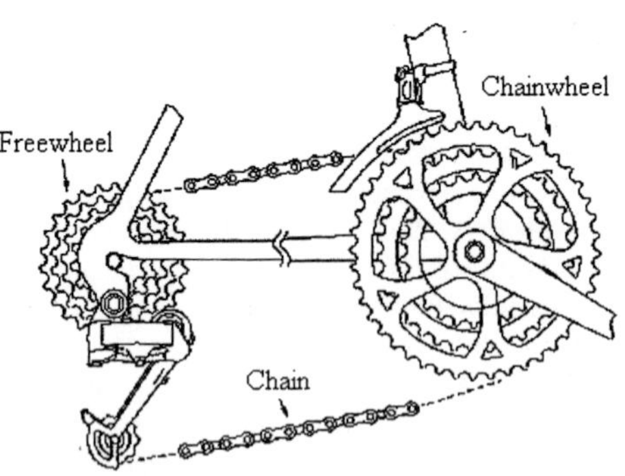

Figure 1. Freewheel-type transmission system

A multi-speed transmission system provides choices for a cyclist to pedal at the fitting cadence and torque under the fitting power output (Kyle, 1988; Whitt and Wilson, 1982). The optimum gear/cadence combination is one in which the legs and cardiovascular system of the cyclist both feel as though they will be able to continue for a long period of time (Marr, 1989). Since the legs and cardiovascular system work well under the fitting cadence (Seabury et al., 1977; Hagberg et al., 1981; Coast and Welch, 1985; Hull et al., 1988; Neptune et al., 1997; Neptune and Hull, 1999), the optimum gear/cadence combination then can be defined as one in which the cyclist will be able to keep pedaling under the fitting cadence.

In this study, a gear shifting occasion calculation method is proposed. Through this method, according to the cycling velocity, a cyclist is able to choose an optimum gear/cadence combination to keep himself/herself pedaling under the fitting cadence while cycling. Furthermore, this method also can be programmed into a microcontroller to build an automatic gearshift transmission. Then a cyclist who rides a bicycle, equipped with the automatic gearshift transmission, can pedal under his/her fitting cadence while cycling.

Characteristics of Discrete Transmission of Bicycles

The discrete bicycle transmission can not exactly satisfy the ergonomic necessity, cadence, of a cyclist since they provide discrete gear ratios. Every gear ratio provided by the discrete bicycle transmission is only able to keep a cyclist pedaling at the cadences close to his/her fitting cadence.

According to the ergonomic studies, a cyclist shall pedal in the same cadence region through every gear ratio. Hence,

$$\begin{matrix} \omega_{m,i} = \omega_m \\ \omega_{M,i} = \omega_M \end{matrix}, \qquad i = 1 \sim N \qquad \text{Eq. 1}$$

where

$\omega_{m,i}$ = Cadence at the lower shifting point of ith gear ratio, rpm

$\omega_{M,i}$ = Cadence at the upper shifting point of ith gear ratio, rpm

ω_m = Cadence at the lower shifting points, rpm

ω_M = Cadence at the upper shifting points, rpm

The cycling velocity can be expressed as

$$V = \frac{\pi \times D_{wheel} \times w \times R_i}{60 \times 100}, \qquad i = 1 \sim N \qquad \text{Eq. 2}$$

where

V = Cycling velocity, m/sec

D_{wheel} = Diameter of the bicycle wheel, m

w = Cadence, rpm

R_i = i^{th} gear ratio

Suppose that external physical conditions does not influence the cycling velocity during a gearshift process, the equality should stand as follows.

$$V_{m,i} = V_{M,i-1}, \qquad i = 2 \sim N \qquad \text{Eq. 3}$$

where

$V_{m,i}$ = Cycling velocity at the lower shifting point of ith gear ratio, m/sec

$V_{M,i-1}$ = Cycling velocity at the upper shifting point of $i-1^{th}$ gear ratio, m/sec

Hence, from Eq. 2 and Eq. 3,

$$w_m \times R_i = w_M \times R_{i-1}, \qquad i = 2 \sim N \qquad \text{Eq. 4}$$

Then,

$$R_i = \frac{w_M}{w_m} \times R_{i-1} = \left(\frac{w_M}{w_m}\right)^{i-1} \times R_1 = (k_w)^{i-1} \times R_1, \qquad i = 2 \sim N \qquad \text{Eq. 5}$$

where

$$k_w = \frac{w_M}{w_m} = \frac{R_i}{R_{i-1}} \qquad \text{Eq. 6}$$

Therefore, only a discrete bicycle transmission whose gear ratios are geometric series is able to keep a cyclist pedaling in the same cadence region through every gear ratio. And such discrete bicycle transmission is called an ergonomically designed transmission.

Different cyclists should shift gear at different points since their fitting cadences are different. However, the cadences at the shifting points of any individual should satisfy the following equation.

$$w_f = \frac{w_M + w_m}{2}, \qquad \text{Eq. 7}$$

where

ω_f = Fitting cadence of a individual, rpm

From Eq. 7 and Eq. 8, the cadences at the shifting points can be obtained as follows.

$$w_m = \frac{2}{1 + k_w} \times w_f, \qquad \text{Eq. 8}$$

$$w_M = \frac{2 \times k_w}{1 + k_w} \times w_f, \qquad \text{Eq. 9}$$

Shifting Points Calculating

Obviously, the numbers of the teeth on the chainrings and the cogs in the discrete bicycle transmission are integers. The integral teeth are almost impossible to exactly generate the geometric gear ratios. Hence, the variance in each gear ratio produces the variances in the shifting points as follows.

Suppose that a cyclist wishes no variation in the upper shifting points, namely w_M do not vary with the variance in each gear ratio, Eq. 4 then can be modified to

$$w'_{m,i} \times R'_i = w_M \times R'_{i-1}, \qquad i = 2 \sim N \qquad \text{Eq. 10}$$

where

$w'_{m,i}$ = Cadence at the lower shifting point of ith varied gear ratio, rpm

R'_{i-1} = $(i-1)^{th}$ varied gear ratio

Hence, the variances in the $w'_{m,i}$ can be represented as

$$\Delta w_{m,i} = w'_{m,i} - w_{m,i} = \left(\frac{R'_{i-1}}{R'_i} - \frac{R_{i-1}}{R_i}\right) \times w_M, \qquad i = 2 \sim N \qquad \text{Eq. 11}$$

Suppose that a cyclist wishes no variation in the lower shifting points, namely w_m do not vary with the variance in each gear ratio, Eq.4 then can be modified to

$$w_m \times R'_i = w'_{M,i-1} \times R'_{i-1}, \qquad i = 2 \sim N \qquad \text{Eq. 12}$$

where

$w'_{M,i-1}$ = Cadence at the upper shifting point of $(i-1)^{th}$ varied gear ratio, rpm

Hence, the variances in the $w_{M,i-1}$ can be represented as

$$\Delta w_{M,i-1} = w'_{M,i-1} - w_{M,i-1} = \left(\frac{R'_i}{R'_{i-1}} - \frac{R_i}{R_{i-1}} \right) \times w_m, \qquad i = 2 \sim N \qquad \text{Eq. 13}$$

In order to produce no variance in an upper shifting point, the variance in the lower shifting point may be too violent for a cyclist, and vise versa. So, it is better to averagely disperse the variances to the upper and the lower shifting point than limit the variances to the upper or the lower shifting point. To improve the cycling performance and comfort, a cyclist needs a shifting points calculating algorithm to overcome the integral teeth problem.

In this study, a shifting points calculating algorithm is proposed as follows. Every two adjacent gear ratios in the transmission are regarded as an independent sub-transmission possessing two gear ratios that are exactly geometric series. Namely, the transmission is treated as a combination of several sub-transmissions whose gear ratios are exactly geometric series. Hence, the lower shifting point of the higher gear ratio can be calculated through Eq. 8. Similarly, the upper shifting point of the lower gear ratio can be calculated through Eq. 9. The cycling velocities at the shifting points also can be calculated through Eq. 2.

Figure 2 shows a calculating example under the fitting cadence at 80 rpm. Firstly, the first and the second gear ratios are separated from the transmission and treated as an independent sub-transmission. Then the cadences and the cycling velocities at the shifting points are calculated. This process will be continued until every gear ratio is processed and the cycling velocity at every shifting point is calculated.

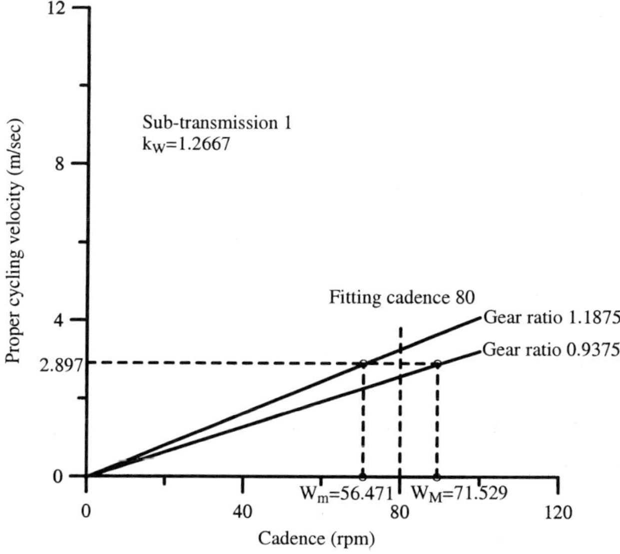

Figure 2. Calculating example

Simulations

In this section, several simulations are implemented to a commercial discrete transmission with 12 gear ratios (US05,261,858) as follows:

Table 1 show the cadences and the cycling velocities at every shifting point under the fitting cadence at 70, 80 and 90 rpm. Obviously, the higher a fitting cadence is, the wider the cadence regions and the cycling velocity regions are. A cyclist who feels the cadence regions are too wide for him should adopt another transmission with more gear ratios.

Table 1. Shifting points of a commercial discrete transmission

(a) Fitting cadence at 70 rpm

Gear ratio $i \rightleftharpoons i+1$	Upper bound cadence of lower gear ratio ($w'_{M,i}$) (rpm)	Lower bound cadence of higher gear ratio ($w'_{m,i+1}$) (rpm)	Cadence region width (rpm)	Cycling velocity (m/sec)
$0.938 \rightleftharpoons 1.188$	78.24	61.77	16.47	2.54
$1.188 \rightleftharpoons 1.304$	73.28	66.72	6.57	3.01
$1.304 \rightleftharpoons 1.5$	74.88	65.12	9.77	3.38
$1.5 \rightleftharpoons 1.652$	73.38	66.62	6.76	3.81
$1.652 \rightleftharpoons 1.765$	72.31	67.7	4.61	4.13
$1.765 \rightleftharpoons 2.087$	75.86	64.14	11.71	4.63
$2.087 \rightleftharpoons 2.235$	72.4	67.6	4.8	5.23
$2.235 \rightleftharpoons 2.5$	73.91	66.09	7.83	5.71
$2.5 \rightleftharpoons 2.824$	74.25	65.75	8.51	6.42
$2.824 \rightleftharpoons 3.167$	74.01	65.99	8.02	7.23
$3.167 \rightleftharpoons 4$	78.14	61.86	16.28	8.56

(b) Fitting cadence at 80 rpm

Gear ratio $i \rightleftharpoons i+1$	Upper bound cadence of lower gear ratio ($w'_{M,i}$) (rpm)	Lower bound cadence of higher gear ratio ($w'_{m,i+1}$) (rpm)	Cadence region width (rpm)	Cycling velocity (m/sec)
$0.938 \rightleftharpoons 1.188$	89.41	70.59	18.82	2.9
$1.188 \rightleftharpoons 1.304$	83.75	76.25	7.5	3.44
$1.304 \rightleftharpoons 1.5$	85.58	74.42	11.16	3.86
$1.5 \rightleftharpoons 1.652$	83.86	76.14	7.73	4.35
$1.652 \rightleftharpoons 1.765$	82.63	77.37	5.27	4.72
$1.765 \rightleftharpoons 2.087$	86.69	73.31	13.39	5.29
$2.087 \rightleftharpoons 2.235$	82.75	77.26	5.49	5.97
$2.235 \rightleftharpoons 2.5$	84.47	75.53	8.94	6.53
$2.5 \rightleftharpoons 2.824$	84.86	75.14	9.72	7.34
$2.824 \rightleftharpoons 3.167$	84.58	75.42	9.17	8.26
$3.167 \rightleftharpoons 4$	89.3	70.7	18.6	9.78

(c) Fitting cadence at 90 rpm

Gear ratio $i \rightleftharpoons i+1$	Upper bound cadence of lower gear ratio ($w'_{M,i}$) (rpm)	Lower bound cadence of higher gear ratio ($w'_{m,i+1}$) (rpm)	Cadence region width (rpm)	Cycling velocity (m/sec)
$0.938 \rightleftharpoons 1.188$	100.59	79.41	21.18	3.26
$1.188 \rightleftharpoons 1.304$	94.22	85.78	8.44	3.87
$1.304 \rightleftharpoons 1.5$	96.28	83.72	12.56	4.34
$1.5 \rightleftharpoons 1.652$	94.35	85.65	8.69	4.89
$1.652 \rightleftharpoons 1.765$	92.96	87.04	5.93	5.31
$1.765 \rightleftharpoons 2.087$	97.53	82.47	15.06	5.95
$2.087 \rightleftharpoons 2.235$	93.09	86.91	6.18	6.72
$2.235 \rightleftharpoons 2.5$	95.03	84.97	10.06	7.35
$2.5 \rightleftharpoons 2.824$	95.47	84.53	10.94	8.25
$2.824 \rightleftharpoons 3.167$	95.16	84.84	10.31	9.29
$3.167 \rightleftharpoons 4$	100.47	79.54	20.93	11

Conclusions

This study proposes a shifting gear occasion calculation algorithm for a discrete transmission on bicycles. Considering the characteristics of a discrete transmission, this study proposes a transmission separating method to overcome the integral teeth problem and makes shifting points calculation possible. Shifting gears at these shifting points can keep a cyclist pedaling in the suitable cadence region, namely satisfying his/her ergonomic necessity.

Based on this algorithm, a professional cyclist can calculate his/her shifting points. Then, according to the cycling velocity, he/she is able to choose an optimum gear/cadence combination to keep pedaling in the cadence region nearby his/her fitting cadence while cycling. Furthermore, this method also can be programmed into a microcontroller to build an automatic gearshift transmission. Then a cyclist who rides a bicycle, equipped with the automatic gearshift transmission, also can pedal in the cadence region nearby his/her fitting cadence. Therefore, this algorithm is helpful for the professional cyclists and the automatic transmission designers.

Acknowledgement

This project was supported by the National Science Council of R.O.C. under the grant number NSC91-2212-E-009-022. The authors would like to express his appreciation for the financial support.

References

Browning, D. L., 1993. Method and system for computer-controlled bicycle gear shifting, *United States Patent #5,261,858*

Coast, J. R., Welch, H. G., 1985, Linear increase in optimal pedal rate with increased power output in cycle ergometry, *European Journal of Applied Physiology*, **53**, 339-342

Hagberg, J. M., Mulin, J. P., Giese, M. D., Spitznagel, E., 1981, Effect of pedaling rate on submaximal exercise responses of competitive cyclists, *Journal of Applied Physiology*, **51**, 447-451

Hull, M. L., Gonzalez, H. K., Redfield, R., 1988, Optimization of pedaling rate in cycling using a muscle stress-based objective function, *International Journal of Sports Biomechanics*, **4**, 1-20

Kyle, C. R., 1988, The mechanics and aerodynamics of cycling. In: Burke, E. R., Newsom, M. M. (Eds.), Medical and Scientific Aspects of Cycling, *Human Kinetics*, **IL**, pp. 235-251.

Marr, D., 1989, *Bicycle Gearing, A Practical Guide*, The Mountaineers

Neptune, R. R., Kautz, S. A., Hull, M. L., 1997, The effect of pedaling rate on coordination in cycling, *Journal of Biomechanics*, **30**, 1051-1058

Neptune, R. R., Hull, M. L., 1999, A theoretical analysis of preferred pedaling rate selection in endurance cycling, *Journal of Biomechanics*, **32**, 409-415

Seabury, J, J., Adams, W. C., Ramey, M.R., 1977, Influence of pedaling rate and power output on energy expenditure during bicycle ergometry, *Ergonomics*, **20**, 491-498

Whitt, F. R. and Wilson, D. G., 1982, *Bicycle Science*, MIT Press, Cambridge, MA

DRINKING GLUCOSE IMPROVES MOOD AT THE BEGINNING OF THE WORKING DAY

Orsolina I Martino and Neil Morris

University of Wolverhampton, Department of Psychology, Wulfruna Street, Wolverhampton WV1 1SB, UK

Maintaining a positive mood state is desirable in the workplace. Blood sugar level is one major factor that has been linked to changes in mood. Heightened blood sugar levels tend to correspond to increased energy and reduced tension. 24 individuals attending a research methods workshop had their blood sugar levels measured at 10 a.m. They were also assessed on a number of mood dimensions using the UWIST Mood Adjective Checklist (UWIST). 12 participants then received a glucose rich drink and the other 12 participants received a saccharine drink matched for sweetness. At 12 p.m. blood sugar was elevated for the glucose group but not for the saccharine group. Similarly mood was significantly more positive for the glucose group but not for the saccharine group. It is recommended that positive steps be taken to ameliorate the effects of low blood sugar at the beginning of the working day.

Introduction

Mood at the start of the working day is likely to be a major influence on ones' affective response to the rest of the working day. Clearly there can be many factors affecting early morning mood, for example, amount of sleep, personal problems etc. Each of these needs addressing and might require professional counselling. However one factor that makes a substantial contribution to reported mood state is blood sugar level (Thayer, 2001). Thayer has shown that low blood sugar level is associated with increases in tension and reduced arousal. This is *temporarily* ameliorated by sugary snacks..

In healthy, young individuals blood sugar levels are maintained at approximately 5 mmol/l via a negative feedback loop. When blood sugar rises above 5 mmol/l insulin is released from the pancreas, which results in glucose being removed from circulation and immobilised as glycogen in the liver and muscles. When blood sugar levels drop below 5mmol/l the pancreas releases glucagon that releases glycogen from the liver; the glycogen is broken down into glucose and released into the bloodstream, increasing blood sugar until insulin release is triggered. Insulin and glucagon are therefore mutually inhibiting. Deviations from the 'set point' for blood sugar, which tend to be due to the absorption of glucose from the gut, are rapidly regulated, whereas the mobilisation of glycogen means that blood sugar can be maintained even in early starvation. Tight control of blood sugar is essential because the brain uses glucose as its 'fuel' but cannot store it - brain processes rely on a constant supply of glucose from the bloodstream.

Blood sugar levels fluctuate, following a diurnal rhythm, with higher levels observed during the morning compared with the afternoon (e.g. Troisi, Cowie & Harris, 2000). This corresponds to the typical pattern of mood variation throughout the day, as described by Thayer (1989). In most individuals, mood is less positive upon waking and improves gradually throughout the morning, peaking late morning and then becoming more negative in the afternoon. This suggests that mood begins to improve once blood sugar levels approach optimum concentration.

Previous research has indicated that increasing blood sugar levels has positive effects on mood. Benton and Owens (1993) found that higher blood sugar levels corresponded to feelings of increased energy and reduced tension. Gold et al. (1995) demonstrated that it was actually possible to drive energy down and tension up by inducing hypoglycaemia.

Glucose ingestion has also been shown to have beneficial effects on cognitive performance. Lapp (1981), for example, found that higher blood sugar levels resulted in improved recall of word lists. Benton and Sargent (1992) found that memory for spatial material and word lists was better after eating breakfast. More recently Morris and Sarll (2001) showed that imbibing a glucose drink improved listening comprehension in students who had missed breakfast, ameliorating the effects of low blood sugar levels resulting from fasting.

This study examined the relationship between blood sugar and mood, addressing the possibility that increasing blood sugar levels by ingesting a glucose rich drink would have a positive effect on mood in the early part of the working day. Mood was measured using the UWIST Mood Adjective Checklist (UWIST) which is known to be both valid and reliable (Matthews, Jones and Chamberlain, 1990). The UWIST measures Energetic Arousal, Tense Arousal, General Arousal, Hedonic Tone and Anger-Frustration and thus extends the range of mood dimensions measured by Benton and Owens (1993). Detailed descriptions of the properties of these dimensions can be found in Morris, Toms, Easthope and Biddulph, (1998). Positive shifts in mood are represented by increases in Energetic Arousal, General Arousal and Hedonic Tone and *decreases* in Tense Arousal and Anger-Frustration. So, for example, it can be argued that while moderate levels of Energetic Arousal as measured by the UWIST (including items such as 'alert', 'energetic', 'active' and 'vigorous') might have beneficial effects upon performance, it would also be beneficial to reduce Tense Arousal.

The present study measured mood before and after the consumption of a glucose rich drink. Half of the participants imbibed a saccharine drink instead of the glucose drink to neutralise experimenter effects. All participants were told that they were drinking glucose *and* saccharine (this was so that if anyone needed to withdraw from the study because either drink would create medical problems they could do so without the precise experimental manipulation becoming apparent). It was predicted that mood would improve following the glucose drink, but not following the saccharine drink.

Method

Participants

6 male and 18 female participants, aged between 19 and 33 (mean age = 22.61, s.d. = 4.22), took part in this study as part of a research methods workshop. Participants came from a diverse range of occupational backgrounds. Diabetics, pregnant women and any individuals with serious medical conditions were excluded. Informed consent was obtained for procedures vetted by the Ethics and Safety Committees of the University of Wolverhampton.

Materials

Mood was measured using the UWIST Mood Adjective Checklist. This consists of 29 adjectives rated on a 4-point scale (ranging from 1 = 'definitely' to 4 = 'definitely not') and measures mood as 5 distinct categories: three bipolar factorial scales of Energetic Arousal, Tense Arousal and Hedonic Tone, the unipolar Anger-Frustration scale and the General Arousal scale, which does not have a factorial basis and comprises positive and negative items taken from the Energetic Arousal and Tense Arousal scales.

Blood sugar level was tested using BM-Test 1-44 blood sugar test strips, following the manufacturer's procedure, then measured with a Prestige Medical Healthcare Ltd. HC1 digital Blood Glucometer.

The glucose drink contained 50 g glucose in 250 ml of water plus 40 ml sugar-free Robinson's 'Whole-orange squash' and 10 ml of lemon juice (to reduce the sweetness). The saccharine ('placebo') drink was identical, except that 2 g of 'Sweetex' replaced the glucose. A pilot study at the University of Wolverhampton indicated that students could not distinguish between the two drinks.

Procedure

Testing took place in a large teaching laboratory at the University of Wolverhampton, where participants were randomly assigned to either the glucose or saccharine group. They were not aware that there was a distinction. After completing a questionnaire that required them to itemise and quantify their breakfast that day blood sugar levels were tested at 10 a.m. Participants tested their own blood sugar levels under strict supervision. Two blood samples were taken from each participant and the average of these two measures was used in the analysis.

The first UWIST was then administered; participants were required to complete it quickly, choosing the response that best described their mood at that particular time. They were then given either a glucose or saccharine drink and were requested to drink it rapidly. This was followed by an interval in which participants engaged in a research methods workshop. At the end of this interval, a second UWIST was administered and blood sugar tested again at approximately 12.00 p.m.

Results

All participants had eaten a light breakfast (roughly calculated to be < 500 kcal), none had fasted. The means and standard deviations for blood sugar readings and UWIST scores before and after the glucose and saccharine drinks are shown in Table 1. A two-way mixed design analysis of variance with Group (glucose/saccharine) as the between-subject factor and Test (before drink vs after drink) as the within subject factor showed a significant effect of Group ($F(1,22 = 16.970, p < 0.01)$ with higher blood sugar levels in the glucose group, a significant effect of Test ($F(1,22) = 23.926, p < 0.01$), showing that blood sugar levels were higher at second test and a significant interaction ($F(1,22) = 44.642, p < 0.01$). Simple effects analysis showed that the two groups did not differ in terms of blood sugar level at the beginning of the study but by the second testing the glucose groups' blood sugar level had significantly increased while the saccharine groups blood sugar readings remained unchanged (p<0.01).

Table 1: Means and standard deviation (in brackets) for blood sugar levels (mmol/l) and UWIST scores before and after a glucose or saccharine drink. The range of possible scores are shown in parentheses.

	Glucose (n=12)		**Saccharine (n=12)**	
	Before drink	***After drink***	***Before drink***	***After drink***
Blood sugar (mmol/l)	**5.21 (1.46)**	**7.93 (1.25)**	**5.18 (0.82)**	**4.75 (0.77)**
Energetic Arousal (min. = 8, max. = 32)	**18.58 (3.34)**	**23.08 (3.26)**	**19.42 (4.06)**	**20.92 (5.25)**
Tense Arousal (min. = 8, max. = 32)	**17.67 (4.05)**	**15.67 (3.47)**	**12.75 (2.53)**	**14.50 (3.37)**
Hedonic Tone (min. = 8, max. = 32)	**23.17 (5.11)**	**26.00 (4.45)**	**26.00 (3.19)**	**25.33 (3.63)**
Anger-Frustration (min. = 5, max. = 20)	**10.58 (3.12)**	**9.42 (4.21)**	**7.75 (3.08)**	**7.83 (2.48)**
General Arousal (min., = 12, max. =48)	**26.42 (2.43)**	**28.25 (2.77)**	**24.33 (2.53)**	**26.00 (4.77)**

As there were between-groups differences in baseline UWIST scores, the percentage by which each participant's score changed from before to after the drink was calculated for each of the 5 dimensions. Percent changes for the glucose and saccharine conditions were then analysed using independent samples t-tests. Percent change in Energetic Arousal was significantly greater for the glucose group ($t(22) = 2.125, p < 0.05$), as was Tense Arousal ($t(22) = 2.655, p < 0.02$), indicating in this instance a greater reduction in tension in the glucose group (see table 1). Percent change in Hedonic Tone was also significantly greater for the glucose group ($t(22) = 2.290$, $p < 0.05$). There were no significant between-groups differences in percent change for Anger-Frustration ($t(22) = 1.607$, $p > 0.05$) or General Arousal ($t(22) = 0.098$, $p > 0.05$).

To summarise, both blood sugar levels and positive mood increased following a glucose drink but not following a saccharine drink.

Discussion

The results of the study showed that consuming a drink rich in glucose resulted in an increase in both blood sugar levels and positive mood. The findings support those of Benton and Owen (1995), who showed that higher blood sugar levels were linked to increased energy and reduced tension, and also those of Gold et al. (1995), which showed that it was possible to directly manipulate mood by altering blood sugar levels. Thus we replicate and contribute to the establishment of the reliability of these findings. However, the findings in this paper also extend these earlier findings by showing increases in hedonic tone. This dimension is a very significant mood dimension that is associated with internal feelings of well-being, for example, feelings of nausea or other somatic discomfort are associated with low levels of hedonic tone. Elevated hedonic tone is associated with increased somatic comfort which is likely to be beneficial to work performance.

In conclusion, the results of the study indicate that raising blood sugar levels has immediately beneficial effects on mood at the start of the working day. However, one should interpret these findings with caution with regards to the extent and duration of these effects - as Thayer (1989) points out, 'sugar-snacking' can rebound as the initial energising and tension reducing effects can induce later fatigue and tension. Whilst future studies might take this factor into account by measuring mood at regular intervals following glucose ingestion, the present results nonetheless indicate that positive mood can be enhanced rapidly, albeit temporarily, by raising blood sugar levels.

Recommendations

(1) The main recommendation is to start the working day by eating a nutritious breakfast (see Morris and Sarll, 2001).
(2) However, the effects of low blood sugar can be ameliorated when breakfast is missed by consuming a convenient form of glucose.
(3) Consuming large quantities of sugary snacks at the beginning of the working day is not recommended as a regular practice.

References

Benton, D. & Sargent, J. (1992). Breakfast, blood glucose and memory. *Biological Psychology, 24,* 95-100.

Benton, D. & Owens, D. (1993). Is raised blood glucose associated with the relief of tension? *Journal of Psychosomatic Research, 37,* 723-735.

Gold, A.C., MacLeod, K.M., Frier, B.M. & Deary, I. (1995). Changes in mood during acute hypoglycaemia in healthy subjects. *Journal of Personality and Social Psychology, 68,* 498-504.

Lapp, J.E. (1981). Effects of glycaemic alterations and noun imagery on the learning of paired associates'. *Journal of Learning Disorders, 14,* 35-38.

Matthews, G., Jones, D.M. & Chamberlain, A.G. (1990). Refining the measurement of mood: The UWIST Mood Adjective Checklist. *British Journal of Psychology, 81,* 17-42.

Morris, N. & Sarll, P. (2001). Drinking glucose improves listening span in students who miss breakfast. *Educational Research, 43,* 201-207.

Morris, N., Toms, M., Easthope, Y., and Biddulp, J. (1998) Mood and cognition in pregnant workers. *Applied Ergonomics, 29, 377-381.*

Thayer, R.E. (1987). Energy, tiredness, and tension effects of a sugar snack versus moderate exercise. *Journal of Personality and Social Psychology, 52,* 119-125.

Thayer, R.E. (1989). *The Biopsychology of Mood and Arousal.* Oxford: Oxford University Press.

Thayer, R.E. (2001) *Calm energy.* Oxford: Oxford University Press.

Troisi, R.J., Cowie, C.C. & Harris, M.I. (2000). Diurnal variation in fasting plasma glucose: Implications for diagnosis of diabetes in patients examined in the afternoon. *JAMA, The Journal of the American Medical Association, v284, i24,* p3157.

INTERFACE DESIGN

GENERATING INTERFACE METAPHORS: A COMPARISON OF 2 METHODOLOGIES

Mark Howell[1], Steve Love[1], Mark Turner[2] and Darren Van Laar[2]

[1]*Department of Information Systems and Computing, Brunel University, Uxbridge, Middlesex, UB8 3PH*
[2]*Department of Psychology, University of Portsmouth, King Henry Building, King Henry I Street, Portsmouth, PO1 2DY*

Interface metaphors have not been successfully applied to the design of automated telephone services. The aim of this study was to address this discrepancy by extending previous methodologies and models from GUI design to the generation and articulation of interface metaphors for such services. The study compared the usability and productivity of a card sorting and a sketching methodology. A questionnaire provided usability data, and a structured GUI metaphor design technique (Cates, 2002) was adapted to provide productivity data. It was found that the card sorting methodology was more usable overall, and more productive for generating 'Sound' features relevant to the chosen metaphors. Card sorting can therefore be recommended as a user-centred means of linking system functions with potential metaphors for telephone-based interfaces.

Introduction

This paper describes an experiment to compare 2 methodologies for generating interface metaphors for voice-based automated mobile phone services. The methodologies are card sorting and sketching, and provide a visual means of choosing an appropriate metaphor-based structure for the auditory information that must be navigated to complete a task using an automated telephone service. The use of metaphors has been successfully applied to the design of a range of graphical user interfaces (GUIs), for example the computer desktop and the web browser, but has not been commercially applied to the design of voice user interfaces (VUIs) for automated telephone services. However, Dutton *et al* (1999) found that experimenter generated interface metaphors improved the usability of a telephone home shopping service. The aim of the present experiment was to follow a user-centred approach to the generation, selection and articulation of potential interface metaphors, in order that they be understood by users, and be congruent with their real world experiences. The following sections discuss the structure and usability problems of automated telephone services; how they may be addressed by using interface metaphors; current methodologies for generating, selecting and articulating metaphors; and how the 2 new methodologies (card sorting, sketching) can be integrated and used to extend these existing methodologies specifically for automated telephone services.

Automated telephone services

The mobile telephone is being used to access an increasing number of automated telephone services, including World Wide Web sites. These services are often structured hierarchically, and consist of lists of spoken menu options arbitrarily assigned to numbers. The user selects a menu option by remembering the number corresponding to the desired option, and then saying the number. The main difference between GUIs and VUIs is that GUIs are able to present multiple channels of information simultaneously using menus, graphics, and audio. The user can quickly orient themselves by scanning this information for navigation cues. Interaction in VUIs is limited to speech and simple sounds, which are slow to produce and serial in nature (Schumacher *et al*, 1995), and which do not allow for scanning. As a result, when interacting with automated telephone services users may suffer from a poor mental model of the service structure and an overloaded short-term memory. This can lead to navigation problems (Rosson, 1985), user frustration, and a consequent reduction in usability. One possible solution to this problem is through the use of an interface metaphor, which may provide navigation information, and a better mental model of the service. However, it has been postulated by Pirhonen and Brewster (1997) that navigation information should not be given because it gets in the way of the information the user is trying to access, and may make the interface slow and clumsy. An interface metaphor would undoubtedly increase the duration of system messages, but if it was designed with high levels of user participation and resulted in fewer errors being made, this could lead to improved usability and acceptance.

Interface metaphors

Metaphors are an integral part of human language (Lakoff and Johnson, 1980), and are commonly used to explain unfamiliar concepts by means of comparison with familiar concepts. Metaphors have been widely employed within human computer interaction (HCI) as an important and successful technique in the design of GUIs, helping to reduce the time and effort necessary for new users to learn to use a system by bridging the conceptual gap between the user and the system (Carroll & Mack, 1985). However, the metaphor must be appropriate and explicit or it may corrupt the user's mental model of how the system works, causing semantic confusion. Opposition to the use of metaphors is most commonly the result of designers imitating or simulating real world objects and processes, rather than highlighting the parallels between seemingly mismatched domains. Designers must be creative to avoid simulation and to encourage user exploration of the system, and also adopt a user-centred approach to avoid imposing inappropriate metaphors on users. But according to Alty *et al* (2000) 'there is very little guidance for software designers on how to select, implement, and evaluate interface metaphors.'

Generating interface metaphors - brainstorming

Alty *et al* (2000) have proposed a 6-step framework for the generation and description of metaphors which recommends brainstorming as a useful technique for initial elicitation. The key features of brainstorming are the fast generation of many new ideas, and the absence of analysis and judgemental evaluation. Alty *et al* (2000) suggest writing the system functionality on a board, selecting related items of functionality, and then mapping real world processes onto them. According to Palmquist (2001) however, users are not good at generating their own metaphors, but can choose and articulate metaphors from a selection provided. It was therefore decided that for the current study the experimenter and 3 other HCI experts would use brainstorming to generate a list of potential metaphors whilst retaining the possibility for participants to develop their own.

Selecting interface metaphors – card sorting and sketching

In order to help users to choose the best metaphors for the system, and to adhere to a user-centred design process, it is necessary to provide suitable techniques to allow them to structure the way in which they naturally conceive of the system. Rather than giving the users a diagram of the structure of a system, and thereby imposing a model onto them, the users were asked to create a structure themselves by means of both card sorting and sketching. Both methodologies have been used within HCI for simple prototyping during the development of the design model, but in this case were used to develop the user's conceptual, metaphor-based model. The strength of card sorting is that it allows the discovery of the latent structure in an unsorted list of items. The strength of sketching is that it has been shown to be a valid technique to represent internal cognitive maps or mental models of the relative locations of objects in a spatial environment (Billinghurst and Weghorst, 1995). It may therefore be possible to extend this ability to represent the spatial structure of items within an electronic environment such as an automated telephone service. Both methodologies yield an overall spatial structure for the telephone service that will form a reference point to facilitate the choice of metaphors sharing a similar overall structure with it.

Articulating interface metaphors – the POPITS model

Once a metaphor has been chosen, the key features of that metaphor need to be isolated. Whereas Alty *et al* (2000) proposed 'fruitful conversation' as a means of identifying additional components related to the underlying metaphor, Cates (2002) proposed a more structured model for developing GUI metaphors, called POPITS. Having identified the underlying metaphor, the POPITS model proposes systematically identifying the related **P**roperties, **O**perations, **P**hrases, **I**mages, **T**ypes, and **S**ounds arising from the metaphor which are relevant to the system. Ideas generated within any of these 6 areas may also act as a catalyst to generate additional metaphors. Within this study the POPITS model was the framework within which users were required to develop their chosen metaphors, and as a result to both test the applicability of the model to VUIs, and select features of the metaphor that corresponded with their own experiences.

Experimental procedure

A 2-condition within subjects design was used with the methodology type as the independent variable. All participants used both the card sorting and the sketching methodologies in counterbalanced order. Eighteen participants took part, and comprised 10 females and 8 males with ages ranging from 18 to 39.

The study was conducted over 9 separate experimental sessions lasting for 1 hour each, with 2 participants taking part in each session. Each session was split into 2 parts, both of which were video recorded, and observed by the experimenter from behind a two-way mirror. Within the Observation Suite participants were seated at tables opposite each other, but separated and obscured by a large Velcro board.

As an introduction participants were given a brief explanation of automated telephone services, and then played a recording of someone performing a task using an automated telephone service (Unified Messaging Service). For part 1 of the experiment, participants were given a set of Velcro backed cards, on each of which was printed a menu option from a Telephone Internet Service (TIS), for example 'Shop'. They were then asked to arrange the cards on the Velcro board into a potential structure for the TIS. The participants were then given a sheet containing a table of 60 'real world systems', which were the metaphors generated in the brainstorming session by the experimenter and the HCI experts. They were asked to choose and rank 5 of these systems, or to additionally generate metaphors of their own, that they considered to be most similar to the structure of the TIS they had produced. Participants were then required to take the metaphor they had ranked as being most applicable to the system structure, and to develop it according to the 6 areas of metaphor features proposed by the POPITS model. Finally participants were asked to explain to each other how to perform two tasks using the service they had constructed, but in terms of language related to their 'number 1' ranked metaphor choice. The rationale underlying these explanations was to evaluate whether participants were able to use the metaphors in their explanations, and whether they incorporated features arising from their POPITS analysis into the explanations.

The second part of the experiment was similar to the first. The only differences were that menu options from a telephone city guide service were used instead of those from a TIS, and that participants were required to sketch a structure rather than sort cards.

Data collection

Data was collected in a number of ways. A ranking sheet was used to record the participants' choice of their top 5 metaphors. A POPITS sheet was used to gather data about the salient features of the metaphor that participants ranked as 'number 1'. Video recording was used to record participants' natural language explanations of their metaphor, and a usability questionnaire was used to record the participants' attitudes towards the usability of each experimental methodology.

Results

Usability of the methodologies

Paired samples t-tests were performed on the usability questionnaire scores to compare participants' evaluations of the 2 methodologies. There was a significant difference between the 2 methodologies for each aspect of usability tested (Table 1). The results indicated that card sorting was more usable than the sketching methodology for all 5 rated aspects of usability.

Table 1. Usability questionnaire results

Usability	Card sorting		Sketching		t	df	Sig.
	Mean	SD	Mean	SD			(2-tailed)
Efficient	5.67	1.03	4.17	1.47	4.75	17	0-001
Easy to use	5.94	1.00	4.06	1.73	4.88	17	0-001
Easy to learn	6.39	0.85	5.22	1.40	3.97	17	0-001
Comfortable	5.50	1.04	4.33	1.61	2.81	17	0.01
Productive	5.78	1.06	3.50	1.62	6.03	17	0-001
Mean	5.86	0.82	4.26	1.32	5.81	17	0-001

Productivity of the methodologies

The data generated from the POPITS section was quantified. Each word or phrase that referred to a specific feature of the chosen metaphor was quantified as one feature. Paired samples t-tests were then performed to compare the number of metaphor features generated using each methodology. Only one significant difference was found between the two methodologies, which was for the POPITS feature 'Sounds' (t(17) = 2.38; p = 0.029). There was no overall difference between the card sorting and sketching methodologies in their abilities to stimulate the productivity of POPITS features relevant to a metaphor (Table 2).

Table 2. POPITS results

POPITS	Card sorting		Sketching		t	df	Sig.
	Mean	SD	Mean	SD			(2-tailed)
Properties	3.67	2.43	3.06	2.13	1.83	17	0.09
Operations	2.33	1.53	2.22	1.63	0.33	17	0.74
Phrases	2.17	1.65	2.72	2.42	-0.97	17	0.34
Images	2.27	1.74	1.67	2.00	1.26	17	0.23
Types	1.61	1.42	1.72	1.84	-0.33	17	0.74
Sounds	1.44	1.25	0.78	0.88	2.38	17	0.03
Mean	2.25	1.09	2.03	1.22	1.13	17	0.27

Using metaphors to explain an automated telephone service

In order to investigate whether participants were capable of using their chosen metaphors to explain tasks, the explanations given were analysed. The effective use of metaphor was defined as that which provided an overall structure for the service, provided depth and detail, and finally was comprehensible to the other participant. On the basis of these criteria, for the card sorting methodology, 15 effectively used metaphor and 3 did not. For the sketching methodology, 10 effectively used metaphor and 8 did not. A McNemar test showed no significant change in the number of effective metaphorical explanations between methodologies (p = 0.063). This result suggests that the methodology used made no difference to the number of participants who were able to produce effective metaphorical explanations.

POPITS features used in the explanations

Another measure of how effectively the participants used metaphors in their explanations was the number of POPITS features they incorporated. Of the participants who gave effective metaphorical explanations, an association would be expected to exist between the number of POPITS features they generated, and the number of POPITS features they incorporated into their explanations. Bivariate correlations were therefore performed to measure the association between the number of POPITS features initially generated, and the number of POPITS features used within explanations, for each methodology. A significant positive correlation was found for participants using the card-sorting methodology (r = 0.601; n = 18; p = 0.008). There was no significant correlation found for the sketching methodology (r = -0.019; n = 18; p = 0.941). It may therefore be suggested that participants using the card sorting methodology were able to provide more detailed metaphorical explanations by incorporating a higher number of the POPITS features they had previously generated.

Conclusion

The main research objectives of this study were to compare the usability and productivity of card sorting and sketching as methodologies for facilitating the generation, selection, and articulation of interface metaphors for automated telephone services. It can be concluded that the card sorting methodology was more usable. There was no overall difference between methodologies in their ability to stimulate metaphorical thought within the POPITS framework. However, participants using card sorting generated significantly more 'sound' related features, which are particularly salient to the construction of metaphor-based telephone services.

The relationship between the number of POPITS features generated, and the number of POPITS features used in explanations for the card sorting methodology suggests that the card based structure produced by participants enabled them to initially generate POPITS features that were more relevant to the metaphor, and consequently easier to incorporate into coherent metaphorical explanations. Due to experimental constraints, participants were given a limited time period to produce an initial structure for each phone system. From observations made throughout the experiment, reactions to the 'time up' signal, and an informal interview given on completion of the experiment, it was clear that the participants using the card sorting methodology were able to design faster and more iteratively.

Card sorting can therefore be recommended as an important visual means of stimulating metaphorical thought about telephone-based interfaces: structuring a telephone service using cards enabled participants to construct more meaningful associations between the structure and the potential metaphor.

References

Alty, J.L., Knott, R.P., Anderson, B. and Smyth, M. 2000, A framework for engineering metaphor at the user interface, *Interacting With Computers*, **13**, 301-322.

Billinghurst, M. and Weghorst, S. 1995, The use of sketch maps to measure cognitive maps of virtual environments, *Proceedings of IEEE 1995 Virtual Reality Annual International Symposium*, Los Alamitos, CA, 1995, 40-47.

Carroll, J.M. and Mack, R.L. 1985, Metaphor, computing systems, and active learning, *International Journal of Man-Machine Studies*, **22,** 1, 39-57.

Cates, W. M. 2002, Systematic selection and implementation of graphical user interface metaphors, *Computers and Education,* **38,** 2002, 385-397.

Dutton, R.T., Foster, J.C. and Jack, M.A. 1999, Please mind the doors – do interface metaphors improve the usability of voice response services? *BT Technology Journal*, **17**, 1, January 1999, 172-177.

Lakoff, G. and Johnson, M. 1980, *Metaphors we live by,* University of Chicago Press.

Palmquist, R.A. 2001, Cognitive style and users metaphors for the web: an exploratory study, *The Journal of Academic Librarianship,* **27**, 1, 24-32.

Pirhonen, A. and Brewster, S. 2001, Metaphors and imitation, *Workshop Proceedings of PC-HCI 2001*, 7-9 December 2001, Patras, Greece, 27-32.

Rosson, M. 1985, Using synthetic speech for remote access to information, *Behaviour Research Methods, Instruments and Computers*, **17**, 2, 1985, 250-252.

Schumacher, R.M., Hardzinski, M.L. and Schwartz, A.L. 1995, Increasing the usability of interactive voice response systems, *Human Factors,* **37**, 2, 1995, 251-264.

HOW TO HANDLE NOTEBOOK INPUT DEVICES: AN INSIGHT IN BUTTON USE STRATEGY

Christine Sutter and Martina Ziefle

Christine Sutter
Martina Ziefle
Department of Psychology, Aachen University
Jägerstraße 17-19, 52056 Aachen
Germany

This study investigated the usability of notebook input devices, trackpoint and touchpad, with regard to characteristics of performance as well as effectiveness of button use strategy. Therefor, 20 participants handled device and mouse button two-handed, one hand controlling the cursor and the other operating the button for click-, drag- and drop actions, as well as one-handed, doing both with one hand simultaneously. Performance was measured by movement time and accuracy of button usage. The touchpad outperformed the trackpoint in speed and accuracy. Performance rose dramatically with the bimanual device handling. The interactive effect of device and button use strategy in drag- and drop errors proved clearly the superiority of the touchpad, starting at a lower level of key errors and an additional advantage of the bimanual button use strategy.

Introduction

The use of portable computers has grown continuously, especially since the nature of work environment is changing and many companies profit from the advantage of mobile workforce. This development places more emphasis on functional and effective notebook technology as well as a more natural and easy communication with interactive systems. Currently, two different input devices integrated in notebooks are available. The trackpoint is a small, force sensitive isometric joystick, and the touchpad which is a touch sensitive panel. Since technical design, location and size of internal input devices are restricted to the chassis, ergonomic requirements and suitability of interaction distinguish integrated and optional input devices (e.g. mouse, trackball, tablet, pen). In the discussion about usability of input devices two important factors have to be named first: The performance potential of input devices which is limited by technical and physical features and the users themselves with their input technique as the decisive factor for the degree of utilization.

The research has hitherto focused on performance of input devices. Most studies are mainly directed towards optional input devices and have neglect the small integrated input devices in notebooks so far (for an overview see e.g. Douglas and Mithal, 1997; Sutter, 2001). One of the first direct comparisons of notebook input devices on different types of tasks was conducted by Batra *et al.* (1998). The conclusion of their broad analysis referred to a superiority of touchpad and trackball over trackpoint.

Moreover, only few studies on user characteristics, specifically regarding their interactive style, were done. There are basically two ways of accomplishing cursor-control actions. Either cursor adjustment and button usage proceed in concert in a unimanual movement or it is divided into a bimanual task, with one hand operating the input device and the other one pressing the button. A comparison of one-handed touchpad interaction techniques (MacKenzie and Oniszczak, 1998) revealed a speed accuracy trade-off between a special tactile selection technique, lift-and-tap and using the physical button. So one-handed selection with a button was the slowest but most accurate style of dialogue. Typically, a unimanual usage of input devices is assumed, although the advantage of labour division in bimanual input compared to the unimanual was already confirmed by Buxton and Myers (1986). Since motor behaviour and control is a broadly discussed research topic (e.g. Schmidt, 1988), the specific topic of asymmetric division of labour in skilled bimanual actions is singled out now. Guiard (1987) pronounced that the possible role-hand assignment in bimanual asymmetric activities follows a lateral preference: "…it seems likely that a macrometric-micrometric specialization of the left and right hands conforms to the requirements of gestures in which one manual role is postural and the other manipulative" (p. 500). As bimanual cursor-control actions can be understood in the framework of Guiards kinematic chain model, their cooperative effect contrary to the unimanual action is in the centre of the current study. Additionally, both interaction styles were applied to object selection and –manipulation. The latter is holding the most extensive requirement on button usage. This study goes for an improvement on quality and efficiency in communication with interactive systems which may lead to more satisfying working conditions of notebook users.

Method

Variables

The study was based on a multiple factorial design with repeated measurements. As within-subject factors two independent variables were defined, the input device (touchpad and trackpoint) and the button use strategy (one-handed and two-handed use of device and mouse button).

The dependent variables included movement time which described the interval of time from onset of cursor movement to final button click or release, the defined end of task. The accuracy was calculated by click or release errors. A click error occurred when participants selected the target while the cursor was outside its boundaries. While highlighting a target, possible click and release errors (e.g. start to highlight at wrong position, release cursor at wrong position) were cumulated as highlight errors. Drag- (cursor outside square target by picking up the object) and drop errors (button release outside drop box) were as well calculated.

Apparatus and materials

Two notebooks, a Toshiba Satellite 1700-300 with a trackpoint and a Dell Inspiron 7500 with a touchpad integrated input device, were used in this study and connected to an external TFT flat screen. The trackpoint is a small isometric joystick that is sensitive to strain gauges, placed between the "G", "H" and "B" keys on the keyboard with two mouse buttons in the wrist rest. The touchpad, a flat 2" by 1.5" touch-sensitive panel, is integrated into the wrist rest beneath the keyboard with two mouse buttons underneath.

In a preliminary examination both devices had been tested on their ratio of force-to-cursor motion (trackpoint) respectively finger motion-to-cursor motion (touchpad) (Sutter, 2001). The cursor velocity of the devices was matched and adjusted to a medium range of cursor speed of about 1500 pixels per second.

Participants

The user group investigated consisted of ten university students and ten employees of the services sector (eight women, twelve men). All of them had been fulltime notebook users for more than 2 years (sd (standard deviation) = 1.4) with an estimated uninterrupted notebook session of 2.6 hours per day (sd = 1.8). The participants were experts in one input device but worked in this study also with the other input device which they were inexperienced in. This design of mixed experience level was chosen for generalisation of results, to compare performance beyond expertise. All of them but one were right handed and 21 to 33 years old (m (mean) = 27; sd = 3.86).

Tasks

Two control-cursor-actions, object selection (Task 1) and object manipulation (Task 2) were presented. In Task 1 each trial consisted of one target selection, as illustrated in Figure 1, left side. The trial started by pressing the space bar. A square target appeared on the screen centre that had to be selected with a cross-hair cursor. Correct cursor positioning inside the target was noticed through visual feedback. The selection confirmation with the left mouse button completed one trial and a new display was presented immediately.

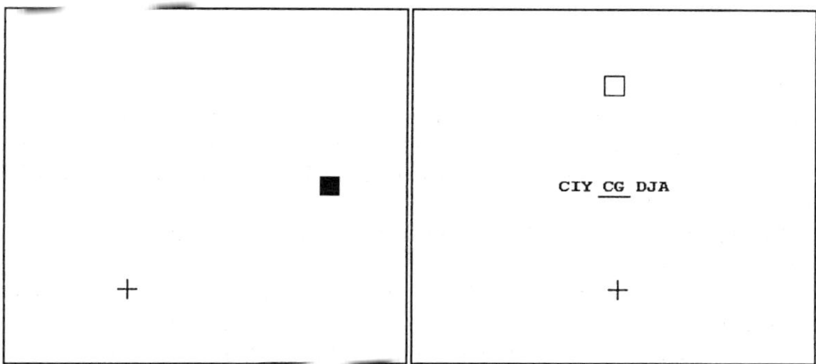

Figure 1. Screenshot of Task 1 (left) and Task 2 (right)

Task 2 focused on object manipulation. The trial started as described in Task 1, but now an alphabetic string was presented as shown in Figure 1 on the right side. First, in a point-drag action the cursor had to be moved from the start point to the leftmost or rightmost side of the target strings, the underlined strings in the middle. They were highlighted by dragging the cursor over the target strings with the left mouse button pressed. After correct button release a target box for a drag-drop action appeared. With the left mouse button pressed the target strings were picked up and dragged to the target box. The target strings were dropped by release of the left mouse button which completed the action. Each correct action was confirmed visually.

Each task included 36 trials with 4 only for training purposes. To avoid confounding effects, target distance, target size and direction of cursor movement were controlled, as they were not objectives of the current study.

Procedure

All participants took part in four conditions, randomly changing, and were instructed carefully to work as fast and accurate as possible.

The screening questionnaire at the beginning was followed by the first part, starting with trackpoint or touchpad. Both tasks were presented in a fixed order and were handled uni- or bimanually as instructed (first condition). After finishing, participants were admitted to the other button use strategy and processed again the two experimental tasks (second condition). The second part of the experiment followed the same procedure as described before. So each participant re-handled the tasks with both button use strategies, now with the second input device (third and forth condition).

Time, spatial information of the cursor movements (x-, y-coordinates) and key actions were recorded online with a logging software tool that was developed at Aachen University.

Results

The results in terms of speed (see Table 1) and accuracy performance (see Figure 2) are reported now for both control-cursor-actions, object selection (Task 1) and object manipulation (Task 2). A three-way repeated-measures analysis of variance (ANOVA) was run comparing input device and button use strategy as well as user expertise (the latter will be discussed in detail on the congress) for each independent variable: movement time and key errors. Follow-up t-tests were done on significant results.

In Task 1 the ANOVA shows a significant effect of input device on mean movement time, $F(1,18) = 8.83$; $p < 0.01$. For an object selection done with a trackpoint the average movement time is about 2.2 s. The touchpad user is 0.6 s faster, so cursor movement increased dramatically. Statistically, the effect of button use strategy on movement time is also significant, $F(1,18) = 13.22$, $p < 0.01$. If a control-cursor-movement is done one-handed, the average movement time needed was 2.1 s in contrast to 1.8 s two-handed. There is a significant difference of 367 milliseconds between the two button use strategies investigated, which is equivalent to a 21 % loss of performance from the bimanual to the one-handed strategy.

Table 1. Means and standard deviation (in parentheses) for movement time as a function of input device or button use strategy (n = 20)

	Trackpoint	Touchpad	One-handed	Two-handed
Object selection	2.2 s (1.1 s)	1.7 s (0.4 s)	2.1 s (0.8 s)	1.8 s (0.6 s)
Object manipulation	8.1 s (3.2 s)	6.2 s (1.8 s)	8.0 s (2.6 s)	6.2 s (2.4 s)

For object manipulation the effect of input device on mean movement time to highlight, drag and drop an object was significant, $F(1,18) = 19.34$, $p < 0.01$. With a trackpoint the whole manipulation took on average nearly 2 s longer than with the touchpad. This is equivalent to a 31 % inferior performance of the trackpoint. The effect of button use strategy is as well significant, F (1,18) = 58.91, p < 0.01. The advantage of 1.8 s is enormous, if the device is bimanually handled. Neither in Task 1 nor in Task 2 an interaction between the two factors was proven to be significant. One reason could be the slight increased standard deviation because of the varying levels of experience. But similar to Task 1, for object manipulation there was also a decrease in performance for the trackpoint and a big advantage in the two-handed button use strategy.

Accuracy was measured by inaccurate click and release actions. Generally, the participants worked accurately, on average only 18 % error trials were observed throughout the whole experiment. Neither the input device nor the button use strategy shows any significant effect on click errors. With both input devices only a few click errors (9 %) occurred. More evidence is given by highlight, drag and drop errors. Highlight errors are influenced significantly by input device, F (1,18) = 11.02, p < 0.01 as well as button use strategy, F (1,18) = 10.31, p < 0.01. No interaction between the two was proven to be significant. The trackpoint (36 %) provoked 1.8 times more highlight errors than the touchpad (20 %). So highlighting an object accurately with a trackpoint was shown to be rather difficult. The same is also valid for the one-handed button use strategy which produces 1.5 times more highlight errors than the two-handed one. The analysis of drag errors shows a significant main effect of input device, F (1,18) = 9.05, p < 0.01, button use strategy, F (1,18) = 41.53, p < 0.01 and a marginally significant device by strategy interaction, F (1,18) = 3.96, p = 0.06. As gradient and percentage of drag- and drop errors are very similar only the latter is visualised in Figure 2.

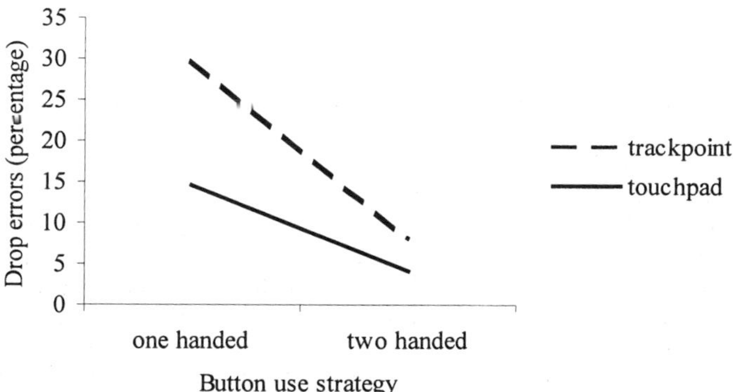

Figure 2. Percentage of drop errors as a function of input device and button use strategy (n = 20)

Drop errors (see Figure 2) show significant effects of input device, F (1,18) = 15.17, p < 0.01, button use strategy, F (1,18) = 49.27, p < 0.01 and an interaction, F (1,18) = 7.03, p < 0.05. So drag- and drop errors decrease dramatically for both input devices when handled bimanually but the touchpad, which starts at a lower error level, still outperforms the trackpoint in accuracy.

Summary and Discussion

Comparing touchpad and trackpoint the observed speed of both cursor-control actions favours the touchpad which is about 25 % faster. Focussing the button use strategy, it is a big advantage to handle the input device bimanually. For accuracy of input devices click errors carry no weight. But the action to highlight, drag and drop an object is clearly more accurate with a touchpad or with the two-handed strategy. Moreover, accuracy of drag and drop actions increases dramatically for both input devices when handled bimanually. But beyond the button use strategy the touchpad still outperforms the trackpoint.

As already shown by Batra *et al.* (1998), the performance potential between trackpoint and touchpad differs significantly. In general, the touchpad is faster and more accurate than the trackpoint. A specific way to direct the utilization degree of device potential is the efficiency in the operation of input devices. In analogy to Buxton and Myers (1986) the superiority of the two-handed input was confirmed. This input technique proves to be the more natural communication style as speed and accuracy increase evidently. Especially in cursor-control actions with intensive button usage, such as drag-and-drop tasks, the two-handed input improves the quality of input device operations. Although MacKenzie and Oniszczak (1998) found the touchpad to be rather slow but accurate used one-handed, results in the current study clearly reveal that both accuracy and speed increase dramatically with the bimanual button use strategy.

On the basis of the current results a two-handed input is highly recommended. Especially for tasks that require an intensive button use a natural way of communication seems relevant. Moreover, differences in input devices refer to the touchpad as the faster and more accurate integrated input device although both input devices profit from the two-handed input.

References

Batra, S., Dykstra, D., Hsu, P., Radle, K. and Wiedenbeck, S. 1998, Pointing device performance for laptop computers. *Proceedings of the Human Factors and Ergonomics Society 42nd Annual Meeting*, (Human Factors and Ergonomics Society, Santa Monica), 535-540

Buxton, W. and Myers, B.A. 1986, A study in two-handed input. *Proceedings of the CHI '86 Conference on Human Factors in Computing Systems*, (ACM, New York), 321-326

Douglas, S.A. and Mithal, A.K. 1997, *The Ergonomics of Computer Pointing Devices*, (Springer Verlag, Heidelberg)

Guiard, Y. 1987, Asymmetric division of labor in human skilled bimanual action: The kinematic chain as a model, *Journal of Motor Behavior*, **19**, 486-517

MacKenzie, I.S. and Oniszczak, A. 1998, A comparison of three selection techniques for touchpads. *Proceedings of the CHI '98 Conference on Human Factors in Computing Systems,* (ACM, New York), 336-343

Schmidt, R.A. 1988, *Motor control and learning,* (Human Kinetics Publishers, Champaign)

Sutter, C. 2001, *Auswirkungen alternativer Eingabegeräte am Notebook auf motorische Leistungsmaße* [Effects of alternative input devices in notebooks on motor performance measures], unpublished master's thesis, (Aachen University, Germany)

BIOMETRIC VERIFICATION AT A SELF SERVICE INTERFACE

Lynne Coventry, Antonella De Angeli and Graham Johnson

NCR- FSD Advanced Technology and Research,
Discovery Centre, 3 Fulton Road,
Dundee, UK, DD2 4SW

The term biometrics refers to a variety of identification techniques, based on a physical, or behavioural user characteristic. Traditionally, access to ATMs, has been controlled by possession of an artefact (card) and knowledge of a Personal Identification Number (PIN). Biometric verification is a potential alternative. At NCR, investigations into biometrics have been undertaken over a number of years. These investigations have used a variety of methods to gradually acquire a sound understanding of consumers' issues. Our research has revealed a number of non-trivial issues with the introduction of this type of technology to the general public. We will present some of our findings and general understanding of the public's attitudes towards and behaviour with, biometrics verification in general and specifically fingerprint at the ATM.

Introduction

Biometrics can be defined as the use of anatomical, physiological or behaviour characteristics to recognise or verify the claimed identity of a person. It requires the collection, processing and storage of details of person's physical characteristics. Biometric technologies are used to confirm that the person is present rather than their token (e.g. bankcard) or identifier (e.g. PIN). A name, password, card, PIN, or key does not confirm the presence of the legitimate person. They rely on the assumption that the card is always with the owner and owners do not reveal their PIN/password to others, however loss and sharing of cards are big security weaknesses in a system.

Our usability research, based on consumers and their behaviour and attitudes, has been instrumental in directing developments of these technologies and their use in self-service settings. It is too easy to forget that advanced technologies *per se* will not succeed without easy adoption by the intended users. Biometric technologies are a good example of advanced technology that impresses many, without understanding the impact users may have on successful implementation.

Biometric Techniques

Physiological biometrics includes those based on verification of fingerprints, hand and/or finger geometry, eye (retina or iris), face and wrist (vein). Behavioural techniques include those based on voice, signature and typing behaviour. The performance of these systems varies greatly.

Performance metrics
The performance of a biometric system must be excellent if users are to trust and accept it. Performance metrics should indicate how well a system performs but it is difficult to get reliable data on particular systems and more independent testing is required. Performance is measured in terms of false accept rates (FAR) – the likelihood that the wrong person will be able to access the system, and false reject rates (FRR) – the likelihood that a legitimate person will be denied access. The problem is the interconnection between these two measures, as one improves the other worsens.

Two other measures are important, failure to enrol (FTE) and failure to acquire an image (FTA). These effectively identify those people who will not be able to use the system (outliers) and those who may have difficulty using the system. It is also necessary to consider the effect of ageing on the chosen technique – for instance facial and fingerprint templates may need to be updated over time.

The method by which these figures are gathered can seriously impact the performance achieved. Performance estimates are often far more impressive than actual performance (Phillips et al 2000). Systems tested in laboratory conditions with a small homogenous set of "good", trained, young, co-operative users may generate completely different results than testing in a live environment with inexperienced and less co-operative users.

Biometrics and the ATM

On the surface there appears to be many valid reasons to replace PIN at the ATM (Automated Teller Machine): the PIN does not prove the identification of the cardholder; PINs can be forgotten leading to user frustration; using the wrong combination of card and PIN may result in retention on cards; people write PINs and disclose to others increasing the risk of fraud (Hone et al 1998); and PINs can be stolen by observation.

Clarke (1994) presents the desirable characteristics of an ideal biometric. The characteristics are that it be universal, unique and exclusive, permanent through life, indispensable, digitally storable, precise, easy and efficient to record and acceptable to contemporary social standards. It seems that not all these objectives can be met by any current method. Many systems do not live up to expectations because they prove unable to cope with the enormous variations among large populations or fail to take into account the needs of people. (Davies 1994)

One of the big issues with the ATM environment is the potential size of the user base – even a small financial institution could have millions of customers. Ultimately the system would have to deal with the entire banking population of the world. This affects a number of factors associated with biometrics including use of verification rather than recognition, enrolment, template storage and handling of outliers.

Verification
Given the potential size of user base, the processing time alone required to identify a person would make it prohibitive to use in a real time application. Thus the use of verification – a one to one match with template for claimed identity is used. The does not preclude back office identification testing for fraudulent identity. The users must collaborate with the system to produce a biometric for instance placing a finger on the device or looking at a camera.

Enrolment
Any person wishing to use a biometric system must be enrolled. This is a key opportunity for learning as well as providing the biometric images from which a template is produced. A good enrolment template is key to efficient and accurate verification. This is also the point where the user can be trained how to use the system and any misconceptions resolved.

Template storage
This has size, bandwidth, privacy and security issues to balance. If fingerprint is adopted, this could be done locally on a smart card with the user carrying the template on their card, even processing their fingerprint on the card to enable use of the card. Thus eliminating the complications of remote storage and processing.

Handling outliers
Extreme examples are those people who do not have the required characteristic, for instance no eyes or no fingers. However, some people may be unable to use the system through illness, e.g. tremors, glaucoma, traumas or injury e.g. cut to the fingers, broken hand. The ageing process can also cause problems and with some biometrics the template would require updating. Outliers must be accommodated without causing discrimination.

Understanding the consumer

To fully understand the role of the consumer in the performance and acceptance of biometrics, we have utilised a number of research methods through different stages of development of new biometric technologies. The most extensive research has been carried out on iris verification. This is reported in Coventry and Johnson (1999) and Coventry *et al* (2003) Our methods include focus groups to identify consumers' understanding, misconceptions, and barriers to acceptance of biometric techniques. This technique elicits attitudes, which may or may not correlate with actual behaviour. Specific studies help to understand how well a specific technology works with the general, untrained public and if it can be adapted to a self service environment. An iterative design and evaluation process is followed to build a self-service application and field trials are used to fully test performance and acceptance. We have found that experience with the actual technology can change people's attitudes in both positive and negative directions.

The remainder of this paper will present our results about biometrics in general, with focus on recent fingerprint studies.

Consumer attitudes to biometrics
Our research has shown that there is a general lack of public understanding of how a biometric or even a PIN works. This is often expressed in terms of a level of suspicion or distrust. Some key findings are that:

- There is little perceived need for the addition of biometrics.
- Consumers have difficulty believing some "futuristic" technologies can work well.
- There is a general concern about the potential misuse use of personal data and is seen as potentially violating privacy and civil liberties. These views vary between countries and cultures and are extensively reviewed in Woodward (1997).

- There are concerns about hygiene with touching such devices and health risks for more advanced technologies such as iris or retina, or even a fear of criminals killing to steal their eye or finger. This view is not helped by films such as Mission Impossible (Davies, 1997)!

With reference to fingerprint, our focus group studies have found different attitudes to those reported elsewhere. We found that people believed fingerprints would work well because it was extensively used in criminology. They did not believe that its association with crime deemed it socially unacceptable. However, a UK based survey of 500 people found that it was deemed to be most reliable biometric but least socially acceptable when compared to signature, facial and PIN.

A user study of fingerprint technology

We are currently involved with the development of a new fingerprint technology that requires the user to move their finger across a thermal sensor rather than placing their finger on a optical or capacitive sensor. This swiping action ensures no fingerprint image is left behind and the technology removes some of the problems with other fingerprint technology dealing with fine, aged or damaged skin.

This study looked at the base usability of the sensor and assessed the amount of support a user would require in order to provide an adequate image. The trial was conducted over 2 days in Edinburgh. Participants were required to enrol and authenticate using the system. A total of 82 people participated in the evaluation. The evaluation included a representative sample range of height, sex, and age. People were spread across three different enrolment conditions:

- Level 0: No *instruction and limited feedback*. Participants were instructed to "swipe their finger" and told whether it was a good/bad swipe. Unsuccessful participants progressed to the next level after 8 attempts.
- Level 1: Instruction *and limited feedback*. Participants were given clear instructions on how to swipe their finger and told whether it was a good/bad swipe. Unsuccessful participants then progressed to the next level after 8 attempts.
- Level 2: Full *instruction and feedback*. Participants were given clear instructions on how to swipe their finger and they were shown their fingerprint image on a laptop. Additionally they were told whether the image was good or bad. Unsuccessful participants were required to repeat the enrolment in the same condition.

Each participant chose which finger to enrol and then used that finger for the duration of the trial. Each participant had to achieve a minimum of four good fingerprint images in order to pass the enrolment procedure. The experimenters subjectively assessed the quality of the fingerprint image and allocated it as "pass" or "fail". Once the interviewer has submitted three good images, the fourth one was automatically compared to the other three and if it matched the participant had successfully completed the enrolment procedure.

During the authentication trial, all participants were given the same instruction. A total of six authentications were required from each participant. At the end participants were asked to complete a questionnaire.

Results

The average age of participant was 34.6 years (SD 14.44), with a range of ages from 16 to 73 years, almost equal numbers of men and women (N = 82, 42 men and 40 women).

The majority of participants (78%) used their right index finger to swipe with. 20% used their right middle finger and 2 participants used their left middle finger. 7 participants (8.5%) failed to enrol. The next analyses are conducted on the sample that succeeded to enrol (N=75).

Figure 1 reports the percentage of participants who successfully enrolled after 1, 2 or 3 enrolment sessions as a function of experimental conditions. The effect of feedback is evident: there is a consistent increase in the number of participants enrolling first time as the initial level of instruction and feedback increases. In Level 0, when no instruction and limited feedback was provided, only 40% of participants succeeded to enrol at the first attempt. In Level 1, when participants were provided with clear instruction on how to swipe their feedback, this percentage raised to 64%. In Level 2, when participants were given instruction and visual feedback, the majority of the sample (80%) enrolled at the first attempt. However, it is also evident that some people had problems with the systems, and needed more practice with the system, independent of the amount of instruction or feedback received.

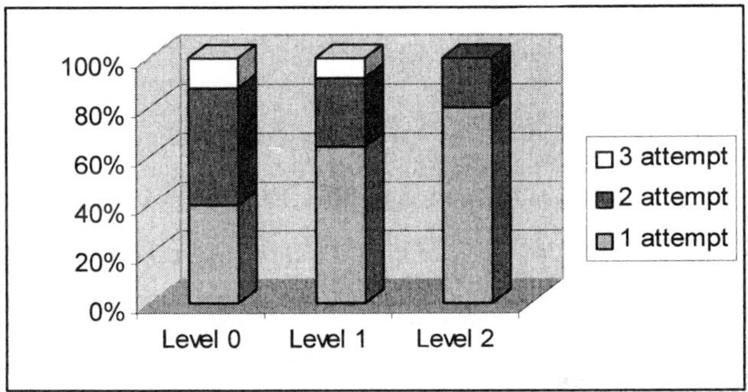

Figure 1: Percentage of successful enrolments

After enrolment people were required to authenticate 6 times. This part of the experiment highlighted new problems with the system. Less than half of the sample (47%) achieved 6/6 successful verifications. Some 28% could be verified 5 out of 6 times and 8% 4 times. The remaining 17% were successful only 3 times or less out of 6 (with 1 participant never being successful).

Because of its severely skewed distribution, the number of successful verifications were analysed by the Kruskal-Wallis H tests, the non-parametric analogue of one-way analysis of variance. Two analyses were conducted to test the effect of feedback received and number of enrolment sessions on the likelihood of successful verifications. Number of enrolment sessions was considered as an indicator of individual difficulties in using the system.

To test the effect of feedback, the final level of enrolment was entered as grouping variable. Results demonstrated that the likelihood of successful verification is not affected by type of instruction and feedback received ($\chi^2_{(1)}$=.16 p = ns).

To test the effect of number of enrolment sessions, two groups of people were compared: those who succeeded at the first attempt and those requiring 2 or more. This analysis showed a significant effect ($\chi^2_{(1)}$= 8.46, p = < .01). Independent of the level of enrolment, people who completed their enrolment after one attempt were much more likely to authenticate that people who needed 2 or more different enrolment sessions. The first group succeed on average 5.2 times, the second group 4.2. This finding confirm the hypothesis that some people had individual problem with the system which could not be solved by instruction, training or feedback.

In the questionnaire people reported believing the fingerprint system was acceptable and would be more secure than PIN.

Discussion

Although biometric technologies are still improving, there are inherent performance limitations which remain and are extremely difficult to work around, except perhaps by combining multiple technologies. These limitations are unique to each kind of biometric technology. We found that some participants simply could not use the system and we were unable to identify the reasons for the failure. This situation would be unacceptable at an ATM if it prevented people from accessing their money.

Our evaluations with the diverse and untrained general public push technology to its limits and see past the hype. Work is still required to ensure biometric technologies are universally usable but resistance to this approach is reducing. Our pluralistic approach has ensured both a broad and deep understanding of the issues to be resolved and understanding the impact of the users on the success of new technologies.

References

Clark, R. 1994, Human Identification in information systems: Management challenges and public policy issues, *Information Technology and People*, 7,4,6-37.

Coventry, L. and Johnson, G.I. 1999. More than Meets the eye! Usability and Iris Verification at the ATM Interface. In S. Brewster et al (eds) *Proceedings of the IFIP TC 13 International Conference on Human Computer Interaction – Interact 99*, Vol 2 (IOS Press/IFIP)

Coventry, L., Johnson, G.I. and De Angeli, A. 2003. Usability and Biometrics at the ATM. *Proceedings of the ACM Human Factors in Computer Systems- CHI'03* (ACM Press)

Davies, A. 1997, The Body as password, *Wired,* July.

Davies, S.G. 1994, How biometric technology will fuse flesh and machine, *Information Technology and People*, 7,4.

Deane, F., Barrelle, K., Henderson, R. and Mahar, D. 1995, Perceived acceptability of biometric security systems. *Computers and Security*, 14, 225-231.

Deane, F.P., Henderson, R.D., Mahar, D.P. and Saliba, A.J. 1995. Theoretical examination of the effects of anxiety and electronic performance monitoring on biometric security systems, *Interacting with Computers*, **7**, 395-411.

De Angeli, A., Coutts, M., Coventry L., Johnson, G.I., Cameron D., and Fischer M. 2002. VIP: a visual approach to user authentication. *Proceedings of the Working Conference on Advanced Visual Interfaces AVI 2002*, 316-323 (ACM Press)

Hone, K.S., Graham, R., Maguire, M.C., Baber, C, and Johnson G.I. 1998. Speech technology for automatic teller machines: an investigation of user attitude and performance, *Ergonomics*, **41**, 7, 962-981.

Phillips, P.J., Martin, A., Wilson, C.L. and Przybocki, M. 2000. An introduction to evaluating biometric systems. *Computer*, February, 56-62.

Woodward, J.D. 1997. Biometrics: Privacy's Foe or Privacy's Friend? *Proceedings of IEEE*, **85**, 9, 1480-1492.

USABILITY AND USER AUTHENTICATION: PICTORIAL PASSWORDS VS. PIN

Antonella De Angeli, Lynne Coventry, Graham Johnson & Mike Coutts

NCR-FSD Advanced Technology & Research,
Discovery Centre, 3 Fulton Road
Dundee DD2 4SW

This paper presents the design and evaluation of the Visual Identification Protocol (VIP) an innovative solution to user authentication based on pictures and visual memory. Three authentication systems were prototyped and compared with the PIN approach in a longitudinal evaluation. The study revealed important knowledge about attitudes towards, and behaviour with, different authentication approaches. VIP was found to be easier to remember and preferred by users, but its usability can be easily disrupted by an inappropriate solution. A detailed error analysis is presented to help understand the limits and constraints of visual memory. This knowledge is instrumental in designing innovative authentication systems of this type.

Introduction

User verification is a crucial component of secure systems that provide access to valuable information or offer personalised services. You cannot withdraw money at the ATM (Automated Teller Machine), log in to your computer, or place a call on your mobile phone without remembering a sequence of numbers or letters. Despite such a wide diffusion, Personal Identification Numbers (PINs) and passwords have a number of well-known deficiencies reflecting a difficult compromise between security and memorability (Adams and Sasse, 1999). Secure codes correspond to random selection of alphanumeric strings being as long as the system allows but humans struggle to remember meaningless strings. Thus, people choose passwords that are related to their everyday life and are often lax about the security of this information, writing it down or deliberately sharing it.

The approach to user authentication so far has been very technical in nature, concentrating on encryption methods and transmission protocols. Few proactive actions have been proposed to raise security awareness and drives secure behaviour (Adams and Sasse, 1999). Yet, current systems still suffer from a neglect of human factors. Biometrics techniques have been proposed as a solution to memory limitations, but there are still many open issues with respect to adopting them in the ATM environment (Coventry *et. al.*, 2003).

This paper reports on the design and evaluation of the Visual Identification Protocol (VIP) a concept aimed at improving user authentication in self-service technology by replacing the precise recall of a numerical code with the recognition of previously seen images, a skill at which humans are remarkably proficient. The limits and potential of the approach are discussed based on a literature review and our experimental findings. In particular, the paper concentrates on errors and mnemonic interference, proposing guidelines to exploit the advantage of visual memory as a means for user authentication.

Exploiting visual memory

Humans have a vast, almost limitless memory for pictures. Classic cognitive science studies have shown that images are usually remembered far better than words or numbers (Madigan, 1993). Furthermore, it has been argued that visual memory is less affected by the general decline of cognitive capabilities associated with ageing than other types of memory (Park, 1997). Picture superiority over alphanumeric material has been attributed to encoding, storing and retrieval differences (Madigan, 1993). In general, free recall suits alphanumeric stimuli; recognition suits visual ones.

The idea of exploiting visual memory in user authentication is not new. Blonder (1996) patented a graphical password requiring users to touch predetermined areas of an image for authentication. The theory was advanced and implemented on a PDA, thus exploiting the input capabilities of graphical device (Jermyn *et al*, 2000). Drawings proved to be harder codes to break than passwords and were assumed to be easier to remember. An user evaluation demonstrated that when an exact match is required, drawings are more difficult to replicate than passwords (Goldberg *et al*, 2002).

Graphical codes are becoming increasingly popular in personal technology. A common design solution requires the user to select target pictures among a set of distractors. An example is Passfaces by Real User Corp., based on face recognition. Users are given 'five faces', which represent their visual code. Each 'face' is displayed on a separate screen amongst different distractors. A field evaluation of this scheme revealed controversial results and did not fully support the expected superiority of faces against password (Brostoff and Sasse, 2000). Following a similar paradigm, Dhamija and Perrig (2000) investigated the memorability of abstract and photographic pictures against passwords and PINs. They showed that creating passwords and PINs is much faster than selecting an image portfolio, with photographic pictures requiring the longest time, but pictures are less error prone after a week interval.

A different approach to graphical authentication requires the user to simulate familiar actions on a graphical interface. An example is V-go by Passlogix, where users can 'mix a cocktail' or 'cook a meal' and the authentication code corresponds to the sequence of objects they clicked on. Despite the growing interest generated by different approaches to visual authentication systems, we are still far away from robust solutions, which could stand up to the challenge of public technology. Our current proposal concentrates on maximising security of personal technology and may overestimate visual-memory potentialities. In particular, little attention has been devoted to understanding memory failures and use of this knowledge to improve graphical authentication systems. This paper is a preliminary contribution in this direction.

The Visual Identification protocol

VIP is intended to improve user authentication in self-service technology, supporting easy, fast and secure interaction. The objective is to find the best compromise between security and usability, following a user centred design process. VIP consists of a self-enrolment and an authentication. At enrolment, the user is automatically given an image portfolio, without the need for a printout. Different authentication processes were designed and evaluated (Table 1). All of them displayed detailed, colourful and meaningful photos of objects on a touch screen interface.

Table 1. Visual authentication systems

Prototype	Type of code	Location	Security scoring
VIP1	4 fixed order images from 10	Fixed	Same as PIN
VIP2	4 fixed order images from 10	Random	Intermediate
VIP3	Portfolio based	Random	Maximum

VIP1 is the pictorial equivalent of the PIN (Figure 1). It requires memorising a sequence of 4 pictures, which are always displayed in the same location on the visual keypad and must be entered in a fixed order. The interface resembles the PIN keypad but a new set of distractors is extracted from the database whenever the user makes an authentication attempt. Images in the visual database are clustered in semantic categories sharing common subject matter (e.g., flowers, animals, rocks, landscapes, etc). Each picture from the authentication code belongs to a different semantic category and the distractors are selected from the remaining categories. From a security point of view, VIP1 add to the traditional PIN all of the intrinsic advantages of pictures. Images are more difficult to describe verbally which should prevent users from revealing their code; they are easier to remember that should decrease the need to write down codes.

VIP2 differs from VIP1 in that the 4 pictures forming the authentication code are displayed in new random positions around the set of 10 locations of the visual keypad at each authentication attempt. Because of that, VIP2 is more secure than VIP1. It minimises the risks related to shoulder surfing since the position of the code is always different. VIP3 is a different concept, designed to investigate the limits of the visual paradigm. The user is assigned a portfolio of 8 pictures. At every authentication attempt, four of these pictures are randomly displayed in the challenge set together with 12 distractors (Figure 1). The distractors are selected from the database, avoiding the categories of the targets currently displayed in the challenge. To authenticate, the users select their images in any order. A new code is presented at each authentication trial. This minimises the risks related to several types of PIN thefts at the ATM.

Figure 1. VIP1 and VIP2 (left hand side); VIP 3 (right and side)

The evaluation

A longitudinal evaluation addressed attitudinal, cognitive, and usability issues related to the VIP approach in comparison with PIN. The experiment involved 61 ATM users (29 males and 32 females), covering a broad range of ages (from 16 to 66 years, mean = 30), and education levels. A touch screen implementation of the traditional PIN approach was used as a control condition. Participants were randomly assigned to one of the four experimental conditions corresponding to different prototypes (PIN, VIP1, VIP2, and VIP3). Data were collected at three stages: learning, test1 and test2. After swiping an ATM card participants underwent the automatic enrolment and performed 10 authentication trials (learning). The first memory test took place 40 minutes later, after a unrelated task. The second memory test took place a week later. During both tests, participants had to swipe their card and entering the code 10 times in a row, as fast and accurately as possible. In case of erroneous entry, they were automatically given up to 3 attempts, as in a normal ATM transaction.

The evaluation provided an insight into cognitive constraints of visual and numerical memory in the context of self-service, addressing both performance criteria (number of errors and entry speed) and subjective evaluation (satisfaction relative to current PIN). Principal findings are summarised below, a more detailed report can be found in (De Angeli et al. 2002).

VIP was found to provide a promising and easy-to-use alternative to the PIN approach. The comparison between VIP1 and PIN demonstrated that pictures are less error prone than numbers after a week interval without compromising the speed of the transaction. The users' reaction to the VIP concept was promising. Overall, users liked VIP better and perceived it as more secure and easy to remember than PIN. Although these results need to take into account the novelty factors, they suggest potential acceptance of the VIP paradigm.

No difficulties emerged with respect to sequence retrieval in visual recognition. As a matter of fact, sequence errors were more frequent when participants had to retrieve sequences of numbers than when they had to recognise sequences of pictures. According to follow-up interviews, users supported sequence retrieval in visual configurations, by incorporating them into a mnemonic story.

The 'best' visual condition as regards objective and subjective evaluation was VIP1. VIP1 differed from VIP2 in that code retrieval was supported by an implicit memory related to the hand movement and to the position of the objects. These cues appeared to be important: they decreased the amount of errors and speeded up the action of entering the code. Participants were also more positive when evaluating VIP1 than any other visual conditions. The 'worst' design condition as regards performance criteria was VIP3. Participants who used VIP3 constantly achieved the slowest and more error prone performance. Nevertheless, VIP3 is the more secure solutions. The following error analysis can help us in the challenge of improving the usability of secure solutions.

Error analysis

None of the participants forgot their visual code throughout the duration of the experiment. However, in almost 6% of the authentication trials (100/1661) the users could not enter the correct code. These errors were not homogeneously distributed among the three conditions, $\chi^2_{(2)} = 46.95$, $p < .001$. Rather, they concentrated in condition VIP3 that accounted for 62% of all the errors (Figure 2). A slight advantage of VIP1 over VIP2 was also observed, $\chi^2_{(1)} = 3.42$, $p = .07$, supporting the idea that fixed location improves code retrieval.

To better understand the factors triggering errors, every wrong selection was tabulated according to its type. Multiple errors occurring in the same selection were coded as independent entries, leading to a sample of 120 errors. The following four error categories emerged from the analysis.

- *Sequence*: the correct code is retrieved but entered in a wrong order.
- *Double click*: the same item is unintentionally selected two consecutive times (the prototype did not allow corrections).
- *Double selection*: the same item is selected twice in non-consecutive positions.
- *Wrong selection*: one or more of the selected items does not belong to the authentication code.

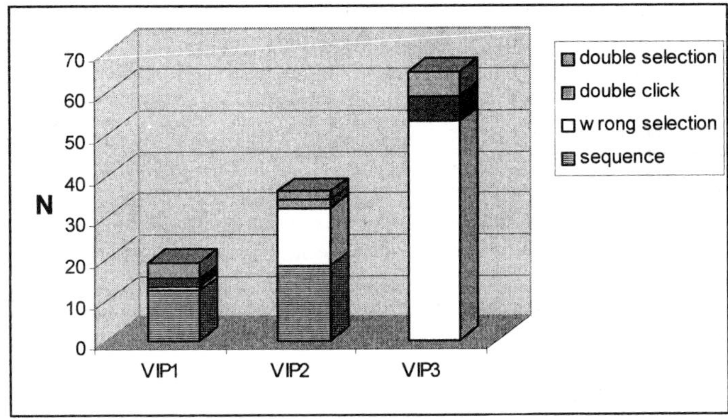

Figure 2. Error types as a function of experimental conditions

Analysing the graph in Figure 2, it emerges that different system configurations trigger specific error types. The poor performance of VIP3 was mainly due to wrong selections. All but four of these errors were due to intra-category confusions: participants tended to falsely recognise distractors belonging to the same category of items from their portfolio, which were not displayed in the current challenge set. Note that this type of errors could not occur in conditions VIP1 and VIP2 because of the selection algorithms implemented. It is worth noting, that some 63% of intra-category errors involved flowers, so that if participants had a flower in their portfolio, they were very likely to identify other flowers in the challenging set as 'their flower.

The wrong-selections occurred in condition VIP 2 (N=14) can be explained by physical proximity to the target and/or similarity. Physical proximity appears to reflect a slip in movement planning, whereas similarity reflects a different type of visual interference. It occurred particularly when the target was visually complex and difficult to label verbally (e.g., rocks and minerals, or skies view). In this case targets were confused with distractors which had very similar visual configurations (same shape or colour) even if they belonged to other semantic categories (e.g., a yellow flower mistaken for a yellow mineral).

VIP1 and VIP2 were mainly affected by sequence errors. However, 95% of them occurred at the very beginning of the interaction, when the user had to build an understanding of the way the system worked.

Conclusion

Visual memory is sensitive to interference (Goldstein and Chanche 1970). This is a natural consequence of the constructive nature of visual memory: a general concept rather than details is stored. This caused problems with users of the prototype systems.

This research has shown that there are potential usability advantages of pictorial authentication over PIN but that there are design features which may have hampered the picture superiority effect. We found that those pictures in our system, sharing common subject matter (i.e., belonging to the same semantic category) interfered with each other. Fine discrimination among items of the same category was difficult. Some problems were also found with pictures which belonged to different semantic categories but looked visually similar when presented in a small picture, for instance a group of yellow flowers was confused with a yellow mineral.

Applying this knowledge gained from this first experiment resulted in two guidelines for future systems design: Firstly use concrete, nameable, clear, coloured, and distinctive images, since they are easier to remember and preferred. Secondly control the visual configuration of the challenge set. In particular, avoid displaying distracters from the same semantic category of items belonging to the user portfolio. These guidelines have been applied to the generation of new visual images which are currently being evaluated as part of a longitudinal study into the memorability of PINs, passwords and pictorial codes.

References

Adams A. and Sasse M.A. 1999, Users are not the enemy, *Communications of the ACM*, 42, 41-46

Blonder, G. E. 1996, Graphical password. *United States Patent 5559961*

Brostoff, S., & Sasse, A. 2000, Are Passfaces more usable than passwords? A field trial investigation. In S. McDonald (ed.) *People and Computers XIV - Usability or Else*! Proceedings of HCI 2000, (Springer) 405-424

Coventry, L., De Angeli, A. and Johnson, G. 2003, Usability and biometric verification at the ATM interface. *Proceedings of the ACM Human Factors in Computer Systems CHI'03 Conference*, Fort Lauderdale, (ACM Press).

De Angeli, A., Coutts, M., Coventry, L., Johnson, G.I, Cameron, D. and Fischer M. 2002. VIP: a visual approach to user authentication. *Proceedings of the Working Conference on Advanced Visual Interfaces AVI 2002*, (ACM Press), 316-323

Dhamija, R. and Perrig, A. 2000, Déjà vu: A User Study Using Images for Authentication. In *Proceedings of 9th USENIX Security Symposium*, 45-58.

Goldberg, J., Hangman, J. and Sazawal, V. 2002, Doodling our way to better authentication. Poster presented at *ACM Human Factors in Computer Systems: CHI'02*, Minneapolis (ACM Press)

Jermyn, I., et al. 2000. The design and analysis of graphical passwords. *Proceedings of the 9th USENIX Security Symposium.*

Madigan, S. 1993, Picture memory. In J.C. Yuille (ed.) *Imagery, memory, and cognition: Essays in honor of Allan Paivio*. (LEA Hillsdale, NJ), 66-89

Park, D.C. 1997, Ageing and memory: Mechanisms underlying age differences in performances. In *Proceedings of the 1997 World Congress of Gerontology.*

A COMPARISON OF DIRECT USER TESTING IN DIFFERENT ENVIRONMENTAL CONTEXTS

Courtney Grant [1] and Steve Cummaford [2]

University College London, 26 Bedford Way, London, WC1H 0AP
[1] now at Atkins, 650 Aztec West, Almondsbury, Bristol, BS32 4SY
E-mail: courtney.grant@atkinsglobal.com
[2] now at TWIi, London www.twii.net
E-mail: steve@cummaford.com

Previous studies have compared the effectiveness of user testing with other usability evaluation methods. Such comparisons have shown that whilst user testing is more costly to conduct than expert evaluations, it also tends to identify new problems, which are not identified by the other techniques. However, these studies examined only lab-based testing, and did not consider the effectiveness of testing in other contexts, e.g. the user's home or workplace. This paper reports a small study that evaluated the relative effectiveness of conducting user testing of a commercial website in different environmental contexts (user testing in a formal usability lab versus user testing in an informal home setting). Effectiveness was expressed in terms of the amount of severe, moderate and minor usability problem types identified in each environmental context, in addition to the amount of usability problem instances found. The main finding was that user testing in the home identified the largest number of severe and moderate usability problems, whereas user testing in the lab identified the largest number of minor usability problems. The paper concludes with some recommendations for effective testing practice, based on the finding.

Introduction

The purpose of conducting a usability evaluation of an interface is to assess whether a system meets specified user performance objectives, and to identify usability problems that prevent these objectives from being achieved (Karat, 1994). Practitioners apply usability techniques in order to complete these evaluations; these techniques can be classified under two different categories; direct user testing and usability inspection methods (i.e. heuristic evaluation, cognitive walkthrough) (Nielson, 1994).

In order to ensure that practitioners select and apply the most effective technique for evaluating a given interface, there is a need to understand how effective each of these techniques are for identifying usability problems. Previous researchers have tended to focus on comparing user testing with various inspection methods, e.g. Jeffries, Miller, Wharton & Uyeda (1991), Karat, Campbell & Fiegel (1992). These studies have focused on the effectiveness of direct user testing for identifying usability problems with an interface, as compared to usability inspection methods.

Current research objective

The aim of the current research is to build on the previous findings by looking at the importance of the environmental context in which user testing is actually conducted. This is an issue that, especially in terms of the world-wide web, has been neglected by previous research. Comparisons focusing solely on comparing user testing in different environmental contexts are virtually non-existent, as user testing research tends to be conducted in a formal usability lab. Rubin (1994) has suggested that when carrying out direct user testing, the actual work environment (the environment in which interactions occur in non-test conditions) should be represented as closely as possible. It can be argued that web-based user interactions predominately occur in an informal environment i.e. at home, which is an environment in which the users will be extremely familiar and comfortable with. The features of this environment are significantly different to those of a formal usability lab.

There is a need to compare user testing performance in different environmental contexts, as important psychological factors, which are affected by context, could have a significant impact on user performance. For example, users may feel less comfortable carrying out a test in an environment that they are not familiar with. Users who travel to a lab to take part in a test may try to hide their confusion, in order to not appear 'stupid' in front of the testing team. This phenomenon, known as the Hawthorne effect, is characterized by the findings of the study being negatively affected by environmental factors, e.g. the context in which the test is conducted. Importantly, it can be argued that the Hawthorne effect is more likely to occur in a formal environment. The rationale for this argument is based on the fact that previous research has shown that users are less likely to exhibit their natural behaviour in an alien environment under test conditions. This is because they are constantly aware of the fact that their performance is being monitored (Preece, Rogers, Sharp, Benyon, Holland, Carey, 1994). This in turn can affect user performance. Therefore, an informal, familiar environment could yield markedly different user performance than a formal, unfamiliar environment.

Furthermore, Nielson (1997) conducted research which supports the argument that the typical internet user has zero-tolerance when it comes to encountering usability problems. The findings suggest that internet users are unlikely to persevere with an unusable interface during real interactions with a site (i.e. not under test conditions). It can be argued that an informal environment that is familiar to the user is more likely to yield natural performance with a reduced chance of the Hawthorne effect occurring, therefore making it more likely for this 'zero tolerance' behaviour to emerge. This could in turn give a better insight into which design features of a given interface cause usability problems.

This forms the basis for the argument that an informal environment that is familiar to the user may possess a higher level of contextual validity than a formal environment that is less familiar to the user, bearing in mind that web-based interactions normally occur in the former. Therefore, data derived from this environmental context could be more representative of actual user performance, and thus form the basis for a more effective prescription for interface re-design.

Description of the website used in the current research

The Boots Opticians website was used in the current research. This website allows the user to view information about the products and services offered by the company; therefore it can be described as an informational website (Kanter & Rosenbaum, 1997). At the time in which the research was conducted, there were no e-commerce facilities available on the website. (Please note that the Boots Opticians website has been significantly re-designed since this research was completed).

Performance Setting
Prior to carrying out the user testing in the different environmental contexts, it was necessary to specify the target users of the system, and their motivations for interacting with the site. The rationale for doing this was to ensure that each user task had a defined goal. In order to complete this, it was necessary to make key assumptions about the likely users of the Boots Opticians website. It was assumed that the users had significant internet experience, and that these users had a genuine motivation for using the Boots Opticians website to perform web-based interactions. This motivation centred upon the desire to access information about products such as glasses and contact lenses.

The users were given a list of tasks, which involved completing key user and business goals on the site – learning about eye-care, browsing product information and booking an appointment for an eye test and fitting.

In order to set desired performance, it was necessary to capture the correct action sequences required to successfully complete each task, in addition to the interface responses to these actions. A cognitive walkthrough of the system was conducted in order to achieve this. The output was a specification of the user and computer behaviour for each of the tasks, which represented how each task could be completed with optimum task quality and minimal resource costs. Usability problems were then defined in terms of deviations from this ideal task specification.

Method

Overall Research Design
The research adopted a between-participants design, and so different users were involved in the two different user testing contexts. The independent variable was the usability context, which had two different contexts (user testing in a home environment and user testing in a Laboratory environment). The dependent variable was the amount of usability problems identified in each context.

Participants & Materials
5 participants were used in each of the two different usability contexts. The participants were all experienced Internet users but had no prior experiences with the Boots Opticians website. In the informal environment, a Camcorder was used for recording user interactions. The users were recorded in their own homes in locations across London.

The lab-based testing was conducted at a formal usability lab London. In this condition, a software tool was used to record user interactions. The software enabled playback, pause and rewind user interactions, and combined a video image of the user's face with a clear screen image. This enabled detailed analysis of screen events to be conducted after the tests.

Procedure

Firstly, the users were given an initial instruction sheet that assured them that the aim of the study wasn't to test their own abilities in any way, but that it was to evaluate and ultimately improve the website. They were then given a short set of instructions for each of the tasks they were asked to complete with the interface. They were told to read the instruction sheet for the first task and then complete the first task, then they could move on and read the instruction sheet for the second task and then complete that task, etc. They were also told that if they really felt that a given task could not be completed, they should move on to the next task. They were also asked to verbalise their thoughts and actions as much as possible whilst carrying out the tasks. A sheet with 'made-up' personal details was also provided for the users, as some of the tasks required the user to fill out personal details and send them to Boots Opticians. The duration of the test was approximately 30 minutes.

After the users had completed the tasks on the website, they were asked to rate how comfortable they felt carrying out the tasks, and whether the particular environment that they were in had any impact on this. This part of the study was incorporated to see whether there were any potential psychological differences between the users in these two environments in terms of how comfortable they felt carrying out the tasks. They were asked to read the following statement 'I felt comfortable carrying out the Internet-based tasks in this environment'. They were then asked to rate to what extent they agreed with this statement using a seven-point scale, ranging from strongly agree to disagree.

Results

Data Analysis

The usability problems identified in each of the user testing contexts were classified using an approach adapted from Mack & Montaniz (1994). Each usability problem identified was defined at the high-level, task, and action levels, in addition to a general description of the usability problem.

Using this approach enabled usability problems to be adequately classified in terms of the differentiation between what constitutes two different usability problems, and what constitutes an instance of the same problem. Total usability problem instances refer to usability problem duplicates (i.e. the same usability problem experienced by more than one user, or one user experiencing the same usability problem on different tasks, etc). Total problem types refer to the total number of usability problem types identified by each usability technique. Therefore, problem types can be said to reflect all usability problem instances identified by each technique minus any duplicates (Mack et al, 1994). Two or more usability problems were deemed to be different instances of the same problem if the high-level problem category, action level and general description of the problem were the same. Therefore, the same usability problem type can occur on different tasks.

Usability problems were given a severity rating using a method adapted from Karat et al (1992), known as 'Problem Severity Classification (PSC) Ratings'. This method calculates usability problem severity using two factors: user frequency and task impact, to generate minor, moderate and severe usability problems.

Findings

The findings were generally small between the two different usability contexts. User testing in the lab proved to be the most effective condition for identifying usability problem instances; 57 different usability problem instances were identified in this condition as compared to 55 found in the home environment. User testing in the lab also proved to be the most effective for identifying usability problem types; 30 problem types were found in this condition, compared to 27 identified in the home environment. User testing in the home was the most effective context for identifying severe usability problems; 5 severe problems were identified in this condition as compared to 4 found in the lab. Furthermore, user testing in the home proved to be the most effective condition for identifying moderate usability problems; 7 moderate problems were found in this condition as compared to 6 identified in the lab environment. User testing in the lab proved to be the most effective context for identifying minor usability problems; 20 minor problems were found in this condition as compared to 15 found in the home environment.

Discussion and conclusion

The main finding from the study was that user testing in the home identified the largest number of severe and moderate usability problems, whilst usability testing in the lab identified the largest number of minor usability problems. Although the differences in findings between the two testing contexts were small, they do suggest that testing in both contexts does result in a more comprehensive identification of usability problems than relying on only one context. In particular, the home testing condition identified some moderate and severe problems, which were not identified in the lab. These problems may have come to light because the users were more comfortable about expressing their frustration or incomprehension, when in their own home environment. The implication is that home testing can provide valuable extra data, which can be used to improve the design of consumer-focused websites and products.

The fact that the differences between the two usability contexts were small in terms of the number of problems identified, combined with the lower costs of user testing in the home, indicates that user testing in the home may be more cost-effective than user testing in the lab environment, in terms of finding usability problem types.

This research only constituted a relatively small-scale study. It is suggested that further research is conducted in this area with a larger number of users, in order to further explore and validate these findings. The current research only used five participants in each user testing condition. The rationale for using this number of participants was based on Nielson's (2000) assertion that 'five users is enough'. However, this assertion has been challenged, e.g. Spool (2001) conducted a user test with an online music store, and found it to be the case that there was no significant occurrence of the 'repeat effect'.

To conclude, the study highlighted the fact that user testing in different environmental contexts is an issue which deserves consideration. It is useful to consider the fact that the environmental context in which user testing is conducted can have an impact on human behaviour, and in turn can influence performance.

References

Jeffries, R, Miller, JR, Wharton, C and Uyeda, KM (1991). User Interface Evaluation in the real world: A comparison of four techniques. *Proceedings ACM CHI'91 Conference* (New Orleans, LA, April 28-May 2): 119-124

Kanter, L & Rosenbaum, S (1997). Usability Studies of WWW Sites: Heuristic Evaluation vs. Laboratory Testing. *SIGDOC Proceedings*, Association for Computing Machinery, Inc

Karat, CM (1994). A Comparison of User Interface Evaluation Methods. In Nielson, J & Mack, RL (Eds), *Usability Inspection Methods*. John Wiley & Sons, Inc

Karat, CM, Campbell, R & Fiegel, T (1992), Comparison of Empirical Testing and Walkthrough Methods in User Interface Evaluation. *Proceedings ACM CHI'92 Conference* (Monterey, CA, May 3-7): 397-404

Mack, R & Montaniz, F (1994). Observing, Predicting and Analyzing Usability Problems. In Nielson, J & Mack, RL (Eds), *Usability Inspection Methods*. John Wiley & Sons, Inc

Nielson, J (2000). Why you only need to test with 5 users. *http://www.useit.com/alertbox/20000319.html*

Nielson, J (1997). Report from a 1994 Web Usability Study. *http://www.useit.com/papers/1994_web_usability_repart.html*

Nielson, J (1994). Heuristic Evaluation. In Nielson, J & Mack, R (eds), *Usability Inspection Methods*. John Wiley & Sons, Inc

Preece, J, Rogers, Y, Sharp, H, Benyon, D, Holland, S, Carey, T (1994). *Human-Computer Interaction. Addison*-Wesley

Rubin, J (1994) *Handbook of usability testing: How to plan, design and conduct effective tests.* New York, NY. John Wiley & Sons

Spool, JM (2001). Eight is not enough. Available from: *http://www.uie.com*

THE USE OF COOPER-HARPER TECHNIQUE TO ENHANCE THE MEASUREMENT OF RATED SENSE-OF-PRESENCE IN VIRTUAL AND REAL ENVIRONMENTS

Edith Ng and Richard H.Y. So*

Department of Industrial Engineering and Engineering Management
Hong Kong University of Science and Technology
Clear Water Bay, Kowloon, Hong Kong SAR
Email: rhyso@ust.hk

* *All correspondence should be sent to Dr. Richard H.Y. So*

A recent study has shown that the Presence Questionnaire (PQ) developed by Singer and Witmer (1996) fail to detect a significant difference between the sense-of-presence reported by participants exposed to a virtual environment (VE) and to its corresponding real environment (RE). Two experiments were conducted to examine the reported failure of PQ and study the benefits of using Cooper-Harper presentation technique. In the first experiment, participants were exposed to a virtual, telematic, and real environment showing similar audio and visual content. In contrary to the literature, the reported PQ total scores reported for the RE condition was significantly higher than that of the VE condition ($p<0.05$). This suggests that PQ can be a sensitive tool to measure rated sense-of-presence. In the second experiment, the use of Cooper-Harper technique to present PQ yielded mixed benefits. Actual and potential applications of this research include the development of an enhanced PQ that can be used across virtual, telematic, and real environments.

Introduction

The Presence Questionnaire (PQ) (Singer and Witmer, 1996; Witmer and Singer, 1998) is a set of questionnaire developed to measure the subjective sense-of-presence in a virtual environment. The questionnaire contains thirty-two 7-point Likert scale questions and the answers to nineteen out of the thirty-two questions are summed to make a PQ total score.

In 2000, Usoh and his colleagues compared the PQ with another questionnaire that they had developed on participants exposed to a real environment and a virtual environment showing similar visual content. They found that while their own questionnaire was able to detect a significance difference between the rated sense-of-presence from the two groups of participants, the PQ failed to do so. This suggests that PQ may not be sensitive to small changes of sense-of-presence. Since the real environment is often used as a benchmark for assessing a virtual environment, a PQ that is sensitive enough to detect a difference between a real environment and a virtual environment will be very useful. The aim of this study is to examine the sensitivity of PQ and seek to improve it using Cooper-Harper presentation technique. Experiment one is conducted to retest the ability of PQ to detect the difference between the rated sense-of-presence from participants exposed to a virtual environment and a real environment showing similar visual and audio content. Experiment two will study the benefits of using Cooper-Harper presentation technique.

Experiment one

Objectives and hypothesis

The objective of this experiment was to study the rated levels of sense-of-presence among participants exposed to a virtual, telematic, and real environments showing similar visual and audio content. It was hypothesized that the PQ total score reported for the real environment and that for the virtual environment was not significantly different. This hypothesis was based on the finding reported in Usoh *et al,* (2000).

Method and design

A full factorial randomized design was used with 12 replications. The experiment has one independent variable that was the types of experimental environment with four levels: (i) real environment with participants wearing a head-mounted weight (RE_weight), (ii) real environment with participants wear a head-mounted weight and a goggle that restrict their field-of-view to 48° (horizontal) by 36° (vertical) (RE_restricted), (iii) telematic environment (TE) presented on a VR4 LCD head-mounted display (HMD) manufactured by Virtual Research Systems, Inc., Santa Clara, CA with 48° (horizontal) by 36° (vertical) field-of-view, (iv) virtual environment (VE) presented on the same VR4 HMD. The visual and audio content of these four experimental environments had been engineered to be as similar as possible and both the VR4 HMD and the head-mounted weight were 1kg. Figure 1 compares a view of the RE_restricted and the corresponding view of the VE. It can be observed that the two environments do look similar. Forty-eight male Chinese university students volunteered for the experiment and they were randomly assigned into the four environments so that there were 12 participants for each environment. Before entering the experimental environment, each participant was asked to complete an Immersive Tendency Questionnaire (ITQ) (Singer and Witmer, 1997) and pre-exposure simulator sickness questionnaire (SSQ, Kennedy *et al.*, 1993). After the exposure to the environments, participants were asked to complete the PQ and the post-exposure SSQ.

During the exposure inside the environment, the participants would be asked to perform a series of visual search tasks, visual-motor tasks (e.g., pick-and-place tasks, writing tasks), and sound localization tasks. These tasks were constructed according to the virtual environment performance assessment batteries (VEPAB, Lampton *et al.,* 1994). In the RE_weight, RE_restricted, and TE conditions, participants used their real hands to manipulate objects. In the VE condition, participants used their real hands to control a virtual hand to manipulate virtual objects. The former way involved full tactile interactions while the latter did not involve any tactile interactions. The latter process was facilitated by the use of a Cyberglove™ manufactured by Immersion Corporation, San Jose, CA.

(a) a view of the virtual environment

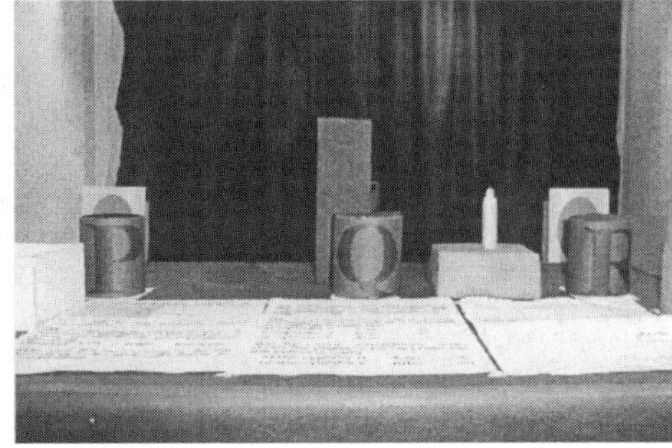

(b) a view of the real environment

Figure 1. Snap shots taken in the real environment and the virtual environment specially constructed to have the similar visual and audio content (the two snap shots are taken from approximately the same view-point) (adopted from So and Ng, 2003)

Results and discussion

This paper will only present the PQ total score results and the SSQ and performance data will be presented later. There was no significant difference among the ITQ scores obtained from participants of the four conditions ($p>0.3$). This suggests that the participants had similar tendency of immersion and any changes in PQ scores were due to the types of experimental environment that they had been exposed to. Table 1 summaries the results of an ANOVA conducted to test the main effects of experimental environment on the PQ total scores. Inspections of the table indicate that the PQ total scores are significantly affected by the changes of environments. Further analyses using Student-Newman-Keuls test indicated that, in this experiment, the PQ was able to detect a significant difference (p<0.05) between the rated sense-of-presence as measured by PQ total scores collected from participants exposed to a real environment (i.e., the RE_restricted) and a corresponding virtual environment (i.e., the VE) showing the similar visual and audio content and with the same field-of-view (i.e., 48° horizontal by 36° vertical) and the same head-mounted weight of 1kg. This disproves the hypothesis and contradicts the previous findings by Usoh *et al.* (2000). Before concluding that the PQ is sensitive, one has to verify that the significant difference founded between the PQ total scores collected from participants in the RE_restricted and the VE was not caused by the fact that the two environments were very different in this experiment. While there is no standard objective judging criteria to assess the different between the RE_restricted and the VE environment, one could examine this issue in various ways. Firstly, a visual inspection of the snap shots taken from the two environments (Figure 1) suggests that the visual content is similar. Secondly, results of further ANOVAs indicated that the PQ sub-scores on auditory collected from the RE_restricted and the VE were not statistically different ($p>0.15$). This is not surprising because the audio content was the same in both environment (i.e., the same set of real speakers were used to present the sound cues in both environments). This suggests that audio content of the two environments were the same. Thirdly, results of further ANOVAs indicated that the PQ sub-scores on interface quality collected from the RE_restricted and the VE were not statistically different ($p>0.2$). At first glance, this result does not seem logical. Further investigations indicate that the interface quality PQ sub-scores were the sum of the answers to three questions asking about the restriction of field-of-view; the interference of control devices (e.g., the use of real and virtual hands) on performance; and the distraction of concentration due to the control devices, respectively. The lack of a significant difference in the interface quality PQ sub-score is consistent with the fact that participants in both RE_restricted and VE conditions view their environment with the same field-of-view (48° by 36°) and also suggest that the participants felt that the virtual hand interface were similar to the real hand interface. Fourthly, results of further ANOVAs indicated that the three PQ sub-scores on natural, haptic, and resolution collected from the RE_restricted and the VE were significantly different ($p<0.02$). This is also consistent with the fact that a real environment is more natural than its virtual counter-part, real objects allow haptic interactions, and real objects can be viewed with higher resolution. All in all, there are evidences to suggest that the real environment (namely RE_restriced) and the virtual environment (VE) were truly showing similar visual and audio content and yet the PQ was able to detect significant differences in rated sense-of-presence in the areas of natural, haptic, and resolution.

Table 1. ANOVA summary table to illustrate the main effects of experimental environments on the measured PQ total scores

Dependent variable	Sources	Sum of Squares	Df	Mean Square	F-value	Sig.
PQ total score	Experimental Environments	1593.729	3	531.243	6.119	.001
	Error	3820.250	44	86.824		
	Total	5413.979	47			

Although the PQ was demonstrated to be able to detect the differences between rated sense-of-presence obtained between participants exposed to a real environment and its corresponding virtual environment, further investigation suggest that participants exposed to the RE_weight condition had been “over conservative” in answering 18 of the 32 questions in the PQ. In other words, participants were reporting that they were in a less-not-natural environment while they were in fact in a real environment. Interviews with the participants revealed three possible reasons: (i) some participants misunderstood the questions, (ii) some participants did not understand the questions and, therefore, selected the middle rating of a 7-point Likert scale, and (iii) the RE_weight environment in this experiment was, in fact, a less-than-natural environment due to the cables associated with the head-mounted weight. Those cables, hanged from the head-mounted weight to the ground, were added as part of the head-mounted weight because the VR4 HMD also had similar cables. Examples to illustrate the first reason include the interpretation of the third question in the PQ. This question asks "how natural did your interactions with the experimental environment seem?". Many participants interpreted that because they were taking part in an experiment, therefore, all of the interactions were unnatural because they were the results of specific instructions from the experimenter.

Experiment two

Objectives and hypothesis

The objective of this experiment was to study the benefits of using Cooper-Harper presentation technique to reduce the level of misunderstanding of the questions in PQ (Cooper and Harper, 1969). The measure of benefits used in this experiment was the answers to the 18 questions identified in Experiment one that had yielded "over conservative" answers. It was hypothesized that the PQ ratings to the selected 18 questions would be significantly higher (i.e., more natural) after the use of Cooper-Harper presentation technique.

Method and design

The PQ was modified using the Cooper-Harper presentation technique originally developed for presenting questions to assess the handling quality of aircrafts and simulators. In short, each of the 32 questions in PQ was presented first in its original form followed by a few long questions with multiple choice answers. Each choice was a complete statement aimed to provide better anchoring so as to avoid misunderstanding of the answer. Further details will be explained in the full version of this paper (So and Ng, 2003). The design and procedure of Experiment two was similar to Experiment one. Again, there was only one independent variable in this experiment which was the types of experimental environment. In this experiment, the four levels of the experiment environments were: (i) the VE, (ii) the RE_restricted, (iii) the RE_weight, and (iv) the real environment (RE) without head-mounted weight. The replacement of TE with RE was done so that the influence of the head-mounted weight on rated levels of sense-of-presence can be studied.

Results and discussion

Results of ANOVA and the Student-Newman-Keuls tests indicated that the PQ total scores obtained from participants exposed to the RE condition were significantly higher than those obtained from participants of the RE_weight condition ($p<0.05$). This suggests that adding a head-mounted weight with cables hanging from the head to the ground could significantly reduce the rated sense-of-presence. The answers to the 18 selected questions of PQ obtained in the RE_weight condition in this experiment was compared to those obtained in Experiment one. Because of the reduced numbers of data samples, nonparametric statistical tests were used in case the distribution of the data no longer confirmed to the normal distribution. Mann-Whitney tests were conducted to compare the answers obtained for each of the 18 selected questions in PQ presented in their original format (Experiment one) and PQ presented in the Cooper-Harper format. Results of 18 Mann-Whitney tests indicated that there were significant differences in the answers to 12 out of the 18 selected questions with higher scores obtained after the use of Cooper-Harper technique. This suggests that the use of Cooper-Harper presentation technique to present PQ could significantly affect the answers to 12 questions within the 32 questions in the PQ.

Conclusions

Evidence is found to suggest that Presence Questionnaire (PQ) is sensitive enough to detect a significant difference between the rated sense-of-presence obtained from participants exposed to a real and virtual environment showing similar visual and audio content.

The addition of a head-mounted weight with cables hanging from the weight to the ground to participants of a real environment could significantly reduce the rated PQ total scores. Some benefits were found with the use of Cooper-Harper technique to present the PQ.

Acknowledgement

This study is funded by the Hong Kong Research Grant Council through HKUST6007/98E.

References

Cooper, G. E. and Harper Jr, R. P. (1969) The use of pilot rating in the evaluation of aircraft handling qualities. AGARD-567, Paris, France: advisory group for aerospace research and development (DTIC No. AD689722)

Kennedy, R.S., Lane, N.E., Berbaum, K.S. and Lilienthal, M.G. (1993). Simulator Sickness Questionnaire: An enhanced method for quantifying simulator sickness. The international journal of aviation psychology, 3(3), 203-220.

Lampton, D.R., Bliss, J.P., Moshell J.M., and Blau B.S. (1994) The virtual environment performance assessment battery (VEPAB): Development and evaluation. Presence, Vol. 3, No. 2, Spring, pp. 145-157.

Singer, M.J. and Witmer, B.G., and Goldberg, S.L. (1996) Presence measures for Virtual Environments: Background and Development. ARI Research Note (Interim Draft). United States Army Research institute for the Behavioral and Social Sciences, Simulator Systems Research Unit, 12350 Research Parkway, Orlando, FL 32826-3276.

So, R.H.Y. and Ng, E. (2003) Measuring rated sense-of-presence in virtual and real environments: the benefits of using Cooper-Harper presentation technique. Awaiting publication.

Usoh, M., Catena, E., Arman, S. and Slater, M. (2000). Using Presence Questionnaires in Reality. Presence, Vol.9, No.5, pp.497-503.

Witmer, B.G. and Singer, M.J. (1998) Measuring Presence in Virtual Environment: a Presence Questionnaire. Presence, 7:3, pp.225-240.

BREAKDOWN ANALYSIS

Andrée Woodcock and Stephen AR Scrivener

The Design Institute, School of Art and Design, Coventry University, Gosford Street, Coventry, UK. email: A.Woodcock@coventry.ac.uk

This paper considers the use, by the authors, of breakdown analysis in the evaluation of computer supported co-operative working environments. In documenting a number of studies we consider the manner in which the method has been refined and extended to provide information that can be used in both formative and summative evaluation. In conclusion we believe that breakdown analysis provides a rapid means of extracting usability problems from audio and visual transcripts and provides more detailed, useful information for system design than many other methods.

Introduction

This paper describes the development, use and refinement of a CSCW (computer supported co-operative working) evaluation method based on communication breakdowns. The origin of the method goes back to early research in CSCW in the context of design and the debate concerning the evaluation of this technology. Hughes (1991) argued that the paradigm shift from HCI to that which considers the group work, communication, social, political, and physical aspects within which a CSCW system is used presents numerous design and development problems. While recognizing the value and potential of the traditional iterative HCI system design approach, including both formative and summative evaluation activities, we also came to realise that the particular methods used to evaluate a CSCW environment required new methods to accommodate the paradigm shift. In the first section, we describe the origin of breakdown analysis. This is followed by examples of our use of breakdown analysis over the past 10 years.

Breakdown analysis

The general aim of a formative evaluation is to produce recommendations for design improvement. Numerous evaluation methods have been developed such as the interview, usability checklist, focus group, expert walkthrough. The selection of one or more methods depends on issues such as time, finance, resources, the type and quantity of data produced, the reliability and validity and the interaction of these techniques.

CSCW system design supports complex group activities. This complexity is reflected in the amount of data required to adequately evaluate a system supporting group activities. Consequently, we felt there was a need for a method that provided reliable diagnostic information from rich data such as audio and video recordings of task focussed system use. Analysis of existing methods led us toward an approach based on "breakdowns".

A breakdown is defined as the moment when a user becomes conscious of the properties of a system and has to mentally break down or decompose his or her understanding of the system in order to rationalise the problem experienced (Winograd and Flores, 1987). Wright and Monk (1989) evaluated the use of a bibliographic data base system using both critical incidents (defined as user behaviour that is sub-optimal with respect to the functionality of a system and the intention of the user) and breakdowns. They found that both critical incidents and breakdowns provided effective evaluation data, but noted that important problems were uncovered via breakdowns and that breakdowns were easier to detect than critical incidents. In conclusion they advised for the use of breakdowns because a critical incident is usually accompanied by a breakdown, whereas the converse is not necessarily the case.

In the next part of the paper we will explore the application of breakdown analysis (BA) and our adaptations to it through a series of studies of computer supported co-operative working. The projects concern looking at the use of a shared drawing application (ROCOCO workstation: Scrivener *et al.*, 1992 and 1993), the effects of bandwidth manipulation on co-operative working (FashionNet: Woodcock and Marshall, 1997, Woodcock and Scrivener, 1997), iterative interface development (DETECT: Woodcock, 2000, Woodcock and Galer Flyte, 2001) and the identification of managerial problems arising during distance working (Lee, Woodcock and Scrivener, 2002). These are used to show the utility of breakdown analysis and how we have extended its use.

ROCOCO Workstation: Applying the TUTE framework to BA

We first used BA in the ROCOCO project (Remote Cooperation and Communication) in which a workstation was developed to allow designers in the UK and Australia to work remotely on a design task using a telephone link and shared drawing application, the ROCOCO Station. Two design dyads each took part in three sessions, each involving a separate design task. At the end of the first and last session each designer completed a questionnaire relating to the fitness for the task and system usability. These results were generally positive. However, it was obvious to observers that communication problems had arisen which were not captured by the questionnaire, that these affected usability and had implications for system redesign. Consequently, we needed a formative evaluation method that would capture these problems using data from the audio and video recordings of the design sessions.

In collaborative tasks verbal discourse is unprompted. Participants are not required to "think aloud" but do so naturally in order to cooperate with their partners. Hence, the collaborative context overcomes the criticisms leveled at the use of concurrent protocols.

Furthermore, we concluded that the breakdowns reported in the concurrent protocols of collaborative tasks were likely to be more reliable than those arising in single user system interactions as they would be reported as a means of bringing the usability problems being experienced into team awareness. Hence, we concluded that BA could be a useful tool for formative evaluation of the performance of CSCW systems.

Shackel (1991) argues that usability depends upon the design of the tool (*i.e.*, the computer system) in relation to the users, the task and the environment, and the success of the user support provided. In general, a breakdown in human-computer communication is an indicator of usability failure. The decomposition of a system into its four principal components (i.e., Task, User, Tool and Environment, or TUTE) provides a framework for usability evaluation which could be used to describe in more detail the breakdowns which occurred. The TUTE model was extended as indicated in Figure Scrivener *et al.* (1996) to accommodate the interactions in the ROCOCO Station. Here, 6 types of breakdown are defined relating to user-user, user-task, user-tool, user-environment, task-tool and tool-environment interaction and refer to vectors 1, 2, 3, 4, 5 and 6 represent respectively.

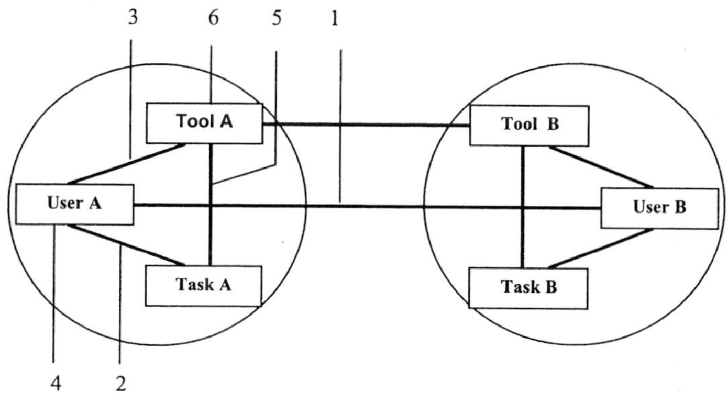

Figure 1: Interaction between two human-computer subsystems. Adapted from Scrivener, *et al.*, 1996, p.163.

In a CSCW environment the user is directly involved in four interactions (user and task, user and tool, user and environment, and user and user) each of which may be subject to breakdown. Although it is not necessary to relate breakdowns to TUTE interactions by doing so it is possible to gain an understanding of which aspects of the system are particularly problematic. For example, when we applied the method to the ROCOCO data, we found 59% of the breakdowns attributable to user-task interactions. The results of BA were fed into iterative system development following a method outlined by Urquijo *et al* (1993), encompassing three stages: 1), transcription and categorisation of breakdowns, 2), causal diagnosis; and 3), remedy prescription.

Fashion-Net Temin: Products, severity of problems and BA

Fashion-Net/Temin was an EU, TEN-IBC funded project looking at the usability of a Multimedia Network Application, (MNA)comprising an online database, shared whiteboard and videoconferencing system- ShowMe) for supporting the European fashion industry, over a trial ATM network running over 2, 4 and 8 MbSec bandwidths. The MNA was tested using fashion design dyads working together in real time, in different countries, over different bandwidth conditions. Summative and formative evaluation methods were used to consider a number of issues including cost benefits for the fashion industry, optimum application performance and comparative MNA performance under the different conditions. Usability was measured through standard questionnaires and BA. For the purposes of this CSCW evaluation BA was extended in a number of ways:

- Analysis software was developed to facilitate transcription and coding
- TUTE was extended to include the outcomes of the interactions i.e. products (e.g. design drawings, videos) as these became degraded with bandwidth. For this purpose product was defined as an entity purposefully generated either for communication or as a goal directed activity.
    - The quality of the product can be effected by the tool e.g. the quality of the sketch is affected by the hardware (pen and graphics tablet) and the functionality provided by the software (e.g. lack of airbrush).
    - Production itself can be impeded. If speech is seen as product, the distortions introduced by ShowMe required users to talk in a very unnatural manner
    - Distortion of the product by the system e.g. colours of fabrics might change with different hardware configurations, echo and distortion of speech with bandwidth restrictions.
- The seriousness of the breakdown was assessed as low, medium or high in terms of the effects it had on the session, the recovery and overall usability. The severity of the breakdown can be measured by the long terms effects and type of solutions used.
- The user was modelled as an active problem solver, reflecting on their experiences and adapting behaviour accordingly. This allowed the solutions which spontaneously arose to breakdowns to be considered

For example, 'bad echo' was classified as a Tool - Environment breakdown caused by running the application over ATM (due to latency). For the user, this created a problem in so far as they could not hear what was being said. The short term effect of this was broken speech, disrupted by delayed feedback, the long term effect was silence. Solutions chosen by the users were to either change the pattern of speech production (slower and louder), change the communication channel (write comments on the graphics tablet), or abandon the trial. Inappropriate solutions caused by false models of system performance may result in behaviour which is inappropriate and adds to the problem (e.g. shouting made the problem worse). It was concluded that when solutions lead users to behave unnaturally, reduce interaction or hinder the effectiveness of the work, application developers should re-address system usability.

Breakdown analysis provided reliable data for all bandwidth conditions which allowed us to differentiate problems to an extent not possible with the other usability instruments and also provided information of sufficient detail for the designers (the target audience of the evaluation method). BA clearly identified the nature of the problem, the overall impact on the quality of the interaction and the ability of the subjects to perform the task using the tools at their disposal. The extension of the framework, to include 'product', focuses attention on the users output, through which communication is mediated.

ADECT: Usability issues working on a shared interface

Woodcock (1999) applied breakdown analysis to the investigation of usability and usage issues to the iterative development of ADECT, a decision support system to assist designers working together (either synchronously or asynchronously) in their application and discussion of ergonomics during concept design. The initial phases of the work of design dyads and triads using ADECT and the Internet to inform a the design of a driver information system were video recorded with a view to understanding the problems designers had with the task, in working together and the tools at their disposal.

BA provided a rapid means of identifying and categorizing the problems found in 1722 minutes of activity, using a one pass through the video. For iterative system development, breakdowns were plotted against the different screens as part of formative usability evaluation. This showed that different screens had different usability issues associated with them. For example, 50% of the problems related to just one screen were attributable to problems with the insertion of data and accidental deletion. This was clearly a user-tool issue.

As a tool to support user centred design, ADECT starts with a product strategy screen - basically a series of questions to get users thinking about the product rationale and functionality. As well as providing information on User -Tool breakdowns, BA was also used to shed light on other CSCW (or co-operative working problems) in particular relating to user - task and user - user problems. For example,

- the first screen of ADECT produced a number of user- task related problems in which the design dyad was required to think through their strategy and what they were designing, prior to commencing with the design. Although a 'problem ' this was what ADECT was designed to do, i.e. bring ergonomics issues to the surface and allow misunderstandings to be articulated earlier in the design process.
- In terms of co-operative working, user -user conflicts were also noticeable in some dyads more than others.

Managerial problems in CSCW: BA and user-user breakdowns

Small design teams often work synchronously and asynchronously across continents, and are supported in their activities by a variety of tools, such as email, videoconferencing and the web. These short duration projects may be prone to many problems relating to language, cultural and technological issues which are difficult for small inexperienced partnerships to predict, overcome and put right in projects of short duration. We set up an international trial between a designer and client to understand the problems which typically arise. All exchanges between participants were recorded. BA was used as a means of identifying where problems occurred. Such problems were subsequently discussed with participant, allowing problems to be traced through the project, and managerial strategies developed to overcome them. A second study, performed on a similar international design project used BA to show the effectiveness of the managerial strategies in reducing all types of TUTE interactions.

Conclusions

BA provides a useful and flexible way of looking at interactions in CSCW environments providing information that can be used by system designers through iterative development. It provides a richer picture of the type of issues which occur and when they occur that might not be gained in questionnaires. Additionally the analysis of breakdowns can be used retrospectively with participants either in focus groups or in depth interviews to develop a fuller picture of interactions and issues which affect performance in CSCW.

References

Hughes, J. Randall, D. and Shapiro, D. 1991, CSCW: Discipline or Paradigm? A Sociological Perspective. *Proc of 2nd European Conference on CSCW* (Kluwer Academic Publishers: The Netherlands), 325-336

Lee, L-C, Woodcock, A. and Scrivener, S.A.R. 2002, Intervention strategies for alleviating Problems in International Co-operative Design Projects, Common Ground Conference, September 2002. 99

Scrivener, S.A.R., Harris, D., Clark, S., Clarke, A., Rockoff, T. and Smyth M. 1992, Designing at a Distance: Experiments in Remote Collaborative Design. *Proc. of OZCHI'92: CHISIG Annual Conference*, Australia, November 1992. 44-53

Scrivener, S.A.R., Clark, S., Clarke, A., Connolly, J., Garner, S., Palmen, H., Smyth, M. and Schappo, A. 1993, Real-time Communication Between Dispersed Work Groups via Speech and Drawing. *Wirschaftsinformatik*, No. 25, Vol. 2. 149-156.

Scrivener, Stephen A.R., Urquijo, S.P., and Palmen, H.K. 1996. "The Use of Breakdown Analysis in Synchronous CSCW System Design." In P. Thomas (ed) *CSCW Requirements and Evaluation* (Springer, London)

Shackel, B. 1991, Usability-Context, Framework, Definition, Design and Evaluation. In: B. Shackel and S. Richardson (eds) *Human Factors for Informatics Usability* (CUP: Cambridge, UK)

Urquijo,S.P., Scrivener, S.A.R. and Palmen, H. 1993,The Use of Breakdown Analysis in Synchronous CSCW System Design. *Proc. of the 3rd European Conference on CSCW* (Kluwer Academic Publishers: Dordrecht, The Netherlands). 281-293

Winograd, T. and Flores, F. 1987, *Understanding Computers and Cognition: A New Foundation* (Ablex: Norwood, NJ, USA)

Woodcock, A. and Galer Flyte, M.D. (2001), An ergonomics decision support tool for designers, in M.A. Hanson (ed),*Contemporary Ergonomics, 2001* (Taylor and Francis, London) 377-382

Woodcock, A. and Marshall, A. 1997, *Fashion-Net/Temin : Evaluation, Internal Project Deliverable, D5.* Available on request from authors

Woodcock, A. and Scrivener, S.A.R. 1999, "Multimedia Network Applications in the Fashion Industry." In B. Jerrard, R. Newport and M. Trueman (eds) *Managing New Product Innovation*, (Taylor and Francis, London)

Wright, P.C. and Monk, A.F. 1989, Evaluation for Design. In: A Sutcliffe and L. Macauley (eds.) *People and Computers V* (CUP: Cambridge, UK). 345-358

WHAT CAN YOU GET FROM A DRIBBLE FILE?

Hugh David

R+D Hastings,
7 High Wickham, Hastings,
TN35 5PB, U.K.

A 'dribble file' records the inputs to a system, showing which key was clicked or where the mouse was positioned and when. Additional data, describing the activities of the system may also be available. This data can be analysed on different levels, depending on the aims of the study. An initial level may concern the response times associated with specific keys. A second level may involve complete sequences of keys, forming a specific input instruction. A third level may examine the strategies used, in terms of sequences of orders. Although these analyses may be carried out for individual exercises, they may equally be summarised and compared between different exercises, to identify individual operating styles, the evolution of expertise or the effects of different operating conditions.

Introduction

It is increasingly easy to develop computer models of human-computer interfaces. The evaluation of such interfaces is often neglected at the initial design stage, because the design of the interface itself quickly becomes complex and demanding. Equally, it is neglected during the implementation of the model because this is proverbially more difficult than expected. Often the evaluator is thrown back on subjective measures, justifiable but less than convincing to the 'objectively' oriented. In a reasonably well-designed interface, the classical experimental psychology paradigms of error rate and failure rate should be so low that they rarely reach meaningful levels.

The 'dribble file' is a simple approach to recording human-computer interface activity. Basically, it is a record of which key is pressed, or where the mouse is positioned and when. In practice, some additional records will be necessary, for example, recording when a data display is updated.

As an example, the analysis of TROTSKY, a small digital Air Traffic Control simulator will be described; showing how successive levels of analysis can be built up.

TROTSKY

TROTSKY is a simple simulator program, written in QuickBASIC, a precursor of Visual BASIC. It demonstrates a type of ATC interface differing significantly from current conventional ATC systems, or the proposed computerised versions of these. The original principles were outlined in Dee (1996) and presented in more detail in David (1997). A fundamental principle of the revised system was that displays should be simplified and present information in graphical rather than numeric or text forms. Although general design principles suggest that a three-dimensional image would be most appropriate, a co-planar display is in practice preferable (Wickens 2000). An initial evaluation was reported at the 2001 Annual Conference of the Ergonomics Society, (David and Bastien, 2001) and a second evaluation (David and Bastien, 2003) is presented at this conference. TROTSKY uses, for demonstration purposes, a conventional low-precision colour display and a standard (QWERTY) keyboard.

The basic record stored consisted of the key being pressed (stored as a one-character 'string'), the time since the start of the simulation, and the time since the last record. It was quickly apparent that some other events needed to be recorded, such as the time at which a new display was presented. Air Traffic Control systems traditionally showed images produced by a rotating radar 'sweep', but this was replaced by a complete image update at a time interval corresponding to a complete sweep of the screen when analog radars were replaced by computer generated synthetic images. TROTSKY presents potential future conflicts superimposed on this image, as new aircraft are generated. The system also displays the flight profile of one of the two aircraft involved in the most urgent conflict.

In the second evaluation, 26 subjects carried out two runs of a nominal one-hour length each, handling about 200 aircraft on random direct routings. The data derived from the dribble file of each exercise was summarised, and analyses of variance were carried out on summarised parameters of each run to identify differences between first and second runs, and between individuals.

Analysis Levels

Keystrokes

The mean times between keystrokes can be analysed, overall or for each key. The times for individual keys will differ depending on their place in the construction of an order. (Only a subset of keys is acceptable at any point in the control process. Keys without meaning were not recorded. In retrospect, this was a mistake, since they may represent either physical or conceptual errors of importance to the ergonomist. All inputs should have been recorded.)

Orders

The basic unit of control for the simulator is the order. This is a sequence of keystrokes, which define how the trajectory of an aircraft is to be modified. In this simulator, as potential orders are input, their consequences are shown immediately on the screen. Thus an order may begin with **F**, **H** or **S**, defining the type of change (Flight Level, Heading, or Speed), followed by **+** or **-** signs indicating increments or decrements in the relevant parameter. As these are input, the display shows the effect on the traffic problem. Once a solution has been found, the **T** key, followed by a series of **+** keys specifies for how long the alteration should be maintained before returning to normal navigation to the planned exit point. A series of orders separated in time can be given to one aircraft, although it is rarely necessary in practice. In addition, a delay can be specified before the initial change of trajectory. A complete order may be accepted using the **Y** key, or rejected using the **N** key, or the last step in an order can be deleted using the **X** key. Finally, the **R** key can be used to cancel the order currently being prepared, and transfer to the other aircraft in the conflict being examined. An accepted order can be specified in a variation of Backus-Naur normal form as shown below.

[()/T+<+>] +<[[F+ [()/.<+>/<->] / [[H/S] + [<+>/<->]]+T + [<+>]]>+Y

Where **()** indicates no key **<a>** indicates a sequence of at least one "**a**". **[A/B/C]** indicates one of the sequences **A, B** or **C**. **A+B** indicates **A** followed by **B**.
(**H++T+++** means "Turn right 10 degrees for three minutes, then go to your exit point")
(At this stage, it becomes apparent that some keystrokes are simply wrong - several possible reasons can be deduced.)

Resolutions
Sequences of orders related to the solution of one conflict can also be identified, usually consisting of an accepted order, preceded by rejected orders to the same aircraft and switches, using the 'R' key from one to the other aircraft involved. If **O+Y** represents an accepted order as above, and **O*** represents an incomplete order then a resolution can be represented by:

[() / < [O*/O]+[N/X/R]>/ R]] + O + Y

Keystroke Analysis

The mean number of keystrokes recorded was 523.1 per exercise, a total of 27,201 keystrokes. The overall mean time (measured from the previous key input or screen update) was 1.05 seconds. This figure was significantly higher for the first run (1.14 sec) than for the second (0.96 sec). Participant means were also very variable, from 0.62 sec to 1.31 sec. There were considerable differences between the times involved for different keys, corresponding to the roles they played in input orders.

*Manoeuvre Keys (**F**, **H**, **S**, and **T**)*
F (Flight Level change) is the most common start for an order, as should be expected. The average time to decide to try a level change was 3.6 seconds, decreasing from 4 to 3.2 seconds for successive runs. **H** (Heading Change) was used on about one fourth as many occasions, and the time to decide to try a heading change was 6.2 seconds, reducing from 7 to 5.4 seconds in successive runs. **S** (Speed Change) was too rarely tried to produce meaningful results. **T** (Time) although sometimes used at the start of an order, usually followed immediately after the establishment of a satisfactory conflict resolution manoeuvre, where it specified the time for which the manoeuvre should be maintained. The mean time therefore is an approximate measure of the recognition that a satisfactory conflict resolution had been achieved.

Increment/Decrement Keys (+/-)
The + and – keys were used to specify how many Flight Level in units of 10 FL, how much Heading in 5 degree units, Speed in 10 Knot units, or Time in 1 minute units were required. It was often possible, particularly for Time intervals, to keep the key depressed, producing a rapid update, at the cost of possible overshoots (see below).

The (+) key is the most commonly used key of all, representing 60% of all keystrokes. The short average time (0.41 Sec) reflects the common use in series to specify time intervals in particular, where the participant could hold down the key and see the successive solutions in sequence.

The (-) key is less used, often to correct an overrun with the (+) key, or to indicate a height decrement or left turn. The longer mean time reflects the need to appreciate the effect of a change before repeating the key.

*Order Control Keys (Y, N, X and **R**)*
The **Y**, **N**, **X** and **R** keys control the disposition of input orders.

The **Y** key accepts an order. Usually, this solves the conflicts in which the selected aircraft is involved. The time taken (1.83 sec) suggests that this is a relatively easy decision, which becomes faster with time, and is variable between individuals. The **N** key rejects an order completely. There are relatively fewer of these, since most conflicts are solved at the first attempt. The time taken (3.42 sec) suggests that participants took longer to decide that a particular possible order would not work than to accept a solution, as might be expected. Again, the task became faster in the second run, and was very variable between participants. The **X** key undid the most recent part of an order. This order was rarely used, except by one participant, who used it in place of the **N** key throughout. (This is an example of a 'superstition', where a system user became convinced that an inefficient interaction is preferable to the one planned by the designer.) The **R** key switches from one aircraft to the other, discarding any incomplete order on the way. It was frequently used, although individual use was, again, very variable. Four controllers used it only once or less per exercise, while one controller used it about once per conflict. It was used more often in the second run.

Order Analysis

In all the 26 participants introduced a total of 2552 orders. Table 1 shows how these are distributed by initial and final keys, with the overall mean time for orders in seconds.

Sixty percent of orders were accepted. Of these 76 % involved height changes. All the accepted orders which began with a time delay (**T**) continued with **F**, so that 82% of accepted orders solved the conflicts by a change in Flight Level. Only Heading changes (**H**) made a significant contribution to the accepted solutions.

Note that accepted orders beginning with **F** are the fastest to input (12.3 sec), while accepted orders beginning with **H** take on average 20.4 seconds. This confirms the idea that Flight Level changes can be pre-planned by reference to the along-track profile, while the map display must be consulted after each increment in heading.

The 2552 orders took 771 different forms, defined as sequences of keys. Examination of these forms showed some redundant keystrokes. For example, changes of direction, represented by **+-** or **-+** occurred 605 times. Some of these were certainly deliberate, for example, when exploring potential heading changes, but others represented corrections of overruns of time before returning to track. For whatever reason, a string of the form **F+--** is functionally equivalent to a string of the form **F-**.

Table 1 – Orders by Start and End Characters

Order	Start									
End	**F**	**H**	**S**	**T**	**+**	**-**	**N**	**Y**	**R**	Total
Y No.	1165	244	18	89	6	1	0	3	0	1529
Mean	12.3	20.4	17.9	16.6	18.0	35.2	.	7.5	.	14.0
N No.	286	122	37	51	2	2	21	0	0	521
Mean	13.7	17.4	14.4	14.8	15.2	20.3	3.2	.	.	14.3
X No.	13	6	0	11	1	0	0	0	0	45
Mean	16.5	26.7	.	16.0	15.4	.	.	.	.	13.1
R No.	8	5	1	1	0	0	0	0	442	457
Mean	12.2	20.8	8.1	5.8	.	.	.	.	4.2	4.5
Total	1472	377	56	152	9	3	21	3	442	2552
Mean	12.6	19.5	15.2	15.9	17.1	25.2	3.2	7.4	4.15	12.3

Repeated keys other than **++** or **--**, which represent nervousness or over-enthusiastic keying, occurred 100 times. **T-** sequences, left after cancelling out sequences of **+++----** occurred on 13 occasions. A few well-formed orders (40) were preceded by + or – sequences, probably because the initial **F, H** or **S** key had not been struck hard enough to register. Finally, sequences where, for example, the participant inserted an **H** immediately followed by an **F**, indicating a change of mind or keying error, occurred 22 times. If these errors are omitted, the 771 different orders are reduced to 536 different sequences. Sequences ending in **X** are too rare to merit analysis, and those ending in **R** are almost all composed of that character alone.

Resolution analysis

The mean number of Resolutions (some of which resolved two or more conflicts) per exercise was 29.4, corresponding to a total of 1529 conflict resolutions. The mean time per resolution was 18.46 seconds.

The mean number of different resolutions, including redundant keying, was 25.84 per exercise, with a mean length of 15.2 keystrokes.

Eliminating redundant keying, there were 24.3 different resolutions per exercise, with a mean length of 14.2 keystrokes.

If these resolutions were condensed by eliminating the + or – elements, so that **F++T+++Y** and **F-T++Y** both reduce to **FTY**, there were 9.9 different resolutions per run, with a mean length of 4.3 keystrokes. By far the most common resolution was **FTY** - the simple flight level change. The mean number of these resolutions was 16.81 per exercise, 57% of all resolutions. This percentage varied from 40.6 to 73.0 for different participants.

Table 2 shows the percentages of resolutions involving one, two, three, or four or more orders, with the corresponding mean times for the complete resolution. Although the total time increases with the number of orders, the time for four or more orders appears to be less – these orders usually consisted of several 'false starts' involving a few keystrokes immediately cancelled before an acceptable solution, usually involving a flight level change, was found.

Table 2 – Number of Orders and Switches between Aircraft

Orders	Per Cent	Mean Time (sec)	Switches	Percentage	Mean Time (sec)
1	78.3	14.2	0	81.4	16.7
2	13.6	34.5	1	11.6	28.2
3	5.6	39.0	2	5.1	37.7
4/+	2.5	18.6	3+	1.9	12.9

It will be remembered that the **R** key switched to the flight profile of the other aircraft in the most urgent conflict. Table 2 also shows the percentage and mean times for resolutions involving no, one, two, or three or more switches, with the corresponding mean times for the complete resolution. A similar pattern appears to apply – where there are three or more switches, these usually appeared at the start, usually followed by a simple flight level change.

Conclusions

- A dribble file presents useful information on the interaction of the user and the computer.
- It should record **all** input to the computer by the user, and sufficient information to place these inputs in context.
- Analysis may be carried out on successively higher levels, making use of knowledge of the operation of the interface.
- Failures, errors and redundancies of input are particularly important in the development of interfaces. Even where they do not affect the operational performance of the system, they may warn of possible poor design.

References

David, H. (1997) *Radical Revision of En-route Air Traffic Control.* EEC Report No. 307. EUROCONTROL Experimental Centre, Bretigny-sur-Orge, France

David, H. and Bastien, J-M. C. (2001) *Initial Evaluation of a Radically Revised En-Route Air traffic Control System,* In Hanson, M.A. (Ed.) *Contemporary Ergonomics 2001*(pp 453-458)*: Taylor* and Francis

David, H. and Bastien, J-M. C. (2003) *Second Evaluation of a Radically Revised En-Route Air traffic Control System,* In McCabe, P.T. (Ed.) *Contemporary Ergonomics 2003* (in press)*: Taylor* and Francis

Dee, T.B. (1996), Ergonomic Re-Design of Air Traffic Control for Increased Capacity and Reduced Stress, In Robertson, S.A. (Ed.), *Contemporary Ergonomics 1996* (pp 563-568): Taylor and Francis.

Wickens, C.A. (2000) The When and How of using 2-D and 3-D Displays for Operational Tasks. In *Proceedings of the IEA2000/HFES 2000 Congress*, .San Diego California USA.

SITUATION AWARENESS IN VIRTUAL ENVIRONMENTS: THEORY AND METHODOLOGY

Ungul Laptaned, Sarah Nichols, John Wilson

Virtual Reality Applications Research Team
School of Mechanical, Materials, Manufacturing Engineering and Management
University of Nottingham, University Park
Nottingham, NG7 2RD, UK

The aim of this research is to develop a new approach and methodology for assessment of situation awareness (SA) in virtual environments (VEs). This paper presents an experiment to examine presence, task/user performance and SA with three different types of virtual environment (VE) display technology and gender. The interrelationship among SA, presence, performance, and sickness is also investigated. The results indicated there was no significant main effect of display type on SA or presence questionnaire, however, there was a significant main effect of display type on simulator sickness questionnaire under total scores ($F = 3.72$, $df = 2(30)$, $\rho < 0.035$). There was also a significant correlation between VRSARM and performance ($r = 0.39$, $\rho < 0.05$), and sickness and recall ($r = -0.377$, $\rho < 0.05$). There were no significant correlation between presence and SA, sickness and performance, sickness and presence, as well as presence and performance. It was therefore concluded that presence does not lead to enhance SA and performance. Nevertheless, this experiment is used to generate design implications for future directions for SA in VE research in which associations between variables might be found.

Introduction

With the rapid development of new technologies in complex and dynamic systems, there has been recently shown an increased concern over the cognitive functioning of system users. This concern has been manifested through increased efforts to consider the cognitive components of performance as new systems are designed. To establish benchmarks for evaluating the degree to which such new design affects user cognitive functioning, it is necessary to evaluate based on performance measurement and selection of metrics. Endsley (1995) pointed out that “a measure of cognitive components is valuable in the engineering design cycle because it assures that prospective designs provide operators with the critical commodity.” This cognitive measure used in explaining performance and decision-making emphasizes the concept of situation awareness (SA) in the decision process. Endsley (1988, p. 97) defined SA as “a person’s perception of the elements of the environment within a volume of time and space, the comprehension of their meaning and the projection of their status in the near future.” It is an understanding of the state of the environment.

SA is a good subjective measure of the quality of any systems such as aircraft, air traffic control operations, tactical and strategic systems, etc. The user of these systems must have knowledge of the current process state at all times, and the ability to use this knowledge effectively in predicting future process states and controlling the process to attain operational goals (Endsley, 1995a). In operations of the complex-systems, Endsley (1995a) has suggested that the concept of SA should be used to investigate a host of new technologies and designs being considered for future systems, including three-dimensional visual and auditory displays, voice control, expert systems, helmet-mounted displays, and virtual reality (VR). In this research, SA is examined and discussed with the advent of VR technologies.

One simple meaning of VR was defined by Loeffler and Anderson (1994) as "a three-dimensional, computer-generated, simulated environment that is rendered in real time according to the behaviour of a user." VR is considered to have an ability to promote the feeling of being there or the sense of presence. Curry (1999) suggested that the concept of presence deals with the degree to which you feel a part of some virtual space; that the space exists and you are occupying it. In the context of SA, presence might provide the sense for a user to perceive the elements in the VE, comprehend their meaning, and project their status. In addition, other types of VR such as spatially immersive systems expressly have been found to have the capability to promote/provoke high SA (Prothero et al., 1995). Therefore, it is essential to determine how the degree of being present could lead to enhanced SA. By assessing a user's SA, it may be able to predict user performance.

To investigate the problem, the research here uses a series of VEs to examine presence and then relate to a measure of its effect on task/user performance, and to explore the concept of SA and the relationship between SA and being present and immersed in the VE. Of particular significance is the three-level model of SA by Endsley, 1995b (perception of the elements in the environment, comprehension of the current situation, and projection of future status), and the way in which VE design might affect these different levels. Subjective SA measurement tool is developed which refers to Virtual Reality Situation Awareness Rating Metric (VRSARM). Further, it is argued that the real world model of SA that proposes three-levels of SA may not be entirely appropriate to VEs, where we may only be concerned with perception of elements and comprehension of the current situation.

Method

The experimental design used in this study was a 3×2, factorial design with 2 within subject variables. The independent variables were Display Type (V8 HMD, Glasstron, and Projection Screen Display) and Gender (Female and Male). Participants in the experiment were recruited by advertisement from the University of Nottingham students. 12 participants (6 females and 6 males) volunteered to partake in this study. They were attended three separate test sessions on different weeks (one for each display type).

The input devices were a standard 2-D mouse and a joystick that used to interact with objects in the VE. Navigation through the environment and object manipulation is achieved using Superscape 3-D control device. The experiment took place in a virtual factory environment. The assigned task entailed entering the factory, moving around, and picking up materials within the factory. There were five tasks that needed to be completed.

Measurement of Situation Awareness

SA measurement included objective and subjective measures. Objective measures were object recalled and performance (search time and errors). Subjective measures were VRSARM (devised by the author, 2000), simulator sickness (Kennedy, 1993), presence (Witmer & Singer, 1998), and transitions to reality (Slater and Steed, 2000).

The VRSARM questionnaire comprises 10 subscales: Situational awareness, Attention, Memory, Workload, Side effects, Immersion/system technology, Involvement, Interaction,

Spatial presence, and Information, as well as a total score. It was developed to investigate the relationship among each factor aforementioned.

The presence questionnaire comprises 4 subscales: Involvement, Natural, Interface, and Resolution, as well as a total score. It was chosen to predict how a sense of presence could lead to enhance SA.

The simulator sickness questionnaire comprises 3 subscales: Nausea, Oculomotor, and Disorientation, as well as a total score. It was chosen to identify that simulator sickness may have an affect on SA and presence.

The transitions to reality questionnaire was selected to evaluate how the participants are distracted by the reasons given for transitions to the real world (shown in Table 1.), which result in a break in the sense of presence during VE use. It is anticipated that the cause of a break in presence may be associated with SA.

Table 1. Reasons Given for Transitions to Reality (Slater and Steed, 2000)

External: Sensory information from the real world intruded into or contradicted the virtual world, either in the form of noises or people talking, or else the touch or feel of interactions with real solid objects.
Internal: This is where something wrong in the virtual world itself is noticed, for example, the laws of physics not being obeyed, objects looking unreal, the absence of sounds, display lag.
Experiment: Some aspects of the experimental set-up itself, or the instructions intruded.
Personal: Some personal feeling intruding, such as embarrassment or consciousness of being observed from the outside.
Attention: A loss of attention to what is happening in the virtual world, or some aspect of the virtual world that results in a loss of presence.
Spontaneous: No apparent reason.

Results

Data from 12 participants was analysed by using a factorial analysis of variance and T-tests. Due to space restrictions, only significant results are reported here.

Objective Measures

1. Recall Measure

The means number of items recalled for V8 HMD, glasstron, and projection screen display were 8.33, 5.75, and 7.83, respectively. The means number of items recalled for female was 8.39 and male was 6.06. There was no significant main effect of display type, but there was a significant main effect of gender on recall (F = 5.59, df = 1(30), $p < 0.02$). Therefore, it was concluded that female recall was better than male participants. There was no significant interaction.

2. Performance Measure

The means search times for V8 HMD, glasstron, and projection screen display were 947.75, 1195.5, and 1271.34, respectively. A significant main effect of display type was not found. The means search time for female was 1154.44 and male was 1121.95, which was not significant. There was no significant interaction. For the means number of errors of display type and gender, no significant main effects or interaction were found.

Subjective Measures

1. Virtual Reality Situation Awareness Rating Metric

The means VRSARM scores on information subscale for female was 25.06 and male was 31.33. There was a significant main effect of gender (F = 11.39, df = 1(30), $p < 0.002$), therefore it was concluded that females experienced lower SA than males for this subscale. There was no significant interaction. There was no significant main effect of display type and gender, and interaction on other subscales and total scores.

2. Presence Questionnaire

The means presence scores on involvement subscale for female was 31.61 and male was 35.17. A significant main effect of gender was found (F = 4.07, df = 1(30), $\rho < 0.05$), therefore it was concluded that male participants experienced higher levels of presence than females. There was no significant main effect of display type on natural, interface, resolution subscale, and total scores. There was also no significant interaction.

3. Simulator Sickness Questionnaire (due to space restrictions, only statistics related to total scores presented here)

The means SSQ scores of total scores during post VE exposure for V8 HMD, glasstron, and projection screen display were 31.79, 15.27, and 7.48, respectively. A significant main effect of displays type was found (F = 3.72, df =2(30), $\rho < 0.035$), and therefore it was concluded that the participants exposed to a V8 HMD had the highest sickness symptoms (on overall scores) compared with those exposed to other display types. The means SSQ scores for female was 24.93 and male was 11.43, but despite this large difference no main effect of gender was found due to individual difference. There was also no significant interaction.

The sickness questionnaire also shows the pre and post exposure symptom levels for V8 HMD, glasstron, and projection screen display. The results found indicated that the change from pre to post exposure of SSQ scores (under nausea, oculomotor, disorientation subscales, and total scores) on a V8 HMD was higher than other display types.

4. Transitions to Reality (n is number of participants who gave responses in the corresponding category) - Transitions particularly interesting for V8 HMD are shown in Table 2.

Table 2. Example of Reasons given for Transitions to Reality

Cause	n	Some Examples
External – Sound	2	-I heard the mouse being clicked when objects were being picked up -Was getting explanations, directions, and navigation help from the experimenter
External – Touch or Force	1	-The experimenter assisted me in using HMD and joystick
Internal	2	-No, because the objects looked unreal -I walked through the walls when I tried to exit
Experiment	3	-HMD is too heavy. I got dizziness -The weight of HMD pressed on my nose. It made me nausea
Personal	2	-Conscious feeling -Suddenly I have started to feel bad, headache, and to have other symptoms and this causes me a transition
Attention	0	
Spontaneous	1	-No reason
None	1	-I did not make any transition

Correlation Matrix

The relationship between the dependent variables was examined using Pearson's correlation. The results were presented in Table 3.

Table 3. Pearson's Correlation

Dependent Variables	Correlation Analyses
VRSARM and Performance	-Search time was positively associated with situational awareness subscale (r = 0.39, $\rho < 0.05$) -Search time was positively associated with side effects subscale (r = 0.34, $\rho < 0.05$)
Within VRSARM	-Situational awareness was positively associated with memory subscale (r = 0.37, $\rho < 0.05$) -Situational awareness was positively associated with total scores (r = 0.52, $\rho < 0.01$)
Sickness and Recall	-Number of items recalled was negatively associated with oculomotor subscale (r = -0.377, $\rho < 0.05$)
Non Significance	
Presence and Recall	-No significant correlation
Presence and Performance	-No significant correlation
Sickness and Performance	-No significant correlation
Sickness and Presence	-No significant correlation

From results shown in Table 3, the correlation between VRSARM and performance indicated that the longer the participants were exposed in VE, the more likely they were to get sickness symptom. In addition, however, those who took a longer search time also experienced higher levels of symptoms. These apparently contradictory results suggested that search time may not be a good direct indicator of SA. This is supposed by previous researches (Nichols et al., 2000). The sickness symptom was also a factor to degrade the recall ability of a participant in VE use. Further, it is suggested that those other subscales in VRSARM are relevant to SA, but this should be examined by the reliability analysis.

Discussions

The present analysis of the impact of VE on user's SA has provided evidence that there is no relationship between presence and SA. This was interpreted that presence does not provide the sense for a user to perceive the elements, comprehend their meaning, and project their status in the VE. It is suggested that presence is also not a factor to support performance.

Search time and number of errors were found not to be significantly different for display types. Males experienced higher levels of presence than females, especially on the involvement subscale. It was concluded that male participants paid more attention to the VE stimuli, and therefore became more involved in the VE experirence. The VRSARM questionnaire indicated that females experienced lower SA than males, based on the information subscale. This is not consistent with the results from looking within VRSARM and recall measure, as SA is positively associated with memory. The higher use of memory on recall, the higher SA, as well as female recalled better than male participants. These results, on the other hand, lead to a conclusion that females experienced higher SA than males. As a result, the reliability analysis for this case is required to investigate the relationship between these factors.

The analysis indicated that the V8 HMD promoted the highest sickness symptoms. It appears that sickness symptoms increase with immersion time. The HMD design might have an effect on the level and type of symptoms they are experienced (Nichols, 1999). Finally, the questions of transitions to reality indicated the reasons for transitions to the real world, resulting in a break in the sense of presence. There were number of reasons given such as external cause (noise in a lab) or experiment cause (VE). These causes may be associated with SA provided that the relationship between presence and SA existed. In the future research, it is necessary to investigate how transitions to reality affects level of SA, and further investigation of the relationship between presence and SA is required.

Acknowledgements

This work was a part of Ph.D. research carried out at the University of Nottingham. The research study was partially funded by School of Mechanical, Materials, Manufacturing Engineering and Management, and International Student Office. Support from researchers at the Virtual Reality Applications Research Team (Dr. S Cobb, Dr. M. D'Cruz, and Dr. R. Eastgate) and Mixed Reality Laboratories (Dr. D. Lloyd and Dr. T Glover) is gratefully acknowledged.

References

Bytrom, K., Barfield, W., & Hendrix, C. (1999, April). A Conceptual Model of the Sense of Presence in Virtual Environments. *Presence: Teleoperators and Virtual Environments*, 8(2), 241-244.

Curry, K. M. (1999, May). Supporting Collaborative Awareness in Tele-Immersion. *International Immersive Projection Technology Workshop, Centre of the Fraunhofer Society Stuttgart IZS*, 10 (11), 253-261.

Endsley, M. R. (1988). Situation Awareness Global Assessment Technique (SAGAT). In *Proceedings of the National Aerospace and Electronics Conference* (*NAECON*) (pp. 789-795). New York: IEEE.

Endsley, M. R. (1995a). Toward a Theory of Situation Awareness in Dynamic Systems. *Human Factors*, 37(1), 32-64.

Endsley, M. R. (1995b). Measurement of Situation Awareness in Dynamic Systems. *Human Factors*, 37(1), 65-84.

Kalawsky, R. S. (1999). VRUSE - A Computerised Diagnostic Tool for Usability Evaluation of Virtual/Synthetic Environment Systems. *Applied Ergonomics*, 30, 11-25.

Kennedy, R. S., Lane, N. E., Berbaum, K . S., & Lilienthal, M. G. (1993). Simulator Sickness Questionnaire: An Enhanced Method for Quantifying Simulator Sickness. *The International Journal of Aviation Psychology*, 3(3), 203-220.

Loeffler, C. E. & Anderson, T. (1994). *The Virtual Reality Casebook*. New York, NY: Van Nostrand Reinhold.

Nichols, S. (1999). *Virtual Reality Induced Symptoms and Effects (VRISE): Methodological and Theoretical Issues*. Unpublished Ph.D.'s Thesis, University of Nottingham, Nottingham, England.

Nichols, S., Haldane, C., & Wilson, J. R. (2000). Measurement of Presence and Its Consequences in Virtual Environments. *Journal of Human-Computer Studies*, 52, 471-491.

Prothero, J., Parker, D., Furness, T., & Wells, M. (1995). Towards a Robust, Quantitative Measure of Presence. In *Proceedings of the Conference on Experimental Analysis and Measurement of Situational Awareness*, Daytona Beach, FL, USA, 359-366. [On-line]. Available: http:// ww.hitl.washington.edu/publications/p-95-8.

Slater, M., Linakis, V., Usoh, M., & Kooper, R. (1996). Immersion, Presence, and Performance in Virtual Environments: An Experiment Using Tri-Dimensional Chess. *ACM Virtual Reality Software and Technology (VRST), Mark Green (Ed.) ISBN: 0-89791-825-8*, pp. 163-172.

Slater, M. & Steed, A. (2000, October). A Virtual Counter. *Presence: Teleoperators and Virtual Environments*. PDF 9.2, In Press.

Witmer, B. G. & Singer, M. J. (1998, June). Measuring Presence in Virtual Environments: A Presence Questionnaire. *Presence: Teleoperators and Virtual Environments*, 7(3), 225-240.

A VIEW OF THE FUTURE? THE POTENTIAL USE OF SPEECH RECOGNITION FOR VIRTUAL REALITY APPLICATIONS

Alex W. Stedmon, Harshada Patel, Sarah C. Nichols, & John R. Wilson

Virtual Reality Applications Research Team (VIRART)
School of MMMEM
University of Nottingham
Nottingham NG7 2RD

Virtual and Interactive Environments for Workplaces (VIEW) of the Future is a project funded by the European Union within the Information Society Technology (IST) program. As part of the project a Special Interest Group has been established to facilitate discussion and research surrounding the potential use of Automatic Speech Recognition (ASR) in VEs. The VIEW Speech Group represents a number of areas of expertise ranging from the technical integration of ASR hardware into VEs and potential application areas of ASR; through to the potential human performance issues associated with using ASR technologies within VEs. This paper builds on the knowledge and theory of using real world ASR and considers its transfer into the virtual world. Speech offers the potential to liberate the user and allow a greater degree of freedom to interact with VEs. However, it is only when the underlying Human Factors issues are considered that the usability of ASR can be enhanced.

A VIEW of the Future

Virtual and Interactive Environments for Workplaces (VIEW) of the Future is a project funded by the European Union within the Information Society Technology (IST) program. The overall goal of VIEW is to develop best practice for the appropriate implementation and use of Virtual Environments (VEs) in industrial settings for the purposes of product development, testing and training. Underpinning this is a clear emphasis on incorporating Human Factors throughout the conception and implementation of Virtual Reality (VR) applications by considering user- and applications-needs in developing initiatives that are fundamentally usable.

As part of the VIEW project a Special Interest Group has been established representing key areas ranging from the technical integration of Automatic Speech Recognition (ASR) hardware into VEs and potential application areas of ASR, through to the potential Human Performance issues associated with using ASR technologies within VEs. This paper builds on the knowledge and theory of using real world ASR in order to consider its transfer into VEs.

Speech as an Input Modality

Speech is the most natural form of human communication; it is our primary medium of communication, which most of us are able to employ in our daily lives from an early age. It is still not fully understood how we learn the subtle rules of syntax and grammar and this is perhaps why it is so difficult to develop such a framework artificially. What is clear, however, is that speech is a "familiar, convenient, [and] spontaneous part of the capabilities the human brings to the situation of interacting with machines" (Lea, 1980).

Speech is also the "human's highest-capacity output communication channel" that offers the highest potential capacity for human-to-computer communication (Lea, 1980). Speech also has other inherent advantages over other more conventional interaction modes. Whereas untrained users may find reading, writing, keyboard skills or manual input difficult without training or practise, speech requires no training except for learning the vocabulary that the machine is able to recognise.

One of the benefits of ASR is that it can be used in situations where other input devices do not perform so well, for example in the dark or around objects or obstacles. As a medium of communication (sound), and through the medium in which it travels (air), speech travels omni-directionally without light, in a way that conventional writing, typing and button pressing are unable to do. As Lea (1980) states, in "using switches, typewriters, cathode ray tube displays, and even the more unusual graphical input devices ... and joysticks, the user must either be in physical contact with the computer console or terminal or must be orientated in a fixed direction to produce input commands and monitor computer outputs". ASR systems therefore permit remote Human-Machine Interaction (HMI) procedures whilst also possibly conducting additional tasks. For example, air traffic controllers may use ASR systems to relate information about position, heading and altitude of aircraft whilst simultaneously looking at displays or keeping an eye out for external stimuli (Lea, 1980).

ASR For Real World Applications

Popular applications of ASR technology include systems used by blind or handicapped operators to control wheelchairs, domestic appliances or use telephones (Baber *et al*, 1996); or industrial systems by personnel unskilled in computer programming (Lea, 1980). Other applications include situations where the operator is suited up and sealed within an environment (such as under the sea, in outer space or perhaps in a highly radioactive environment) where speech would be suited to remote locations and would appear to be unaffected by weightlessness (Lea, 1980).

With the advent of the computing age and Artificial Intelligence (AI) ASR systems have been developed, refined and applied to more advanced applications allowing more flexible systems for user interaction. In military aircraft cockpits and infantry systems where there is little room to employ extra input devices, ASR interfaces may offer benefits as speech would appear to be less affected by high gravitational and acceleration forces encountered in aircraft and vehicular manoeuvres, than typewriting, button pushing, twisting knobs, thumb wheels or handwriting (Lea, 1980).

ASR in the driving domain relies upon the careful design of the interface to match the expectations, preferences and abilities of various user groups (Stedmon & Bayer, 2001). As standard in-car systems generally require drivers to operate visual displays and manual controls whilst driving, ASR may improve the usability and safety of in-car systems including mobile phones, entertainment systems and Intelligent Transportation Systems (Graham & Carter, 2000). However, what might be a potential aid could just as easily prove to be hazardous if it distracts drivers from safe control of their vehicles (Stein, Parseghian, & Wade Allen, 1987). The driver's ability to attain an acceptable level of performance in terms of the ASR process (commands, menus, vocabulary, etc) may be constrained, as the car can be considered a 'hostile environment' for ASR (Baber & Noyes, 1996). Such hostile environments are typically characterised by high levels of noise, stress and workload affecting the way people talk and how their speech is recognised by the interface (Baber, *et al*, 1996).

Speech as an Input Device

Of particular importance in the VIEW project is the need to examine which input devices are best suited to conducting specific applications and tasks. There is also a strong interest for investigating how participants navigate around VEs, as well as looking at how VE content can support usability.

Input devices can be defined as the medium through which users interact with a computer interface and, more specifically in the context of the present research, a VE. Currently, there is an increasing variety of input devices on the market that have been designed for VR use, which range from tradition mouse and joystick devices; to wands, data-gloves, speech, etc. With such a variety, this may lead to individuals selecting an inappropriate input device which could compromise the overall effectiveness of a VR application and the user's satisfaction in using the application.

Kalawsky (1996) indicates that the design of an input device should match the perceptual needs of the user. As such, the integration of input devices should follow a user needs analysis to map their expectations onto the attributes of the overall VR system. Jacob *et al*, (1993) recommended that the device should be natural and convenient for the user to transmit information to the computer and further, Jacob *et al*, (1994) suggest that input devices should be designed from an understanding of the task to be performed and the interrelationship between the task and the device from the perspective of the user. Building on a sound understanding of user needs, it is important, therefore, to analyse the task in the correct level of detail so that the VR system and the VE that is developed supports user interaction and overall application effectiveness. As such, Barfield *et al*, (1998) indicate that a VE input device should account for the type of manipulations a user has to perform and that the device should be designed so that it adheres to natural mappings in the way in which the device is manipulated, as well as permit movements that coincide with a user's mental model of the type of movement required in a VE.

Despite such recommendations, there are very few established guidelines that detail what is required of an input device in terms of the parameters that pertain to a user and user performance within a VE (Kalawsky, 1996).

ASR for VR Applications

There is still much to be understood about the Human Factors issues involved with both ASR and VR, and the use of ASR *within* VEs is an exciting integration of the two that is not without its challenges!

One of the (subtle) benefits of ASR technology is that speech can be used in the dark and around obstacles. Within VEs, users are not necessarily in 'darkness' but immersed in a 'virtual reality' where typical 'obstacles' are not necessarily concrete concepts (as in the real world) but anything from the interface and menus they are using to the actual environment they are interacting with. Speech offers the potential to liberate the user and allow a greater degree of freedom to interact with VEs. For the purpose of the VIEW project, speech has been identified as a novel input device which can be used in isolation or in conjunction with other input devices such as a mouse, joystick, data-gloves, etc.

Within the VIEW project a number of potential issues have been highlighted, such as:

- which commands/tasks can be triggered by speech;
- integrating speech into existing toolkits;
- how useful/comfortable is voice control within different VE systems;
- which combination of voice control and other input devices is suitable.

Within VIEW, speech has been primarily implemented to navigate and manipulate objects in VR training applications. Speech can be used to select menu commands, where, for example, specific commands can turn a gravity simulation on or off, and in terms of manipulation, objects can be selected, grabbed, or released. Speech in a text-to-speech application has been used to provide task instructions and explain concepts to users in addition providing feedback.

If a speech metaphor is well designed then it can be intuitive, taking the burden of navigation or interaction control away from a physical input device, and increasing opportunities for multi-modal interaction.

Speech allows hands-free operation and maybe particularly useful for navigating through menus and short cut commands. In future applications speech can be used as a replacement for other input devices such as mouse or keyboard. Furthermore, speaker independent ASR systems support multi-user sessions in collaborative VEs allowing many user to interact with each other and the VE at the same time.

There are, however, technical problems that may compromise the successful use of speech in VEs. The integration of ASR software into the building of VEs is still a problem and using microphones which are too sensitive may pick up background noise and electrical interference which could affect the accuracy of processing speech commands. Hand-held microphones or head-mounted microphones can be heavy and obtrusive but blue-tooth technology can offer more versatility in terms of wireless mobility and integration of devices.

In addition, anecdotal evidence and the findings of Stedmon *et al*, (in press) suggests that speech may not be best suited for specific actions such as navigation and so the best use of speech interfaces might be in combination with other input devices for a more integrated approach.

General HF Issues of ASR

Speech offers the systems designer a ready input modality for HMI, however, much of the impetus for the development of ASR technology has come from the belief that speech exists as an untapped resource. This is a highly contentious issue, for as Linde & Shively (1988) argue, speech is already an active or semi-active mechanism in most working environments. Another important issue which comes out of the work conducted by Baber *et al*, (1996) is that using an ASR system is demanding in itself and as such it may be necessary to "review the notion of 'eyes free/hands free' operation as [sole] justification for ASR" (Baber *et al*, 1996). It is only when the underlying Human Factors issues begin to be understood that the usability of ASR can be enhanced.

No strict guidelines exist for the design and development of ASR interfaces. It seems that various developers adopt their own best practise strategies and designs. Within any system application Human Factors variables can be divided into four groups as represented in Figure 1 below.

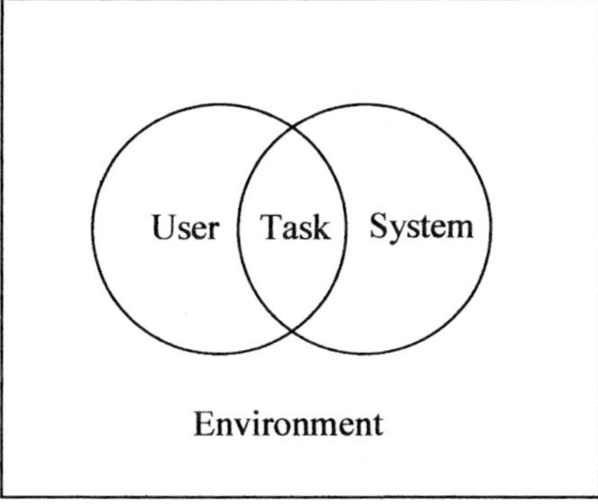

Figure 1. The principal components of a person-machine system

Within this representation a **person** (representing user variables) interacts with a **machine** via displays and controls (representing system variables) to perform a **task** (representing task variables) within an operational **environment** (representing environmental variables). Thus, any ASR application should be both designed and evaluated, using:

- **representative users** - according to variables such as age, gender, experience, accent, etc., and taking into account the variation between users. Within VR applications, users may come from many backgrounds and with collaborative VEs users may interact using different languages. As such, these variables need to be understood if the overall VR application and integrated ASR system is to perform properly;
- **representative systems** - with realistic accuracies, response times, dialogue structures, vocabularies, etc. Within VR applications, ASR systems will generally be led by the real-world applications market. As such, the systems variables are generally defined by what the conventional market dictates and what technology can deliver;
- **representative tasks** - including any concurrent or competing tasks, and realistic levels of mental workload, stress, fatigue, etc. Whether in the real world of a virtual world, ASR applications are very much application specific. As such, detailed analysis of the task needs to be carried out to address how and where speech input may support the user over more conventional input devices;

- **representative environments** - including physical attributes such as noise and vibration, and psychosocial aspects such as the presence of others. Within VR there are many technical environments that are used to present VEs. These can range from traditional headset or desk-top presentations; through to projection walls and VR CAVEs. In these situations variables such as acoustics and ambient temperature, as well as the presence of any other users, may affect the overall system performance and use acceptance of ASR in VR applications.

Overall, the success of speech recognition depends on the user, system, task and environmental variables, all of which have significant effects on overall system performance, as outlined above.

It should be remembered that, despite all the published guidance on ASR system design, even human factors experts will not produce the perfect system first time. Therefore, the development of any ASR application should be an iterative process of design-test-redesign, involving both system experts and end-users.

Conclusion

Within the VIEW project a programme of research includes the exploratory testing of existing applications and interaction devices as part of an iterative design process in order to provide guidelines for the development of new interfaces; as well as more traditional experiments that will ultimately feed into the creation of an interactive design support tool for VE developers. This will integrate the knowledge gained from the project (e.g. guidelines, code of practice, best use guidance, examples, etc.) for the specification, development and use of VEs in general, particularly workplace oriented VE applications.

This research will not only be an exploration of the key issues of using ASR technology in a VE but will hopefully lend itself to the development of an ASR protocol for VE applications, through the lifespan of the VIEW Project, and beyond.

Acknowledgements

The work presented in this project is supported by IST grant 2000-26089: VIEW of the Future.

References

Baber, C., Mellor, B., Graham, R., Noyes, J.M., & Tunley, C., (1996) "Workload and the use of Automatic Speech Recognition: The effects of time and resource demands". In, Speech Communication, 20, (1996), pp.37-53.

Baber, C., & Noyes, J. (1996) "Automatic Speech Recognition in Adverse Environments". In, Human Factors, 38(1), pp.142-155.

Barfield, W., Baird, K., & Bjorneseth, O. (1998). Presence in virtual environments as a function of type of input device and display update rate. *Displays*, 19: 91-98.

Finch, M.I., & Stedmon, A.W. (1998). The Complexities of Stress in the Operational Military Environment'. In, M. Hanson (ed). *Contemporary Ergonomics 1998.*

Proceedings of The Ergonomics Society Annual Conference, Cirencester 1-3 April, 1998. Taylor & Francis Ltd. London.

Graham, R., & Carter, C. (2000). Comparison of Speech Input and Manual Control of In-Car Devices whilst on the Move. *Personal Technologies* 4, 155-164.

Jacob, R., Leggett, J., Myers, B., & Pausch, R. (1993). An agenda for human-computer interaction research: interaction styles and input/output devices. *Behaviour and Information Technology*, 12(2): 69-79.

Jacob. R., Sibert, L., McFarlane, D. & Mullen Jr., P. (1994). Integrality and separability of input devices. *ACM Transactions on Computer-Human Interaction*, 1(1): 3-26.

Kalawsky. R. (1996). Exploiting virtual reality techniques in education and training: Technological issues. *Prepared for AGOCG.* Advanced VR Research Centre, Loughborough University of Technology.

Lea, W.A. (1980) "The Value of Speech Recognition Systems". In, W.A. Lea (Ed.). *Trends in Speech Recognition.* (1980) Prentice Hall: USA.

Linde, C., & Shively, R. (1988). Field study of communication and workload in police helicopters: implications for cockpit design". In, *Proceedings of the Human Factors Society 32nd Annual Meeting.* Human Factors Society, Santa Monica, CA. 237-241.

Stedmon, A.W., & Bayer, S.H. (2001). Thinking of something to say: workload, driving behaviour & speech. In, M. Hanson (ed). *Contemporary Ergonomics 2001. Proceedings of The Ergonomics Society Annual Conference, Cirencester. 2001.* Taylor & Francis Ltd. London.

Stedmon, A.W., Patel, H., Nichols, S.C., & Wilson, J.R. (in press). Free Speech in a Virtual World: Speech Recognition as a Novel Interaction Device for Virtual Reality Applications. In, P. McCabe (ed). *Contemporary Ergonomics 2003. Proceedings of The Ergonomics Society Annual Conference, Cirencester. 2003.* Taylor & Francis Ltd. London.

Stein, A.C., Parseghian, Z., & Wade Allen, R (1987) A Simulator Study of the Safety Implications of Cellular Mobile Phone Use. *Proceedings of 31st Annual Meeting of the American Association for Automotive Medicine.* New Orleans, USA.

FREE SPEECH IN A VIRTUAL WORLD: SPEECH RECOGNITION AS A NOVEL INTERACTION DEVICE FOR VIRTUAL REALITY APPLICATIONS

Alex W. Stedmon, Sarah C. Nichols, Harshada Patel & John R. Wilson

Virtual Reality Applications Research Team (VIRART)
School of MMMEM
University Park, University of Nottingham
Nottingham NG7 2RD

This paper considers the potential use of Automatic Speech Recognition (ASR) in Virtual Reality (VR) by analysing data from an Input Device Usability (IDU) Questionnaire developed for the Virtual and Interactive Environments for Workplaces of the Future (VIEW) project. Data from two VR trials and an Expert User Group were collected and analysed over a number of parameters, such as: interaction, distraction, ease of use, user comfort, frustration, enjoyment, error correction and overall input device usability. Analysis of the data illustrates that participants who believed they were using an ASR system to control their interaction in the VE, rated the usability of speech higher than the participants instructing another person or an expert user group. This points to an element of user perception and experience upon the overall usability of ASR as an input device in VR applications.

A VIEW of the Future

Virtual and Interactive Environments for Workplaces of the Future (VIEW) is a project funded by the European Union within the Information Society Technology (IST) program. The overall goal of VIEW is to develop best practice for the appropriate implementation and use of Virtual Environments (VEs) in industrial settings for the purposes of product development, testing and training. As part of the project a Special Interest Group has been established representing key areas ranging from the technical integration of Automatic Speech Recognition (ASR) hardware into VEs and potential application areas of ASR; through to the potential Human Performance issues associated with using ASR as novel input devices within VEs.

Speech as a Novel Input Device

Of particular importance in the VIEW project is the need to examine which input devices are best suited to conducting specific applications and tasks. Input devices can be defined as the medium through which users interact with a computer interface and more specifically in the context of the present research, a VE. Currently, there is an increasing variety of input devices on the market that have been designed for VR use. However, this may lead to individuals selecting an inappropriate input device which could compromise the overall effectiveness of a VR application.

The potential of ASR for VR applications is considered in Stedmon *et al* (in press). In many ways, Human Factors understanding of both ASR and VR are still being pioneered and the use of ASR *within* VEs is an exciting integration of the two that is not without its challenges! Only by incorporating Human Factors throughout the conception and implementation of VR applications, by considering user- and applications-needs, can such applications be designed that are, fundamentally, usable.

Input Device Usability Questionnaire

This paper considers the potential use of ASR in VR by analysing data from an Input Device Usability (IDU) Questionnaire that has been developed for the project by the University of Nottingham. The IDU Questionnaire is part of a post-immersion evaluation set that comprises a number of existing and new questionnaires that have been designed to assess factors associated with VR/VEs experience such as Simulator Sickness Questionnaire; Stress Arousal Checklist; Presence Questionnaire; Usability Questionnaire; Input Device Usability Questionnaire; Behaviour During VE Viewing; Post Immersion Assessment of Experience; and an Enjoyment Questionnaire. The pre- and post-immersion evaluation sets have been developed by the University of Nottingham over a number of years, based on established research in relation to VR Induced Symptoms and Effects (Cobb *et al*, 1999). The pre- and post-immersion evaluation questionnaires, when used together, offer an insight into the differences between users' VR experiences before and after using a VR application.

The IDU questionnaire contains 15 questions that investigate a number of parameters concerning input device use in VR applications, such as: interaction, distraction, ease of use, user comfort, frustration, enjoyment, error correction and overall input device usability. The questionnaire has been developed from previous usability research at the University of Nottingham (Barone, 2001) and established sources (Kalawsky, 1996) which were then formulated through expert review and developed specifically for input device usability issues.

VR Trials and Expert User Group

Two complementary experiments were been conducted to investigate the use of speech for navigation and object manipulation within a training VE developed by the University of Nottingham. The VE was developed to assess the effectiveness of VR as a training tool for computer network card replacement.

In both experiments the pre- and post-immersion evaluation questionnaires were administered, although only results from the IDU questionnaire are reported here. In addition an expert user group assessed the potential for ASR use in VR applications by independently completing the IDU questionnaire.

The primary aim of the experiments was to evaluate users' behaviour and actions when interacting with a computer via speech, but examining how this behaviour is affected by the users' beliefs about whether they are speaking to a computer or another human. Therefore in one condition participants believed they were speaking to a computer with an ASR system, and in the other participants believed they were speaking to a person, who was interpreting their actions and interacting with the computer via a standard input device. In fact, in both conditions, a human was interacting with the computer, but in the first condition, the human operator was concealed behind a screen, and participants were informed that they were speaking to a computer, with the human operator only being revealed at the end of the experiment.

Therefore the three sets of respondents to the IDU questionnaire were as follows:

- **VR experiment 1: speech recognition** – 12 participants conducted the VR task using speech input to control interaction, and believed they were talking to a computer.
- **VR experiment 2: instructing another person** – 12 participants conducted the VR task using speech to instruct another person (who, in turn, used a standard input device to interact with the computer).
- **Expert User Group:** 12 participants conducted a stand-alone assessment for the potential of ASR in VR applications.

IDU Questionnaire Comparisons

The IDU questionnaire is rated across a 5 point Likert scale (where 1 = strongly agree and 5 = strongly disagree). Data were collected and compared between the two experiments and the expert user group. Mean data scores for each questionnaire were obtained and analysed using T-tests.

The mean scores for the different user groups were: Speech – mean = 3.35 (SD = 0.32); Person – mean = 3.05 (SD = 0.50); Expert – mean = 3.06 (SD = 0.31). The data were analysed using a one way ANOVA, and no significant differences were observed ($F=2.34$; $df=2,33$; $p>0.05$).

The data scores for each question are represented in Graph 1 below, and illustrate a consistent pattern between questions for each participant group, with an overall higher set of ratings for those who believed they were talking to a speech system.

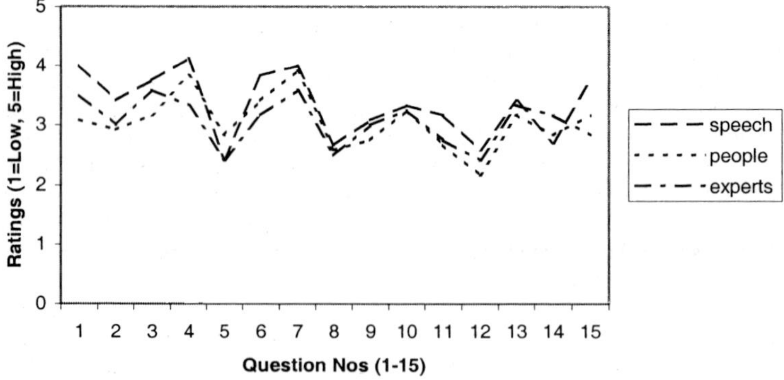

Graph 1. IDU Profiles for different groups

Discussion

From an initial analysis of the data the participants who believed they were using an ASR system to control their interaction in the VE, rated the usability of speech higher than the participants instructing another person or the expert user group. Whilst the results of the ANOVA were not significant, if a larger participant sample had been used then it is likely that these differences would have been significant. Even so, the data would seem to support the an element of user perception and experience upon the overall usability of ASR as an input device in VR applications.

Graph 1, above, illustrates a similar pattern in the IDU profiles for the questionnaire data for the different participant groups. Considering the questions in more detail participants felt that:

- they would enjoy using speech
- it would be comfortable to use speech
- speech would make it easy to interact with the VE
- that using a different input device would not make it easier to move around the VE
- it would be not be easy to move and position themselves in a VE using speech
- it would not feel natural to use speech to control movement in a VE

These finding suggest that participants generally felt that speech was an easy and enjoyable input device if it is used for the right sort of user interactions. The finding that using a different input device would not make it easier to move around the VE might indicate that the VE itself is a difficult environment to navigate around and thus highlights the need for careful integration of input devices and VE design.

Participants were invited to contribute their own comments which further illustrate some of the findings made. These are summarised in Table 1, below. The comments from the participants with first hand experience of the speech interface relate more to actual system use than the participants who instructed another person or even the expert user group. The comments from the speech group illustrate that the speech interface made interaction easier and quicker, than instructing another person or potentially using another interaction device. However, comments were also made about how relevant speech might be for tasks such as navigation or trivial and repetitive tasks.

These findings support the argument that the design of an input device should match the perceptual needs of the user (Kalawsky, 1996). From this study, users' perceptions would appear to be moulded by their direct experience of using an ASR interface. Jacob *et al*, (1993) recommended that the device should be natural and convenient for the user to transmit information to the computer and thus, a device should account for the communication-relevant characteristics of humans. Again, the findings support this, as users felt that speech is a natural input device to use. Further, Jacob *et al*, (1994) suggest that input devices should be designed from an understanding of the task to be performed and the interrelationship between the task and the device from the perspective of the user. Barfield *et al*, (1998) indicated that a VE input device should account for the type of manipulations a user has to perform and that the device should be designed so that it adheres to natural mappings in the way in which the device is manipulated, as well as permit movements that coincide with a user's mental model of the type of movement required in a VE. With respect to this the findings highlight how some tasks (menu selection, object manipulation) might be suitable for speech input, whereas other tasks (such as navigation) might be better suited to other input devices.

Table 1 Summary Comments from IDU Questionnaires

	Advantages of Speech	Disadvantages of Speech
Speech	• Good for interacting with objects within the environment • More natural form of command than typing (or moving a mouse) • Didn't have to think as much when deciding how to interact with the environment • Understood quite complex commands • Made control of VE more relaxing and less intense • It was quick once you knew the command • It was simple to use	• Not the most natural method for moving in a VE • Microphones can be too large, bulky and intrusive • Commands can take time to perfect • Joystick/keyboard easier for controlling movement • Felt self-conscious • Not sure what commands to use
People	• Could handle several instructions in one command • Less strenuous than using a mouse/joystick, etc • No real learning process required • Natural language - the ultimate user interface	• Too easy to say one thing when you mean another • Not good for trivial, simple repetitive tasks • Have to add instructions when initial instructions are not carried out • Not sure what are acceptable commands • Microphones can be too large, bulky and intrusive
Experts	• Good for selecting menus • Good for simple instructions • Good for menus and settings • Natural provided it works! • Good for multitasking • Adds to already available interactions when with existing input devices, especially in a 'busy' VE • Single word can initiate a complex automated procedure • Hands free, allow other tasks to be performed	• Could be disturbed by other people • Might feel self conscious • Fine adjustment manipulation may prove difficult • Might be difficult for navigation • Interaction metaphors not as precise as using joystick or mouse • Inaccurate if user loses concentration • Dislike using for locomotion • Could lead to side effects

Conclusion

Overall, the success of speech recognition depends upon the user, system, task and environment variables (Stedmon *et al*, in press), all of which have significant effects on overall system performance. At present there are very few established guidelines that detail what is required of an input device in terms of the parameters that pertain to a user and user performance within a VE. Therefore, the development of any ASR application should be an iterative process of design-test-redesign, involving both system experts and end-users.

Acknowledgements

The work presented in this project is supported by IST grant 2000-26089: VIEW of the Future.

References

Barone, A. (2001). *A Usability Comparison of Two Virtual Environments and Three Modes of Input.* Unpublished MSc Human Factors in Manufacturing Systems. University of Nottingham.

Cobb, S.V.G., Nichols, S.C., Ramsey, A.R. & Wilson, J.R. (1999). Virtual Reality – Induced Symptoms and effects (VRISE). *Presence: Teleoperators and Virtual Environments,* 8(2) pp.169-186

Barfield, W., Baird, K., & Bjorneseth, O. (1998). Presence in virtual environments as a function of type of input device and display update rate. *Displays*, 19: 91-98.

Jacob, R., Leggett, J., Myers, B., & Pausch, R. (1993). An agenda for human-computer interaction research: interaction styles and input/output devices. *Behaviour and Information Technology*, 12(2): 69-79.

Jacob. R., Sibert, L., McFarlane, D. & Mullen Jr., P. (1994). Integrality and separability of input devices. *ACM Transactions on Computer-Human Interaction*, 1(1): 3-26.

Kalawsky. R. (1996). Exploiting virtual reality techniques in education and training: Technological issues. *Prepared for AGOCG.* Advanced VR Research Centre, Loughborough University of Technology.

Stedmon, A.W., Patel, H., Nichols, S.C., & Wilson, J.R. (in press). A View of the Future? The Potential Use of Speech Recognition for Virtual Reality Applications. In, P. McCabe (ed). *Contemporary Ergonomics 2003. Proceedings of The Ergonomics Society Annual Conference, Cirencester. 2003.* Taylor & Francis Ltd. London.

AIR TRAFFIC CONTROL

SECOND EVALUATION OF A RADICALLY REVISED EN-ROUTE AIR TRAFFIC CONTROL SYSTEM

H. David[1], J.M.C. Bastien[2]

[1]*R+D Hastings,*
7 High Wickham, Hastings,
TN35 5PB, U.K.

[2]*Université Paris V – René Descartes*
45 Rue des Saints-Peres
75270 Paris, France

26 fourth-year students (native French speakers) were given an overall briefing on air traffic control, including a video description of an advanced ATC system, followed by individual coaching using 20 selected training examples. They were instructed that conflict resolution was their main task, but that aircraft should leave at the predetermined point if possible.

They were then presented with two exercises presenting traffic corresponding to an entry rate of 200+ aircraft per hour in random direct flight for one-hour nominal duration. (When no action was required, the simulation rate was accelerated by a factor of six, so that the average length of one exercise was about 14 minutes.)

24 participants controlled this traffic correctly, with no unresolved conflicts. Of the 1720 potential conflicts only two were not resolved correctly. Of the 10468 aircraft, only 5 left at a time, place or height different from that planned.

Introduction

An examination of the 'En-Route' Air Traffic Control (ATC) system, summarised in Dee (1996) and in more depth in David (1997a), led to the publication of EUROCONTROL Experimental Centre (EEC) Report 307, (David, 1997b). This report describes how En-Route ATC can be re-designed to provide a human-oriented task (primarily conflict resolution) supported by computer-based facilities, employing satellite communication and navigation systems. Briefly, after a careful task definition, a set of task analyses assuming different levels of human involvement were carried out, and a semi-automated system was adopted. Sub-tasks were allocated to define a satisfying, functional human controller task. Displays and control operations were then devised to provide simple (for the controller) displays and control options. The system was adapted, for demonstration purposes, to use a conventional low-precision colour display and a standard (QWERTY) keyboard. Although general design principles suggest that a three-dimensional image would be most appropriate, a co-planar display is in practice preferable (May et al 1996,Wickens 2000). An initial evaluation was reported at the 2001 Annual Conference of the Ergonomics Society, (David and Bastien, 2001a) and was more fully described in EEC Report 360 (David and Bastien, 2001b). Some minor but functionally important features, an additional 'swap' feature and enhanced error-checking and recording facilities were added for a second evaluation, as described below.

The RADICAL Interface

The 'co-planar' display provides a conventional 'map' display, with an additional height/ time and distance profile for a selected aircraft. These two windows are the only ones provided in the demonstrator – additional information would be made available in any real system. In normal operation, only aircraft involved in potential future conflicts would be shown on the map display, linked to show which aircraft are in potential future conflicts.

Aircraft are shown by icons representing their relevant characteristics, Colour indicates flight level, size indicates size, position and angle represent position and direction of flight, and wing sweep represents speed. The system presents the profile for an aircraft in the most urgent conflict, although an experienced controller may choose to over-ride this choice. The profile shows all conflicting aircraft by blocks representing the duration of the conflict, in solid colour for conflicts with the present flight plan, or in outline for aircraft that would be in conflict if a level change were made. The full trajectory of the aircraft is shown on the map display, with the trajectories of conflicting aircraft up to the point of closest approach – which may be after the tracks have crossed. A white line, indicating the distance and direction of the problem links the positions of the target and problem aircraft.

All displays are entirely symbolic, using no letters or digits. The extensive use of colour to differentiate aircraft and other information makes it impossible to provide a useful monochrome illustration. Interested readers are referred to EEC Report 307 (David 1997b) or to the EEC website (www.EUROCONTROL.fr) or to the first author's website (www.hughdavid.com - currently under construction).

The controller constructs a potential solution to the problems of the target aircraft by specifying manoeuvres via the keyboard. As each keystroke is entered, the display shows its consequences. Additional conflicts may appear because of a proposed manoeuvre, or the aircraft may leave at a wrong position, height or time. Red lines joining the actual and required exit indicate these differences. Some manoeuvres may be unacceptable, for example, requiring excessive speeds to exit correctly. They are indicated by a red final track segment. The controller may cancel all or part of a set of manoeuvres, or may use the 'R' key to transfer to the other aircraft in the most urgent conflict. The system will query a solution which leaves conflicts unsolved or requires an aircraft to leave at a wrong position, height or time, but the controller may over-ride the system. (Such solutions are usually the result of a keying error, but may sometimes form part of a larger strategy.) (It is assumed that aircraft communication is by data-link. The manoeuvres decided by the controller are coded and sent to the relevant aircraft for action, and to adjacent aircraft for information. A speech message would be generated at the controller's position, and in the pilots' cockpits, in their language of choice. A 'message repeat' facility would be available at each position.)

Aims

The aim of this study was to demonstrate that the RADICAL interface could be learned and operated efficiently without the expert skills required for conventional ATC operations.

Method

Twenty-six fourth-year students (All French-speaking, none familiar with air traffic control) were given an initial general briefing on Air Traffic Control, followed by twenty individual short coached training examples. They were briefed that the principal aim was to resolve future potential conflicts, but that aircraft should leave at the point, time and height originally planned. They were advised to seek a solution in terms of flight level change first, and to use the 'R' control to display the flight profile of the other aircraft if no immediate solution to the problem presented itself.

They were then presented with traffic building up to 40 aircraft simultaneously present, in random direct flight in a space representing 120 x 90 NM (Nautical Mile) corresponding to an entry rate of 200+ aircraft per hour for a nominal one hour. (The system replaced each aircraft as it left, providing new aircraft with an initial and final 2 minutes of conflict free flight. Since different solutions provided slightly different flight paths, samples diverged during successive exercises.)

Because the system was set to show only aircraft in conflict, it was completely blank when there were no conflicts. To save time, the simulation rate was increased to six times real time during these periods, and automatically returned to real time when a new conflict was detected.

Results

Table 1 presents the mean performances of the 26 participants. (These are unweighted means of individual performance means for each exercise, not overall mean values.)

Twenty-four of 26 participants correctly resolved all the conflicts presented to them. Two participants each failed to solve one conflict in time. In one instance, the participant became pre-occupied with solving one problem, and failed to leave enough time to solve a second conflict. In the other, the participant produced a nearly satisfactory solution, but hit the "Y" key twice, over-riding the system warning.

The mean time taken to resolve conflicts was 22.8 seconds, measured from the first appearance of the conflict on the screen to the acceptance of an order resolving the conflict. The mean time to construct an order was 14 seconds, and a mean of 30.4 orders was required to solve a mean of 33.8 conflicts. (One order may solve several conflicts, but a less urgent conflict will not be considered until all more urgent ones have been solved.) There were few modified orders – virtually all attributable to one participant who preferred to use the 'modify' procedure rather than the 'cancel' procedure. Ten orders were cancelled per run. Cancelled and modified orders do not necessarily imply errors by the participant. They usually represent the exploration of a potential solution that did not in fact work. (Solutions involving changes of flight level only can be checked visually on the height profile display before entry, but those involving height or speed changes can only be checked for conflicting traffic by entering the solution and observing any consequences.)

Table 1. Mean Performance

Parameter	Arithmetic Mean	% Nil Scores
Aircraft	201.3	
Conflicts	33.08	
Resolved (correctly)	33.04 (=99.88%)	
Unresolved	0.04 (= 0.12%)	96%
Time to resolve Conflicts	22.80 seconds	
Time before Conflict start	29.4 Minutes	
Number of Control Orders	30.35	
Time to Construct Orders	13.95 seconds	
No. of Modified Orders	0.87	48 %
Time to construct	13.06 seconds	
No of Cancelled orders	10.02	
Time to construct	14.33 seconds	
Total Exercise Duration	13.89 Minutes	
Position Exit errors	0.019	98 %
Height Exit errors	0.038	96 %
Time Exit Errors	0.096	90 %

The mean time taken to resolve conflicts was 22.8 seconds, measured from the first appearance of the conflict on the screen to the acceptance of an order resolving the conflict. The mean time to construct an order was 14 seconds, and a mean of 30.4 orders was required to solve a mean of 33.8 conflicts. (One order may solve several conflicts, but a less urgent conflict will not be considered until all more urgent ones have been solved.) There were few modified orders – virtually all attributable to one participant who preferred to use the 'modify' procedure rather than the 'cancel' procedure. Ten orders were cancelled per run.

Five aircraft left at the wrong height. (Participants were instructed to allow an aircraft to leave at a wrong height if it would otherwise be in conflict at the exit position.) One of these aircraft also left at the wrong place and time, and one at the wrong time. These errors are not enough for serious analysis, but it is significant that only 5 of 10,468 aircraft left at unplanned positions, which would have required coordination with the adjacent airspace, corresponding to one every ten hours of running time.

Discussion

The capacity increase provided by the RADICAL interface is such that no formal experimental comparison can be made – the least efficient novice performs better using this interface than the most experienced team of professional controllers using conventional methods. Such a team would consist of two experienced controllers, requiring three years of initial training and six months of training for each sector involved. The aircraft are constrained to fly on predefined routes, so that separation can be verified over fixed points. In favourable conditions, with the most modern equipment, some fifty to sixty aircraft per hour could be handled. A few aircraft might be given direct routes, at considerable cost in additional monitoring effort.

The capacity of the demonstrator was limited by the software available to 40 aircraft at any time. (A higher-capacity version is now available. Although this is limited by data retrieval delays, samples involving up to 500 aircraft simultaneously present have been handled successfully in preliminary trials.) The operational capacity of the revised system is defined by the frequency of potential conflicts. Since it takes approximately 15 seconds to solve a conflict, a 75% workload would imply the solution of 180 conflicts per hour. The frequency of potential conflicts in a direct-route environment would depend on the density of traffic, particularly at choke points around reserved airspace. As in the current system, potential conflicts could be reduced by adopting flight level allocation conventions. It appears probable that, if this system were adopted for flights above, say, FL200 (20,000 feet), conflict-free routings could be provided for most climbing aircraft before they reached that level, and for cruising aircraft before reaching the boundary of the controlled airspace. Aircraft on direct routings to the same airport would probably not converge significantly before descending below FL200, but some special procedures would be needed to avoid problems around the edges of restricted airspace.

That the proposed interface is easy to learn does not imply that it removes the need for skill in its use. During their two nominal hours of control activity, the participants developed measurably different methods of treating traffic. It can reasonably be expected that controllers would 'learn the traffic' as they do at present, although it would not be necessary for them to do so to provide an efficient control service.

Finally, it should be recognised that the practice of real Air Traffic Control, as opposed to more-or-less elaborate simulation, requires mental and moral stamina, patience and professional conscience, which, in the long run, may be more important that the short-term cognitive and motor skills examined here.

Conclusions

1. The RADICAL interface for En-Route Air traffic control can be learned and efficiently operated with one morning's training by non-ATC personnel.

2. The interface is independent of the language spoken by controllers and/or pilots, and can be organised to enhance pilots' mutual awareness. (The participants in this study were French-speaking.)

3. The improved efficiency of the revised system removes most of the coordination effort needed in the contemporary system.

4. The capacity of the interface was limited by the available software to 40 aircraft simultaneously present, corresponding to about 200 aircraft in random direct flight through a 90 by 120 NM area. The true capacity is better expressed in terms of potential conflicts per hour, as approximately 180 potential conflicts per hour.

Acknowledgements

The first author gratefully acknowledges the vision and generosity of the second author in undertaking this study without financial support. Both authors gratefully acknowledge the co-operation of the participants of the Laboratoire de l'Ergonomie de l'Informatique. They also acknowledge the encouragement of the director of the EUROCONTROL Experimental Centre, M. J-C Garot. This study does not form part of the official EUROCONTROL work program.

References

(EEC Notes and Reports are available at the EEC Web-site (www.eurocontrol.fr))

David, H., (1997a) Deep Design – Beyond the Interface, In K-P. Holtzhausen *Advances in Multimedia and simulation. Human-Machine-Interface Implications* (pp 65-77) Bochum, Germany: Fachhochscule Bochum

[1]David, H. (1997b) *Radical Revision of En-route Air Traffic Control.* EEC Report No. 307. EUROCONTROL Experimental Centre, Bretigny-sur-Orge, France

David, H. and Bastien, J-M. C. (2001a) *Initial Evaluation of a Radically Revised En-Route Air traffic Control System*, In Hanson, M.A. (Ed.) *Contemporary Ergonomics 2001*(pp 453-458)*: Taylor* and Francis

David, H. and Bastien, J-M. C. (2001b) *Initial Evaluation of a Radically Revised ATC Interface*, EEC Report No. 360, EUROCONTROL Experimental Centre, Bretigny-sur-Orge, France.

Dee, T.B. (1996), Ergonomic Re-Design of Air Traffic Control for Increased Capacity and Reduced Stress, In Robertson, S.A. (Ed.), *Contemporary Ergonomics 1996* (pp 563-568): Taylor and Francis.

May, P.A., Campbell, M. & Wickens, C.D. (1996) Perspective displays for air traffic control: Display of Terrain and Weather, *Air Traffic Control Quarterly* 3(1) 1-17

Wickens, C.A. (2000) The When and How of using 2-D and 3-D Displays for Operational Tasks. In Proceedings *of the IEA2000/HFES 2000 Congress*, .San Diego California USA

---

[1] This report includes a 3.5"disc with compiled code, source code and documentation for demonstrations of traditional (DEMOLD), intermediate (DEMON) and fast interfaces (DEMFAST)

Compiled code, source code and documentation for the improved version (TROTSKY) and for analysis programs are available from the first author.

TOWARDS A NEW TASK BASE FOR AIR TRAFFIC CONTROLLERS IN THE FUTURE ATC/ATM

Arnab Majumdar and Washington Yotto Ochieng

Centre for Transport Studies
Imperial College London
South Kensington
London SW7 2AZ

The demand for air travel in the World continues to grow at a rapid rate, especially in Europe. The ever-growing demand continues to stretch the available resources for air traffic management (ATM) and control (ATC) with serious consequences for the society in terms of passenger safety, the environment and the economy. Various initiatives have been proposed to increase airspace capacity in Europe. These rely upon a combination of new technologies and procedures. This paper considers the literature on proposed new controller tasks and their modelling. The first section describes the airspace capacity problem in Europe and the proposed solutions. This is followed by a section outlining the various possible tasks of controllers in a future ATC/ATM environment. The final section considers an example of how to model these new tasks in a fast-time simulation model of air traffic controller workload.

Introduction

The demand for air travel worldwide continues to grow at a rapid rate, especially in Europe and the United States. In Europe, the demand exceeded predictions with a real annual growth of 7.1% in the period 1985-1990, against a prediction of 2.4%. By the year 2010, the demand is expected to double from the 1990 level (ATAG, 1996). Within the UK international scheduled passenger traffic is predicted to increase, on average, by 5.8 per cent per year between 1999 and 2003 (DETR 2000). The demand has not been matched by availability of capacity. In Western Europe many of the largest airports suffer from runway capacity constraints. Europe also suffers from an en-route airspace capacity constraint, which is determined by the workload of the air traffic controllers, i.e. the physical and mental work that controllers must undertake to safely conduct air traffic under their jurisdiction through en-route airspace (Majumdar et al., 2002).

In 1989, the annual cost to Europe due to air traffic inefficiency and congestion in en-route airspace was estimated to be 5 billion US Dollars (Lange, 1989). This was mainly due to delays caused by non-optimal route structures and reduced productivity of controllers due to equipment inefficiencies. This figure has been confirmed recently by the European Commission and major delays remain e.g. in 1999, 25% of the total delays were caused by a lack of en-route ATC capacity. Therefore, to in order to decrease the total delay, an increase in en-route capacity is of paramount importance. Note that whilst the demand figures have been revised downwards since the unfortunate incidents of 11

September 2001, the belief in the industry is that demand will rise again to the original predicted levels within a few years.

At a global scale and in the early 1980s, the International Civil Aviation Organisation (ICAO) recognised that the traditional air traffic control (ATC) systems would not cope with the growth in demand for capacity. Consequently, it established the Special Committee of Future Air Navigation Service (FANS) to study, identify and assess new technologies to enable ATC to cope with this demand, as well as to make recommendations for the future development of navigation systems for civil aviation. This led to the development of a satellite-based system concept to meet the future civil aviation requirements for communication, navigation and surveillance/ air traffic management (CNS/ATM). The concept involves the application of state-of-the-art technologies in satellites and computers, data links and advanced flight deck avionics to cope with air traffic service requirements. This should remove the need for relatively expensive ground-based equipment, which use line-of-sight technology and has inherent limitations. It is expected to produce benefits in efficiency, economy and safety. More importantly it is expected to be an integrated global system with consequential changes to the way air traffic services are organised and operated.

In Europe, the problem of airspace capacity has been exacerbated by the lack of an integrated ATM system. Since 1990, the organisation EUROCONTROL (established in 1960 to co-ordinate European ATM) has been charged by the ECAC (European Civil Aviation Conference) Member States, to address the issues of airspace usage (ECAC, 1990). In the 1990s, EUROCONTROL launched EATCHIP (European Air Traffic Control Harmonisation and Integration Programme) for en-route airspace, and the APATSI (Airport/ Air Traffic Services Interface) programme for the airports (EUROCONTROL, 1991). Notwithstanding the measured success of EATCHIP in addressing the airspace usage issues (ECAC, 1998), the programme has been superseded by the EATMP (European Air Traffic Management Programme). The aim of EATMP is the implementation of ATM policies, processes and measures, identified by the ATM Strategy for 2000+ initiative, which involve the development of ATC/ATM strategies based on new technology within the CNS functions (as defined by ICAO) of ATM (EUROCONTROL, 1998). Two key changes envisaged for en-route airspace will be the introduction of real time 4D navigation and the increasing delegation of responsibilities for control to flight crew, by the use of airborne separation assurance between aircraft, leading eventually to 'free flight' airspace.

EATMP proposes a broad range of measures and technology, with the focus on the progressive introduction of a number of operational improvements to keep pace with the anticipated demands for capacity, efficiency, safety and environment. In particular, in the future ATM/ATC service for Europe, aircraft will be controlled accurately and with high integrity in four dimensions with the aid of on-board and satellite navigation and communications technologies. Each aircraft will negotiate and re-negotiate a 4D flight plan in real time with the ground based ATM system. This will provide airborne autonomous separation to give conflict-free tracks between departure and destination in the form of 4D profiles to be accurately adhered to by aircraft. To achieve this will require enhanced surveillance methods and automated communications will be needed, which will also have profound impacts on ATM functions and working methods of pilots and air traffic controllers.

The concept of *free flight* has emerged in the USA, which will result in pilots being free to choose their preferred route between the airports of departure and destination. Free flight envisages each aircraft communicating with both ground controllers and proximate aircraft to negotiate and establish conflict free flight paths. Most of the flight plan negotiation will be carried out semi-automatically via one of a number of new data links. In line with the US, the promotion of free flight is also defined as one of the aims of EUROCONTROL's ATM 2000+ programme (EUROCONTROL, 1998), though this is envisaged to occur in stages with increasing degrees of delegation of responsibilities to flight crew. This move toward free flight will lead controllers and pilots to face several changes in their operations. For example, traffic will no longer be constrained to a fixed route structure with predictable congestion points and controllers may need to be familiar with the whole airspace (as opposed to fixed flight profiles) to allow flexibility in the dynamic assignment of airspace to traffic. Crucially in free flight, pilots will be responsible for aircraft separation. Using the appropriate satellite-based technology, such delegation will have a profound influence on the roles and consequent tasks that the pilots and controllers, e.g. monitoring, must carry out (The CAST Consortium, 1998). The following section considers possible new controller tasks.

Possible New Controller Tasks

The pervious section has noted that new technologies and procedures have been proposed to increase airspace capacity in Europe. These will have profound impacts on the roles and tasks of air traffic controllers. A good description of the current tasks of air traffic controllers in the en-route environment comes from a report by Rodgers and Dreschsler (1993) on the controller task base for the US Federal Aviation Authority (FAA). Their detailed description of every task of the air traffic controller sub-divided into detailed levels of sub-activities, etc. had six major categories of ATC activities:

- Perform Situation Monitoring.
- Resolve Aircraft Conflicts.
- Manage air traffic sequences.
- Route or Plan Flights.
- Assess Weather Impact
- Manage Sector/Position Resources.

Whilst various studies have been undertaken in Europe to consider the duties and future tasks of air traffic controllers, the study with the greatest depth of detail is that undertaken by the CAST Consortium (CAST Consortium, 1998), to consider the future training needs of air traffic controllers. The CAST Consortium's report identified the nine major categories that would require either a modification of tasks, or new tasks or both, as a consequence of increasing automation and new procedures in ATC/ATM:

- Anticipation and planning
- Situation Awareness
- Monitoring
- Detection
- Evaluation
- Resolution
- Verification
- Communication
- Decision Making

Table 1 indicates the tasks involved in these CAST categories, and identifies them in the categories of Rodgers and Dreschsler (1993) FAA task category.

Fast-time Simulation Modelling

It has been mentioned that in Europe, en-route airspace capacity is primarily determined by controller workload. Consequently, airspace planning and management often uses fast-time (i.e. computer based) simulation modelling exercises of controller workload. A well used such model is the Reorganised ATC Mathematical Simulator (RAMS), which together with its precursor has had over 25 years of use in airspace planning in Europe. RAMS is a discrete-event simulation model, and requires an appropriate task base which assigns a particular task to an event in the simulation. There are five major task categories in RAMS (EUROCONTROL, 1995):

- Coordination tasks, both external and internal;
- Flight data management tasks;
- Planning conflict search;
- Routine radio/telephony communications tasks;
- Radar tasks associated with conflict resolution.

Table 1 also indicates how the future tasks outlined in the CAST Consortium Report may be incorporated in the current RAMS task categorisation. A key finding Table 1 is that whilst many future controller tasks can be incorporated n to the current RAMS task base, a considerable amount of new controller tasks involve automation based tasks. This can involve checking on the resolution solution proposed by a controller tool, implementing the solution and verifying it. Currently the task base in RAMS cannot adequately capture such tasks. Hence there maybe a need for a future fast-time simulation model to incorporate automation tasks as a separate category, though this may require further discrimination to adequately capture such tasks.

An example of such future tools and their effects on fast-time simulation modelling can be seen with the use of the Medium Term Conflict Detection Tool (MTCD) (EUROCONTROL, 1998), which aims to help the planning controller (PC) aid the tactical controller (TC) in the sector by allowing the PC to solve in-sector conflicts. When the MTCD is used in conjunction with a Conflict resolution Advisory (CORA), then the TC's workload can be greatly reduced. Therefore this technology will lead to major changes in working methods for PC and TC. Whilst such a change will require considerable alterations in RAMS, e.g. conflict detection and resolution algorithms, with regards to controller tasks, new tasks in the resolution and automation category will be required for PC conflict resolution; decision to use the resolution advisory and monitoring the advisory resolution.

Table 1. Consequences of future ATM systems for air traffic controller

Anticipation / Planning	**FAA Category**	**RAMS task-base**
Differentiate between actual ATC situation and predicted state	Perform Situation Monitoring/ Route or Plan Flights.	*AUTOMATION*
Gauge confidence in computer-generated plan	Perform Situation Monitoring	*AUTOMATION*
Commit planned route for aircraft to memory	Route or Plan Flights.	*AUTOMATION*

Situation Awareness	**FAA Category**	**RAMS task-base**
Build up and maintain situation awareness using computer assistance	Perform Situation Monitoring (Housekeeping)	Flight data management (change needed)
Ensure situation awareness does not fall below a critical level	Perform Situation Monitoring	*AUTOMATION*
Seek possible alternatives to computer- generated projections or solutions	Perform Situation Monitoring/ Manage air traffic sequences? Route or Plan Flights?	*AUTOMATION*
Maintain awareness of what system is doing	Perform Situation Monitoring	*AUTOMATION*
Maintain awareness of system	Perform Situation Monitoring	*AUTOMATION*
Monitoring	**FAA Category**	**RAMS task-base**
Maintain "global" perspective of the sector	Perform Situation Monitoring	*AUTOMATION* Flight data management?
Monitor for system faults	Manage Sector/Position Resources/ Manage air traffic sequences	*AUTOMATION*
Ensure visual channel does not become overloaded	Manage air traffic sequences/ Route or Plan Flights	*AUTOMATION* Radar surveillance?
Detection	**FAA Category**	**RAMS task-base**
Detection of conflicts in the medium- term	Resolve Aircraft Conflicts/ Route or Plan Flights	
Manage track deviations detected by system	Manage air traffic sequences	
Evaluation	**FAA Category**	**RAMS task-base**
Evaluate ATC situation based on system- processed information	Perform Situation Monitoring/ Manage air traffic sequences? Route or Plan Flights?	
Resolution	**FAA Category**	**RAMS task-base**
Select conflict resolution alternatives with the assistance of system tools	Resolve Aircraft Conflicts	
Communication		
Seek source of information presented automatically by system	Route or Plan Flights/ Manage Sector/Position Resources	R/T Communications/ Co-ordinations?

Manage electronic message exchange (between controllers and between controllers and pilots)	Route or Plan Flights	R/T Communications/ Co-ordinations?
Ensure rapid, unambiguous communication in emergency situations	Route or Plan Flights	R/T Communications/ Co-ordinations?
Ensure datalinked message has been received accurately by pilot	Route or Plan Flights	R/T Communications/ Co-ordinations?
Verification	**FAA Category**	**RAMS task-base**
Obtain information on acceptability of plan and actions from computer tools	Route or Plan Flights	
Decision Making	**FAA Category**	**RAMS task-base**
Ensure that controller decision making ability is not prejudiced by manual reversion	Route or Plan Flights/ Manage Sector/Position Resources	AUTOMATION
Take account of system- generated information when making decisions	Manage Sector/Position Resources/ Perform Situation Monitoring	AUTOMATION
Decide when and how to use computer assistance	Manage Sector/Position Resources/	AUTOMATION
Consider mode of operation	Manage Sector/Position Resources	AUTOMATION

References

ATAG (1996) *European traffic forecasts 1980-2010,* ATAG: Geneva, Switzerland.

The CAST Consortium (1998) Consequences of future ATM systems for air traffic controller Selection and Training, WP1: Current and Future ATM Systems, Contract No.: AI-97-SC.2029, The European Commission

DETR (2000) *Public private partnership for the National Air Traffic Services Limited,* DETR, London

ECAC (1990) *ECAC Strategy for the 1990s,* ECAC, Paris

EUROCONTROL (1991) *European ATC Harmonization And Integration Programme (EATCHIP)* - Report Phase I, Brussels.

EUROCONTROL EXPERIMENTAL CENTRE (1995) RAMS system overview document, Model Based Simulations Sub-Division, EUROCONTROL, Bretigny-sur-Orge, France

EUROCONTROL (1998) *Air Traffic Management Strategy for 2000+,* Volumes 1&2, EUROCONTROL, Brussels.

Majumdar, A., Ochieng W.Y. and J.W. Polak (2002), Estimation of European airspace capacity from a model of controller workload, *The Journal of Navigation*, **55** (3), 381-403

Rodgers, M. and G.K. Dreschsler (1993) Conversion of the CTA, Inc., en route operations concepts database into a formal sentence outline job task taxonomy, Federal Aviation Administration Civil Aeromedical Institute report, DOT/FAA/AM-93/1.

CULTURAL DIFFERENCES IN THE MEANING OF COMPETENCY IN AIR TRAFFIC CONTROL

Deirdre Bonini[1] and Barry Kirwan[2]

[1]Department of Psychology, Trinity College Dublin, Dublin 2, Ireland
dbonini@tcd.ie
[2]Eurocontrol Experimental Centre, B.P. 15, 91222 Bretigny sur Orge CEDEX France
barry.kirwan@eurocontrol.int

The safe and efficient control of aircraft is the result of a air traffic controllers collaborating with each other, with an increasing reliance on automation. Competence is believed to be an important variable in an efficient collaboration and use of technology. This paper describes a tripartite methodology and the results of a study that collected information on the characteristics considered important by French, Irish and Italian air traffic controllers in attributing competence. Differences were found in both controller and technology descriptors of competence, not only in terms of the characteristics provided but also in the value that air traffic controllers attributed to them.

Introduction

Air traffic controllers accomplish the safe and efficient air traffic control of aircraft by collaborating with colleagues and pilots through technology, and by using information provided by complex automated systems. As reliance on technology is perceived as increasingly necessary, so is an understanding of the mechanisms that optimise the interaction of controllers with technology.

There are a number of factors that influence the acceptance and use of technology, such as its reliability and its perceived need by the users (Muir, 1994). One of the most interesting and less understood of these mediating elements is trust. In differing degrees, trust mediates all our interactions with both people and technology (Golembiewski & McConkie, 1975; Lee & Moray, 1992).

A decision to trust a colleague and delegate a task to him/her, or to base ones action on the information provided by technology is effectively the choice of a control strategy. When we choose to trust another we are deciding if they are "good enough" to carry out the task we are delegating to them, or if the automation is "accurate enough" to give us the right information to make an informed and correct decision. This judgement of appropriateness can be conceptualised in terms of an assessment of competency.

In order to understand air traffic controllers' (henceforth controllers) trust in other controllers and automation it is thus necessary to identify the characteristics that are relevant in a controller's attribution of competence.

Competence is a complex combination of attitudes, values, knowledge and skill that allow for satisfactory performance in actual working situations (Gonczi &

Athanasou, 1996). Satisfactory performance may be understood according to standards in that occupational area, but also as defined by the working culture of those in the domain. The model of competence that is most often referred to is the component model (Sptizberg and Cupach, 1984), that includes knowledge, skill, and motivation, and describes the concept in terms of knowing what behaviour to carry out when, and being motivated to do so.

In the literature the concept is treated either as an underlying characteristic of an individual, assessed by psychometrically derived tests, or as a socially situated ability to perform tasks and roles to an expected standard, which varies with experience and responsibility (Eraut, 1998). The latter understanding best suits an analysis of a competent controller, as someone able to "apply his appropriate knowledge, skills and experience to provide air traffic control services as notified in his air traffic controller licence" (EUROCONTROL, 2000: 49).

Storey (2001) describes competence as a dynamic process that changes with experience. What changes with time is the repertoire of behaviours available, but also the ability to recognise salient characteristics more readily. These indicators can be attributes of a competent controller or features of an efficient and useful technology. Holmes (1994) argues that social processes need to be considered as the concept of competence is culturally defined.

Given that competence appears to be dependent on experience and social processes, it is likely that competence may have different 'markers' in different countries or cultures. This is particularly important for Air Traffic Management (ATM), because it transgresses national boundaries, and future automation (to support the controllers) will be implemented in many different countries, and there is a general European goal of enabling future mobility of controllers throughout Europe. However, the cultural impacts on trust and competence, both for human-machine and human-human interactions, have received little general study, and no specific studies were found in the domain of ATM.

The aim of this study was therefore to identify descriptors of controller and technology competence, according to controllers from three different European countries. If such differences exist, then they need to be taken into consideration when introducing automation into ATM, or in general controller training.

Competence Descriptors

The methodology used to collect controller and technology descriptors was the same for all three nationalities and encompassed three tasks: an information collection task, a card sorting task and a paired comparisons task. The results of each task served as the basis for the next stage. Ireland, Italy and France agreed to participate in the study.

Information collection task

The first task aimed at collecting a representative list of characteristics that controllers use to describe competent colleagues and technology. The information was collected through a brief interview or through a questionnaire sent by email. The questions are reported in Table 1 and were administered in the respondents' mother-tongue. Competent technology was defined as technology that works well and is considered helpful by a controller.

Table 1. Questions asked in the information collection part of the study

Question 1.	Think of a controller whom you regard as competent. What characteristics does he/she have?
Question 2.	Please list the characteristics that describe technology that works well. If possible include examples of efficient and inefficient technology.
Question 3.	Please list the characteristics that describe technology that helps a controller work well. If possible include examples of useful and useless technology.

A total of 68 controllers replied to the three questions. Replies were catalogued using a very simple content analysis. The lists resulting from French controllers' replies contained 32 controller descriptors and 20 technology descriptors. Irish controllers provided a list of 24 controller descriptors and 20 technology items. Italian controllers were more articulate in talking about technology. Their lists consisted of 12 controller descriptors and 30 technology descriptors.

To understand the relative saliency of each item the next step was to run a paired comparisons task. However, with the exception of the Italian list of controller descriptors, the other lists were too long for this exercise. Thus, a group of controllers of each nationality were first asked to carry out a card sort task to organise the words, in order to reduce the number of items to compare.

Card sorting task

A card sorting task is a technique to explore how people group items. It has been used for many purposes, from designing an interface to investigating the educational needs of patients (Lewis, 1991; Luniewski, Reigle, and White, 1999).

Controllers were given a pack of randomly sorted cards containing the characteristics derived from the information collection task. They were asked to organise them in a way that made sense to them.

A total of 49 controller participated in the sorting tasks. Although it was not possible to have the same group of controllers carry out both sorting tasks, an effort was made to achieve a representative group of participants in terms of age, experience and gender.

A hierarchical agglomerative cluster analysis that sequentially merges all cases into one group was used to analyse the results (Aldenderfer & Blashfield, 1984). The output of this method is represented in the form of a tree structure (a dendogram), that portrays a hierarchical organisation of the relation between items.

Controller and technology descriptors for French and Irish controllers clustered into 4 groups. The Italian participants sorted technology items into 5 clusters.

Paired Comparisons Task

The paired comparisons method is a psychological scaling technique (Seaver and Stillwell, 1983; Hunns, 1982) and has been most often used in the human reliability domain to elicit expert judgements on likelihood of error. The technique is based on the idea that experts are better at comparing one item with another and then deciding which is higher or lower on a scale, rather than making absolute judgements for each item (Kirwan, 1994). The judgements made by experts are transformed into interval scales that represent a 'psychological continuum' of perceived characteristics. The results are represented graphically as 'rulers'.

During a break from their work, participants were given pairs of characteristics and asked to choose which of the two was most important to be considered either a competent controller or technology. These pairs were derived from matching each expression in the lists, with every other item with which it had been grouped in the word sort task.

A total of 66 controllers participated in the paired comparison tasks for controller and technology descriptors. The exercise resulted in 23 scales of competence describing the characteristics attributed to competent colleagues and technology according to French, Irish, and Italian controllers.

Results

The characteristics that controllers gave to describe a competent controller and technology varied greatly between countries. French and Italian controllers did not share any characteristic that was not also shared with Ireland.

In describing a competent controller Irish and French controllers agreed on the importance of 'self-confidence', 'experience' and having a 'sense of humour'. Irish and Italian controllers shared 'flexibility' as a characteristic. All nationalities agreed that 'being calm' and able to 'keep ones cool' were necessary to attribute competence. With regard to technology, French and Irish controllers both agreed on 'user friendly', 'easy to set up' and 'simplifies the job'. Irish and Italian controllers shared 'flexible', 'simple to use' and 'reliable'. All three nationalities agreed on 'fast', 'reduces workload' and 'saves time' as key attributes.

Looking at the clusters into which controllers grouped items, it was possible to read the underlying strategy for most rulers. For the French controller, for example, two of the rulers represented character and personal style, and show the way in which work was carried out, their 'technique'. Italian respondents grouped the technology descriptors in terms of the interface, presentation of information, time, efficiency and the consequence of the technology on their work (e.g. reduces stress, workload, etc.).

With regard to the distribution of the contents of the rulers, from least to most important, Irish and Italian controllers ranked items more closely than French controllers.

The highest positions on the technology rulers of French controllers were held by 'gives you more time', 'fast', 'simple', and 'effective human machine interface'. The fact that the technology 'helps in planning', is 'intuitive', 'user friendly' and 'allows you to make precise measurements' were seen as less important.

Irish controllers ranked 'accurate', 'safe', 'user friendly' and 'essential in my job' in the highest positions in their groups of items describing technology. The four characteristics 'fast', 'easy to understand', 'gives you more time', and 'well maintained', were less important.

Respondents from Italy rated technology that was 'reliable' and 'increases the quality of work' as most important attributes. The fact that the technology is 'fast', 'easy to use' and 'makes you work more' were not as important.

Discussion

Analysing the words in terms of the competence component model (Spitzberg and Cupach, 1984) it was found that French and Irish controllers focused more on characteristics describing skills and motivation, whereas Italian controllers focused less on motivation and more on knowledge. French controllers felt that 'communication' was one of the most important factors, but that 'anticipation', 'being rigorous' and 'being able to quickly recover from mistakes' were also salient. Irish controllers felt that the most important characteristics were 'communicating clearly', 'being part of the team', 'helpful', 'not afraid to admit mistakes' and 'able to maintain a cool head under pressure'. Italian controllers focused on 'knowledge of English'.

An example of a ruler is provided in Figure 1. A few points can be made about the common characteristics. The item 'fast', for example, was the highest characteristic on one of the French rulers. The same item was the lowest item in both Irish and Italian rulers. Irish and Italian controllers ranked clarity of display of information high on their rulers.

Lowest					**Highest**
0	**36**	**38**	**53**	**61**	**66**
FAST	**GAIN TIME**	**OF HELP**	**SIMPLE TO USE**	**USEFUL**	**CLEAR DISPLAY**

Figure 1. A competence ruler for Italian controllers with respect to technology. The numbers refer to the distance between the characteristics, from the item considered least to that considered most important

With regard to 'easy to set up', French respondents ranked it highest and Irish lowest. The item was not mentioned by Italian controllers. Some of these respondents may be soon having a system up-grade bringing more functionality. Italian controllers, on the other hand, only recently changed over to one of the most sophisticated systems in Europe and may now take setting up the added features as part of their routine. It is essential to read the information in the context of the system that is used by respondents.

The controller characteristics that resulted identify attributes that are important for team-mates to be considered competent. Working well together is essential in ATM, and the results indicate importance is attributed to different characteristics. On the other hand, the technology attributes are interesting for the development and introduction of future technology. Their diversity emphasises that it is important to consider the salient features of present systems before introducing change. The differences between cultures identified relate not only to the actual characteristics, but to the value given to them.

The scales of competence derived from this study will be used in a number of scenarios that aim at testing the hypothesis that trust depends not only on a judgement of competence, but also on the degree of reliance of the truster on the trustee or on automated tools. This work can then help to inform the development of automated support tools for controllers in future European air traffic management.

Acknowledgements

The authors wish to thank all the controllers who participated in the study, we are grateful for their help, time and enthusiasm. We would also like to thank the EUROCONTROL project managers and Irish, Italian and French control centre staff who kindly organised our visits and found time to include us in their busy simulation schedules and operational rosters.

References

Aldenderfer, M.S. & Blashfield R.K. 1984, *Cluster Analysis*, (Sage University Press, Beverly Hills, California)

Eraut, M. 1998, Concepts of competence, *Journal of Interprofessional Care*, **12** (2): 127-139

EUROCONTROL 2000, *European Manual of Personnel Licensing – Air traffic controllers*, EATCHIP, HUM.ET1.ST08.10000-STD-01

Golembiewski, R.T. and McConkie, M. 1975, The centrality of interpersonal trust in group processes. In C. Cooper (ed.) *Theories of Group Processes*, (Wiley and Sons Ltd.), 131-179

Gonczi, A. and Athanasou, J. 1996, *Instrumentación de la educación basada en competencias. Perspectiva de la teoría y la práctica en Australia*, (Ed. Limusa)

Hunns, D.M. 1982, Discussions around a Human Factors Data-Base. An Interim Solution: the Method of Paired Comparison. In A. E. Green (ed.) *High Risk Safety Technology*, (Wiley and Sons Ltd.)

Kirwan, B. 1994, *A Guide to Practical Human Reliability Assessment*, (Taylor & Francis, London)

Lee, J. and Moray, N. 1992, Trust, control strategies and allocation of function in human-machine systems, *Ergonomics*, **35** (10): 1243-1270

Lewis, S. 1991, Cluster analysis as a technique to guide interface design, *International Journal of Man-Machine Studies*, **35**: 251-265

Luniewski, M. Reigle, J. and White, B. 1999, Card Sort: An Assessment Tool for the Educational Needs of Patients with Heart Failure, *American Journal of Critical Care*, **8** (6): 297-302.

Muir, B.M. 1994, Theoretical Issues in the study of trust and human intervention in automated systems, *Ergonomics*, **37** (11): 1905-1922

Seaver, D.A. and Stillwell, W.G. 1983, *Procedures for using expert judgement to estimate human error probabilities in nuclear power plant operations*, **Nureg/cr-2743** (U.S. Nuclear Regulatory Commission)

Spitzberg, B.H., and Cupach, W.R. 1984, *Interpersonal communication competence*, (Beverly Hills, CA: Sage)

DRIVING

THE USE OF TACTILE AND AUDITORY LANE DEPARTURE WARNINGS IN A FIELD OPERATIONAL TEST

John Bloomfield[1, 2]
Kathleen A. Harder[2]
Benjamin J. Chihak[2]

[1] *University of Minnesota and University of Derby*
[2] *University of Minnesota*

The Driver Assist System (DAS), that was developed to help specialty vehicle operators deal with the very low visibility conditions that can occur in winter, provides a virtual representation of the roadway ahead on a Head-Up Display mounted in front of the driver. A Field Operational Test (FOT) of the DAS, involving 21 subjects (eight snowplow operators, one highway patrol officer, and 12 ambulance drivers), was conducted on a 45-mile section of highway in Minnesota from December 2001 until March 2002. Unfortunately, this proved to be the state's mildest winter on record—compromising the planned evaluation of the DAS. However, it was possible to compare lane departure durations when the DAS was *On* and when it was *Off*, in good visibility conditions. The results of the three-and-a-half-month long FOT illustrate the perils of field testing.

Introduction

Low visibility conditions caused by falling, blowing, and/or drifting snow—as well as fog—often occur in winter, in Canada and the Northern tier states of the U.S.A. These conditions are of particular concern to specialty platform drivers—i.e., snowplow operators, ambulance drivers, and highway patrol officers. A Driver Assist System (DAS) was developed in order to provide assistance to them in low visibility conditions. The DAS integrates information acquired using several technologies, including the Differential Global Positioning System, digital geo-spatial databases, and forward-looking radar. It presents a virtual representation of the roadway ahead on a Head-Up Display mounted in front of the driver. The DAS also provides warnings of vehicles ahead on the Head-Up Display, as well as lane departure warnings that are presented in three modalities—visual, auditory, and tactile.

To test the efficacy of the DAS, we used the back-to-back research strategy suggested by Gopher and Sanders (1984)—conducting a series of studies, first in the laboratory, then in the field.

In the first study of the series (Harder, Bloomfield, and Chihak, in press), we carried out two simulation experiments—both with very low visibility. In both, we investigated various lane departure warnings that could be used in the DAS. Visual lane departure warnings were examined in the first simulation experiment, in which 15 subjects (all members of the general public) drove in 30 meters of fog—a visibility so poor that, in the real world, it would make most drivers stop driving.

In the second simulation experiment, we investigated the effectiveness of presenting lane departure warnings in three modalities—visual, auditory and tactile. The 55 participants were 24 snowplow operators, 17 state patrol troopers, and 14 ambulance drivers. This time the visibility was 90 meters and the roadway was completely snow-covered—so that, effectively, the participants were driving in whiteout conditions. Without exception, when the 55 participants drove *without* using the DAS, *none* of them could stay on the road—even driving as slow as 5 mph (8.05 km/h). In contrast, when they did use the DAS, they were able to drive at normal speeds. As a result of this experiment, we recommended that lane departure warnings should be presented in all three modalities simultaneously

Following the simulation experiments, we conducted a field test, with 13 snowplow operators driving a snowplow that was fully equipped with the DAS on a 4.5-mile (7.2-km) long closed track (Bloomfield and Harder, in press). The field test was conducted after the first snows of the 2000-2001 winter, in temperatures that varied between 30 deg F (minus 1 deg C) and minus 18 deg F (minus 28 deg C)—consequently, a mixture of ice and snow covered the course throughout the test.

Each participant drove the course once in clear visibility conditions. Then, he or she drove the course several times in zero visibility conditions—curtains were used to occlude the view through the windshield and side windows of the snowplow. The most important result of the field test was that all 13 participants were able to drive the snowplow, using the DAS, in zero visibility conditions. In the straight sections of the course, they were able to drive using the virtual representation of the road ahead that was presented on the Head-Up Display. However, the 4.5-mile (7.2-km) track included three 90-degree turns and, when the snowplow operators negotiated these turns, the virtual lane-markings were not shown on the Head-Up Display. They were not shown because the view provided by the Head-Up Display coincides with the normal view through the windshield, and because *all drivers* when negotiating sharp turns—like the 90-degree turn encountered in this field test—have to look through the side windows. In spite of having *no* outside view of the world and *no* lane information on the Head-Up Display, all 13 operators were still able to negotiate the turns. They did this by driving very slowly—at 3 mph (4.83 km/h), or less—and by using the combined auditory/tactile warning as a *turn advisory*.

The field test demonstrated that it was possible to use the DAS to drive a snowplow in zero visibility on a closed track. The next step was to test the system more extensively during winter weather in a Field Operational Test.

The Field Operational Test

The Field Operational Test (FOT) was conducted between December 2001 and March 2002. It involved 21 specialty vehicle operators—eight snowplow operators, one highway patrol officer, and 12 ambulance drivers. They drove in one of six vehicles that were all fully equipped with the DAS—four snowplows, one ambulance and one state patrol vehicle. The test was conducted on a 45-mile (72.4 km) section of Minnesota Trunk Highway 7 (MNTH-7) between Hutchinson and Hopkins, Minnesota.

Traveling east from Hutchinson on MNTH-7, the first 28 miles of the route are of straight horizontal alignment, with a flat to slightly rolling profile, and pass through farmland; the next 12 miles are relatively hilly, with a number of curves, and with trees along the side of the highway; the final five miles are also relatively hilly and have curves and trees but, in addition, are urban. There are four types of highway in the 45-mile FOT route—(1) for approximately 33.5 miles, there is one lane in each direction, with no controlled intersections or thru-stops; (2) there is a 1.5-mile section with a third passing lane; (3) there are approximately 5 miles with two lanes in each direction; (4) there is an approximately 5-mile section that is a four-lane divided highway with a median barrier, and with eight non-coordinated traffic signals.

Six meteorological sites recorded local atmospheric visibility measurements at five-minute intervals throughout the FOT. The sites were situated between five and twelve miles apart along MNTH-7.

We have already described some of the problems that occurred in the FOT (Bloomfield and Harder, 2002).

The Mildest Winter on Record

Unfortunately, the winter of 2001-2002 was the mildest winter on record in Minnesota. Although it snowed sometimes, there were only two occasions when there were relatively heavy snowfalls. They occurred near the end of the FOT, on March 8-9 and March 14-15. The visibility data obtained at the six meteorological sites during these snowfalls was examined—it is aggregated in Table 1.

Table 1. Time visibility was in the 200-299, 100-199, and 0-99 meter ranges aggregated across the six meteorological sites in the two snowfalls

	Visibility		
Snowfall	200-299 meters	100-199 meters	0-99 meters
March 8-9	95 minutes	—	—
March 14-15	190 minutes	15 minutes	—
Total	285 minutes	15 minutes	

As Table 1 shows there were no times during the FOT when the visibility was as low as in either the simulation experiments or the field test. There was a 15-minute period, at one site, when the visibility was in the 100-to-199-meter range, which includes the range within which Hawkins (1987)[1] found the greatest reductions in speed. However, no specialty operators were driving at that time. There were several periods, totaling 285 minutes, when the visibility was in the 200-to-299-meter range. Briefly, at different times, three snowplow operators were driving when the visibility was in this 200-to-299-meter range. However, none of these operators was in the vicinity of the particular meteorological site at which low visibility was recorded, and the visibility was clear enough that none of the three needed to use the Head-Up Display during this time.

[1] In an extensive observational study conducted on British motorways, Hawkins (1987) reports the following effects of limited visibility, caused by fog, on traffic speed: (a) for traffic in the fast lane, reductions in speed occur when the visibility drops below 300 meters; (b) for traffic in the slow lanes, reductions in speed occur when the visibility drops below 250 meters; (c) in all lanes the greatest reductions in speed occur when the visibility drops into the 150-180 meter range; and (d) traffic speed is reduced by 25 % to 30 % when the visibility approached 100 meters.

Although vast amounts of driving performance data were collected during the three and a half months during which the FOT was conducted, the lack of low visibility conditions resulted in all these data being obtained in clear visibility.

Lane Departure Warnings

During the FOT, the operators did not need to use the Head-Up Display. Typically when they were driving—whether the DAS was switched *On* or *Off*—they looked directly through the windshield, with the Head-Up Display pushed up out of their line of sight. Because some of the time the DAS was *On* and some of the time it was *Off*, it was possible to examine the effect of the DAS's auditory and tactile lane departure warnings—but only in good visibility conditions.

First, we eliminated lane departures that occurred when the operators used the turn signal, indicating a turn or a change of lane change—since it was clear in these cases that the operators did not need a lane departure warning. Then, we determined the number of lane departures where the turn signal was not used for each of the 21 specialty operators who took part in the FOT. Table 2 shows the number of these lane departures that occurred when the DAS was switched *Off* and when it was switched *On*.

Table 2. Number (and percent) of lane departures that occurred with the DAS *Off* and with the DAS *On*—in good visibility conditions

Subject	Number (& percent) of lane departures (DAS *Off*)	Number (& percent) of lane departures (DAS *On*)
201	2,390 (57.9 %)	1,735 (42.1%)
202	2,991 (74.7 %)	1,013 (25.3 %)
203	12 (5.7 %)	199 (94.3 %)
204	409 (29.3 %)	985 (70.6 %)
205	47 (30.1 %)	105 (69.1 %)
206	2,983 (98.8 %)	35 (1.2 %)
207	472 (65.8 %)	245 (34.2 %)
208	799 (88.7 %)	102 (11.3 %)
301	639 (80.4 %)	156 (19.6 %)
401	100 (66.7 %)	50 (33.3 %)
402	301 (87.2 %)	44 (12.8 %)
403	108 (85.0 %)	19 (15.0 %)
404	275 (89.3 %)	33 (10.7 %)
405	90 (95.7 %)	4 (4.3 %)
406	340 (98.6 %)	5 (1.4 %)
407	233 (100.0 %)	— (0.0 %)
408	48 (100.0 %)	— (0.0 %)
409	22 (100.0 %)	— (0.0 %)
410	48 (100.0 %)	— (0.0 %)
411	56 (100.0 %)	— (0.0 %)
412	121 (100.0 %)	— (0.0 %)

Table 2 shows the number and percent of lane departures for the eight snowplow operators (subjects 201-208), the state highway trooper (subject 301), and the 12 ambulance drivers (subjects 401-412). The table also shows that there were large differences in the numbers of lane departures both between conditions (DAS *On* and DAS *Off*) and between the operators. For ten of the operators—two snowplow operators (subjects 203 and 206) and eight ambulance drivers (subjects 405-412), it was not appropriate to compare the duration of the lane departures in these two conditions, because of the very large discrepancies in the number of lane departures when the DAS was *On* and *Off*—with less than 10 % of the lane departures in one of the conditions.

For the remaining 11 operators, although there were still considerable discrepancies in the number of lane departures when the DAS was *On* and *Off*, we did compare the durations of the lane departure. The distributions of these durations were positively skewed. Although skewness can be addressed with the use of a logarithmic transformation, so that data can be tested with parametric statistics, the inequalities in numbers cannot be addressed by any transformation. Because of this, we used the nonparametric Kolmogorov-Smirnov two-sample test to compare the degree of similarity between the cumulative distributions of lane departure durations obtained when the DAS was *Off* and when it was *On* for each of the remaining 11 operators. The results of this comparison are shown in Table 3.

Table 3. Result of using the Komogorov-Smirnov test to compare the durations of lane departures that occurred with the DAS *Off* and those that occurred with the DAS *On*—in good visibility conditions.

Subject	Test result
201	Not significant
202	Departure durations *longer* with DAS *On* (p<0.05 level)
204	Not significant
205	Departure durations *longer* with DAS *On* (p<0.01 level)
207	Not significant
208	Not significant
301	Not significant
401	Not significant
402	Departure durations *shorter* with DAS *On* (p<0.01 level)
403	Not significant
404	Not significant

The first six subjects (201, 202, 204, 205, 207, and 208) in Table 3 were snowplow operators; subject #301 was a state highway trooper; and the last four subjects (401-404) were ambulance drivers. The table shows that, for five snowplow operators, the state highway trooper, and three ambulance drivers, there were no significant differences in the durations of the lane departure. There were statistically significant differences for the remaining three operators—for the two snowplow operators the lane departure durations were significantly longer (at the p<0.05 level for subject 202, and at the p<0.01 level for subject 204) when the DAS was *On* than when it was *Off*; while, in contrast, for the remaining ambulance driver (subject 401) the lane departures were significantly shorter (at the p<0.01 level) when the DAS was *On* than when it was *Off*. It is difficult to make much of these results—especially since the comparisons were made using data that was obtained in conditions of good visibility, and not in the poor visibility conditions for which the DAS was designed.

Conclusion

The results obtained from the three and half month long FOT illustrate the perils of field testing—where the testing conditions are beyond the control of the investigators. The winter of 2001-2002 did not provide the poor visibility conditions required to determine whether the DAS was effective.

Contradictory results were obtained when we looked at the differences in lane departure durations with the DAS *Off* and the DAS *On* in good visibility conditions—with no statistically significant difference for eight of the 11 operators for whom a comparison could be made, significantly longer durations with the DAS *On* for two snowplow operators, and significantly shorter durations with the DAS *Off* for one ambulance driver.

It is worth mentioning that, when responding to a questionnaire, several snowplow operators said they would like the lane departure warnings to be tied to a lane width that is wider than the standard 12-foot lane, so they would not get lane departure warnings when they plow across the centerline and right edge line of the highway.

Also, when the lane departure warnings were tested under the poor visibility conditions for which they were intended, in the previously-mentioned second simulation experiment (Harder, Bloomfield, and Chihak, in press) and in the field test conducted on a closed track (Bloomfield and Harder, in press), the operators said that they thought it was useful to have lane departure warnings presented in three modalities—visual, auditory, and tactile.

References

Bloomfield, J.R. and Harder, K.A. 2002, Ergonomics applied: Taking a system designed to assist drivers in poor visibility conditions from the laboratory to the field. Paper presented at *The Ergonomics Society Annual Conference*, Homerton College, Cambridge, 3-5 April 2002.

Bloomfield, J.R. and Harder, K.A. in press, Using a head-up display in poor visibility conditions II: Knowledge acquisition from unstructured interviews in a field study. In Gale, A.G. (editor) *Vision in Vehicles IX*, (Elsevier Science Publishers B.V., Amsterdam, Holland)

Gopher, D. and Sanders, A.F. 1984, S-Oh-R: Oh stages! Oh resources! In Prinz, W. and Sanders, A.F. (editors), *Cognition and Motor Behaviour*, (Springer, Heidelberg, Germany)

Harder, K.A., Bloomfield, J.R., and Chihak, B.J. in press, Using a head-up display in poor visibility conditions I: Investigating lane departure warnings in two simulator experiments. In Gate, A.G. (editor): *Vision in Vehicles IX*, (Elsevier Science Publishers B.V., Amsterdam, Holland)

Hawkins, R.K. 1988, Motorway traffic behaviour in reduced visibility conditions. In Gale, A.G., Freeman, M.H., Haslegrave, C.M., Moorhead, I., & Taylor, S. (editors) *Vision in Vehicles II*, (North-Holland Elsevier Science Publishers, B.V., Amsterdam, Holland), 9-18.

VISUAL DEMANDS ASSOCIATED WITH RURAL, URBAN, & MOTORWAY DRIVING

Terry C. Lansdown

Heriot Watt University,
Edinburgh, EH14 4AS

This paper discusses the visual attentional demands imposed on the driver in three commonly experienced road environments. It aims to benchmark normative distributions of visual behaviour and identify any significant differences between rural, urban and rural driving. Findings indicate these road types impose significantly difference visual demands on the driver with resultant changes in recorded distributions of visual behaviour.

Introduction

Advanced driver information systems are becoming more common in road vehicles. Rockwell (1988) stated that: "Since, most electronic displays being introduced into vehicles today have little positive safety benefits, they could be construed as increasing the accident potential if drivers attend to them to the detriment of roadway visual sampling.", page 317. The need for additional information regarding driver visual behaviour has been highlighted (Rockwell 1988; Wierwille, Hulse et al. 1988). Visual behaviour can be empirically investigated by examining the driver's spare visual capacity, the distribution of their scanning and the impact of changes in the driving environment on his/her fixations. This paper is concerned with the environmental influences on driver visual demand.

Spare Visual Capacity

In most situations the driver retains considerable spare visual capacity to deal with additional cognitive and visual demands (Rockwell, 1972). However, under some circumstances the driver may experience maximal demands from the driving situation. Hughes & Cole (1986) report experimental work with video tapes of road scenes and suggest that between 30% to 50% of visual attention may be spare resource. Suzuki, Nakamura, & Ogasawara (1966) measured eye movements and also propose that approximately 50% of available visual attention is not related to driving, supporting Hughes & Cole's (1986) proposition.

Environmental Influences

It has been reported that visual behaviour is influenced by changes in environmental workload (Senders, Kristofferson et al. 1967). For example, driving through a complex junction or busy road may be considered to impose more workload on the driver than a quiet country drive. Experimental manipulation of the road environment has resulted in changes in the drivers' visual behaviour (Rockwell, 1972), that is: "Using spatial and temporal analysis it was found that drivers in open road driving at 65 mph on the same test section give essentially the same pattern of eye movements. Changing the test section landscape background introduced statistically significant changes in spatial density plots even though the road geometry and the task remained the same.", page 324. Hughes & Cole (1986) and Spijkers (1992) provide further evidence of different visual demands imposed by the type of road, i.e., residential streets, motorway or urban shopping streets.

This study aimed to determine inherent differences in visual demands imposed by normative driving in three road environments (i.e., rural, urban and motorway driving). It was hypothesised visual demand would be highest in the urban condition.

Method

The experiment had a one factor repeated measures design. The independent variable was the road type driven: rural, urban or motorway. Dependent variables were glance frequency, glance duration and percentage time per region. Presentation of conditions was pseudo-randomised. A SAAB 9000i was used for the experiment. The vehicle was equipped with recording equipment to combine views of the drivers' face and forward view and time-code the resultant images.

Participants

Eight licensed drivers with greater than two years experience participated in the trial, six males and two females. Unfortunately, data for two of the males was lost due to equipment error. The ages ranged from 21 to 28 years (mean = 24.67, SD = 2.42). All had normal or corrected to normal vision and were naive to the research aims.

Video Tape Transcription

The participant's visual behaviour was analysed post-hoc from videotape records. The visual scene was divided into five regions: forward view, driver mirror, right region, left region and into vehicle (including the instrument panel), see Figure 1. Visual behaviour was transcribed in terms of the number and duration of visual deviations from the forward view. To illustrate: the time in minutes and seconds when an event occurred, (for example, the participant looking to the interior driver mirror); the location of the view (interior driver mirror) and the duration of the glance (e.g., 0.8 secs) were recorded. Resolution for the transcription was restricted by the video recorder's twenty frames per second limit. Glance frequency per minute (gf/min) to each region was obtained and percentage time per region calculated.

Experimental Route

The route undertaken (rural, urban and motorway driving) was balanced in terms of the time required to drive (15 minutes per region) and the order of presentation. The experiment took one hour per participant including familiarisation and transfer sections. Peak travel periods were avoided to minimise traffic variability.

Results

Glance Duration

The duration of participants' glances to the driver mirror were significantly different ($F(2,5) = 8.5$, $p < 0.01$) between regions, see Table 1. Post-hoc analysis revealed glance duration to the driver mirror in the urban condition was significantly shorter than in either the rural or motorway conditions. There were no significant differences between the other regions of the visual scene.

Table 1. Mean glance durations and Tukey's HSD post-hoc comparisons* (SD in parenthesis)

	Urban	Rural	Motorway
Driver mirror	0.77 (0.03) A	0.93 (0.01) B	0.92 (0.11) B
Right region	1.26 (0.14)	1.36 (0.40)	1.17 (0.11)
Left region	1.34 (0.21)	1.44 (0.32)	1.20 (0.24)
Instrument panel	1.09 (0.61)	0.95 (0.16)	0.91 (0.18)
Overall mean	1.16 (0.19)	1.17 (0.10)	1.05 (0.14)

• *Means with the same letter are not significantly different (α = 0.05). For example, mean glance duration to the driver mirror was significantly lower in the urban condition than either the rural or motorway driving.*

Figure 1. Visual regions looking to the forward view

Glance Frequency per Minute

Glance frequency was significantly different for the right region ($F(2,5) = 6.373$, $p < 0.05$), left region ($F(2,5) = 5.374$, $p < 0.05$), instrument panel ($F(2,5) = 8.176$, $p < 0.01$) and the overall mean glance frequency ($F(2,5) = 9.619$, $p < 0.005$), see Table 2. Drivers glanced to the right significantly less frequently during the rural drive than either motorway or urban conditions. Participants were found to glance more frequently to the left region during urban driving than in rural or motorway driving. During urban driving glance frequencies to the instrument panel were significantly lower than during motorway driving. Analysis of overall mean glance frequency revealed participants glanced away from the forward view significantly more often on the motorway than the rural drive.

Table 2. Mean glance frequency per minute (gf/min) and Tukey's HSD post-hoc comparisons* (SD in parenthesis)

	Urban	Rural	Motorway
Driver mirror	3.06 (1.67)	3.53 (2.47)	5.47 (2.34)
Right region	3.28 (0.74) A	1.80 (0.58) B	3.44 (1.10) A
Left region	1.30 (0.27) A	0.81 (0.43) B	0.64 (0.32) B
Instrument panel	0.36 (0.14) A	0.79 (0.76) A & B	1.49 (0.78) B
Overall mean	2.01 (0.50) A & B	1.76 (0.61) A	2.74 (0.46) B

* *Means with the same letter are not significantly different (α = 0.05).*

Percentage Time per Region

The percentage of time allocated to the regions of the visual scene was significantly different for the forward view ($F(2,5) = 8.789$, $p < 0.01$), driver mirror

(F (2,5) = 4.165, p < 0.05), left region (F (2,5) = 5.588, p < 0.05) and the instrument panel (F (2,5) = 5.909, p < 0.05) for all conditions, see Table 3. The percentage of time looking to the forward view was significantly greater in the rural condition when compared to motorway driving. Visual allocation to the driver mirror was found to be significantly different. However, the conservative nature of Tukey's HSD revealed no individual differences between means. Drivers' total percentage of time spent glancing to the left region was significantly higher during the urban condition than during motorway driving. In motorway driving participants glanced to the instrument panel for a significantly greater proportion of time than urban driving.

Discussion

This experiment has attempted to identify normative differences in attentional demand associated with commonly recognised driving environments. Significant differences were established for visual behaviour experienced by participants during rural, urban and motorway driving.

Table 3. Mean percentage of time per region and Tukey's HSD post-hoc comparisons* (SD in parenthesis)

	Urban (%)	Rural (%)	Motorway (%)
Forward view	85.0 (3.2) A & B	87.0 (3.2) A	81.0 (3.3) B
Driver mirror	4.0 (2.3)	5.2 (2.6)	8.2 (3.9)
Right region	7.0 (2.1)	4.0 (1.8)	6.8 (2.3)
Left region	3.0 (0.6) A	2.3 (1.2) A & B	1.3 (0.5) B
Instrument panel	0.7 (0.5) A	1.5 (1.4) A & B	2.5 (1.5) B

* *Means with the same letter are not significantly different ($\underline{\alpha}$ = 0.05).*

Rural Driving

Rural driving in the U.K. involves negotiation of predominantly single lane roads, a de-restricted speed limit of 60 mph (97 kph) and a typically low traffic density. The driver may reasonably expect to encounter sharp turns and a wide variety of junction types. Participants spent more time glancing to the forward view during rural driving (87%) than the motorway condition (81%). Rural driving may have provided less driving relevant visual distracters than motorway driving and therefore, drivers could have been more able to concentrate on the forward view in this condition than during motorway driving. Motorway driving requires greater situational awareness because the vehicle must safely negotiate three lanes of traffic rather than one.

On average, participants glanced to the right region (mirror and window) less frequently during rural driving (20 gf/min) than motorway (35.17 gf/min) or urban (38.5 gf/min) conditions. Urban driving was a more visually rich environment with respect to potential driving hazards than the rural condition. Increased right region checking in the urban environment may have been caused by more traffic at junctions than the rural drive. Hence, the delays to emerging traffic would require additional visual checking particularly from the right region (i.e., the immediate direction of oncoming traffic). Motorway driving requires right and left mirror and window checking prior to lane changing or overtaking manoeuvres.

Motorway Driving

The motorway network may be characterised as typically having three lanes, a speed limit of 70 mph (113 kph) and a high traffic density (dependent on the time of travel). The roads tend to have long gentle geometry requiring gradual steering adjustments. There are no roundabouts or traffic-light junctions, entry or exit is by slip road. Driver glance frequency may be considered as a measure of the perceived need to extract information from the environment. Participants' average glance frequency was found to be higher during motorway driving (27.88 gf/min) than rural (18.95 gf/min). Thus, increased glance frequency during motorway driving may have reflected the greater complexity of the motorway visual environment and more demanding driving.

Interestingly, the percentage of time glancing to the instrument panel was greater during motorway driving than urban. No significant difference was found between the time checking the instrument panel between the rural and motorway conditions. It may have been the driver's self-regulation of speed that was responsible for this behaviour. During rural driving maintenance of speed may be limited by the complexity of the road layout, forcing the driver to slow the vehicle to negotiate bends, etc. In urban driving the close proximity of junctions and the traffic density precludes high speeds. Hence, in both urban and rural driving speed is effectively regulated for the driver. During motorway driving the road layout encourages speed in excess of the regulated limit, traffic density permitting. It is therefore the responsibility the driver to maintain 70 mph (113 km). Consequently, without geographical restrictions the driver must sample the instrument panel more frequently to obtain the required information.

Urban Driving

Towns and cities may include driving on from one to four or more lanes. In this study only urban roads with a single carriageway were used. The speed limit is 30 mph (48 kph) or 40 mph (64 kph) in some regions and traffic density is widely variable. Road geometry will also vary with many types of junction in use. The majority of the significant findings of this study relate to changes in driver visual behaviour during urban driving.

Participants' glance durations to the driver mirror were significantly shorter in urban driving. This may have reflected the fact that in the urban environment hazards in the forward view were more frequent and the time to potential incident would be shorter (e.g., a child running in front of the vehicle). This may have made drivers reluctant to glance away from the forward view for longer than necessary. Glances to the left region were more frequent during urban driving. This possibly reflected greater caution on the part of the driver when emerging from right turn junctions. Similarly, the percentage of time spent looking in the left region was significantly greater than during motorway driving. The number of times participants checked the instrument panel during urban driving (0.36 gf/min) was significantly lower than during motorway driving (1.49 gf/min). The driver could have either been reluctant to glance away from the forward view in this complex environment or not require speedometer information, as they knew they were not exceeding the speed limit.

Spare Visual Capacity

Post-hoc videotape transcription provides general information regarding the location of driver fixations. Consequently, the technique cannot determine if the driver is glancing to a road sign (driving relevant) or an advertisement (driving irrelevant). The transcriber is unable to state the specific location of fixation, only the general direction of regard (i.e., forward view, left (mirror and window) & right (mirror and window) regions and the instrument panel). Spare visual capacity was suggested to be between 30% and 50% (Hughes & Cole, 1986), and 30.2% (Mourant et al., 1970). Although not directly comparable, results from this study (percentage of time allocated to the forward view: urban 85%, rural 87%, and motorway 81%) suggest that the allocation is somewhat lower, more in line with the 90% of fixations being within ± 4° of the focus of expansion suggested by Rockwell (1972).

The study was limited in several respects. The results should be treated tentatively as the participant group was small, with the corresponding effect on the power of statistical tests employed (n = 6). The results would be most appropriately considered as an informative exploration of visual workload to establish general trends relating to the distribution of the features of driver visual behaviour. There were inherent differences in the length and geometry of the road types. For example, the motorway contained no roundabouts. However, the findings of this study suggest that the visual demands associated with road type and environment do differ and therefore warrant further investigation.

Conclusions

This paper has described an experimental investigation into the visual demands imposed in three environments: rural, urban and motorway driving. The study established significant differences in normative driver behaviour in the experimental conditions. Generally, urban driving was found to impose greater visual attentional demands on the driver than rural or motorway driving, although other condition specific effects were observed. Drivers typically exhibited spare visual capacity. The findings suggest that results from experimental work may not be easily generalised to all road types, as the visual demands inherent in a particular road may be significantly different in another. Selection of experimental routes should therefore be carefully considered to encompass a broad range of road types and junctions or specify and indicate the specific road type findings appropriately relate to.

References

Hughes, P.K. & Cole, B.L. (1986). "What attracts attention when driving?" *Ergonomics* **29** (3) 377-391.

Mourant, R.R., Rockwell, T.H., & Rackoff, N.J. (1970). "Driver's eye movements and visual workload." *Highway Research Record* **292**: 1-10.

Rockwell, T.H. (1972). *Eye movement analysis of visual information acquisition in driving: an overview.* Australian Road Research Board.

Rockwell, T.H. (1988). *Spare visual capacity in driving-revisited: new empirical results for an old idea.* Second International Conference on Vision in Vehicles, Nottingham, U.K., Elsevier Science.

Senders, J.W., Kristofferson, A.B., Levision, W., Dietrich, C.W., & Ward, J.L. (1967). "The attentional demand of automobile driving." *Highway Research Record* **195**: 13-15.

Spijkers, W. (1992). "Distribution of eye-fixations during driving." *Journal of the International Association of Traffic & Safety Sciences* **16** (1): 27-34.

Suzuki, T., Nakamura, Y., & Ogasawara, T. (1966). "Intrinsic properties of driver attentiveness." *The Expressway and the Automobile* **9**: 24-29.

Wierwille, W.W., Hulse, M.C., Antin, J.F., & Dingus, T.A. (1988). Visual attentional demand of an in-car navigation display system. *Vision in Vehicles-II.* A. G. Gale, M. H. Freeman, C. M. Haslegrave, P. Smith and S. P. Taylor, Elsevier Science Publishers: 307.

ON-ROAD ASSESSMENT OF THREE DRIVER IMPROVEMENT INTERVENTIONS

B Ng'Andu and Tay Wilson

Psychology Department, Laurentian University
Ramsey Lake Road
Sudbury, Ontario, Canada, P3E 2C6
tel (705) 675-1151 fax (705) 675-4889

Drivers subjected to self-behaviour monitoring, behavioural analysis, safety propaganda, or no treatment were successfully distinguished by means of discriminant analysis of their on-route rated driving. Examination of these variable patterns suggests that relatively the specific social skills regarding pedestrian/driver interactions taught to the behavioural analysis group (one hour discussion of individual's own profile on driver/pedestrian interactions) were most effective on the targeted behaviour, followed by the self-behavioural monitoring group (on road check list of specific pedestrian/driver interactions), and the safety propaganda group in that order. All three treatment groups could be readily distinguished from the no-treatment group which had a relatively high weighting on what might be considered "high aggression/danger variables" (ignoring pedestrian right of way at crossings, changing lanes abruptly, and exceeding the speed limit).

Introduction

It is argued that the extent to which a driver modification programme will favourably influence driver behaviour depends upon how well it teaches specific skills to the driver; moreover these skills are primarily social and locale specific as opposed to perceptual/motor (Wilson, 1991; Wilson & McArdle, 1992). Welsford (1968) suggests that perceptual/motor skills but not necessarily social skills develop primarily as a result of intentionally repeated performance of the task. Consequently, one might argue that repetition would improve perceptual motor aspects of driving which could lead to the driver being less rather than more careful while social skills might rather develop vicariously in the form of an individual driving style. The question of interest here is whether drivers subjected to self-behaviour monitoring, behavioural analysis, safety propaganda, or no treatment can be distinguished by means of their on-route rated driving.

Method

On the basis of several days of on-site out-of-car and in-car observation periods, some sixty social driver variables were chosen and operationally defined so that raters could use them effectively: ignoring pedestrian right of way, obstructive stops, mutineer stops and engine revving at pedestrian crossings; fast approach to non-traffic light, traffic light, straight, and turning junctions; racing green and amber lights or entering junction from stop during amber; accelerating through, tailgating through, omitting sideways visual inspection light at, stopping obstructively or in motion at controlled and uncontrolled junctions; cutting junctions, omitted or late signals, and crossing path of oncoming traffic while turning; average running speed, speed variability, number of times speed limit exceeded, headway maintenance and variability, abrupt, no or late signalled lane change, wandering lanes, lane straddling, fast or cutting curves, passing stationary vehicles or cyclists closely or by crossing centre lane in face of oncoming traffic; frequent or junction lane change; omitting rear mirror use before junction approach or turn or lane change; infrequent use of mirror, overtaking frequently, in face of oncoming traffic, sharing lanes or cutting in; increasing speed while being overtaken; one handed, hand crossing or slipping steering wheel behaviour; insisting on right of way against pedestrian crossing away from pedestrian crossings and finally trip time.

Eight female and thirty-two male drivers ranging in age from 19-53 of which 20 were accident free, 12 had one accident and 8 two or more accidents in the last five years were recruited and assigned randomly to one of four groups and driving day. All drivers brought their own cars and met outside the Mornington Crescent underground station in Camden, London, England where they were informed that the purpose of the study was to determine the reliability of two raters, one in front and one in back, assessing driver performance and asked to do two 24 km runs in the Camden area on separate days.

The behavioural analysis group were additionally told that a separate meeting to discuss their driving profile would be necessary in order allow them to help judge whether the profile was correct. At these one-hour meetings, discussion was restricted to the driver's individual profiles on driver pedestrian interaction (ignoring right of way, braking harshly, stopping obstructively or in motion, revving engine at pedestrian crossing and insisting on right of way against pedestrians crossing away from pedestrian crossings). First, specific discussion of the driver's behaviour in these variables was held. Second, drivers were presented with data on accidents in general in Camden and pedestrian involved accidents in particular. Finally, discussion of the accident generating potential of the above variables was carried out in detail with an attempt to get the driver to acknowledge responsibility for helping pedestrians to cross safely.

The self-behavioural monitoring group drivers were instructed in and asked to use a self-rating form over the next seven days which consisted of daily estimates of frequencies of ten undesirable pedestrian interaction variables: ignoring pedestrian priority, harsh or mutineer braking, obstruction, engine revving, overtaking, hurrying pedestrians, failing to give way at pedestrian crossings and passing pedestrians too closely or giving unclear intentions. The safety propaganda group drivers were asked to read at home a safety propaganda sheet containing information about the numbers of road injuries and deaths, that humans are solely responsible for about two-thirds of all accidents and that accidents can be reduced by conscious acts of safety by each driver ending with the exhortation: "Make safety your business. Do at least one conscious act of safety and provide a good model for other drivers." The no-treatment group drivers were thanked for their participation.

After treatment, all drivers were then asked to volunteer for a further study aimed at assessing accuracy of estimating journey time while driving along a 9-km route through Camden. On this third drive, one rater assessed all of the driver variables and additionally asked drivers to estimate trip time.

Results

Inter-rater correlations for all but two of the variables (approaching traffic light junctions fast and obstructive stops at junctions were above .9 and generally .95 justifying further analysis of only the first raters data throughout. Discriminant analysis on the pre-treatment variable scores for all drivers in the four groups was carried out. No significant discriminant functions separating the groups were found suggesting that the random assignment of drivers to groups was successfully carried out. Moreover, out of 57 variables having non-zero means and variances, only three had significant F ratios ($p < .05$): tailgating during junction turn (the no-treatment group was higher than the other three groups), lane wandering (the no-treatment group was again higher), and passing parked vehicles close (the behavioural analysis group was higher) which is not inconsistent with chance results (binomial probability $p = .23$). The pre-treatment groups were thus accepted as equivalent.

Discriminant analysis on the 52 post-treatment variable scores having non-zero means and variances for all drivers in the four groups was then carried out. Unilabiate F tests showed that three out of these 52 variables (accelerating through non-traffic light controlled junctions, harsh braking at junctions and mutineer stops at junctions were significantly different ($p < 0.05$) across groups with the behaviour analysis group drivers scoring lower (better driving) than the other two groups. The binomial chance probability of three significant F ratios out of 52 is 0.22. However, under the assumption of independence of treatment group from significant variable, the probability of three significant variables all from the behavioural analysis group is 0.027. Two discriminant functions accounted for a total of 94% of the variance (81%, Canonical correlation .99, $p < .0001$ and 13%, canonical correlation .96, $p < .056$, respectively). The group centroids (100% correct classification of drivers to group) on the two discriminant functions respectively for the four groups were no-treatment group (14, 2), safety propaganda group (-9, 5), behavioural analysis group (-6, -1), and behavioural monitoring group (-1, -5).

The discriminant functions were simplified by taking the highest loading variables and weighting them by their standardized scores (^{r}= .6 and .3 respectively for correlations calculated between simplified and full discriminant functions). For the first discriminant function: high positive scores involved approaching non-traffic light controlled junctions fast, ignoring pedestrian rights of way at pedestrian crossings, revving engine while waiting at pedestrian crossings, changing lanes abruptly and exceeding the speed limit while high negative scores involved approaching junctions fast, mutineer stops at pedestrian crossings, omitting sideways visual inspection through traffic light controlled intersections, rushing into junction as soon as traffic lights change and accelerating through non-traffic light controlled intersections. For the second discriminant function, high positive scores involved omitting sideways visual inspection through non-traffic light controlled junctions, racing against green traffic lights, approaching straight junctions fast, cutting junctions during turn, harsh braking at junctions and mutineer stops at pedestrian crossings while high negative scores involved approaching non-traffic light controlled junctions fast, racing against amber lights, accelerating through traffic light controlled junctions, approaching traffic light controlled junctions fast and accelerating through non-traffic light controlled junctions and to a much lessor extent ignoring pedestrian right of way at crossings.

Discussion

Interpreting the simplified discriminant functions, one might picture the no-treatment groups as relatively approaching non-traffic light controlled junctions fast, ignoring pedestrian rights of way at pedestrian crossings, revving engine while waiting at pedestrian crossings, changing lanes abruptly and exceeding the speed limit. The self-behavioural monitoring group would be pictured as lacking these attributes but additionally as relatively approaching non-traffic light controlled junctions fast, racing against amber lights, accelerating through traffic light controlled junctions, approaching traffic light controlled junctions fast, accelerating through non-traffic light controlled junctions and to a much lessor extent ignoring pedestrian right of way at crossings. The behavioural analysis group would be pictured as relatively approaching junctions fast, having mutineer stops at pedestrian crossings, omitting sideways visual inspection through traffic light controlled intersections, rushing into junction as soon as traffic lights change and accelerating through non traffic light controlled intersections. The safety propaganda group would be pictured as being relatively even more pronounced in the variables typifying the behavioural analysis group and additionally as relatively omitting sideways visual inspection through non-traffic light controlled junctions, racing against green traffic lights, approaching straight junctions fast, cutting junctions during turn, harsh braking at junctions and mutineer stops at pedestrian crossings. Examination of these variable patterns suggests that relatively the specific social skills regarding pedestrian/driver interactions taught to the behavioural analysis group (one-hour discussion of individual driver profile on driver/pedestrian interactions) were most effective on the targeted behaviour, followed by the self-behavioural monitoring group (checklist of specific pedestrian/driver interactions) and the safety propaganda group in that order. All three treatment groups could be readily distinguished from the no-treatment group which had a relatively high weighting on what might be considered "high aggression/danger variables" (ignoring pedestrian right of way at crossings, changing lanes abruptly, and exceeding the speed limit). These results provide support for the contention that driver improvement programmes aimed at improving specific social skills in specific locales can be relatively effective.

References

Welsford, A.J. 1968, *Fundamentals of Skill.* Methane.

Wilson, T. 1991, Locale driving assessment – A neglected base of driver improvement interventions. In *Contemporary Ergonomics*, (Praeger, London), 388-393.

Wilson, T. and McArdle, G. 1992, Driving style caused pedestrian incidents at corner and zebra crossings. In *Contemporary Ergonomics*, (Praeger, London),388-393.

HAS SAFETY ENGINEERING ON THE ROADS WORKED? COMPARING MORTALITY ON ROAD AND RAIL

Anthony H Reinhardt-Rutland

Psychology Department, Room 17J15, University of Ulster, Shore Road, Newtownabbey, Northern Ireland BT37 0QB

Road deaths have steadily diminished over many years, suggesting that engineering initiatives, such as seat-belts and grippier road-surfaces, are effective. Yet there are anomalies. Pertinent road deaths in the period around the introduction of any engineering initiative often suggest an effect that is at best short-lived; such an outcome is often attributed to changed behaviour in response to the engineering initiative. It is argued in this paper that mortality must be considered in conjunction with injury statistics. Unlike road deaths, injuries have not tended to diminish over the years. Yet - at least with regard to seat-belt introduction and car occupants - the pattern of injuries parallels the pattern of deaths. Analysis of another transport mode - the railways, where the behaviour of users cannot substantially affect the casualty outcome - shows that the proportions of deaths in relation to overall casualties has substantially reduced over the years. A major reason for this is the improved medical care of trauma injury. The same must apply to the roads. Therefore, with the possible exception of traffic-calming, the evidence that safety engineering has reduced motorist casualties is weak. Furthermore, the deterioration in driver behaviour that is inherent in this conclusion argues that the development of less intrusive means of transport may not be attainable.

Introduction

Engineering initiatives on the roads have had mixed effects. Some have perhaps addressed incorrect assumptions about the information-processing necessary for driving. Thus, initiatives directed at conspicuity - for example, high-intensity fog lamps and retroreflective material - do not address the nature of human perception of space and motion and so are unlikely to have much effect (Reinhardt-Rutland, 1992). Other initiatives seem to have been a success, at least in the vicinity of their application. The obvious example here is traffic-calming - the installation of road-humps and chicanes (Davis and Coffman, 2001).

Many more initiatives have obvious potential. An excellent example is provided by seat belts: comparison between belted and unbelted occupants of the same vehicles shows that the fatalities for the belted occupants is lower by 40% (Evans, 1990). Of course, to carry predictive power, such research must assume that there are no untoward effects occurring over time. Most obvious in this regard is change in behaviour. The inherently poor ride associated with traffic-calming measures such as road-humps is not vitiated over time - although if vehicle suspension improves substantially this may need revisiting - so no change in behaviour is likely. In contrast, seat-belts reduce the discomfort associated with near-collision - the driver is not thrown towards the steering-wheel and windscreen (Reinhardt-Rutland, 2001) - so change in behaviour is possible over time.

The issues concerning changes in behaviour are contentious, since much of the argument is likely to depend on making comparisons in data collected over a number of years. The opportunities for full experimental control are therefore restricted: many factors - changed performance of automobiles, new legislation and road building - can introduce nuisance variables during the course of a long study. Despite this, such evidence is crucial and useful conclusions can be drawn.

Arguments for the failure of safety engineering

Critics of safety engineering argue that engineering initiatives have often had the unfortunate effect of increasing speeding, close-following, overtaking and other dangerous driving behaviours. The argument is made most explicit in Gerald Wilde's (1994) well-known theory of *risk homeostasis*. This asserts that drivers habitually behave with some desired level of risk, which is determined by a combination of social and individual factors; when this level is disturbed - particularly because of a new engineering initiative - the driver responds by bringing risk back to this desired level. A controlled study regarding antilock ABS brakes provides support: in matched groups of drivers operating either ABS-fitted or conventionally-braked taxis, collisions over three years turned out not to be statistically different (Aschenbrenner and Biehl, 1994).

However, the issue of long-term controlled studies in general has already been noted. For example, many jurisdictions now have legislation enforcing seat-belt use, so the remarkable potential of seat belts demonstrated by Evans (1990) is not easily open to testing over several years. Nonetheless, evidence suggests that seat-belt use has engendered speed increase (Janssen, 1994; Reinhardt-Rutland, 2001). British mortality rates concerning compulsory belt use tell a complementary story, although it should be borne in mind that drink-driving legislation was introduced at the same time as compulsory belt use at the beginning of 1983 (Adams, 1985). The rates for drivers and passengers were 2443 in 1982, 2019 in 1983, 2179 in 1984 and - some years later - 2426 in 1989 (Transport Statistics, 1990).

Although risk homeostasis has received strong attention over the years, there are other conceptualisations of the behavioural changes that can occur: examples invoke zero-risk (Summala *et al*, 1988) and learning involved in the procedures for classical and operant conditioning (Fuller, 1984, Reinhardt-Rutland, 2001).

The latter theory avoids one of the most intractable problems of risk homeostasis: the invocation of a precise balance between the technical effectiveness of an engineering initiative and the subsequent more dangerous behaviour. This is important: one might want to dismiss the precise predictions entailed by Wilde's theories, while still subscribing to a link between the introduction of engineering initiatives and change in driver behaviour.

Arguments for the success of safety engineering

The case that behaviour change has not seriously undermined the potential benefits of safety engineering has recently been made by Robertson and Pless (2002). They identify two major points. First, humans have difficulty in assessing risk objectively (Pidgeon, 2001). There are many examples of this: the risk of road casualties is often overlooked in comparison with (a) the Northern Ireland "Troubles" and (b) rail casualties. Yet, as is shown later, (a) and (b) have elicited far less casualties than the roads. One of the contributory factors to poor risk assessment is the relative paucity of the threatened event: even on the road, Robertson and Pless (2002) note that the "average" driver has few, if any, collisions over a period of time from which to assess the risk of any given action.

However, the assumption that *only* actual collisions contribute to the assessment of road risk seems unwarranted. For example, any driver will judge that overtaking on a wide, straight road with little traffic is safer than making the same manoeuvre on a narrow, windy road with heavy traffic. Risk perception may be complex, but the supposition that we have no realistic conceptions of risk on the road is at best contentious.

The second point offered by Robertson and Pless is much more telling: overall mortality rates have steadily dropped over the years in just about every jurisdiction for which there are data. Take examples from Northern Ireland: there were 288 road deaths in 1978, 178 in 1988 and 150 in 1998 (Flanagan, 1999). The accumulated effects of safety engineering would seem to be clear, so we can rely on engineering solutions to eventually bring road casualties to acceptable levels. We may not be able to pinpoint in time when a new engineering initiative affects the casualty statistics, but that must be for further research.

Road deaths and road injuries

The trends in the GB mortality data over that 1980s suggest a short-lived effect for seat-belts, yet trends over many years in many jurisdictions suggest that the cumulative effects of safety-engineering measures are substantial. How is the paradox to be explained?

Mortality does not tell the complete story. Collisions cause a range of injuries, only a small percentage of which lead to death. There have often been worries about statistics for non-fatal injuries, for example, with regard to their completeness: under-reporting may affect injuries that cannot be witnessed by others (Adams, 1985). Nonetheless, much can be gleaned from injury statistics. In particular, if an engineering initiative reduces deaths, it should also reduce injuries: a crash that would have killed now leads to injury, while a crash that would have injured now leads to no injury.

Corroborating this point are the UK injury statistics for drivers and passengers associated with seat-belt legislation. These were 149,878 (1982), 130,425 (1983), 143, 624 (1984) and 170,705 (1989). As with the related mortality data described earlier, the effectiveness of seat-belts appears short-lived; indeed injuries have increased. That an interpretable trend is present argues that injury statistics are not too flawed to be useful.

Regarding the longer-term Northern Irish injuries, these were 8080 (1978), 10,789 (1988) and 12,294 (1998). In comparison with the related Northern Irish mortality data, the main point in the present discussion is that road deaths as a proportion of casualties have reduced from 3.5% to 1.2% over 20 years: deaths and injuries have not remained in step. The small percentages of deaths suggests that there is a fine line between non-fatal injury and death: a given severity of injury is now less likely to lead to death.

Following an earlier point suggesting that risk perception does not relate well to actual casualties, note that political deaths due to the Northern Irish "Troubles" have been about half the road deaths - averaging 75 per year during the 1980s - and political injuries have been less than 10% of road injuries (Flanagan, 1999).

Railway statistics: evidence consistent with improved medical care

Reduced road deaths in relation to injuries might reflect a number of changes. Examples include the reduction in vulnerable means of travel such as walking and cycling. This paper is mainly concerned with another change: the improvement in emergency medical care in the postwar years (Walsh, 2000).

Corroboration of the latter assertion comes from collisions on the railways. A first point to note is - as with the Northern Irish political casualties - the difference between the casualty rates on railways and roads. During their 175 years' existence in Great Britain, railway collisions are estimated to have caused a total of about 3000 deaths (Kichenside, 1997) - less than the current 3500 road deaths *each year*.

Because of their paucity, examination of the data for the railways year-by-year is less fruitful than for the roads: for example, the seeming rash of major railway crashes in each of the last few years is in contrast with the lack of crashes in the early 1990s. Specific collisions provide interesting contrasts. Two collisions over nearly fifty years can be identified as leading to similar casualty totals: Harrow in 1952 (261 casualties) and Ladbroke Grove in 1999 (276 casualties). However, Harrow led to 112 dead and 150 injured (Kichenside, 1997), while the Ladbroke Grove figures are 31 dead and 245 injured (Milner and Pigott, 1999): the reduction in deaths in relation to injuries is striking and borne out more generally. Kichenside supplies death/injury data from other past collisions - for example, 90/109 (St. Johns, 1957) and 49/80 (Hither Green, 1967). Recent data include 13/70 for Great Heck in 2001 (Marsden *et al*, 2001) and 7/76 for Potters Bar (Milner and Pigott, 2002).

A summary of data calculated from serious UK railway crashes over fifty years is presented in table 1. The trend over many years is clear enough, even if the figure for the 1980s fails to conform.

Table 1. Proportion of railway deaths relative to all railway casualties in serious collisions over the last fifty years

Year range	1950-9	1960-9	1970-9	1980-9	1990-9	2000-
Proportion of deaths (%)	40.2	36.9	30.0	33.3	13.0	12.0

In drawing conclusions from the rail data, some comments are in order. Rail transport, like other transport modes, has changed markedly over the last fifty years. On the railways, rolling stock is stronger and more fire-resistant, but it travels much faster. The Harrow smash entailed an estimated closing speed of 60mph, while the corresponding figure for Ladbroke Grove may have been as much as 145mph. Despite modern attention to fire-resistance, a major fire accompanied the Ladbroke Grove smash. The implication is that the smaller number of deaths at Ladbroke Grove compared with Harrow cannot be explained by reduced severity at the former. Safety-engineering on the railways is largely directed towards avoiding collision: Harrow, St. Johns and Ladbroke Grove drew attention to the visibility of signals and automatic warning systems, while Potters Bar raised issues of maintenance supervision. There has been no recourse to the likes of seat-belts and airbags on the railways: the implication is that collisions are never likely to be as endemic as on the roads.

It is safe to assume that the reduction in rail deaths relative to injuries is due in no small part to modern medical practice. Important components of this are the greater involvement of the fire service in facilitating the release of casualties and the paramedical training of ambulance crews: treatment of trauma-injury and patient stabilisation is possible at the site of the crash, rather than at a distant hospital (Walsh, 2000).

Concluding remarks

If conclusions are to be drawn from mortality statistics over a short and clearly defined period of time with regard to the introduction of seat-belt legislation, it is reasonable to ask that they by corroborated by injury data. They are: death and injury rates show broadly matching trends. However, Robertson and Pless' belief that the long-term reduction in road deaths indicates the effectiveness of safety engineering is highly dubious and is not corroborated by injury statistics: deaths reduce, while injuries increase or remain relatively constant. The effectiveness of medicine provides a more plausible explanation for the long-term trends: previous "goners" would now have a good chance of survival and restoration to health.

The effects of changing behaviour are mainly - if not exclusively - centred on motorists: one might argue that little is lost for motorists if an engineering initiative becomes ineffective, provided that the changes in behaviour do not go beyond merely negating the ameliorative effect of the engineering initiative.

However, the more dangerous behaviour that is entailed must have an adverse effect on non-motorists. Most obviously this concerns pedestrians and cyclists. Users of public transport should also be considered: they must access bus-stops and railway stations, most likely on foot or by cycling. In light of the increasing congestion on the roads, it is clear that public transport needs to be expanded: for a number of reasons, systematic expansion of the road system is no longer an option. However, the expansion of alternatives to driving is going to be seriously impeded if the road environment is regarded as too aggressive for non-motorists. The issue is particularly pronounced with regard to school-age children, whose mobility is frequently curtailed because of the dangers of the road system (Davis *et al*, 2001).

References

Adams, J. G. U. 1985, *Risk and Freedom: The Record of Road Safety Legislation*, Transport Publishing Projects, Cardiff

Aschenbrenner, M. and Biehl, B. 1994, Improved safety through improved technical measures? Empirical studies regarding risk compensation processes in relation to anti-lock brake systems'. In: R.M. Trimpop and G .J. S. Wilde, (eds.) *Changes in Accident Prevention: The Issue of Risk Compensation*. Styx, Groningen, Netherlands

Davis, R. and Coffman, A. 2001, *Safe Roads for All*. Institute for Highway Incorporated Engineers, York

Evans, L. 1990, Restraint effectiveness, occupant ejection from cars and fatality reductions, *Accident Analysis and Prevention*, **22**, 167-175

Flanagan, R. 1999, *Chief Constable's Annual Report*. Royal Ulster Constabulary, Belfast

Fuller, R. 1984, A conceptualization of driving behaviour as threat avoidance, *Ergonomics*, **27**, 1139-1154

Janssen, W. 1994, Seat-belt wearing and driving behaviour: an instrumented-vehicle study, *Accident Analysis and Prevention*, **26**, 249-261

Kitchenside, G. 1997, *Great Train Disasters*, Parragon, Bristol

Marsden, C., Milner, C., Piggott, N. and Longman, J. 2001, Thirteen killed in ECML disaster at Heck, *Railway Magazine*, **147/1200**, 4-7

Milner, C. and Pigott, N. 1999, HST and Turbo in 120 mph head-on collision, *Railway Magazine*, **145/1184**, 4-7.

Milner, C. and Pigott, N., 2002, Potters Bar tragedy, *Railway Magazine*, **148/1215**, 4-6.

Pidgeon, N. F., 2001, Crisis? What crisis?, *The Psychologist*, **14**, 230-231

Reinhardt-Rutland, A. H. 1992, Poor-visibility road accidents: theories invoking "target" risk-level and relative visual motion, *Journal of Psychology*, **126**, 63-71

Reinhardt-Rutland, A. H. 2001, Seat-belts and behavioural adaptation: the loss of looming as a negative reinforcer, *Safety Science*, **39**, 145-155

Robertson, L. S. and Pless, B. 2002, Does risk homeostasis theory have implications for road safety. Against, *British Medical Journal*, **324/7346**, 1149-1152

Summala, H. and Naatanen, R. 1988, Risk control is not risk adjustment: the zero-risk theory of driver behaviour and its implications, *Ergonomics*, **31**, 491-506

Transport Statistics Great Britain 1990, HMSO, London

Jones, G. 2000, Introduction: Nursing in accident and emergency. In B. Dolan and L. Holt (eds.), *Accidents and Emergency: Theory into Practice*, Bailliere Tindall, London

Wilde G. J. S. 1994, *Target Risk*, PDE, Toronto

HOW DOES A SPEECH USER INTERFACE AFFECT THE DRIVING TASK?

Elisabet Israelsson[1] & Nina Karlsson[2]

[1]*Linköping University, Linköping, Sweden*
[2]*Department of Human-System Integration, Volvo Technology Corporation, Gothenburg, Sweden*

The main objective of this study was to investigate the effect that speech user interfaces have on the drivers' mental workload in a real driving situation, and to compare this to manual interfaces.

Introduction

In recent years, speech user interfaces have become increasingly popular in many different areas. In the automotive area, under the slogan "eyes on the road – hands on the steering wheel", the notion of speech has often been viewed as a savior to compensate for poorly designed user interfaces with deep menu structures. However, few studies have been done that investigate the impact that speech user interfaces have on the driving task, and to confirm the high expectations, especially in real driving environments (a few studies have been made in driving simulators). Driving is typically divided into primary and secondary tasks. The primary task of driving pertains to actions related to navigating and controlling the vehicle. The secondary tasks are those not directly related to the primary task, but are nonetheless performed while driving.

Due to the state of the technology, most speech user interfaces used in vehicles today are command-based and not particularly conversational, hence adding a load on the user's memory. Are these type of speech user interfaces still safer in the sense that they demand less mental workload from the driver compared to manual interfaces?

Results from a study in a driving simulator suggest that mental workload decreases when dialing a phone number on a cellular phone using voice rather than pressing buttons (Graham *et al*, 1998). Another study by Carter *et al* (2000) compares the impact different input devices have on the driving task. The respondents performed a phone dialling task using voice control, steering wheel buttons and the conventional buttons located on the centre stack. Voice control proved to have the least impact on the driving task and was most preferred by the respondents. However, using the conventional buttons on the centre stack was the fastest mode of interaction.

The main objective of this study was to investigate the effect that speech user interfaces have on the drivers' mental workload in a real driving situation, and to compare this to manual interfaces.

Method

The general method of the study is described below.

Respondents
Nineteen respondents participated in the study, thirteen men and six women.

All respondents were Volvo employees, none of them having worked with anything related to navigation systems or speech user interfaces. Age range was from 23 to 44 years old, with a median of 30 years. All respondents were experienced drivers, i.e. they drove more than 5000km/year. None of them had any prior experience with speech user interfaces; eight had prior experience of Volvo's navigation system.

Equipment
A vehicle equipped with a voice controlled navigation system was used in the study. The navigation system's display is situated on top of the dashboard, and is controlled by using a set of buttons placed on the backside of the steering wheel. The buttons can be reached without letting go of the steering wheel when holding a regular "ten-to-two" grip. Pressing a push-to-talk (PTT) button on the front side of the steering wheel activates the voice control system.

A peripheral detection task (PDT) was used with the visual stimuli placed on top of the dashboard in a manner that the 16 diodes were reflected on the windshield. A video camera was used to record the test sessions.

Experimental set-up
The respondents were seated in the driver's seat. They were informed of the procedure of the test, and were given a chance to familiarize themselves with the navigation system for approximately 10 minutes. Additionally, they were given a map of the route. Each respondent drove the route three times. The first lap was considered a baseline-lap. During the first lap, no tasks were performed except the PDT. During the second and third laps, the respondents either performed speech tasks (using the speech user interface) or manual tasks (using the conventional buttons on the steering wheel) in the navigation system. The order of the second and third lap was balanced. After each lap the respondents filled out the NASA-TLX questionnaire. After the last lap they also filled out an attitude questionnaire.

Measures
Four dependent variables were measured: i) hit rate on the PDT, (ii) reaction time on the PDT, (iii) NASA-TLX score, and (iv) time to solve a task.

PDT is a secondary task measurement used to measure workload while driving (van Winsum *et al*, 1999). The instrument consists of a plate with size of 16*5 centimetres, onto which 6 red randomly blinking diods are placed. Each light remained lit for one second. Each time the respondents detected a red light they were instructed to press a button mounted on their left index finger. The number of correct responses, hit rate, and the reaction time is measured.

NASA-TLX is a subjective measurement of workload. It consists of a multidimensional scale with 6 dimensions of factors related to mental workload (Hart, 1999). Since the term mental workload can be interpreted somewhat differently among the respondents, their personal opinion on what mental workload means for them is taken into the final calculation of the NASA-TLX score. This is done by subjective weighting of each scale.

Results

PDT – hit rate

A 3*2 SPANOVA (Split-Plot Analysis of Variance) was performed to find out whether there were any effects of task type (manual or speech) or interaction. There was a significant difference between manual and speech task [F(2,34)=20,753, p<0,0001].

A least significant difference test revealed the hit rate was lower for manual tasks than baseline [p<0,0001] and for speech tasks [p<0,0001]. However, there was no significant difference between the speech tasks and baseline regarding hit rate.

Additionally, there was no effect of experience and no interaction effects.

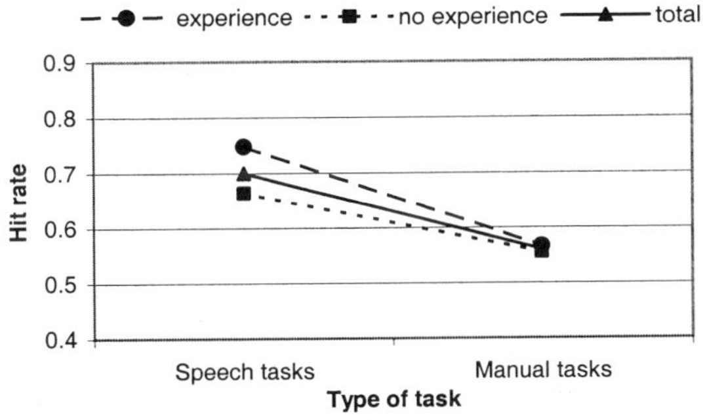

Figure 1. PDT hit rate

PDT – reaction time

No effect of type of task, that is modality, was discovered. Furthermore, there was no effect of experience or any interaction effects. All respondents basically had similar reaction times.

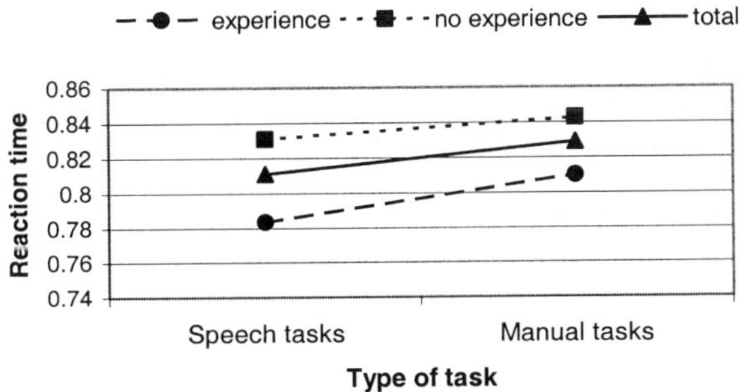

Figure 2. PDT reaction time

NASA-TLX

There was a significant difference between the manual tasks and speech tasks [F(2,34)=53,438; p<0,0001]. Pairwise comparisons revealed that the respondents experienced the manual tasks to have a significantly higher demand on workload compared to the speech tasks [p<0,0001] and baseline [p<0,000]. However, there was no significant difference between the speech tasks and baseline.

Finally, there was no effect of experience or any interaction effects.

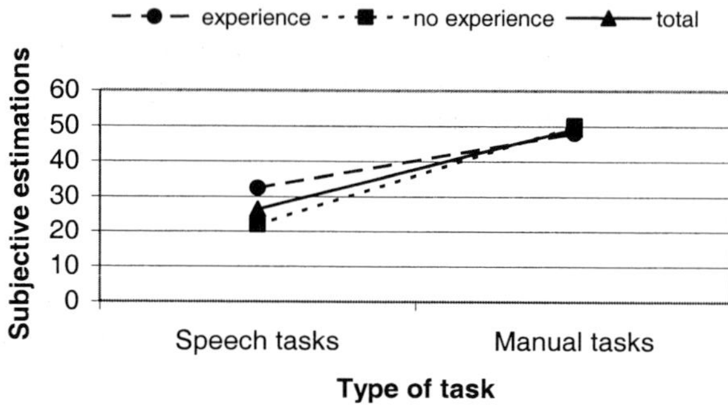

Figure 3. NASA-TLX score

Task completion time

There was a significant difference between the manual tasks and speech tasks [F(1,17)=176,855; p<0,0001]. No effect of experience or any interaction effects was detected.

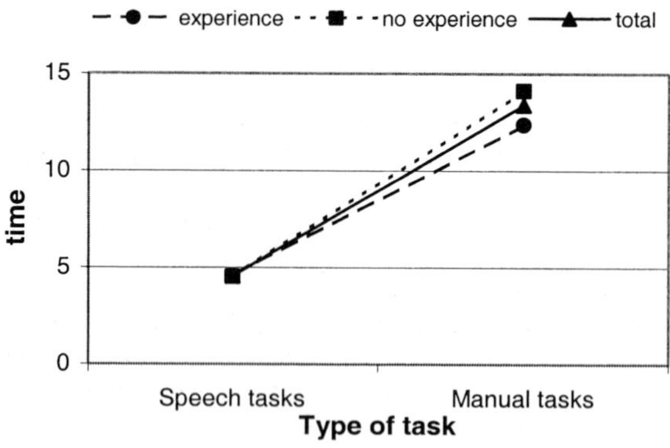

Figure 4. Task completion time

Errors

Even though all respondents successfully completed all tasks, they performed a number of errors along the way. The most frequent types of error when performing the speech tasks were not saying the correct utterance and command, forgetting to press the PTT-button, waiting too long before uttering the command (resulting in that they were timed out), or speaking too unclear.

When performing the manual tasks, the most frequently occurring errors were mixing up the button, that is moving backwards instead of forward, and getting lost in the menus.

Attitudes

Nearly all respondents experienced the speech user interface to be more fun and easier to use than the conventional manual interface, with only one respondent preferring a manual interface. They could also see themselves buying a vehicle with voice control in the future. In addition, they believed that using a speech user interface would actually increase safety.

Discussion and Conclusions

The results from the peripheral detection task (PDT) and the NASA-TLX ratings show that performing tasks manually in the navigation system requires a higher workload than performing the same tasks using voice control. These results were somewhat expected, since prior studies conducted in driving simulators reveal the same tendencies.

Surprisingly however, there was no significant difference between baseline driving (no secondary tasks) and driving while operating the navigation system using voice control. This was unexpected since using voice control requires the user to remember a specific command (e.g. press the PTT-button and utter the command) and to acknowledge feedback from the system in order to make sure he/she was correctly understood. This finding could be a result of the selected tasks being relatively simple. Only one-command tasks were used in the study, due to limitations in the prototype navigation system. If longer dialogs had been used, requiring the user to interact with the system in two or more iterations before one task is completed (such as entering a new street address), the difference might have been greater.

Additionally, the order in which the respondents drove the baseline route was not varied due to safety reasons. The baseline was considered a chance for the respondents to familiarize with the vehicle, route, and PDT-task. Giving the respondents a chance to do this before driving the three laps would have required too much time and might have been considered too tedious for the respondents.

The reaction time on the PDT-task was similar during all conditions for all respondents. This might have been a result of the tasks being difficult. The respondents either saw the red light blink and reacted to it or did not see it at all.

Prior experience of using the navigation system did not significantly affect the respondents' performance on the secondary tasks. This might imply that the experienced users were not really that experienced (cannot be compared to experienced users of a radio for example) or that the interface differ to such an extent that experience of one does not affect the other. However, it is natural to believe that an experienced user is familiar with terms and functionality in a way that he/she will more easily remember for example a command.

A closer look into the general mean scores on the NASA-TLX score shows that the respondents experienced the manual interface as twice as demanding as the speech user interface. This fact is probably related to the positive remarks given in the attitude questionnaire, where almost all of the respondents preferred the speech user interface over the conventional interface.

The time to complete a task was significantly longer for the manual tasks compared to the speech tasks. This is not surprising since using the manual interface requires one to wander around in menus while the speech user interface only requires one to utter a command. However, some of the manual tasks were actually faster than some of the speech tasks. These were the tasks that required only one or two button presses. This could imply that speech user interfaces are best suited for current manual tasks requiring more than a single push of a button.

Further studies are needed in order to fully understand the impact of voice control on the driving task. The next step is to compare simple commands with longer dialogs and to measure vehicle data and eye movement. Is the driver really keeping his eyes on the road?

References

Graham, R., Carter, C., Mellor, B. 1998, *The use of automatic speech recognition to reduce the interference between concurrent tasks of driving and phoning,* in 5th Conference on Spoken language Processing 1998, Australia

Carter, C., Graham, R. 2000, *Experimental comparison of manual and voice controls for the operation of in-vehicle systems,* in Proceedings of the IEA 2000 / HFES 2000 Congress, HUSAT Research Institute, Loughborough University, UK.

Hart, S. G. 1988, *Development of NASA-TLX (Task load index): Results from Empirical and Theoretical Research,* Aerospace Human factors Research Division, NASA-Ames Research center, Moffett Field, California.

van Winsum, W., Martens, M., Herland, L. 1999, *The effects of speech versus tactile driver support messages on workload, driver behaviour and user acceptance,* TNO-report, TM-99-C.

THE IMPACT OF MOBILE TELEPHONE USE AND ACCIDENT HISTORY ON TURN INDICATION DECISIONS WHILST DRIVING

Mark Turner[1] and Steve Love[2]

[1]*Department of Psychology, University of Portsmouth, King Henry Building, King Henry I Street, Portsmouth, PO1 2DY*
[2]*Department of Information Systems and Computing, Brunel University, Uxbridge, Middlesex, UB8 3PH*

Participants completed a simulated urban driving task that required intermittent turn indication decisions to be made. The task was performed with or without the presence of a concurrent mobile phone conversation task. Performance on the driving task was measured by the frequency and duration of turn signalling, by mental workload assessment and by a post-exposure driver attention test. Driver accident history was measured by means of a self-reported questionnaire. Findings indicated that primary and subjective measures of workload were increased during concurrent driving and telephone use compared to driving alone, whilst performance on the telephone task was significantly impaired during driving. Turn indication frequency was reduced by telephone use. No evidence was found to suggest a greater accident history exacerbated concurrent task performance.

Introduction

Advances in vehicle technology have seen an increase in the range of in-car systems which compete for driver attention in addition to the driving task. An example, which has led to widespread public and academic debate is the use of cellular (mobile) telephones whilst driving. Results from experimental studies have indicated that mobile telephone use may lead to clear deficits in driver control such as increasing braking and reaction times (e.g. Alm and Nilsson, 1995; Lamble et al, 1999), increasing variance in lateral positioning of the vehicle (e.g. Brookhuis et al, 1991) and altering the allocation of visual attention (e.g. Wikman et al, 1998). Telephone use whilst driving has also been reported to exacerbate the successful maintenance of vehicle speed and distance from other vehicles more than drinking alcohol equivalent to the UK legal driving limit (Direct Line Insurance, 2002).

Analyses of accident statistics have also shown links between mobile telephone use whilst driving and greater traffic accident rates (e.g. Stevens and Minton, 2001). Increased accident risk is thought to be associated with increases in driver workload resulting from dual-task performance (e.g. Haigney and Westerman, 2001; Royal Society for the Prevention of Accidents, 2002). Rather than performing one task proficiently, drivers may perform both tasks insufficiently leading to increased driver error. Concurrent driving and phone tasks have been found to increase subjective and physiological measures of workload (e.g. Haigney et al, 2000; Garcia-Larrea et al, 2001).

Most experimental studies have focused on the potential impairment of discrete operational components of the driving task specific to vehicle control such as braking or steering. Although such deficits have obvious and critical implications for road safety, they are not wholly representative of the driving task. By comparison, relatively little attention has been paid to the potential latent effects of telephone use on the tactical or decision-making aspects of driving which may contribute to but not directly cause accidents. Concurrent driving and phone tasks have been found to impede driver detection of and reaction to simulated traffic signals (Strayer and Johnston, 2001) and to increase navigation errors in the absence of any observed deficit to vehicle control (Noy et al, 1999). It follows that the effects of concurrent telephone communication and its implications for driving safety may extend beyond simple control aspects of the driving task. The aim of the current study was to examine the impact of telephone use on measures of driver attention and turn indication decisions during a simulated driving and to consider the relationship between concurrent task performance with driver accident history. Turn indications were used to represent driving performance since they require the making of decisions based on constantly changing information about the speed and movement of oncoming vehicles and the driver's own vehicle (Guerrier et al, 1999).

Method

Participants and Tasks

A total of 24 participants (15 female and 9 male) were used in the study. The mean age of participants was 22.6 years (SD, 7.9). All were relatively inexperienced drivers, having held a driving licence for a maximum of 5 years (mean driving experience, 3.4 years; SD, 1.3) and driving on average less than 6,000 miles per year. Each participant was randomly allocated to one of two groups. Group 1 completed a simulated driving task at the same time as performing a concurrent mobile phone task. Group 2 completed the simulated driving task and mobile phone task independently from each other.

The driving task consisted of a video recording of a real car journey viewed from the driver's perspective. The video depicted a typical urban driving scenario of other cars, pedestrians and roadside fixtures. A mock vehicle information display was superimposed onto the bottom left-hand corner of the video display consisting of a serious of circular lights to which participants were required to attend. The vehicle passed a total of 45 road junctions (possible target) and made a total of 13 turning manoeuvres (target event) during the journey. The actual probability of the vehicle making a turn at any one junction (ratio of target events to possible targets) was therefore 29%.

The phone task was the Working Memory Span Test used Alm and Nilsson (1995). Participants were presented with a series of semantically correct or incorrect five-word sentences and asked to judge if each sentence made sense or did not make sense. After five sentences had been presented, the participant was then asked to recall the last word of each of the previous 5 sentences, before a further set of 5 sentences were presented. Eight sets of 5 sentences were used in the study, yielding a maximum test score of 40.

Procedure

Participants in simulated driving conditions were instructed that they would be shown a video of a typical urban driving scenario following which they would be asked questions about the events that occurred in the scenario. Whilst the vehicle was moving, a green light (representing the vehicle's turn indicator) flashed on the mock vehicle information display each time the vehicle performed a turn. Whilst the light was flashing, participants were required to depress a switch linked to a computer that recorded the frequency and duration of responses made. In this manner, the accuracy of participant turn indication responses in relation to the actual manoeuvres made by the vehicle could be examined

The simulation was presented on a monitor located at an approximate viewing distance of 700 mm in front of the participant with the turn signal switch located to be operated by the participant's right hand. For telephone task conditions, a Nokia 3210 handset was used. Participants were instructed to hold the mobile phone with their left hand. Only the effect of engaging in conversation on driving performance was examined, rather than additional activities such as dialling or using other handset functions.

Following the driving simulation, participants completed a questionnaire that measured their attention to and recall of events or objects presented during the simulation. The questionnaire comprised 15 true or false statements, yielding a potential range of scores for driver attentiveness between 0 and 15.

To assess the mental workload associated with concurrent and independent experimental tasks, participants in all conditions completed the NASA-TLX workload measure (Hart and Staveland, 1988). This requires participants to rate each task from low to high on six sources of workload; mental demand, physical demand, temporal demand, frustration, effort, and self-rated performance. Data are then combined to generate a single measure of subjective workload (WWL) for each task ranging from 0 to 100 (higher values indicating greater workload).

Participants also completed a modified accident history questionnaire (Porter and Corlett, 1989) that required them to indicate the frequency and severity of accidents they had experienced on a scale of 0 (not accident prone) to 7 (accident prone) in 5 different settings; driving, playing sports, at home, at work and general.

Results

In-Task Performance Measures

Performance on the turn indication task was compared with the actual number and durations of turn signals during the simulation. Drivers who completed the task whilst performing the concurrent mobile phone task were found to make significantly fewer turn indications than occurred in the simulation ($t(11)=2.21$, $p<0.05$), Of the drivers in the concurrent task condition, 50% (6 out of 12) made the correct number of signals and 42% (5 out of 12) made fewer than the correct number of signals. No significant difference between the number of turn signals made by participants and the true number of turns was found for drivers completing the simulation without the phone task ($t(11)=2.01$, n.s.). Of the drivers completing the driving task alone, 66% (8 out of 12) made the correct number of turn signals with the remainder initiating more than the correct number of signals.

Participants performing the concurrent mobile phone task made significantly fewer turn indications than those not performing the mobile phone task during the simulation, however no significant difference was found in the average duration of turn signals initiated by participants, and no difference was found in performance on the driver awareness questionnaire, as a function of mobile phone use (Table 1). Performance on the concurrent phone (working memory span) task was found to be significantly impaired by the driving simulation compared to when performing the phone task alone.

Table 1. Performance measures as a function of combined or separate driving and telephone task (df=22)

Measure	Combined driving and phone task		Separate driving and phone task		t	p
	Mean	SD	Mean	SD		
Number of signals	11.75	1.96	13.42	0.67	2.79	0.01
Signal duration	10.06	1.76	10.52	2.71	0.50	n.s.
Awareness questionnaire	9.75	1.36	10.00	1.48	0.43	n.s.
Working Memory Span Test	16.75	3.65	23.92	3.70	4.78	<0.001

Subjective Workload Assessment

To examine differences in NASA-TLX scores between participants rating of the combined driving and phone task with the performance of either task alone, MANOVA was performed using the six workload subscales and weighted workload score (WWL) as the dependent variables. A significant difference in workload was found between the three conditions (F(14,30)=4.71, p<0.001, Wilks' Lambda =0.1) which was attributable to differences in the mental demands and effort required to perform tasks (Table 2). Post-hoc univariate comparisons using Tukey's HSD test revealed no significant differences in mental demands (p=0.25 n.s.) or operator effort (p=0.71 n.s.) between the combined driving and phone task and the phone task conducted on its own. The mental demands required to perform the driving task alone were rated as being significantly lower than the combined driving and phone task (p<0.001) and the phone task only (p=0.001). The driving task alone was also rated as requiring significantly less effort than the combined driving and phone task (p=0.04) and the phone task only (p=0.02).

Table 2. NASA TLX workload scores (df=2,21)

Source of workload	Driving and phone task		Driving only		Phone task only		F	p
	Mean	SD	Mean	SD	Mean	SD		
Mental	96.25	5.27	55.83	21.78	85.83	11.58	20.62	<0.001
Physical	19.17	18.20	20.83	13.93	8.33	5.16	1.32	n.s.
Temporal	77.50	10.11	64.17	22.00	57.50	32.06	2.18	n.s.
Effort	77.50	17.65	55.83	18.00	84.17	12.42	4.96	0.017
Performance	74.17	24.85	66.67	13.66	58.33	28.23	0.92	n.s.
Frustration	75.00	23.93	57.50	21.62	75.83	20.84	1.38	n.s.
WWL	82.92	9.74	61.00	12.20	76.50	9.61	8.99	0.002

Accident History

To examine the effects of accident history on mobile phone use whilst driving, driving related items from the accident history questionnaire were totalled and participants divided into two groups that had a relatively low (n=13) or high (n=11) history of driving accidents, determined by a median split of questionnaire scores. The effects of accident history and concurrent phone use were then examined using two-way factorial ANOVA. A significant interaction was observed between accident history and the number of turn indications made with or without the concurrent phone task (F(1,21)=5.83, p<0.05) which suggested less accident prone individuals were more adversely affected by performing the driving and phone task concurrently. Post-hoc analyses revealed no significant difference in the number of turn signals made by more accident prone participants when performing the combined task (mean, 13.0) compared to the driving task alone (mean, 13.0). However, participants with a lower accident history rate were found to make significantly fewer turn signals when performing the phone and driving task concurrently (mean, 11.0) as opposed to turn signals made during the driving task alone (mean, 14.0). No other significant effects of accident group were found.

No significant differences were found between participants with high and low accident histories in the duration of turn signals made ($t(22)=0.67$, n.s.), in attention during the simulated driving task as measured by the post simulation questionnaire ($t(22)=0.98$, n.s.) or in working memory span test performance during phone conditions ($t(22)=1.23$, n.s.). No differences in weighted workload scores (WWL) were evident between participants with high and low accident histories when performing either the phone and driving task concurrently ($t(10)=0.84$, n.s.), or when performing the driving task alone ($t(10)=0.81$, n.s.) or the phone task alone ($t(10)=0.96$, n.s.).

Discussion

Drivers using a mobile phone whilst performing a simulated driving task made significantly fewer turn indications than drivers not using a mobile phone. This may suggest that drivers who engage in phone conversations pay less attention to their environment whilst driving. However, no differences were found on post simulation awareness questionnaires or in average turn indication times (once initiated) between concurrent phone task and no phone task driving conditions. This suggests that it may not be a lack of attention to events but a failure to convert attention into meaningful actions relevant to the driving activity, (e.g. the manipulation of vehicle controls) rather than the failure to complete already initiated actions that may represent the true deficit created by concurrent phone use. This would appear to contradict previous research that has suggested concurrent mobile phone use whilst driving creates deficits due to a 'narrowing of attention' (Alm and Nilsson, 1995).

Turn indication data may also tentatively suggest that drivers who reported fewer previous traffic accidents were more adversely affected by the concurrent phone and driving task than those with a greater accident history. However, no other differences in phone or driving task performance measures or in subjective ratings of workload were found as a function of accident history. Taken as a whole, these data suggest that individual performance whilst driving is just as likely to be affected by phone use regardless of whether a driver has or does not have a history of traffic accidents. It should be noted however, that the accident rates reported by participants in the present study were generally low (even for those in the high accident history group) due to the relatively young age of the drivers examined.

The workload (NASA-TLX) of the driving task was rated as being subjectively more difficult when combined with the phone task, than when performing the driving task alone. However, the subjective difficulty of the combined driving and phone task was not found to be greater than the reported difficulty of the phone task alone. This may suggest that it was the phone task, rather than the interference between tasks that was largely responsible for the reported difficulty of the combined task. The working memory span test used in this study represents a challenging cognitive task requiring a constant and high degree of attention. Whilst such tasks allow a standard phone stimulus to be presented, the net effect of their use as a concurrent task may be an overestimation of the influence of phone use on driving performance compared to phone dialogues more indicative of everyday conversations. The examination of alternative dialogue formats on driving performance may represent a useful way of improving the ecological validity of phone tasks used in this research area.

In conclusion, the current study adds to a growing body of research suggesting that the use of mobile telephones whilst driving is detrimental to driving performance. The current research demonstrates that phone use may impede the driver's ability to respond to, rather than perceive, specific road situations.

References

Alm, H. and Nilson, L. 1995, The effects of a mobile telephone task on driver behaviour in a car following situation, *Accident Analysis and Prevention*, **27**, 707-715.

Brookhuis, K.A., De Vries, G. and De Waard, D. 1991, The effects of mobile telephoning on driving performance, *Accident Analysis and Prevention*, **23**, 309-316.

Direct Line Insurance 2002, *The Mobile Phone Report*, (Direct Line Insurance, Croydon).

Garcia-Larrea, L., Perchet, C., Perrin, F., and Amenedo, E. 2001, Interference of cellular phone conversations with visuomotor tasks: an ERP study, *Journal of Psychophysiology*, **15**, 14-21.

Guerrier, J.H., Manivannan, P. and Nair, S.N. 1999, The role of working memory, field dependence, visual search and reaction time in the left turn performance of older drivers, *Applied Ergonomics*, **30**, 109-119.

Hart, S.G. and Staveland, L.E. 1988, Development of NASA-TLX (Task Load Index): Results of empirical and theoretical research. In P. Hancock and N. Meshkati (eds). *Human Mental Workload*. (North Holland, Amsterdam), 139-183

Haigney, D.E., Taylor, R.G. and Westerman, S.J. 2000, Concurrent mobile (cellular) phone use and driving performance: task demand characteristics and compensatory processes, *Transportation Research Part F*, **3**, 113-121.

Haigney, D.E. and Westerman, S.J. 2001, Mobile (cellular) phone use and driving: A critical review of research methodology, *Ergonomics*, **44**, 132-143.

Lamble, D., Kauranen, T., Laakso, M., and Summala, H. 1999, Cognitive load and detection thresholds in car following situations: safety implications for using mobile (cellular) telephones while driving. *Accident Analysis and Prevention*, **31**, 617-623.

Noy, I., Cassidy, H., Brown, C.M., 1999, *Quality of Driving with Cellular Telephones. Technical Memorandum TME 9901*, (Ottawa, Transport Canada).

Porter, C., and Cortlett, E. 1989, Performance differences of individuals classified by questionnaire as accident prone or non-accident prone. *Ergonomics*, **32**, 317-333.

Royal Society for the Prevention of Accidents 2002, *The Risk Of Using A Mobile Phone While Driving* (RoSPA, Birmingham).

Stevens, A and Minton, R. 2001, In-vehicle distraction and fatal accidents in England and Wales, *Accident Analysis and Prevention*, **33**, 539-545.

Strayer, D.L., and Johnston, W.A. 2001, Dual-task studies of simulated driving and conversing on a cellular telephone, *Psychological Science*, **12**, 462-466.

Wikman, A., Nieminen, T. and Summala, H. 1998, Driving experience and time-sharing during in-car tasks on roads of different width, *Ergonomics*, **40**, 358-372.

DETECTING RULE VIOLATIONS IN A TRANSPORT SYSTEM SIMULATION: EFFECTS OF WORKLOAD DISTRIBUTION

N. Morris, N. Feratu, O. I. Martino, & L. Oki

University of Wolverhampton, Department of Psychology, Wulfruna Street, Wolverhampton WV1 1SB, UK

In a simulation of a transport system 30 pairs of operatives were required to monitor the transactions of a vehicle and detect violations of three rules relating to passenger numbers and routes to be taken. The workload was equally distributed between the two operatives in each pair but for 10 of the pairs each operative was required to monitor alternate messages. Another 10 pairs alternated over blocks of five messages and the remaining pairs alternated randomly. Each operative had a sheet with the messages to be monitored indicated in highlighted boxes and they recorded any violation in these boxes. Alternating monitoring produced > 80% rule violation detection whereas blocking and randomisation both resulted in < 60% violation detection. A questionnaire examining memory for events during the one hour long simulation showed that operatives in the alternate monitoring condition remembered significantly more about the simulation. The results are discussed in terms of the mental workload demands of complex tasks.

Introduction.

The mental workload associated with performing a cognitive task is a limiting factor in the efficiency of performance. Such tasks can be seen as loading up working memory. There is a tendency to consider working memory capacity as being limited by the number of 'chunks' of information held in short-term memory (Miller, 1956) or in terms of the time taken to rehearse items by repetition (the word length effect – Baddeley et al. 1975). However we argue that there is indeed a capacity limitation in terms of the amount of information that can be retained by rehearsal but this is less important than the through put per unit of time. This has been dealt with in a more elaborate form elsewhere (Morris, 1991).

The point that we wish to stress is that working memory capacity is apparently very modest but once information has been used (for example, reasoning has been carried out, mental arithmetic completed, transfer to long-term memory achieved) the capacity of working memory is rapidly freed for further work. Indeed the contents of working memory can be updated, even when full span loads are held, in under one second (Morris and Jones. 1990). Viewed from this perspective, working memory may well appear to be burdened by even small loads when we look at it in a 'snapshot' in time. However if one views the processing that can occur in, say, one hour then it is a formidable processing device.

In a series of studies we have established that working memory updating, the process by which information is processed through working memory and then replaced by new material, is relevant to a range of complex cognitive tasks (Crofts and Morris, 1997; Morris, 1991; Morris and Everill,

1994; Morris and Jones, 1988;1990; Morris, Jones and Milne, 1990; Morris and Lamb, 1993 and Morris, Milne and Jones, 1991).

From this perspective, mental workload problems, in some settings, may be tackled as a time-sharing problem, given that we may not be able to increase the available space in working memory. If memory updating, i.e. real-time processing rate, is formidable but nevertheless limited, it may be possible to improve the amount of information processed per unit of time across the duration of a lengthy task.

Suppose a given task makes heavy cognitive demands and requires a high level of accuracy, for example, some monitoring tasks. If two individuals share the task then there are a large number of options for allotting 50% of the workload to each operative. They might alternate after each event, tackle half of the shift each or use some other allocation. Thompson (2002) has reviewed memory 'collaboration'.

There is no clear theoretical optimal choice derivable from the memory updating perspective. It is an empirical matter. If each operative deals with 50% of the task irrespective of the method of allocation then they always have equal cognitive loads for a whole shift. If performance efficiency is different with different work schedules then this implies that some workload distributions are more compatible with the dynamics of working memory processing than others. The temporal patterning of work deployment will be an important factor in maximising processing efficiency without creating cognitive overload.

In this study we examined this using an imaginary transport system task. The workload distribution appropriate to a given task may vary with the nature of the task and may need to be established empirically. Nevertheless we chose three disparate allocations which should be roughly generalisable because they load working memory in radically different ways. We chose to develop an imaginary transport system because past simulations (for example, Morris, Milne and Jones, 1991) have suggested that participants engage with these better than more arid tasks using number or consonant processing.

In this simulation students working in pairs were asked to monitor a simulation of the University's free shuttle service between campuses. The task had a similar duration to an actual shuttle run. The pairs of students had to identify rule violations that could occur during the journey. Rule violations are less likely to be detected when working memory is overloaded (Baddeley, 1986). A questionnaire at the end of the simulation tested memory retention for events in the simulation.

To vary cognitive load in the short-term but match net loads three conditions were created. In one condition an individual was responsible for detecting violations for alternate messages. This resulted in the amount of information held being equal across all conditions but removed the possibility that two rule violations would have to be detected in 40 seconds. The possibility that multiple violation detections would need to be made when transactions were blocked in fives or randomly assigned was high. Thus we predicted that avoiding accumulating loads that must be processed for rule violations would result in the most efficient performance. Blocking is most likely to overload working memory and thus be most detrimental to performance. We predict that random allocation would produce intermediate performance levels.

Method

Participants

Sixty undergraduate students at the University of Wolverhampton volunteered to participate in the study. The participants were aged between 18 and 35 (mean age = 24.3 years), 29 were male and 31 female.

Materials

A script was prepared relating to the route of two vehicles travelling between the five University campuses (Wolverhampton, Compton, Dudley, Walsall and Telford. An example of a typical message is *'Vehicle A, pick up 15 students, Wolverhampton Campus'*. The messages are structured to designate the vehicle, the number of students picked up or dropped off, and the campus at which the students enter or depart the vehicle. The script contains seventy-five messages: fifty messages relating to Vehicle A and twenty-five messages relating to Vehicle B. The messages were designed so that, for each vehicle, messages alternate between picking up students and dropping off students, i.e.;

'Vehicle A, *pick up* 20 students, Wolverhampton Campus',

'Vehicle B, *pick up* 10 students, Compton Campus',

'Vehicle A, *drop off* 30 students, Dudley Campus'

The vehicles have three rules that must be followed at all times. In the script, each rule was violated four times (twelve rule violations in total). Messages containing rule violations were placed into the script using a random number generating programme. Each message was recorded on tape with random spacing of about 20 seconds between each message. Spacing between messages was randomised to more closely resemble operational settings. At the beginning of the tape, the initial status of each vehicle was announced, for example, *'At the start of the journey, both vehicles are situated at the Wolverhampton Campus and are carrying 0 students'*). After every fifteen messages in the script, a status update message was provided (e.g.; "*Vehicle A is currently carrying 25 students and is situated at the Telford Campus. Vehicle B is currently carrying 15 students and is situated at the Walsall Campus*"). The update messages were then repeated to allow the participant to verify that they had correctly updated their knowledge about the current state of the simulation run.

Procedure

Participants were given a sheet outlining the instructions, and were told that they were to listen to a tape concerning the routes of two buses travelling between the five University Campuses. It was explained that at each campus, the bus picks up or drops off a particular number of students, and travels to the subsequent campus. However, the vehicles have three rules that must be followed at all times:

- Rule 1 - The Vehicle should not carry less than 5 passengers at any time.
- Rule 2 - The Vehicle should not carry more than 60 passengers at any time.
- Rule 3 - The Vehicle should always follow the route: Wolverhampton – Compton – Dudley – Walsall - Telford.

The task of the participant was to continuously update their memory of the status of the vehicles, and detect any apparent rule violations that occurred and to record any rule violations on a check sheet.

It was made clear to participants that, after every fifteen messages, an update message would be announced on the tape (e.g.; "Vehicle A is currently carrying 25 students and is situated at the Compton Campus. Vehicle B is currently carrying 10 students and is situated at the Dudley Campus"). Update messages were used to assist participants in updating their memory upon the current status of the vehicle as, if a participant made a mistake at an early stage in the simulation run, this mistake would have a 'knock on' effect throughout the session.

Participants were asked to read the instructions thoroughly and to read the rules until they had memorised them. The researcher asked the participants to recall the instructions and the three rules of the study, to demonstrate that the participants understood what was expected of them, and understood the rule violations that were to be detected.

Once the task was understood, participants were asked to put on headphones and were instructed to listen to the tape concerning the route of the vehicles. They were instructed to listen to all messages on the tape, but to only record rule violations for vehicle A. As noted above, the script contained seventy-five messages, with fifty of these concerning the route of vehicle A. The remaining twenty-five messages concerned vehicle B. These messages were used in the script as 'distractor traffic', and participants were instructed to ignore these messages during the study. For this reason, the boxes on the response sheets were reserved only for messages relating to vehicle A. When a rule violation occurred, participants were asked to tick the appropriate box in one of the columns entitled either *'rule 1 violation'*, *'rule 2 violation'* or *'rule 3 violation'*. If a rule violation had not occurred, participants were asked to tick the appropriate box in the column entitled *'no rule violation'*. Thus participants ticked a box for every message they were required to monitor. Participants were only required to record rule violations for particular messages, 50% of the Vehicle A messages, depending on the condition they were in. To make this clear to participants, the messages they were to monitor for violations were highlighted, on the check sheet, in yellow ink; alternate messages for condition one, messages in blocks of five for condition two and messages in a random order for condition three. To counter-balance, each condition was further split into two groups. For example, in the alternate messages condition, half of the participants were asked to record rule violations for 'odd' messages, and half of the participants were asked to record rule violations for 'even' messages; in the 'blocks of five messages' condition, half of the participants recorded rule violations for messages 1-5, 11-15, 21-25, etc., and the other half of participants recorded rule violations for messages 6-10, 16-20, 26-30, etc.; and in the 'random messages' condition, half of the participants recorded rule violations for messages 1, 3, 4, 7, etc., and the other half of participants were asked to record rule violations for messages 2, 5, 6, 8, etc.

Following completion of the simulation run, participants received a memory recall test. This was a set of questions relating to the route of vehicle A, and was designed to assess the amount of information about the simulation that had been retained by participants. Participants were instructed to answer all questions, to only provide one answer for each question, and to guess any answers that they were unsure of.

After completion of the memory test, participants were debriefed and thanked for their participation.

Results

Rule Violations

As the rule violations were randomly assigned in the script, there could be slightly different number of violations for participants to detect depending on their particular target messages. For this reason, the scores, in terms of number of rule violations detected, were converted into percent correct detections. The percentages were then used as the data for statistical analysis. The mean number of rule violations detected from the response sheets were calculated and are shown in table 1, along with standard deviations for each of these scores.

Table 1– Means and Standard Deviations For Violation Detection Scores.

Condition	Mean	Standard Deviation
Alternate Messages	**83.13**	**17.81**
Blocks of 5 Messages	**57.78**	**30.93**
Random Messages	**59.86**	**22.89**

A One-way, independent samples analysis of variance (ANOVA) indicated a significant difference between the three conditions ($F(2,57) = 6.61$, $p<0.005$). Tukey HSD tests were carried out on the scores. This showed that, there was no significant difference between the blocked message format and the random message format ($p>0.05$), but performance was significantly enhanced for the alternate message format compared to the blocked message format ($p<0.005$). Performance was also improved for the alternate message format compared to the random message format ($p<0.01$).

Memory Recall

Table 2 shows the mean scores and standard deviations of the three conditions for memory recall.

Table 2 - Means and Standard Deviations for memory recall.

Condition	Mean	Standard Deviation
Alternate Messages	**7.85**	**1.04**
Blocks of 5 Messages	**1.04**	**1.67**
Random Messages	**5.90**	**2.10**

A one-way, independent samples ANOVA was also carried out upon the memory recall scores, where it was found that the three conditions were significantly different ($F(2,57) = 7.23$, $p<0.001$). Tukey's HSD indicated that performance for the alternate messages format was superior to the blocked messages format ($p<0.001$) and better than the random messages format ($p<0.001$). It was also indicated that the random message format resulted in better memory than the blocked messages format ($p<0.01$).

In summary, it is clear that detection of rule violations is significantly affected by the monitoring format used. There was no difference in detection ability between participants detecting rule violations in blocks of five messages and messages in a random order. However, performance is significantly better in the alternate message monitoring condition. The memory recall data show that retention is best with alternating messages and worst with blocked messages. The random format produces intermediate levels of retention.

Discussion

These results clearly show that the rule violations are better detected when participants deal with alternate messages rather than unsystematic patternings of monitoring or blocking. This occurs even though *all* of the target messages must be tracked, by each participant, if violations are to be detected. Such continuous monitoring is essential because a violation occurs as the outcome of a string of transactions culminating in a specific violation finally exceeding the parameters of the rule. This suggests that it is the need to decide that a violation has occurred, not just the requirement to update ones knowledge about the status of the bus, that is burdensome to working memory. Rule violation detection rate was comparable when work allocation was in blocks or randomised suggesting that alternating is by far the best strategy in this sort of scenario.

It is clear that memory updating plays an important role in detecting rule violations. In Morris et al. (1991) it was shown that memory updating efficiency is affected by rule violation detection. Memory updating was selectively improved for transactions involving rule violations. In the present study the condition with most efficient rule violation detection, alternation, memory recall was also best. This implies that the strategy adopted in the alternating condition also leads to improved memory for the events in the simulation. Although strictly speaking post-simulation retention of information is not relevant one should bear in mind that comprehension of verbal material is closely associated with efficiency of processing in working memory (Baddeley, 1986). Thus on job training involving reflecting on performance during a work shift might also benefit from such structuring.

We intend to investigate further the value of collaborative monitor and do not therefore wish to make strong recommendations at this time. It is sufficient to suggest, given the tentative nature of the study, that investigating work pattern allocation from a working memory perspective should prove to be a fruitful line of research into task allocation strategies.

References

Baddeley, A.D. (1986) *Working Memory.* Cambridge: Cambridge University Press.

Baddeley, A.D., Thomson, N., and Buchanan, M. (1975) Word length and the structure of short-term memory. *Journal of Verbal Learning and Verbal Behavior,* **14,** 575-589.

Crofts, L. and Morris, N. (1997) Variable workload effects on monitoring abilities within a freightline simulation. In S.A. Robertson (Ed.) *Contemporary Ergonomics 1997.* p. 351-356.London:Taylor and Francis.

Miller, G.A. (1956) The magical number seven, plus or minus two: Some limits on our capacity for processing information. *Psychological Review,* **63,** 81-97.

Morris, N. (1991). The cognitive ergonomics of memory updating. In E.J. Lovesey (ed.) *Contemporary Ergonomics 1991.* London: Taylor and Francis.

Morris, N. and Everill, J. (1994) Real-time memory processing across the menstrual cycle. In S.A. Robertson (Ed.) *Contemporary Ergonomics 1994.* London:Taylor and Francis.

Morris, N. and Jones, D.M. (1988). The effects of relevant and irrelevant transcription on memory updating in a freight service simulation. In E. Megaw *(ed.). Contemporary Ergonomics 1988.* London: Taylor and Francis.

Morris, N. and Jones, D.M. (1990). Memory updating in working memory: the role of the central executive. *British Journal of Psychology,* **81,** 111-12 1.

Morris, N., Jones, D.M. and Milne, A.B. (1990). Using simulations to examine complex cognitive processes. In E.J. Lovesey (ed.). *Contemporary Ergonomics 1990.* London: Taylor and Francis.

Morris, N., and Lamb, L. (1993) Does memory capability decline with age? In E.J. Lovesey (Ed.). *Contemporary Ergonomics 1993.* London: Taylor and Francis.

Morris, N., Milne, A.B. and Jones, D.M. (1991). Updating one's knowledge about the current status of vehicles in a freight system simulation. *Applied Ergonomics,***22,** 401-408.

Thomson, R. (2002) Are two heads better than one? *The Psychologist.* **15,** 616-610.

RAIL

EVALUATION OF RAILWAY SIGNAL DESIGNS USING VIRTUAL REALITY SIMULATION

Guangyan Li[1], W. Ian Hamilton[1] and Ian Finch[2]

[1]Human Engineering Ltd.
Shore House, 68 Westbury Hill, Westbury-On-Trym, Bristol BS9 3AA UK
[2]Great Western Zone, Railtrack (part of the Network Rail Group),
125 House, 1 Gloucester Street, Swindon, SN1 1GW UK

An experimental study was carried out using virtual reality (VR) technology to evaluate signal designs for the UK railway industry. Eight design schemes were tested using 36 train drivers. Dependent variables were signal sighting distance and reading error rates, and subjective data were recorded regarding how easy or difficult it was to identify the signal aspect and to associate the signal with the line of travel.

The results show that driver responses in signal identification varied significantly with the type of design schemes, with the major differences lying between the staggered and non-staggered signals. Drivers could identify staggered signals further back compared to the non-staggered ones. However, the drivers also tended to make more errors with staggered signals. Backplate shapes designed within the current 3-aspect size did not improve the quality of signal sighting, nor did the bordered signals in the situation where the signals were identified at longer distances (e.g., over 200m).

Introduction

Following a previous study investigating the effect of railway signal backplate design on driver responses using an impoverished computer simulation (Li et al, 2002), further work was carried out to test the railway signal design/layout features in a more realistic VR environment. The aims of this study were to test the effects of different signal backplate design and layout schemes on driver responses, including the distance to the signal when its aspect can be identified, signal reading error rate for the association of signal with the line of travel, and the driver subjective opinions on each signal design scheme. This work was the second part of a 3-phase study aimed at identifying the optimum signal design scheme for the Paddington Throat.

Experimental Studies

Design of the trials

The test scenarios were set up in VR based on the realistic signal settings in the real world, i.e., the signals were mounted on gantries (6 signals on each gantry) and the signals were tested for different types of backplate designs and their horizontal layout patterns. In total eight signal design schemes were tested, including the current standard backplate (3-aspect design, 6 signals in a line); a combination of different backplate shapes along a gantry (6 in a line, Figure 1); a combination of backplate shapes with added furniture (known as 'Theatre box') around the signal; the current standard design with a white border (including 10% and 20% of backplate width, Figure 2), and the current standard design with staggered arrangements along the gantry, including 1-by-1, 2-by-2 and 3-by-3 staggered layouts (all with a 750mm vertical offset between adjacent signals, measured from the centre of signal head, Figure 3). This was based on suggestions that the practical situation only allows signals to be vertically offset by up to 750 mm on a gantry. All schemes were tested in both inbound (3 gantries) and outbound (4 gantries) directions, thus forming 16 test sessions.

The current standard 3-aspect backplate was based on Railway Group Standard GK/RT0031 (1996)(600x990mm), and different backplate shapes were designed within this envelope. The diameter of the signal light was 200mm and was kept constant for all trials. All signal backplates along gantries had a single signal head that was located at the bottom of the backplate (180mm from the bottom edge, measured from the centre of the signal head). This was to represent future implementation of single-headed fibre optic signals in the track sections concerned.

The selection of combined shapes was based on a previous study showing that signals with these backplate shapes performed as well as or even better than the current standard rectangular design (HEL, 2001).

Different routes were used in the trials during which the train would change its line from gantry to gantry. These routes were randomly assigned to different test sessions and the drivers were not informed of which route they were travelling, therefore their responses to the signals ahead would be totally dependent upon the simultaneous identification of the signal for their line. Junction indicators were used to show which direction the train was going to change its line. In order to reduce the possibility that drivers might use junction indicators as a hint to identify their line of travel, more than one junction indicator was installed on the gantry if there was a line change ahead.

For all trial sessions, the background lighting was set as 12:00am, 15th August. This was based on analysis of SPAD data from 1993-2001 in the area concerned, which showed that most SPADs occurred around this time (with a peak also in December).

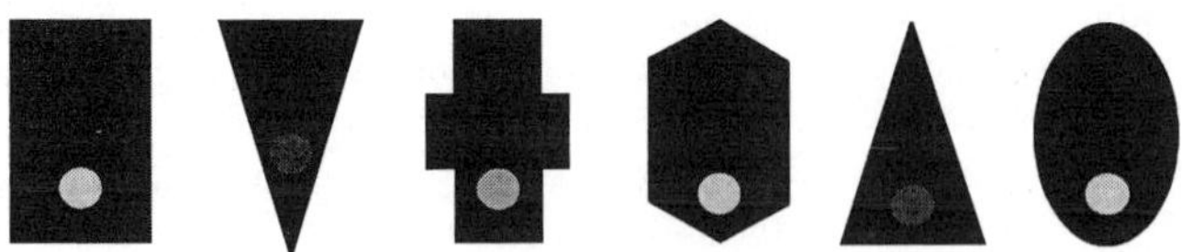

Figure 1. Signal backplate shapes tested in the trial

Figure 2. Signals with bordered backplate (20% of backplate width or 120mm wide)

Figure 3. 1-by-1 staggered layout

Participants

A total of 36 train drivers (34 male and 2 female) took part in the trials. Their average age was 40.9 years (range=22-64.5, SD=10.6), and their average driving experience was 12.8 years (range=0.7-45.0, SD=13.5). All drivers completed a Participant Record Form and a Participant Mood Test Form. Although several participants reported being on various types of medication, all had been passed as fit to drive by their respective doctors, and were on active duty at the time, they were therefore deemed fit to participate in the trail.

Equipment

A Dell Workstation (PWS330, Intel® Pentium®, 4CPU, AT/AT Compatible with 512 MB RAM and 3D graphic card (Geforce2)) was used as virtual reality simulator for the trials. The screen size was 18 inch; screen area (resolution) was set at 1280 by 1024 pixels and the colours were set for 'True Colour' (32 bit). The computer was running Microsoft Windows 2000 system. The VR signal testing software programme used in the trial was based on a driver 'Route Learning' programme which was developed by Infrasoft Solutions Ltd and modified to meet the requirements of the present trials.

Procedure

Following an initial briefing and a subsequent training session, the participants were seated at a height-adjustable chair and allowed to adjust it to their preference, ideally maintaining their eye-height level with the centre of the screen, and their eye-to-screen distance at approximately 450mm, this was to achieve a visual angle which was calculated to present images at realistic angular subtense.

The participant was instructed to start the train by pressing the 'up' arrow key on the computer keyboard and then to control the train speed using the up/down arrow keys. They were requested to control the train speed in the same way as if they were driving in the real situation. They were then to press any appropriately coloured key when they could positively identify the signal aspect for their line of travel (at this point the equivalent sighting distance was recorded). The letter keys on the keyboard were divided into three areas and were visually marked green, yellow and red respectively, thus allowing the participant to tap any one of the appropriate keys to identify the signal aspect. By the end of each 'run' the participant was asked to subjectively 'rate' the signal design scheme using a 5-point scale, in terms of how easy or difficult it was to identify the signal aspect and how easy or difficult to associate the correct signal with the line.

The trials were conducted in an office with its lighting level maintained at approximately 10-13 lux throughout the trials. It took approximately 90 minutes for each participant to complete the trials, and there was a 10-15min break in the middle of the trials during which refreshments were served.

Results

Of the 36 participants, one driver did not complete the trials, thus the results of 35 participants were analysed. The sighting distance data corresponding to the eight design schemes were analysed using ANOVA, which showed a significant condition effect at $p \leq 0.000000001$ level [$F(7,34)=12.819$]. Figure 4 shows the quantitative effects of different designs on sighting distances, overlaid with driver ratings for these designs.
Further analysis with Post hoc tests (Tukey HSD test) showed that the major difference amongst the design schemes tested was between the staggered layouts and the combined shapes. By comparing the three types of staggered layout with combined shapes and also with the current standard design, 3-by-3 layout appeared to be the best as it was also significantly better than the current standard design ($p<0.05$). However, there were no significant differences between the three staggered schemes themselves with respect to their influence on sighting distances, suggesting that any pattern of staggering should be able to increase sighting distance relative to non-staggered patterns, but the amount of change in sighting distances within the three staggered patterns is statistically negligible.

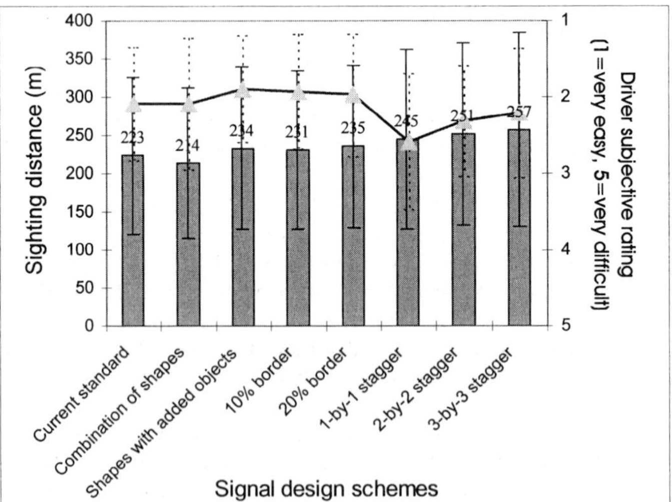

Figure 4. Effect of signal design schemes on sighting distance, overlaid with driver opinions on signal-line association (n=35)

Multiple comparisons with driver ratings on the identification of signal aspect and on the signal-line association showed very similar results, indicating that the major significant differences in driver opinions were with 1-by-1 staggered pattern as compared with all the other 'non-staggered' schemes. In general, the participants regarded these conditions significantly 'better' than the 1-by-1 staggered condition, but they considered the 2-by-2 and 3-by-3 staggered layouts to be as good as all the other non-staggered ones.

Driver signal reading error rates in relation to different designs are given in Table 1.

Table 1. Driver response error rate corresponding to signal design schemes

	Current standard	Combined shapes	Added objects	10% border	20% border	1-by-1 stagger	2-by-2 stagger	3-by-3 stagger
Error	6.1%	8.2%	7.3%	4.9%	4.9%	12.7%	13.1%	11.0%

Multiple comparisons show that the main differences in driver subjective ratings existed between the 1-by-1 and 2-by-2 staggered layouts and the bordered signals, confirming that the driver response errors were significantly lower with both bordered signal designs than that with the two staggered designs.

Discussion

This study demonstrated that VR simulation technology is a robust and economical approach to quickly testing a variety of signal design schemes which would otherwise be very difficult or even impossible to be set up and tested in the real world.

The results showed that signals with staggered layout could be identified at some 30 metres further than the current standard layout (Figure 4). One reason for this might be due to the fact that in the staggered layout, some signals were placed lower than their standard positions thus could be seen at an earlier stage especially from underneath tunnels or bridges.

A combination of signal backplate shapes (designed within the envelope of the current standard 3-aspect backplate) did not aid driver performance on signal sighting. On the contrary, it reduced the distance at which the signals could be identified. This was believed to be due to the small size of the backplate. Combinations of shapes may assist signal reading if a larger backplate envelope size is used, following the '150% rule' (Li et al, 2002).

This study also indicated that signals with 10%-20% white borders worked as well as the current standard design for viewing distances within 200-250 metres. Bordered signals may work better in shorter viewing distances such as 'shunting signals'.

Data analysis indicated that driver opinions were not significantly different towards all but one design, 1-by-1 staggered layout, which was less favoured than the other non-staggered ones. Some drivers commented that they prefer 2-by-2 or 3-by-3 layouts rather than 1-by-1 because they did not like moving their eyes up and down frequently as they went through the 1-by-1 layout.

However, this study indicated that while the staggered signals allowed the drivers to 'identify' the signals at further distances, they were also associated with more reading errors. One of the reasons is possibly due to 'dual reveal' with staggered signals, i.e., the drivers might only have seen some but not all signals on the gantry when a decision was made. This problem can happen especially when signals were obscured by bridges before gantries. This study has therefore suggested that, where signals are likely to be obscured by bridges on approach, it is better to place all signals at the same level along the gantry, as close as practicable to the driver's eye level. Staggered signals may work better in open approaches where drivers are able to see the full array of signals throughout the entire time of approach.

It should also be noted that, due to some limitations of image presentation on a PC screen (eg. the smallest pixel size possible), the simulated environment cannot fully represent seeing signals in the real world. It is anticipated however, considering the sighting distance alone, if the signals can be identified in the VR environment, they should be identifiable in the real world, from at least the equivalent range.

Conclusion

The results from is study suggest that in visually restricted areas, signals on gantries should be placed at the same level, but as low as practicable. Staggered signals may improve signal reading in open approaches where all signals on a gantry can be seen at the same time. A combination of different backplate shapes, as designed within the envelope of the current standard backplate size, did not aid signal sighting, nor did the bordered signal backplate for viewing distances greater than 200 metres.

Following this study, it was suggested that further investigation would be required to evaluate the signal design schemes that have not been tested in the present trials, such as six signals in a line at a lower height and other types of staggering patterns, which was the work of the following phase, as reported in HEL (2002).

Acknowledgements

This work was funded by Railtrack Great Western Zone. Thanks are due to Clive Lovelock at Railtrack for his support, and the ergonomists at Human Engineering for their technical input. Thanks are due also to Infrasoft Solutions Limited for support with VR computer programming.

References

Human Engineering Ltd, 2001, The Influence of Backplate Design and Offset on Railway Signal Conspicuity. Test Phase-1 of Human Factors Support to the Fibre Optic Signal Assessment for Railtrack Great Western Zone. (Report prepared by G. Li and authorised by W Ian Hamilton), HEL/RGWZ/01543/RT2.

Human Engineering Ltd, 2002, Further evaluation of staggered and non-staggered signal designs for the Paddington Throat in a virtual reality environment. Technical Report of Test 'Phase-2 Plus' of Human Factors Support to the Fibre Optic Signal Assessment for Railtrack Great Western Zone. (Report prepared by G. Li and authorised by W Ian Hamilton) (HEL/RGWZ/02702/RT4).

Li, G., Rankin, S. and Lovelock, C., 2002, The influence of backplate design on railway signal conspicuity. In: P. T. McCabe (ed.) *Contemporary Ergonomics 2002*, (Taylor and Francis, London), 191-195.

Railway Group Standard GK/RT0031, Issue Two, December 1996, *Lineside Signals and Indicators.*

DRIVER RECOGNITION OF RAILWAY SIGNS AT DIFFERENT SPEEDS - A PRELIMINARY STUDY

Guangyan Li[1], W. Ian Hamilton[1] and Theresa Clarke[2]

[1]Human Engineering Ltd.
Shore House, 68 Westbury Hill
Westbury-on-Trym, Bristol BS9 3AA

[2]Ergonomics Group, Railtrack (part of the Network Rail Group)
Railtrack House, Euston Square, London, NW1 2EE

Experimental trials were conducted using computer simulation to investigate the influence of train speed on driver recognition of railway signs. Twenty-four professional train drivers participated in the study and thirty-six types of lineside signs were tested under the simulated approach speeds of 100, 200 and 300km/h respectively.
Based on driver performance data, critical train speed was estimated for each type of sign. Beyond these speeds, the signs could no longer be correctly identified within a minimum approach time. The study also indicated that with increasing speed a larger visual angle is required in order to correctly identify a sign. In addition, more reading errors are related to signs containing two or more pieces of information, with each information item containing two or more letters or numbers.

Introduction

With advancing technology and increasing demand on railway transport, trains are travelling faster than they were decades ago. However, a majority of existing railway signs are still based on traditional designs that have been used for many years. Faster approach speed reduces the time that a sign stays in the driver's view, which increases the chance for the signs to be misread. There is thus a need to better understand whether these signs can still be correctly identified by drivers while travelling at different speeds.

The scientific basis for the present study has been reported elsewhere (HEL, 2002), which reveals that, despite much research that has been carried out to date regarding the design and safety issues for traffic signs, most published studies are in connection with road transport. Little is known about the effect of train speed on driver responses to different types of lineside signage, and no satisfactory answers could be found to some of the important questions. For example, how much information can be correctly identified while travelling at different speeds? Are all the current lineside signs, especially those designated for high-speed use, readable for high-speed train operation? If not, what are the speed 'cut-off' points beyond which the current signs are no longer effective? (thus an in-cab information system may be required to present such information); What is the maximum acceptable complexity level of a lineside sign in relation to a particular train speed?

Experimental trials were conducted to test the effects of train speed on driver recognition of lineside signs, using a computer simulation programme which was developed by HEL and successfully used in previous studies (e.g. Li et al, 2002). The tests covered the current operational (or potential) train speeds in the UK (from 100km/h or 62mph up to 300km/h or 186 mph) and included most standard lineside signs.

Experimental studies

Representation of Railway Signs

The lineside signs were shown at full apparent size (based on their current standard design and calculated visual angles). They were presented at an equivalent starting distance of 800 meters and 'moved' (expanded) towards the subject at an apparent speed of 100, 200 or 300 km/h (one at a time, in random order). The starting distance was chosen to ensure that the drivers should not be able to see the signs at the start of the trials.

Thirty-six railway signs were tested, which covered most of the standard lineside signs that are currently used in the UK for this speed range. Figure 1 shows some examples.

Figure 1. Examples to show some of the railway signs tested
(Drawings are for illustration purpose only, simulations were based on standard designs GK/RT0033, 1996)

Testing Theoretical Signs

In addition to testing the current standard realistic signs, a theoretical test session was also conducted to investigate how much information (and presented in what pattern) the drivers can effectively identify while approaching a sign at different speeds. The theoretical signs were designed such that each signboard contained from one to three items of information (combined letters and numbers), and each information item also had a complexity level from one to three (Figure 2). The size of the signboard was kept constant at 750x1050mm which is a standard size for most combined numeric/text signs.

It was hoped that the results would help to develop guidelines for the design of new signs which may have to convey multiple messages.

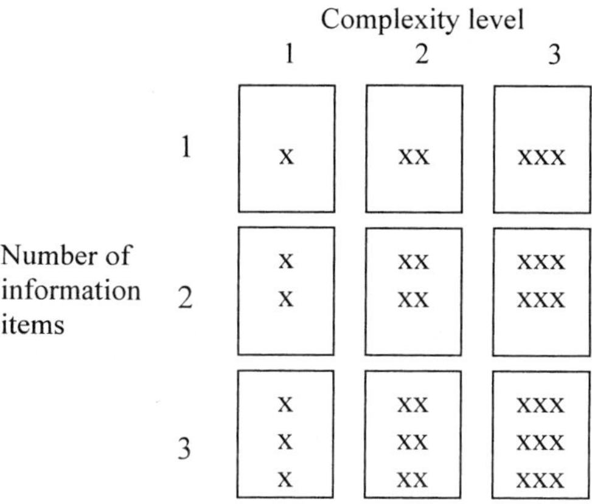

Figure 2. Patters of theoretical signs

Participants

Twenty-four train drivers from nine Train Operating Companies (23 male and 1 female) participated in the trials. Their average age was 40.6 years (Range=27-64, SD=9.3), and their average duration of driving experience was 12.1 years (Range=0-42.0, SD=10.6).

Apparatus

Two Intel Pentium®4 desktop computers with 512 MB RAM and 3D graphic card (Geforce2) were used for the trials. This enabled two parallel trials to be run with two participants at the same time (in two separate rooms). The monitor size was 17 inch; screen area (resolution) was set at 1280 by 1024 pixels and the colours were set for 'true colour' (32 bit). The screen background was set to medium grey (128). The computers were running MS Windows 2000 system.

Procedures

The participants were seated at the simulator and were allowed to adjust the chair to their preference. Their eye-to-screen distance was maintained constant at approximately 50cm. The simulation settings ensured that this distance would achieve a correct visual angle to the eyes such that the simulated signs viewed on the screen were equivalent in size to how they would appear in the real world.

The participants were required to press a key (e.g. the 'space bar') as soon as they could positively identify the sign, and at this point the computer recorded the equivalent sighting distance/time to sign. At the end of each run the participants gave their answers on what they had seen and the results were recorded by the experimenter.

It took approximately 60 minutes for each participant to complete the trials. The room lighting was dimmed to approximately 10-15 lux throughout the trials.

Results

Findings of the Railway Sign Tests

The influence of train speed on driver response to lineside signs was tested by calculating an equivalent visual angle at the eye at the moment when a particular type of sign was correctly identified while approaching the sign at one of the speeds tested. The results show that in order to correctly identify a certain type of sign at a certain approach speed, the sign has to achieve a minimum visual angle at the eye. The pattern of visual angle varied depending on the types of signs. In general, ANOVA and multiple comparisons (Tukey's HSD test) showed a trend which was that the faster the approach speed, the greater the required visual angle (or the larger the sign should be) at the point when its content could be correctly identified ($p<0.00001$ for all three speed levels tested), as shown in Figure 3. This is probably due the fact that subjects need a minimum amount of time to process the sign's information content.

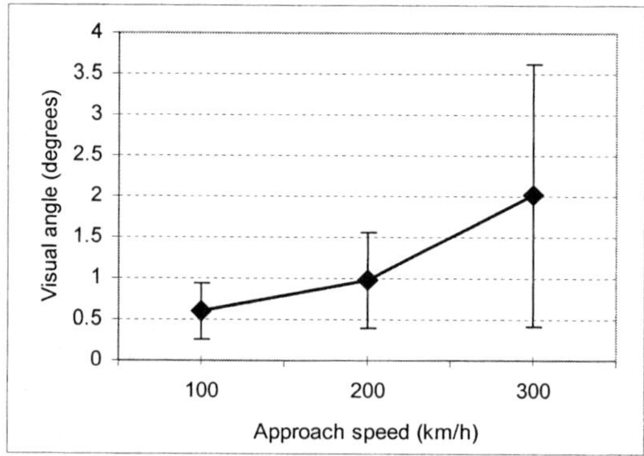

Figure 3. Visual angles required to correctly identify the lineside signs tested

The relationship was also estimated using linear regression between approach speed and time remaining to reach the sign at the moment when it is correctly identified. This gives an equation to estimate a speed limit beyond which the sign can no longer be correctly identified within a set time remaining before it is passed: $V = a - bT$

Where: V is approach speed (km/h); 'a' is the maximum speed limit when the signs can be read correctly at any time reaching/passing the sign (T=0s); 'b' is a constant showing the amount by which a change in time remaining (T, measured in seconds) will cause a reduction in speed limit in order to identify the sign correctly. Table 1 gives the estimated maximum speed corresponding to some of the signs tested, and beyond these speeds the signs are no longer readable.

Table 1. Estimated speed limit based on approaching time remaining

Type of signs	Equation for speed estimation	Max speed km/h (When T=0s)
Whistle board	301.1-65.4 T	301 (187mph)
Cab signalling and AWS gap warning	257.7-97.3 T	257 (159mph)
Countdown marker 100 yards	261.7-175.3 T	261 (162mph)
Countdown marker 300 yards	232.8-184.1 T	232 (144mph)
Level crossing, temp AWS cancelling/cab signalling	273.7-34.2 T	273 (169mph)
Electrification neutral section warning	277.7-77.1 T	277 (172mph)
Emergency indicator	270.3-43.1 T	270 (167mph)
Permissible speed indicator (triangle)	293.0-94.8 T	293 (141mph)
Permissible speed indicator (circle)	320.1-104.9 T	320 (198mph)
Temp differential speeds	275.3-121.6 T	275 (170mph)
Standard differential speeds	297.5-122.3 T	297 (184mph)
Permissible speed–standard differential	253.0-159.2 T	253 (156mph)
Permissible speed–arrow diverging route (red circle)	347.4-130.2 T	347 (215mph)
Permissible speed–arrow diverging route (yellow triangle)	278.1-190.9 T	278 (172mph)
Temp speed restriction–diverging route (warning lights)	247.9-93.1 T	247 (153mph)
Permissible speed – HST or DMU trains (red oval)	270.4-256.8 T	270 (167mph)
Permissible speed – HST or DMU trains (yellow triangle)	196.0-115.2 T	196 (121mph)
Radio channel change markers and SPAD indicators	270.0-150.5 T	270 (167mph)

Findings of the Theoretical Sign Test

Driver response errors were significantly affected by sign complexity at $p<0.000000001$ level, but the effect of speed on reading error was insignificant. However, there was a significant interaction between complexity and speed ($p<0.001$) with respect to their influence on sign reading errors. Multiple comparisons indicated that the major differences lay only between sign designs 3x2 and 3x3 (i.e., 3 pieces of information each having 2 or 3 digits) and the remaining sign patterns. The results suggest that lineside signs should not contain information which is at or above 3x2 complexity level, especially if the sign is to be used for an approaching speed of over 100km/h.

Discussion

This study found a significant influence of approach speed on driver recognition of railway signs in terms of the visual angle required at the eye and the time remaining when the sign could be correctly identified. Using the trial data, a critical train speed can be estimated beyond which the contents of the corresponding sign may not be correctly identified within a given time. It must be understood, however, that these results were based on only 24 drivers' performance data in an impoverished simulated environment, it is not yet known to what extent driver responses/behaviour in this environment represents that in the real world. In addition, the image display ability of a PC screen is limited by the smallest pixel size, therefore, it can be assumed that, as far as the sighting distance is concerned, if a sign can be viewed at a certain distance on the simulator, it should be seen farther in the real world, which also allows a faster approach speed.

However, it should also be understood that the simulated environment in the impoverished simulator is much simpler and visually cleaner than what the drivers see in the real world, in which for example, the information can be cluttered, the signs can be covered with dirt, the windscreen of the train may be smudged or there may be glare from the sun. In these situations, the real speed limits could be even lower if the sign is to be correctly identified.

The present study tested the driver sign reading performance under three speed levels only. Information is thus limited for the development of a more realistic relationship between sign reading performance and approach speed based on the data collected. Further data analysis showed the accuracy of the linear models of around R^2=0.56 on average (Range: 0.22-0.77, SD: 0.12), suggesting that the relationship between speed and driver recognition of lineside signs may be to some extent non-linear.

Drivers commented in the present study that they often do not need to recognise the signs before reaching the signpost. Therefore, it may be reasonable to assume that the remaining time (T) can be less than 4 seconds (the minimum sighting time required by standards) for the recognition of signs, even up to the point of reaching it (i.e., at T=0).

Speed did not significantly affect driver response errors for the identification of most of the railway signs tested. One of the reasons is possibly due to the fact that during the trials, the drivers were given a free choice not to respond until they were sure that the content of the sign had been positively identified, resulting in a low reading error rate, but a shorter sighting distance (or equivalently larger visual angle at the eye).

Conclusion

This study indicated a dynamic relationship between a visual angle required to correctly identify a sign and the approach speed, suggesting that larger signs would be required for higher train speed in order to ensure a minimum reading time. The maximum speed limit relating to each type of lineside signs (current standard designs) can be estimated using the linear models developed on the basis of the experimental data; and beyond the maximum speed the corresponding sign may not be correctly identified. However, due to the limitations of this study, as discussed earlier, the results at this stage should be regarded as being only indicative rather than conclusive.

Beyond speeds of 100km/h more reading errors occur with signs containing two or more pieces of information, with each information item containing two or more letters/numbers. Therefore, when designing new lineside signs, consideration should be given to how to avoid presenting the drivers with more than two pieces of information with each information item containing three letters or numbers. Similarly, for signs containing up to three pieces of information, each information item should not contain more than two digits.

References

Human Engineering Ltd., 2002, *Driver recognition of lineside signals and signs during the operation of high-speed trains – Scientific basis for the study*. Technical report to Railtrack (part of the Network Rail Group), prepared by Guangyan Li and authorised by W. Ian Hamilton, HEL/RT/01651/RT1.

Li, G., Rankin, S. and Lovelock, C., 2002, *The influence of backplate design on railway signal conspicuity*. In: Contemporary Ergonomics 2002, (ed. P.T. McCabe), London: Taylor & Francis, 191-195.

Railway Group Standard GK/RT0033, Issue Three, July 1996, *Lineside Signs*.

DEVELOPMENT OF A RAILWAY SAFETY RESEARCH SIMULATOR

Mark S. Young

Railway Safety
Evergreen House, 160 Euston Road
London NW1 2DX

In response to a series of high profile incidents in the rail industry, Railway Safety has directed a significant portion of its research budget into human factors themes. To complement this research, we have identified the need for a driving cab simulator primarily for research in human factors and operations. This paper provides the justification for a cab simulator in terms of the advantages for the Railway Safety research programme. This is followed by a detailed review of current driver training simulators in the industry, highlighting the differing requirements of a research simulator as opposed to a training aid. Finally, a functional specification for an 'ideal' research simulator is derived. The paper concludes with an outline of the future development programme for the Railway Safety research simulator.

Introduction

The UK rail industry is currently under an intense safety spotlight. Accidents such as those at Ladbroke Grove, Hatfield, and Potters Bar have undermined public confidence in rail travel. To address these problems, Railway Safety was set up at the start of 2001 as an independent safety body to lead research and set standards for the Railway Group.

It is generally perceived that the rail industry, when compared to other transport domains, has lagged behind somewhat in its consideration of human factors. Since its inception, Railway Safety has allocated a significant portion of its research budget towards redressing this balance. As a result, there is now a growing portfolio of research into human factors, covering issues such as communications, train protection and control, risk assessment and reducing the frequency of signals passed at danger (SPADs).

To support this research, Railway Safety has identified the need for a driving cab simulator primarily for research in human factors and operations. Following an initial feasibility study to identify the company's specific requirements, a full specification for both the simulator functionality and its hosting requirements has been developed. This paper provides an overview of how that specification was derived, based on a detailed review of current technology.

Justification

A state-of-the-art driving cab simulator will provide Railway Safety with a dedicated facility for conducting high quality research into human factors and operations issues. Many human factors aspects of train driving lend themselves to simulator research. For example, Railway Safety has previously funded studies into visual and mental acuity, and drivers' eye movements (Merat *et al*, 2002). Both of these projects experienced difficulties in conducting experiments in the field, due to conflicts of interest with scheduled services, train crew availability, and track access. Using a simulator would have avoided many of these problems and reduced the timescales for such research.

Simulator studies have several advantages for research of this nature (Senders, 1991). First, they can be used to put people into situations that would not be ethical in the real environment, particularly where such situations compromise safety. Second, simulators can be used in controlled experimental studies. Differences in driver performance can then be attributed to the manipulated variables with a high degree of confidence, and the influence of confounding variables can be minimised. Finally, we are able to compress experience, to collect data on a whole range of situations unlikely to be encountered in the natural environment in a short time frame.

The use of simulation in research environments is not without controversy. A review of the literature (Stanton, 1996) found that the main issues surrounding simulator use were focused upon the level of fidelity within the simulated environment. The key areas of concern can broadly be divided into physical fidelity (i.e. the degree to which the simulated environment looks like the real environment) and functional fidelity (i.e. the degree to which the simulated environment behaves like the real environment). Simulators have been successfully applied to research in other areas, particularly aviation and driving (e.g. Bloomfield and Carroll, 1996; Harris *et al*, 1999; Nilsson, 1995, Smith *et al*, 1997; Young and Stanton, 2002). The evidence from these domains suggests that functional fidelity is of greatest importance to transfer effects (the degree to which behaviour in the simulator transfers to the real operational environment). Physical fidelity may help convince the experimental participant that the task should be taken seriously, which would be less convincing in a more abstract environment.

In terms of research, it is difficult to define in advance the exact nature of experiments that would take place in the Railway Safety Research Simulator, since each study is specifically designed to answer applied questions when they arise. However, themes which would benefit from simulator research have been identified, including: SPAD reduction and mitigation, communications, abnormal and degraded working, train protection and control, competence, risk assessment, and occupational health.

Review of current simulator technology

To devise a specification for a research simulator, it was first necessary to review the current state of technology in the railway industry. Several train operators use simulators in driver training, and three of the larger simulators were visited for this review. Since the requirements specification for a training simulator differs from those for a research simulator each simulator was assessed for its applicability to research activities. The simulators were rated along a similar pro forma as that adopted by Stanton and Young (2002); the review presented below is an abridged summary of the key points.

EWS Simulator Training Centre

The EWS simulator was commissioned in October 2001 in response to the recommendations from the Southall and Ladbroke Grove formal inquiries. It consists of a full cab on a static base, with a single forward projection screen displaying a 40-degree horizontal field-of-view. Within its 45-mile track database, the software contains a variety of track types, including 2- and 4-track layouts, with both semaphore and colour light signalling. The track also has a variety of gradients, from level to 1 in 39. A quadraphonic sound system provides realistic engine noise, while a subwoofer gives the driver a sense of vibration in the absence of a motion base. A variety of events and environmental conditions are programmable via a simple scenario editor, and a dedicated post-run analyser can be used to examine the data on a range of train variables.

In terms of fidelity, the most obvious limitation is with the visual field, since a wider view would provide greater immersion for drivers. However, drivers who have used the simulator give positive feedback about it, suggesting that it is a good representation of the real task. Furthermore, anecdotal reports suggest that after a period of acclimatisation, drivers do become quite immersed in the simulation.

The absence of motion might also be seen as a disadvantage, although it is the opinion here that this did not greatly detract from the experience. The vibrations of the subwoofer compensate somewhat by giving some proprioceptive feedback.

Regarding its application to research activities, the versatility in scenario development is a major advantage. However, there are also a number of drawbacks with the EWS simulator. These mostly relate to the flexibility of the visual environment. Although extensive, the visual database is limited in terms of its adaptability, since it is not possible to alter objects embedded in the environment. For instance, in a study on signal sighting, it would be desirable to use different positions and configurations of signals; this would not be possible in the EWS simulator. Furthermore, the functionality of other objects is limited, since other trains are the only objects that can be animated. Trackside workers, and vehicles at level crossings, are unfortunately static.

The other major limitation is regarding data manipulation. The post-run analyser is useful for viewing data in its various forms, but empirical studies would require exporting to a statistical package for more powerful analysis. However, the post-run analyser in the EWS simulator is not compatible with such software.

In sum, the EWS simulator provides very high levels of fidelity, and seems ideally suited to its role as a training simulator. However, if it were to be considered as a research tool, some improvements in its flexibility would be necessary.

Eurostar simulator – Holmes House training centre

The Eurostar simulator has been in service since 1993, and consists of a fully functional cab replica with a static base. The digitised image is projected onto a single forward screen, with a field-of-view of approximately 40 degrees in both planes. Digitised engine sound is routed through a four-way speaker system in the cab, and system alarms are sounded through individual speakers. The cab even has dynamic illumination at the side windows to give a peripheral impression of movement.

The visual databases are set within the simulator for five of the Eurostar routes. Scenario development takes place via a simple Windows interface at the instructor's workstation providing 'drag-and-drop' access to a variety of events and conditions. In addition to the scenario display, the instructor has three additional screens, showing simulator data and current track position, the actual forward view, and a TV monitor of the cab interior. Detailed information on train operations can be recorded and time-linked with a video record if specified in advance. However, data files must be downloaded and processed at a separate facility, and it is not possible to export them to external packages.

As a training simulator, the Eurostar facility performs its function well. Although the graphics are perhaps of lesser quality than more modern machines, they still provide a decent sense of immersion and visual flow. Furthermore, the visual field fully covers the smaller windscreen of the Eurostar, thus improving psychological fidelity.

Disadvantages with the simulator are primarily related to the lack of capability for data recording and analysis. There is also limited functionality in altering the visual database. Whilst Eurostar do not need such functionality for their purposes, these aspects are essential criteria for a research simulator, and must be included in the specifications.

Virgin Trains training centre

The Virgin training centre contains three full-task driving simulators. There are two motion-based Voyager simulators for the cross-country routes, and one static Pendolino for the West Coast Main Line. The facility has been operational since mid-2000.

Each simulator consists of a full cab mock-up of the relevant production train, with accurate 3D images of real routes, covering about 60km of railway. Images are displayed on a single forward projection screen with an approximate 50-degree field-of-view horizontally and vertically. Neon light boxes installed at the cab's side windows provide some sense of movement in the visual periphery.

From a research point of view, the major advantage over other training simulators is the capacity for data recording and manipulation. The software for the Voyager simulator writes data directly to an Excel file, making it possible to statistically analyse the results (or even transfer them to a more powerful statistical package).

Flexibility in scenario development is also an advantage, with a simple GUI for designing a variety of situations for drivers. A wide number of events are available, which can either be preset in advance or initiated by the instructor during the simulation. However, the flexibility ends with the scenarios, since it is not possible to alter the visual database for creating new tracks or environments. A track building tool, which can offer such functionality, is available as an add-on package.

Simulator fidelity is very high, and drivers are generally impressed with the experience. Most of this quality can be attributed to the Silicon Graphics imagery and the effective motion base. However, both of these aspects greatly inflate the costs involved in setting up and maintaining the facility.

Generally, then, the flexibility of the Virgin simulator in scenario development and data recording make it most suitable as a research tool. However, it lacks variety in visual environments, and would benefit from the capability to construct new routes.

Summary

Each of the systems reviewed can be classified as *replica simulators*, in that they are a complete, exact duplication of the human-machine interface and a realistic set-up of the system environment. As such, they are all bound to score highly on criteria of fidelity. However, whilst each simulator is ideally suited to its role in a training facility, their applicability to research was generally lacking in the areas of scenario flexibility and data analysis. For properly controlled empirical studies to be conducted, such functionality is essential. The Virgin simulator was rated highest on these criteria, but there was still room for improvement in manipulating the visual databases.

Functional specification for a Railway Safety Research Simulator

So far, this paper has identified the types of research for which a driving cab simulator would be advantageous, and has reviewed the technology available in three large training simulators currently in use by train operators. These activities, in conjunction with previous experience of research simulators, have led to the development of a detailed functional specification for a Railway Safety Research Simulator.

A key question about the simulator relates to its level of fidelity. Any simulator for research use should have a high level of functional fidelity. Whilst physical fidelity seems to be more important for training simulators, it can also improve the feeling of immersion for participants. If drivers feel like they are in a real train, they are more likely to behave as they would in a real environment.

Simulator fidelity is very often associated with movement. Motion systems, though incurring additional costs, can greatly enhance the participants' experience if implemented effectively. However, if there are even slight inaccuracies in synchronisation with the displayed image, subjective validity can be greatly reduced, and rates of simulator sickness are likely to be high. Indeed, the benefits of motion systems are rarely proportionate to the costs associated with them, and are therefore usually the first feature to be sacrificed if budget is limited. A low frequency sound resonator (or simple subwoofer) channelling engine noise can provide enough vibrations throughout the cab to give a sense of movement at a fraction of the cost. However, motion bases can provide crucial performance advantages when perception of acceleration is important (as in railway operations). Furthermore, simulator movement bolsters face validity for participants, as well as representing the state of the art in current technology. The Railway Safety Research Simulator is intended to be a flagship facility for the industry, and will thus incorporate a full motion base.

Therefore, the simulator will feature high quality graphics, sound, and a full cab mock-up on a motion base with up to six degrees of freedom. The image should encompass at least 50 degrees horizontally and vertically, with preference to a multi-projection system providing up to 180 degrees horizontally. Most importantly, though, the software shall provide full versatility in developing experimental scenarios and visual databases, including a number of real routes from the UK network as well as the capability to create entirely new track layouts. Ideally, the simulator will also allow the experimenter to manipulate the design of signals and other trackside objects. Finally, the simulator shall record all relevant performance variables at rates specified by the experimenter. Cursory descriptive results shall be viewable within the simulator software itself, with full compatibility with external statistical packages for deeper analysis.

Railway Safety is also considering investing in an additional, scaled-down version of the simulator, perhaps based on a desktop computer. The reason for this is that some experiments may not need the high levels of fidelity that a full cab simulator offers, or there may be a need for a higher throughput of participants. Quick answers to simple questions may be possible on a desktop simulator.

At the time of writing, there are two simulator contracts out to tender – one for the delivery of the simulator, and another for hosting the facility. Therefore, neither the specification, timelines, nor location have been confirmed. The simulator will be hosted and managed by an external body with a proven history of simulator operation, although the hosting contract does not necessarily guarantee the successful bidder any of the associated research work. It is anticipated that the contracts will be finalised early in 2003, with delivery and commissioning in mid-2004.

Conclusions

- A Railway Safety Research Simulator would clearly add value to the research programme in a number of areas.
- Current simulator technology for driver training provides high levels of fidelity but lacks the flexibility necessary for a research tool.
- The research simulator would address these shortcomings while incorporating high quality graphics, motion, and a full cab environment.
- An additional, scaled-down (i.e. desktop) version of the simulator would be advantageous for smaller experiments.

Acknowledgements

Railway Safety would like to thank EWS, Eurostar and Virgin Trains for their cooperation and assistance in the production of this paper.

References

Bloomfield, J. R. and Carroll, S. A. 1996, New measures of driving performance. In S. A. Robertson (ed.), *Contemporary Ergonomics 1996*, (Taylor and Francis, London), 335-340

Harris, D., Payne, K. and Gautrey, J. 1999, A multi-dimensional scale to assess aircraft handling qualities. In D. Harris (ed.), *Engineering Psychology and Cognitive Ergonomics Volume Three: Transportation Systems, Medical Ergonomics and Training*, (Ashgate, Aldershot), 277-285

Merat, N., Mills, A., Bradshaw, M., Everatt, J. and Groeger, J. 2002, Allocation of attention among train drivers. In P. T. McCabe (ed.), *Contemporary Ergonomics 2002*, (Taylor and Francis, London), 185-190

Nilsson, L. 1995, Safety effects of adaptive cruise control in critical traffic situations. *Proceedings of the second world congress on intelligent transport systems: Vol. 3*, (VERTIS, Tokyo), 1254-1259

Senders, A. F. 1991, Simulation as a tool in the measurement of human performance, *Ergonomics*, **34**, 995-1025

Smith, K. C. S., Lewin, J. E. K. and Hancock, P. A. 1997, The invariant that drives conflict detection. In D. Harris (ed.), *Engineering Psychology and Cognitive Ergonomics Volume One: Transportation Systems*, (Ashgate, Aldershot), 229-236

Stanton, N. A. 1996, Simulators: research and practice. In N. A. Stanton (ed.), *Human factors in nuclear safety*, (Taylor and Francis, London), 117-140

Stanton, N. A. and Young, M. S. 2002, Driving simulators: a frame-of-reference for comparing psychological dimensions. Manuscript submitted for publication.

Young, M. S. and Stanton, N. A. 2002, Malleable Attentional Resources Theory: A new explanation for the effects of mental underload on performance, *Human Factors*, in press.

DEVELOPMENT OF A HUMAN FACTORS SPAD HAZARD CHECKLIST

Claire Turner[1], Rachel Harrison[1], Emma Lowe[2]

[1]Human Engineering Ltd, Shore House, 68 Westbury Hill, Westbury-on-Trym, Bristol BS9 3AA, UK
[2]Railtrack (part of the Network Rail Group), Railtrack House, Euston Square, London NW1 2EE, UK

That there are human factors issues connected with the occurrence of SPADs (Signals Passed At Danger) is well known, but the understanding of these issues in the rail industry is sometimes limited and often based only on subjective opinion. An objective and consistent approach to the investigation of human factors causes of SPADs was therefore necessary to create a generic performance-based understanding of human factors, eliminate the need for subjective opinion, and pinpoint the root human factors causes of SPADs. The development of the Human Factors SPAD Hazard Checklist, outlined in this paper, was intended to address this need. A model of driver behaviour and error, based on established cognitive theory, and a validated database of SPAD hazards, form the basis of this work. The final checklist has been developed for ease of use both proactively, by signal engineers and sighting committees, and reactively, to investigate the root causes of a SPAD incident and to help to prevent their future occurrence.

Introduction

The general trend in the occurrence of SPADs has decreased over the last year. For example, 24 SPADs were reported in the HSE's 'SPADs Report' for September 2002, which is 6 fewer than in September 2001, and 19 fewer than the average September figures for the last 6 years (HSE, 2002). However, the understanding of the influence of human factors on SPAD occurrence remains limited. It has historically been limited to driver issues such as competency, shift patterns, and fatigue, and is often based only on subjective opinion. But as Lord Cullen pointed out in his inquiry into the Ladbroke Grove crash of 1999, focusing solely on driver error fails to identify root causes such as poor infrastructure and management failures.

A rational, objective and coherent approach to the investigation of human factors causes of SPADs was therefore needed to create a generic performance-based understanding of human factors, eliminate the need for subjective opinion, and pinpoint the root human factors causes of SPADs. The Human Factors SPAD Hazard Checklist addresses this need.

Approach

The checklist originates from two main sources of information: -

- A cognitive model of the driver task (Cognitive Task Analysis)
- A database of human factors SPAD hazards

Development of the Cognitive Model

In order to reduce SPADs, it is important to understand driver performance. Through an examination of exactly what is involved in the driving task and of how drivers organize and process information, it becomes possible to identify where mistakes can be made. Cognitive Task Analysis (CTA) was the method used to achieve this. Cognitive Task Analysis allows the train driving task to be analysed from the point of view of the driver, and aspects of the task that place heavy demands on the driver's cognitive resources - their attention, perception and so on – to be identified. By focusing upon the parts of the signal reading task that drivers will find hardest to learn and perform, it becomes possible to see where errors are most likely to be made. The ultimate aim was to use the Cognitive Task Analysis approach either to predict SPAD risk for a given layout or to identify significant human factors within a SPAD event.

The way the driver processes information was modelled in three stages using a combination of established ergonomics techniques:

- Task analysis of signal reading – a step by step analysis of the driver task
- Recognise-act cycle modelling – to uncover driver information processing
- GEMS modelling of human error – to identify the types of errors that drivers could make and their consequences.

Recognise-Act Cycle Modelling

Information processing can be thought of as a series of repetitions of the *recognise-act cycle*. In each cycle, a driver perceives features (e.g., signals) in the surrounding environment, and compares them with rules and procedures stored in memory in order to decide upon the appropriate action (e.g., the signal is at single yellow, therefore anticipate that the next signal will be at danger).

By combining knowledge of the driving task from a task analysis with the recognise-act cycle, it is possible to produce a more detailed representation of the signal-reading task in terms of the information processing involved. This is illustrated in Figure 1.

Modelling Human Error

The more skilled we are in performing a task, the more automatic it becomes, but the very nature of this effort-saving cognitive design leaves us prone to making errors. These can occur at any stage of information processing. The GEMS (Generic Error-Modelling System) model developed by Reason (1990) defines three levels of error, which have been identified in human performance and error analysis in various industries: -.

- Skill-based: slips and lapses - usually errors of inattention or misplaced attention, when task demands outweigh or outrun cognitive capacity.
- Rule-based: mistakes - usually a result of picking an inappropriate rule or procedure, caused by a misjudgment of current circumstances or by making incorrect assumptions based on previous experience

- Knowledge-based: mistakes - due to incomplete/inaccurate understanding of system, or overconfidence.

Our cognitive resources are limited, which explains why skill-based slips and lapses are the most common forms of human error as these occur due to failures in attention and memory. Rule-based mistakes are the second most common form of failure. Because train drivers are highly trained operators performing a very procedural task, knowledge-based mistakes in train driving are rare.

Vigilance	Signal sighting is a vigilance task that requires sustained levels of attention and alertness
Signal Detection	Surface features of a signal are detected in the environment
Signal Recognition	1. Signal perception – the signal's form is identified and discriminated from surrounding objects 2. Association with line – signal is recognised as appropriate to the driver's route
Signal Interpretation	Signal aspect is read and an appropriate response is chosen
Action	Driver responds to the signal

Figure 1. Illustration of driver information processing during the signal-reading task

Developing the SPAD Hazard Database

Driver performance is influenced by a complex environment that contains a number of 'performance shaping factors' (PSFs). These factors can cause disruption at each stage of information processing, which may cause the types of errors described above. An example is passenger distraction: a passenger could disrupt driver vigilance resulting in the driver failing to detect an upcoming signal. This would be classed as a skill-based error – the driver's attentional capacity was exceeded, resources could not be split between the driving task and the distraction, and a lapse in awareness of lineside signals occurred. Possible factors that could influence driver performance were identified as a result of a literature review (of HMRI and Railway Safety documents, academic research projects, etc) and a series of site visits to various signals. A total of 52 signals across the seven Railtrack Zones were assessed, with SPAD histories ranging from 0 to 17. A combination of cab rides and track walks enabled digital video footage of each signal, its approach and surrounding area to be obtained. Interviews with drivers were also undertaken to ratify our understanding of the driver task.

Three validation criteria were developed to enable consistent qualification and categorisation of human factors SPAD hazards. In order to be included in the final database, a hazard needed to be clearly defined, it's effect on human performance and behaviour well understood, and also have evidence (whether real-world or academic) to back it up.

Using these three criteria as a benchmark, 69 PSFs were validated as human factors SPAD hazards. For convenience, these were grouped into factors that better define their effects. These can be related back to the aspect of driver behaviour they impact. The database includes not only infrastructure factors, such as gradients and curved approaches, but also the effects of driver performance, environmental conditions, operational procedures, and train cab design. Some examples are provided in Table 1.

Table 1 – Examples of Hazard Factors

Stage of Driver Task		Hazard Factor	Example
VIGILANCE	A	Personal factors	Fatigue
	B	Sources of distraction	Complex track layout
DETECTION	C	Visibility factors	Foliage
RECOGNITION	D	Perceptibility factors	Dark or complex signal background
	E	Association with line	Irregular signal spacing
INTERPRETATION	F	Reading aspect	Expectation
	G	Interpretation	Flashing yellow aspect sequence
ACTION	H	Action/Performance failures	Poor train cab ergonomics

Qualitative and quantitative analyses of the site visit data were carried out. This revealed a significant relationship between the number of hazards present at a signal site and number of SPADs at that signal.

Outline of the Human Factors SPAD Hazard Checklist

In order to bring the cognitive model and database of hazards together in a practical, usable format, a checklist was developed for use by signal sighting engineers and committees, and SPAD investigators (HEL/RT/02719/RT2, 2002). The checklist follows a simple logic that is rooted in the sequential stages of driver information processing. It is designed to address the human factors SPAD hazards that may be encountered by the driver.

It is assumed that a number of external influences, e.g., train, infrastructure or signaller failures, are ruled out in any investigation prior to using the checklist. Any SPAD without authorisation is considered to be a violation of the driver rulebook, so the starting point for the checklist is "Was the violation deliberate?" This allows a distinction to be made between abnormal or criminal behaviour and excusable violations.

Eight outline questions direct the user to the relevant checklist questions. These are summarised in Table 2.

Questions within each checklist part relate to the presence or absence of hazards and have been worded to enable simple yes/no answers. This eliminates the need for subjective opinion and promotes a precise, performance-based understanding of SPAD hazards. Help is provided for each checklist question, which explains the argument behind the hazard (i.e., the precise description of the hazard and examples of where it could occur), evidence for the hazard, and what it is necessary to do in order to answer the question. Using the help as a guide, the user is required to work his/her way through the checklist, answering yes to any relevant question.

Table 2 – Checklist Questions

Outline Question		Checklist Part	Example Checklist Question
Is there evidence to suggest that personal factors are involved?	**A**	**Personal factors**	Had the driver worked successive night shifts in the week prior to the SPAD?
Is there evidence to suggest that the driver was inattentive?	**B**	**Sources of distraction**	Is the signal positioned soon after an OLE neutral section?
Is there evidence to suggest that the driver could not see the signal?	**C**	**Visibility factors**	Is the signal obscured by station structures or furniture?
Is there evidence to suggest that the driver could not perceive it as a signal?	**D**	**Perceptibility factors**	Is the signal viewed against a dark or complex background, e.g., buildings, foliage, etc?
Is there evidence to suggest that the signal could not be associated with the correct route?	**E**	**Association with line**	Does the number of lines visible to the driver (including sidings) differ from the number of signals visible?
Is there evidence to suggest that the driver could not read the signal aspect correctly?	**F**	**Reading aspect**	Is the signal normally (i.e., >75% of the time) encountered at a proceed aspect?
Is there evidence to suggest that the driver could not interpret the signal aspect correctly?	**G**	**Interpretation factors**	Are flashing aspects used on the approach to the signal ?
Is there evidence that the driver could not perform the correct action?	**H**	**Action/ Performance failures**	Did poor cab ergonomics cause a delay in driver action?

Determining the Level and Source of SPAD Risk

Once the checklist has been completed, SPAD risk at a particular signal location can be easily calculated. Each "yes" answer to a question is assigned one penalty point. These are summed to derive a total for each part of the checklist, and an overall total for the checklist. Very simply, the higher the overall number of resulting penalty points, the higher the SPAD risk at a particular signal. In this way, the signals that require the most urgent attention can be assessed and the investment in SPAD risk reduction measures prioritised. This approach has been designed to complement other risk assessment tools that exist for assisting with decisions about investment and risk reduction.

Identifying Risk Reduction Measures

Risk reduction measures have been suggested for the majority of hazards, intended to prevent, control or mitigate the effect of a SPAD. The measures proposed are based on evidence or experience (a risk reduction measure implemented at a multi-SPAD signal that has not since been SPADed, for example). Alternatively, new measures have been suggested based on good ergonomic principles.

The totals for each part of the checklist direct the user to the most appropriate risk reduction measures. A reference book of possible risk reduction measures is provided.

Numbering is consistent throughout for ease of use, so, for example, by referring back to the relevant queries in the visibility part of the checklist, the associated prevention, control or mitigation measures can be identified.

Conclusions

The human factors SPAD hazard checklist brings together an understanding of the driver task and human error mechanisms. Based on evidence and subject to a rigorous hazard qualification procedure, it provides a coherent structure for analysing the human factors causes of SPADs.

The traditionally driver-centric view of human factors has been expanded to cover the infrastructure, train design, environmental and operational factors that are likely to be the underlying causes of SPAD incidents. However, in order that all possible human error causes of SPADs are identified as part of the investigation process, further work is required on the role of signaller error in SPADs.

A major benefit of the checklist is that it allows signals to be assessed both reactively, in support of a SPAD inquiry, and proactively, by signal engineers and sighting committees to evaluate the SPAD risk at a particular location. The scoring method provides a rational and objective basis for allocating resources to address those signal locations that are the most at risk of a SPAD incident.

Further Validation and Future Development of the Checklist

Use of the checklist is being supported with training of signal sighting professionals and SPAD investigators. A pilot application of the checklist by the Railtrack Zones is the next step. The checklist has been the subject of broad research and development, but as with all new methods, the best way to ensure that it works is to test it in the field. The new risk reduction measures put forward would in particular benefit from this type of assessment.

A structured three-step process for validating human factors SPAD hazards has been specified. This not only ensures that hazard identification is consistent and open to verification, but also can be adopted as a means of capturing and processing any further SPAD hazards that may arise when the checklist is applied in the field. The information generated as a result of the validation process will also be useful in deciding upon appropriate risk reduction measures.

Feedback from initial application will be collated and, if necessary, the checklist modified to increase its usability and scope.

References

Reason, J. 1990, *Human Error* (Cambridge University Press, Cambridge)

Health & Safety Executive, 2001, *The Ladbroke Grove Rail Inquiry Report The Rt. Hon Lord Cullen PC* (HSE)

Health & Safety Executive, 2001, *SPADs (Signals Passed At Danger) Report for September 2001* (HSE)

Health & Safety Executive, 2002, *SPADs (Signals Passed At Danger) Report for September 2002* (HSE)

HEL/RT/02719/RT2, 2002, *Human Factors SPAD Hazard Checklist "User Guide."* (Report prepared by C. Turner and authorised by W.I. Hamilton for Railtrack, Plc.)

SAFE AND UNSAFE BEHAVIOUR AMONGST RAILWAY TRACK WORKERS

Trudi Farrington-Darby, Laura Pickup and John Wilson

Centre for Rail Human Factors
Institute for Occupational Ergonomics,
University of Nottingham,
University Park, Nottingham, NG7 2RD

This paper discusses the main findings of a qualitative piece of research carried out in collaboration with a UK rail infrastructure maintenance company (IMC). A series of group and individual interviews addressing the question of "What makes people unsafe when working out on the track?" were performed. Analysis against a conceptual model devised from safety culture and behaviour literature resulted in a final forty factors that were perceived to be contributing to unsafe working behaviour within this IMC. These were presented to the organisation and disseminated to staff and now form the basis of a strategy to reduce unsafe act precursors and move in the direction of a positive safety climate and culture.

Introduction

In the past the tendency when dealing with accidents in the rail industry, like many others has been firstly, to focus on the people who were directly involved in the incident and blame the individual and secondly to implement disciplinary action, increased training and/or rules and procedures in an attempt to prevent a reoccurrence. These are likely to be short term and ineffective methods for reducing unsafe behaviour and improving safety culture within an organisation (Reason, 1997). Accidents and unsafe behaviour take many forms. Reacting like this to each one is time and energy consuming and fails to anticipate the next accident-creating scenario. The organisation does not necessarily become any stronger and safer as a result. This approach also does not consider the latent and organisational acts that contribute to event or the culture and climate that they occur in.

A more constructive approach is to reduce incidents and accidents through understanding their causes (creating what Reason refers to as an informed and possibly learning culture); part of this involves accepting that the individual who behaves unsafely is not the sole interaction responsible for the event, and that organisational or contextual factors may have put them in a position which made their unsafe behaviour more likely (Reason, 1997).

Precursors to unsafe acts (violations or errors) can be identified and analysed. Additionally the general organisational and safety culture these factors and individuals exist in need to be considered in order to really understand the environment for safety. This underlies the philosophy of organisational safety management being easier to manage than trying to manage people's minds (Reason, 1997)

The literature surrounding safety culture and safety climate is often conflicting, inconclusive and difficult to directly apply to a specific industry. The reasons for this are legitimate in that it is a relatively new concept, that spans various disciplines and is by nature difficult to define and measure. Whilst there are several models and definitions proposed some of the agreement that appears to be emerging suggests safety climate is being considered as a measure of the climate at the time; a snap shot of safety culture that is typically measured though psychometric design and indicating personnel's attitudes and beliefs at the time. Safety Culture is considered to be a more related to the organisational policies and systems in place that suggest practice and set precedent of how safety is regarded and managed (Cox and Flin, 1998;). Models of safety culture and climate typically involve determining its component parts and their relationships (Cheyne et al, 1998; Glendon and Litherland, 2001; Trimpop, 2000; Brown et al, 2000). No published studies for railway infrastructure maintenance were found.

Reason's (1997) approach of addressing the components of safety culture and bringing them together is one proposed way of achieving this. He suggests the components of a safety culture being its five subcultures: learning, reporting, just, flexible and informed.

Whilst the literature will further understanding of safety culture and climate and bring us closer to reliable measurement, for an organisation with a safety culture problem their main concern will be to make sense of their own problem. They will also want to determine the best course of action to address a negative safety culture, perhaps using the five subcultures described by Reason (1997) as top-level goals. They require something tangible. A possible route to achieving this with our current understanding of safety culture, safety climate and unsafe behaviour is to look at what influences attitudes, beliefs and behaviour related to safety.

Study back ground

A railway Infrastructure Maintenance Company (IMC) who were concerned about the death and injury of staff on the track wanted deeper information than from a recently applied safety climate questionnaire survey. They approached the Institute for Occupational Ergonomics at University of Nottingham to conduct a research study to address these concerns. Whilst there is great value in the use of safety climate tools for problem detection, measuring changes over time and after the implementation of management ideas, there are also limitations to their use, such as the lack of freedom for explanation (the IMC required more in-depth information regarding the issues that the questionnaire had highlighted), fear of reprisals when writing responses (of particular importance in the rail industry), overuse without feedback, misinterpretation, and the use of written words for expression in an industry that is traditionally driven by verbal communication. Additionally the questionnaire results were sometimes contrary to what the organisation suspected really was happening in terms of working safely. Accordingly they employed an independent group (the IOE) to collect data directly from their staff.

Method overview

The investigation process is outlined in Table 1. The imperfections of performing practical field research in industry are highly relevant to this study and their resolution a key component to the success of the project along with the collaboration with the main organisation.

Table 1. Outline of investigation process

Stage 1:	Background research and literature review
Stage 2:	Interview preparation
Stage 3:	Semi structured Interviews (6 groups totalling 34 track workers and 8 individual interviews)
Stage 4:	Data gathering and preparation
Stage 5:	Data analysis
Stage 6:	Data presentation
Stage 7:	Dissemination
Stage 8:	Action plan and strategy

As the issue of blame was a concern, the whole study design was considered carefully from the outset. To encourage cooperation and open and honest responses, measures were put in place that included confidentiality agreements, working closely with the trade union representatives throughout, ensuring feedback and providing payment for time and travel.

The methods for analysing rich contextual data in an efficient manner were approached pragmatically in light of the client's needs. The aims were to determine what track workers, their supervisors and managers thought caused unsafe behaviour and to present these factors in a meaningful way to the client to allow them to develop appropriate strategies. The analysis process underwent several iterations and the main steps are shown in Table 2.

Table 2 Process for data analysis

Stage A:	Direct transcription of tapes to database
Stage B:	Initial coding against conceptual model
Stage C:	Inter and intra coder reliability testing
Stage D:	Final framework for coding from reviewed model
Stage E:	Completed and updated coding of data
Stage F:	100 factors identified through thematic coding using framework
Stage G:	Dependence classification to reduce categories
Stage H:	40 factors that influenced unsafe behaviour identified

The statements transcribed in stage A were coded against a conceptual model of the grouped themes of factors that were described in the literature; an initial 100 factors were identified. This level of detail is common when dealing with such rich data but stage G ensured that a number of factors workable for managers of the organisation were presented without losing the original themes. Sorting on the basis of dependence between factors and those that would have far reaching effects on others identified 40 main factors that influence track workers' safety behaviour. These 40 factors were presented with quotes to expand on them, suggestions for ways of dealing with the factors and whether the factors were considered to have a direct and immediate influence on track workers' behaviour (e.g.

weather) or are more medium (e.g. supervisors style of management) or long term (e.g. contradictory rules) in the proximity of their source to the unsafe event. These are not hard and fast classifications but offer a way of looking at factors that influence safety behaviour, in terms of the depth within the organisation's functioning that they occur and the range of functions and systems they affect

Findings

The final influencing factors are summarised in table 3.

Short term or more immediate factors

Some of the factors thought to affect behaviour at the time of performing a job at the track side are: excessive and poor communication; poor behaviour on the part of safety role models such as controllers of site safety (COSS); inconsistent team members either due to poor planning, staff shortages or the heavy reliance upon contract staff. Track workers also reported a lack of systematically performed and regularly performed risk assessments. Various reasons could be found to explain this, that staff become familiar with their jobs and the risks associated with them, that they do not have the skills with which to perform risk assessments or they do not have the resources to perform risk assessments in a dynamic work environment.

Other factors influencing unsafe behaviour at the time of a job were reported as the perception of safety and risk; an operator's knowledge; the equipment available, the condition it was in and the appropriateness of the equipment for that job; operator fatigue and loss of concentration; individual competence to perform a task safely; certification of competency and a sense that things are different and the same rules don't apply when you are working out of the normal hours of work.

Medium term and less direct factors

More medium term factors were related primarily to the immediate supervisors (line managers) and work planners in the organisation. Both of these groups of people provide a link between the track workers and the organisation. Supervisors are seen to be the first line representation of management, organisational beliefs and policies, regulation and discipline, support and communication. Planners provide a link between operational staff and the programme of work, which forms the organisation's main business. They indirectly influence safety at track side through their competence to plan and their operational knowledge. The planners and the supervisors form part of a communication feedback loop and are particularly relevant to the capacity of the organisation to learn and evidence of operational staff that they were being listened to.

Some of the medium to long-term factors relate to managers more distant from the track workers than their direct supervisors. In particular factors such as management visibility, accessibility, communication methods, feedback and messages provided implicitly through decision and actions were suggested to affect behaviour at trackside and culture in general.

Long term and more indirect factors

Some of the factors identified as long term were inherent in the organisation's and industry's functioning as a whole, for instance paperwork volume, relevance and purpose, rules that are contradictory to each other, what is practical and what is expected, poor rules dissemination, the usability of the rule book (on several levels including format, readability, availability and weight) and the perceived purpose of the rules. The blame culture and culture of prosecution has led to sceptism regarding the motives behind some of the rules, procedures and paper work in the rail industry. Rejection on that level may lead to negative culture through a lack of trust.

Safety systems that affect the organisation's functioning throughout such as communication pathways, feedback cycles, and reporting methods were reported as indirect factors influencing the safety behaviour of staff on the track.

Another area of concern highlighted through the study was that of training and recruitment. Staff and skill shortages were seen as directly affecting the ability to perform work safely on the day but had their source in long-term organisational policy.

Table 3. 40 Factors influencing safe and unsafe behaviour

Communication on the job (excessive and poor quality)	Poor and underused real time risk assessment skills	Individual perception of what safe is	Track workers knowledge and understanding
Inconsistent teams/ subcontractors	Safety role model behaviour	Social pressure of home life	Setting up site safety on the day
Rule dissemination	Physical conditions	Peer pressure	Feedback cycle
Competence capability and certification	Working hours: different behaviours out of normal hours	Manager's communication methods	Information / communication route clarity
Pre-job information dissemination	Planners knowledge for job resourcing	Job feedback to planners	Volume of paperwork
Feedback messages from managers	Manager's railway knowledge	Manager's visibility and accessibility	Perceived purpose of paper work
Supervisors style visibility, communication, representation of staff	Supervisors: (technical competencies and assessment of them)	Supervisors presence (visibility, leading by example, opportunity for verbal communication)	Fatigue, concentration, ability to function (alcohol)
Equipment (condition, appropriateness and availability)	Practical alternatives to rules	Perceived purpose of the rule book	Rule book usability and availability
Information pathway flow	Planners competency to plan	Methods for reporting	Information systems use
Contradictory rules	Recruitment methods	Training needs analysis	Training methods

Conclusions and next steps

The factors which influence (un) safe behaviour, perceived by operational staff and their line managers can be classified by their "distance" from their outcomes. Like the latent and organisational errors described by Reason (1997) factors that influence behaviour can be traced back as far as higher management policies as well as including those that occur on the day as a result of something at the time of an event.

Many of the factors related to those that are suggested to affect safety culture and climate in the literature, and to those that are described as performance shaping factors for error and violations.

This work has not aimed to provide a new model of safety culture or even to redefine what causes unsafe acts. It has provided clear identification of the perceptions of what affects behaviour, from the viewpoint of the people that work in a specific organisation. For the purpose of this study the main factors were identified in a way that has formed the basis of an interactive dissemination programme to operational staff and supervisors, through safety briefs and "training trainers" sessions. A strategy for addressing tangible factors contributing to unsafe behaviour and death and injury on the track is being produced from the evidence.

Current management action plans include review of training, and specifically risk assessment training to include human factors considerations, training of safety role models such as COSS on the importance of their roles, review of skills other than technical that these staff may require (e.g. man management), review of incident reporting systems and the ways they can be designed and used to encourage reporting culture and improve organisational learning.

The background philosophy for all dissemination and action was one of continued participation of operational staff and constant consideration of safety culture and the five subcultures defined by Reason (1997). Awareness that every action has a message about safety governs the continuation of activities from this work.

References

Brown, K. A, Willis. G. P, Prussia. G. E 2000, Predicting safe employee behaviour in the steel industry: Development and test of a sociotechnical model, Journal *of Operations Management,* **18**, 445-465.

Cheyne, A., Cox, S., Oliver, A. and Tomas, J. 1998, Modelling safety climate in the prediction of levels of safety activity, *Work & Stress*, **12 (3)**, 255-271

Cox, S. and Flin, R. 1998, Safety Culture: Philosophers Stone or Man of Straw? *Work & Stress,* **12 (3)**, 189-201

Glendon, A. and Litherland, D. 2001, safety climate factors, group differences and safety behaviour in road construction, *Safety Science*, **39**, 157-188

Reason.J. (1997) *Managing the risks of organizational accidents*. Ashgate Publishing Ltd. Hampshire, England.

Trimpop, R. 2000, Safety Culture. In P. Elzer, R. Kluwe and B. Biussoltara (eds) Human Error and System design and Management, (Springer-Verlag, London), 189-199

MENTAL WORKLOAD OF THE RAILWAY SIGNALLER

Laura Pickup[1], John Wilson[1], Theresa Clarke[2]

[1] *Centre for Rail Human Factors, Institute for Occupational Ergonomics University of Nottingham, University Park, Nottingham, NG7 2RD.*
[2] *Head of Ergonomics, Network Rail, Railtrack House, Euston Square, London, NW1 2EE*

This paper outlines the qualitative approach taken to understanding the workload of a railway signaller and highlights the need to develop tools appropriate for the assessment of mental workload within this domain. Findings from signaller interviews are presented leading to a summary of how the term 'workload' should be considered within this domain. This empirical work is placed in the context of the literature to provide an early conceptual model and approach for the development of assessment tools relevant to the rail industry.

Introduction

A variety of studies completed within railway signal boxes have provided numerous opportunities for observation and informal discussions to understand how railway signallers carry out their duties. This approach has provided ergonomics based appreciation of what a signaller is required to do, the culture that exists amongst this group and the qualities considered necessary to make a 'good signaller'. This has led to the understanding of how to develop tools to assess workload in a railway signaling control environment.

The need to assess the workload of a signaller has become an increasing priority for the rail industry with a shift from more physical to mental work as signalling systems move from lever controlled signal boxes to increasing levels of automation involving a change in task demands. Problems associated with workload within this industry tend to be identified from incident reports or from signalling managers and staff representatives about a specific signal box. However, the need to predict the workload prior to introducing a new system is also important and therefore both retrospective and predictive assessment of workload is required.

Unlike aerospace there appeared to be little public domain evidence of ergonomics research or tools to assess rail workload. Interviews with human factors consultants identified limited success had been experienced with the use of tools such as time line analysis and NASA TLX (Hart and Staveland, 1988). The opinion was that these did not capture all of the variables considered relevant to the signaller's workload. The literature reveals little on a clear set of principles or guidance as to how to develop a workload assessment within such an environment. Although some of the methods from aerospace were considered of value none offered an immediate set of tools suitable for the rail industry to use within the field for the different relevant facets of workload (see below). This appeared to be due to the signaller being required to complete multiple activities concurrently, to prioritise and share tasks for a particular situation and working environment, and the nature of effort involved in completing their work. Additionally many of the tools focus on identifying conditions of overload e.g. NASA TLX, SWAT (Reid and Nygren, 1988), time line analysis (Kirwan and Ainsworth, 1992), W/INDEX, (North and Riley, 1989) - with no consideration to underload. The increasing levels of automation have highlighted underload as a potential risk to the railway signalling control environment. For the purpose of this work the concept of workload was proposed to be a spectrum with high and low workload at opposite ends. Conditions that fall into either extreme should be considered as potentially impacting upon personal well-being and system performance (Hancock and Caird, 1993).

The absence of appropriate measurement tools "off the shelf" and uncertainty of how to systematically interpret workload led us to explore the concept of workload in relation to first principles, and therefore direct how it should be considered and subsequently assessed within the rail industry. This work was initiated with minimal assumptions about the concept of workload but with the literature being considered in the light of the understanding of a signaller's work and working environment (through observations, reviews of task and goal analyses, interviews and questionnaires). The aim was to provide a conceptual framework for the concept of workload (Vicente et al, 1987), to identify the areas within the framework that require tools to enable assessment of each dimension.

Workload in the context of rail signalling

A view commonly held by signallers is typified by the comment that a 'good' signaller is 'someone who can think ahead and predict what is going to happen and work around things'. A signaller will describe their responsibility within the organisation as 'to ensure safety on the track'. They are required to control trains without conflict with each other, or the intended timetable, whilst accommodating engineering work on the infrastructure, which involves high levels of communication and vigilance of the system's state. The system itself provides protection against the loss of separation between trains and conflict management. However, the infrastructure, coupled with a large number of rules and organisational instructions limit the options available to the signaller. The signaller's overall goals can be considered as maintaining awareness of the situation, considering strategies, plans and making decisions, and acting to achieve their decisions. This is a basic description of the railway signaller but illustrates the dynamic puzzle that the signaller is presented with.

The outcome of the signaller's ability to complete this work will be judged by the organisation. If a delay occurs, and it is considered to be due to the way that an individual regulated a train, then a code will be assigned to highlight the cause of the delay; this is known as delay attribution. There are many other reasons why delays may occur and these are also recognised, but the feedback to a signaller is presented in terms of loss to the organisation caused by the delay.

Two stages were used to help understand the intuitive meaning of the term workload (Huey and Wickens, 1993). Stage one was to ask signallers through interviews and questionnaires an open question: 'What is workload?' and what it means to them. The responses included a number of phrases and terms such as 'an acceptable level of tasks', safety, stress, 'the amount you have to deal with at any one time', efficiency, 'the ability to switch from relaxed to active at the flick of a switch', 'the amount of work done during shift' and the 'required *effort* to complete the days work'. The term effort was most frequently mentioned. The views tended to consider the tasks and the overall working environment with organisational and system demands all grouped together under the umbrella term 'workload'.

Subsequently human factors experts within the rail industry were asked for their definitions. From this understanding and assessment of workload within the rail industry has been suggested as requiring an insight into:

- the number, complexity and interaction of tasks performed over time or at one point in time
- the load perceived by someone over time or at one point in time
- the number of functions necessary to be performed in different situations
- the compatibility of working arrangements for the functions needing to be completed

Stage two involved the use of semi-structured interviews with signallers to explore the variables considered as relevant to the signaller's workload. This was undertaken as part of investigations into specific signal boxes reporting workload problems. The request was to identify and understand the cause of the reported problems and whether these issues were likely to influence either performance or signaller wellbeing.

Jahn's (1973) early model of workload appeared to make few assumptions and represented the terms described within a number of the definitions of workload mentioned earlier and also frequently cited by many authors with experience of research in this field of work (Borg, 1978, De Waard.D, 1996, Hancock and Caird, 1993, International standard ISO 10075, 2000, Johannsen, 1977, Moray, 1982, Rouse et al, 1993, Vicente et al, 1987). Jahn's (1973) basic model was adapted (figure 1) to avoid using the term 'workload' and to focus the interview to probe for factors considered as influencing the load on, or the effort from, the signaller, followed up by asking what were the potential or actual effects of these factors.

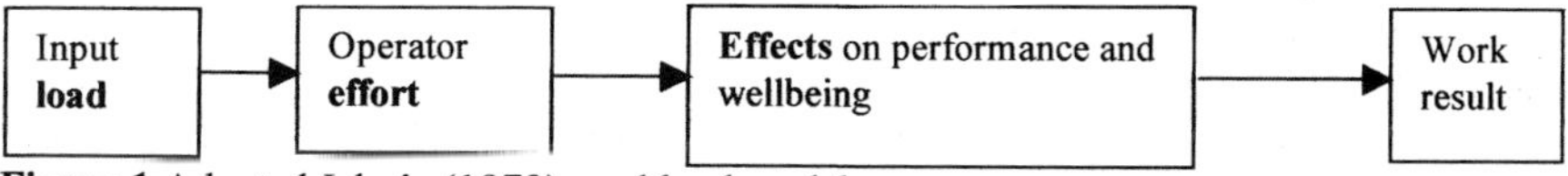

Figure 1 Adapted Jahn's (1973) workload model

The data collected were systematically reviewed to code it under the categories of loading factors, effort and effects. The effects were either directly stated or interpreted from the signaller's comments, and divided into two main groups, which referred to emotional states and more physical states. The findings have been summarised in Table 1 and suggest that the perceptions held by signallers about the sources of load and effort refer to a number of different variables.

The evidence relating to workload dimensions – a conceptual model

A suite of tests was thought the most effective way of assessing the number of relevant dimensions as one tool would seem incapable of assessing all variables relevant to the signaller (Brookhuis and De Waard 2002, Williges and Wierwille, 1979,). The basic model guided further exploration of each dimension, to develop a conceptual framework to represent the key variables.

The understanding of loading factors often appears confused within the literature as they refer to both the source of load and the perception of load. This differentiation is relevant when considering methods of assessment, with the perception of load relying heavily on subjective assessments. Subjective experience of load is very relevant in understanding the

Table 1 Findings of variables influencing signaller's work

Loading Factors
Organisation, environment context (e.g. normal, degraded, emergency), management, individual, team, job design, equipment, social situation feedback on performance
Effort
Complexity of decision making, options available, time sharing of activities, predictability of situation, perception of control on situation, interruptions, distraction from activity, perception of likelihood of achieving intended goal (due to time, experience, organisational or environmental limitations), consequence of actions, blame for actions
Effects - Emotional states
Anger, frustration, guilt, loss of esteem and anxiety, stress.
Effects - Physical states
Distracted, pressured by time, fatigued and high or low levels of arousal.

level of demand experienced from the load on an individual (Borg 1978, Vicente et al, 1987, Rouse et al, 1993), but the source of the load is also relevant in that the load is imposed by a certain number of tasks that require a level of performance within a time available. The internal load created by an individual (the individual's goals and expected or acceptable level of performance) is a further consideration as this will also create a level of demand and influence the perception of load (Borg, 1978, De Waard, 1996, Navon & Gopher, 1979, Rouse et al 1993, Vicente et al, 1987). Load was therefore divided into three sub dimensions to consider each of the relevant factors: imposed load, perceived load and internal load. These have been illustrated within the conceptual model, figure 2.

The mental effort exerted is dependent upon the individual as much as the task being undertaken. The skill, experience, level of arousal and motivation for the task will all determine the degree of effort exerted or experienced by any individual (Kahneman, 1973). Human behaviour theories suggest we are fundamentally goal driven and the effort exerted is dependent upon the priority of the goal and how the goal is achieved (Kahneman, 1973, Rasmussen, 1986). The individual also has a level of effort they deem to be acceptable. Once reached it is this that will influence the increase or decrease in any further effort by adoption of strategies that allow them to work at an acceptable level of effort to achieve their intended goal or level of success (Kahneman, 1973). If the work result is assessed as less than desirable then the individual could moderate their effort as required, suggesting a feedback loop within the conceptual model, figure 2. Demand is implied by the literature as being a consequence of load and a cause of effort for an individual (Borg, 1978, Hancock and Caird, 1993, Rouse et al, 1993) but understanding why a source of load or a perception of load creates a demand is complex. Theories of attention and information processing (Kahneman, 1973, Navon and Gopher, 1979, Norman and Bobrow, 1975, Rasmussen, 1986, Shiffrin and Schneider,1977, Wickens, 1992, Wickens 2002) identify possible factors that influence demand such as: interaction and compatibility of tasks, situation or context, dynamic goals, task difficulty, task automaticity, task priority, task number, time pressure, individual strategies, skill, preferences, expectations and motivations, quality of information.

There appears little evidence to suggest that effects on performance are predictable following an increase or decrease in workload (or levels of load, demand and effort).
There are two main reasons, firstly effort is determined by the strategies adopted but a strategy

that conserves effort may not necessarily impact on performance (Rouse et al, 1993,Wickens, 1991). Secondly trying harder or increasing effort may have little impact on performance

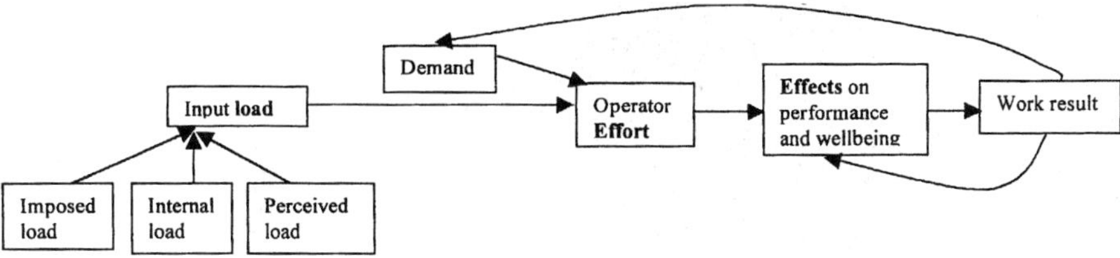

Figure 2 Conceptual model of mental workload

(Rouse et al, 1993). Thus fluctuations in load and demands may impact upon the effort experienced but with no guarantee of an effect on performance. The effects that may influence performance or wellbeing have been suggested from interviews with signallers and include fatigue, arousal, pressure and stress with a number of more emotional states also relevant. All of these effects can direct ergonomists to use subjective assessment tools already developed to highlight unacceptable levels of these 'effects'.

Conclusion – developing new tools

The assessment of the mental workload of a railway signaller needs to consider the whole work situation functionally. Tools to identify imposed and perceived loading factors can direct the management of how to improve a signaller's workplace. Tools that assess individual characteristics of the signaller, system, situation, combinations of tasks and working environment, and suggest conditions of unacceptable levels of demand and effort, can highlight zones of unacceptable workload at both ends of the spectrum.

From this programme of research a new unidimensional scale, multi dimensional scale, laddered decision tool, tools to record combinations of tasks and tools based on relative judgments of operational characteristics within a particular environment or scenario are all under development and will be reported on in the future.

A suite of assessments would enable conditions and perceptions to be revealed that could provide a profile and place the level of mental workload along a spectrum. A zone of acceptability would exist at the centre of the spectrum along a figurative line with conditions of unacceptable levels of mental workload being at either end of the spectrum (high and low). The zone of acceptability should be guided by industry standards, signallers' perceptions of acceptability and the theories and principles associated with mental workload. Judging workload in terms of the acceptability of a particular scenario is not a new idea (Brabazon and Conlin, 2001, Rouse et al, 1993,) and seems to be the most pragmatic way to approach the measurement of such a dynamic concept with a vast number of variables.

References

Borg, G. 1978, Subjective aspects of physical and mental load, *Ergonomics*, **21**, (3), 215-220

Brabazon, P. and Conlin.H. 2001, Assessing the safety of staffing arrangements for process operations in the chemical and allied industries, *Contract Research Report 345/2001,* (Health and Safety Executive, Sudbury)

Brookhuis, K.A. and De Waard, D. 2002, On the assessment of (mental) workload and other subjective qualifications, *Ergonomics*, **45**, (14), 1026 - 1031

De Waard, D. 1996, The measurement of drivers' mental workload, *PhD thesis*, University of Groningen, Haren, (Traffic Research Centre, The Netherlands)

Hancock, P.A. and Caird, J.K. 1993, Experimental evaluation of a model of mental workload, *Human Factors,* **35**, (3), 413-429

Hart, S.G. and Staveland, L.E. 1988, Development of NASA-TLX (task load index): results of experimental and theoretical research. In P.A. Hancock and N. Meshkati *Human mental workload 1988,* (North Holland, New York), 139-178

Huey, M.B. and Wickens, C.D. (1993). *Workload transition: Implications for individual and team performance.* Washington, DC: National Academy Press.

International standard ISO 10075, 2000, Ergonomic principles related to mental workload – part 1: General terms and definitions, (European Committee for Standarisation, Brussels)

Jahns, D.W. 1973, In Johannsen,G.1977, Workload and workload measurement. In N.Moray (Ed) *Mental workloa- Its theory and measurement 1977*,(Plenum Press, New York),3-11

Johannsen,G.1977, Workload and workload measurement. In N.Moray (Ed) *Mental workload – Its theory and measurement 1977*, (Plenum Press, New York), 3-11

Kahneman. D. 1973, *Attention and Effort*, (Prentice Hall inc, New Jersey)

Kirwan,B. and Ainsworth, L.K. 1992, *A Guide to Task Analysis*, (Taylor & Francis, London)

Moray, N. 1982, Subjective mental workload, *Human Factors,* **24**, 25-40

Navon.D.and Gopher.D. 1979, On the economy of the human-processing system. *Psychological Review*, **86**, (3), 214-255

Norman. D.A. and Bobrow. D.G. 1975, On data-limited and Resource-limited processes, *Cognitive psychology*, **7**, 44-64

North, R.A. and Riley, V.A. 1989, W/INDEX:A predictive model of operator workload. In G.R.McMillan, D.Beevis, E.Salas, M.H.Strub, R.Sutton, and L.Van Breda (Eds) *Applications of Human Performance Models to System Design,* (Plenum, New York), 81-89

Reid, G.B. and Nygren,T.E. 1988, The subjective workload assessment technique a scaling procedure for measuring mental workload. In P.A. Hancock and N. Meshkati (Eds) *Human mental workload 1988,* (North Holland, New York), p185-218

Rouse,W.B., Edwards.S.L., Hammer.J.M. 1993, Modeling the dynamics of mental workload and human performance in complex systems, *IEEE Transactions on Systems, Man and Cybernetics* , **23**, (6), 1662-1670

Shiffrin. R.M. and Schneider.W. 1977 Controlled and Automatic Human Information Processing: II, Perceptual learning, automatic attending, and a general theory, *Psychological Review,* **84**, 2, 127-190

Vicente, K.J., Thornton, D.C., Moray.N., 1987, Spectral analysis of sinus arrythmia: A measurement of mental effort, *Human Factors,***29**, (2), 171-182

Wickens, C.D. 1992, *Engineering psychology and human performance*, second edition, (HarperCollins Publishers Inc, New York)

Wickens.C.D 2002 Multiple resources and performance prediction, *Theoretical issues in Ergonomic Science,* **3**, (2), 159-177

Williges, R.C. and Wierwille, W.W. 1979, Behavioural measures of aircrew mental workload, *Human Factors,* **21**, 549 – 574

BASELINE ERGONOMIC ASSESSMENT OF SIGNALLING CONTROL FACILITIES

Joe Capitelli, Fionnuala O'Curry, Mike Stearn & John Wood

CCD Design & Ergonomics Ltd
Golden Cross House,
8 Duncannon Street,
London, WC2N 4JF

Historically ergonomics assessments of Signalling Control Facilities throughout the National rail network have focussed on specific perceived problems and have been carried out as a reaction to an incident or event at a particular location. Furthermore they have been carried out by a range of in-house teams or outside consultancies using different assessment methods and reporting in a way appropriate to the concern raised. Consequently information has tended to be localised and difficult to compile and compare. To enable a uniform overview of ergonomics performance within Signalling Control Facilities across the rail network the 'Baseline Ergonomic Assessment Programme' has been developed and implemented. This is a standardised and systematic ergonomics assessment method which allows a general overview of ergonomic effectiveness and operational demand to be determined. It may be used reactively or as part of a proactive assessment exercise. For the first time differing types and generations of Signalling Control Facilities can be compared on a broad range of ergonomic and operational variables. The data generated can form the basis of prioritised local and system wide action plans to allow a co-ordinated response to the upgrade of Signalling Control Facilities based on ergonomic issues.

Introduction

Signalling Control Facilities are where the regulating and monitoring train schedules throughout the U.K is performed. More than one thousand of these control facilities exist nationally supporting a wide variety of control systems ranging from older 'mechanical lever frames' to 'electronic entry-exit panels ' (NX panels – see Figure 1) and on to the more recent VDU based Integrated Electronic Control Centre (IECC) and Network Management Centre (NMC) systems.

In the past, ergonomics assessments of Signalling Control Facilities have generally focused on specific perceived problems, and have been conducted on a reactive basis. Typically, a variety of different ergonomics consultants and in-house teams have conducted such assessments using different methods and reporting in a way appropriate to the concern raised. As a result, information about signalling issues and actions taken to address such issues have remained localised limiting the ability to gain a comparative overview of ergonomic performance across the rail network. This limitation suggested the need to adopt a more holistic approach to undertaking ergonomics assessments within Signal Control Facilities.

To facilitate this holistic approach to Signal Control Facility assessment Network Rail and UK ergonomics consultancy, CCD Design & Ergonomics Limited, have developed the 'Baseline Ergonomic Assessment Programme'. This Programme provides a standardised ergonomics assessment methodology, analysis model and issues reporting structure that provides a general overview of ergonomics performance and operational demand within railway Signal Control Facilities identifies key ergonomics issues and prioritises areas for attention or more detailed review.

Figure 1: A Signaller Operating an NX Panel

The 'Baseline Ergonomics Assessment Programme' is usable by both in-house and external Ergonomists and local, suitably briefed, managers. The components of the Programme are currently being integrated into a Toolkit, along with a number of others, for use in the rail environment. By assessing signalling control facilities in a standardised and systematic manner, ergonomics performance across different generations of signalling control facilities can for the first time, be compared across a broad range of ergonomic and operational variables. Actions can be prioritised both locally and system wide to ensure a coordinated response to Signalling Control Facility upgrade.

In summary, The 'Baseline Ergonomic Assessment Programme' aims to deliver:

- A standardized Toolkit by which Signalling Control Facilities can be assessed, rated and compared for ergonomics performance on a uniform basis

- An assessment Programme that focuses on interfaces that affect human performance and welfare
- A methodology by which ergonomics information can be consistently and effectively collected within Signal Control Facilities and analysed to identify and prioritise ergonomics issues

Method

Initial Method Development

The Baseline assessment method was shaped by the requirements to take a global view of ergonomics issues, to minimise disruption to operational staff working within operational control facilities (with the initial focus being panel and IECC based facilities) and to allow information to be collected and analysed both by professional ergonomists and by experienced infrastructure production managers. It was also apparent, from previous experience, that variations in ergonomics interfaces between individual Workstations/Panels within multi-manned Signal Control Facilities (i.e. equipment, responsibility, control area, operational demand etc), required the assessment method to focus at a Workstation/Panel level as opposed to a whole Facility level.

A Baseline Ergonomics Assessment method was developed which contained;

- An interview proforma for collection of views from managers, operators and staff representatives.
- A grid template for recording room, panel or workstation dimensions.
- An ergonomics checklist for scoring 140 ergonomics variables.
- A software analysis model for collecting and analysing scores from the ergonomics checklist

The method requires a Baseline Assessor to visit each Signal Control Facility and complete each part of the programme. Ergonomics variables within the checklist were established from requirements within infrastructure owner standards, ergonomics standards, guidance and accepted best practice. Where feasible, objective ergonomics variables were recorded (i.e. dimensions, distances, noise, thermal and light measurements) however fields were left within the checklist and analysis models for recording of subjective views given by on-duty Signallers during each assessment,

Ergonomics variables were scored on a scale of 1 to 3 where 1 equates to ergonomically effective and 3 equates to ergonomically ineffective as judged against standards and best practice requirements. All questions were scored equally i.e. no weighting was applied (however any issues judged to be of major significance were highlighted as Key Issues in the Baseline Report for that workstation).

An Analysis Model was produced using an Excel workbook to score automatically checklist data, record Signaller comments and Assessor observations and log Key Ergonomics Issues established from the assessment. Key Issues were defined as those likely to affect Operator Welfare or Performance. The output of the Analysis Model is a Baseline Report for each Signal Control Facility Panel/Workstation listing scores for each ergonomics variable, Key

Ergonomics Issues, Signaller comments and Assessor observations.

Pilot Study

The Baseline method was initially piloted at 2 Signalling Control Facilities and reviewed before full piloting at a further 16 sites spread (producing in total 30 individual Panel/Workstation Baseline Assessments). These sites covered a range of signalling control systems (Panels, RETB facility and IECC) and Signal Control Facility size. Following the Pilot Study, a detailed review was undertaken of the method with representatives from the railway Ergonomics Team and with Signallers and Signal Box Managers from the Pilot study. The method, scoring system, organisation of ergonomics variables and delivery of the assessment programme were all updated

Main Study

From the Pilot Study, ergonomics variables were agreed for assessment as detailed in Table 1. These variables were analysed within five ergonomics categories also detailed in Table 1. Category scores were calculated by averaging the scores for each of the ergonomics variables within that category.

Table 1: Data Structure

Ergonomics Category	**Ergonomics variables**
Welfare	Environment Comfort, Furniture, Posture, Space, Storage, Access and Facilities, Thermal and Vibration
Environment Performance	Lighting, Glare, Noise
Workstation/Panel Performance	Design, Display and Control Interfaces, Alarms, Level Crossings, Secondary Activity Support
People and Organisation	Signaller Experience Level, Training, Roster & Manning, Rules, Handover, Planning, Supervision and Work Breaks
Operational Demand	Infrastructure Complexity, Traffic Volume, Traffic Speed Variation, Terminus Platform Working, Intervention Frequency

Baseline assessments were conducted at 140 Signal Control Facilities across the rail network. The main study assessed all IECC facilities and the larger Panel based facilities. In total, 264 individual signal Panels/Workstations were Baselined.

Signal Control Facilities assessed in the main study were rated, compared and prioritised based on scores for each ergonomics variable, each ergonomics category and overall ergonomics effectiveness (OEE). Key Issues and comments and observations collected during the on-site assessment describe ergonomics issues and the actions suitable to address them.

Analysis and Deliverables

Baseline Reports

Every panel assessed has a corresponding Baseline Report detailing the Ergonomics Category Scores (i.e. for Welfare, Environment, Signalling Workstation, People and Organisation and Operational Demand) and the OEE score. This report also includes the list of Key Issues for that workstation as well as the Comments and Observations captured during the assessment.

Priority Ranking Lists

By ranking the ergonomic effectiveness scores for workstations on a local or system wide basis, priority ranking lists can been compiled from the OEE score or more specific scoring categories (e.g. Welfare or Workstation/Panel Performance). These can be used to categorise and prioritise workstations to highlight those where improvement might be best focussed. Further prioritisation can be achieved by ranking workstations of similar score by the Operational Demand score. Figure 2 shows an example of a scatter plot of Panel Performance against Operational Demand for a group of workstations.

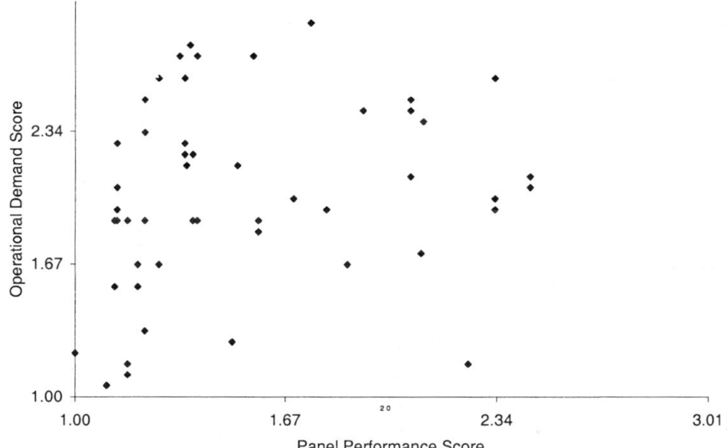

Figure 2: Scatter Plot of Panel Performance Score against Operational Demand

Signalling Control Facility Reports

A report for each facility is generated. These reports identify any Key Issues at each facility and suggest appropriate improvements or highlight areas in need of further investigation. These reports are discussed and agreed at a local level to ensure effective prioritisation of issues to support the programme of continuous improvement to Signalling Control Facilities.

Spreadsheet Analysis

A spread sheet has been compiled containing all the data generated by this work. Potentially the analysis of this can supply a huge amount of data. Preliminary analysis of this data has been carried out and used to support the local and system wide analysis referred to below.

Local and System Analysis

Local and system wide issues and comparisons can be highlighted by considering the data at the relevant level. Based on both the qualitative and quantitative data generated those ergonomic issues most common at a national or local level can be identified.

Limitations

While the Baseline programme has generated much valuable information and made a considerable contribution to the organisations knowledge base, there are limitations to the Baseline programme.

- Each Baseline assessment was developed to last a maximum of 2 hours per workstation. As such, Baseline reports do not provide a detailed ergonomics assessment of the facility visited. Instead they provide a general overview and can highlight the need for a more detailed assessment to be carried out if required.

- Only a limited number of signalling staff at the facilities assessed were available for interview and comment. Due to this any data collected that was subjective in nature may be biased towards the view of a limited number of subjects.

- Due to the quantity of data gathered and the way the data is compiled it was found difficult to develop a sustainable weighting system that would accurately reflect the differing level of importance each variable holds in relation to its associated operational and ergonomic risk. Currently no weighting system is applied to the data. To address this issues a system of weighting variables is currently being developed.

Conclusion

The Baseline Ergonomic Assessment Programme has provided a standardised ergonomics assessment method that allows a general overview of ergonomics performance and operational demand to be determined for a wide range of Signalling Control Facilities. By use of this standard and systematic approach ergonomics performance can be compared across different types and generations of signalling control facilities and action can be prioritised at a local and system wide level to allow a co-ordinated response to the upgrade of Signalling Control Facilities based on ergonomic issues.

THE ROLE OF COMMUNICATION FAILURES IN ENGINEERING WORK

Philippa Murphy and John R. Wilson

Network Rail, Railtrack House, Euston Square, London NW1 2EE and Centre for Rail Human Factors, Institute for Occupational Ergonomics, University of Nottingham, University Park, Nottingham NG7 2RD

This paper outlines the research undertaken to further the understanding of communication failures during engineering work on the railways. It provides an overview of three, complementary pieces of research that examine rail communications and their processes from difference perspectives. These are a root cause analysis tool, a content analysis and an examination of the critical failure points within the rule book. The findings of these are then placed in context of the implications for the industry and how they can be implemented to improve safety communications.

Introduction

In recent years, operations associated with engineering work on the railways have come under scrutiny, to both optimize performance and control risk. As would be expected, the causes of these were multifaceted. Incidents of a lesser magnitude occur every day on the infrastructure, resulting in loss of the infrastructure, time, and delay or personal injury. Recognising the importance of communication and the implication of communication failures, a detailed study commissioned. This paper summarises the work carried out and examines three tools used for assessment, the results will have on operations and procedures.

Methodological considerations

The goals and objectives of this work followed an extended period of participant observation, from being based in the sponsoring company (Railtrack, now Network Rail) for an extended period of time. The first year was spent undertaking participant observation at varying levels (Hammersley et al, 1998) and the flexibility of the method allowed emergent themes to develop. Following attendance at various industry working groups, the importance of engineering work incidents was stressed, with the focus on communication failures within these. The first piece of research was the development of a tool to understand the root causes of incidents. As a result of these findings a second tool was then developed to analyse the content of communications to understand whether there were any significant differences between incident and routine communications. Finally, and again as a result of these findings, a piece of work was conducted into an analysis of the procedures in the rule book, where communication-based violations occur. These tools are discussed in detail in the following section.

Tool development

Root Cause Analysis Tool (RCAT)

Following involvement in industry groups, analysis of formal inquiries and participation in various work streams, it became apparent that further work was required in the area of root cause analysis. This resulted in the development of the RCAT, which had three aims. To understand:

- The types of communication failures which occur
- What other factors are associated with these failures
- Where in the causal chain communication failures occur

Following extensive testing, development and application, in the industry generally and on formal inquiry reports, the results showed that communication failures were more varied than previously thought. A high level description of the results follows, but a more detailed explanation can be found in Murphy (2001). One of the most important results was that communication failures were more complex than previously realised by the industry. The research found that the most common failure types were:

- Communicating an incomplete message
- Communicating an incorrect message
- Not communicating

Archive analysis showed that, in formal inquires, the investigation into root cause finished at 'communication failures'. The RCAT has shown that this is not sufficient for understanding why the failure occurred and is therefore not focused on putting useful mitigation measures in place. To target failures effectively, it is imperative that causes such as those noted above, are properly understood, as each requires different methods to prevent their recurrence.

With regard to the second aim, the most commonly occurring factors were: violations (40%), laziness (24%) and complacency (16%). The most common causes of planning failures were 'incorrect information in the Weekly Operating Notice' (50%), not having the correct information to hand (38%), and poor planning due to a lack of local knowledge, by the operational staff (6%). The last of these categories was most commonly exposed by the placement of detonator protection in the wrong place or using the incorrect limits for the work. The mitigation measures for these are complex and require a long-term initiative by the entire industry to change.

Finally, with regard to the place communication failures took in the causal sequence, they were most commonly noted to be the final failure before the incident. When considered in terms of Reason's (1990) organisational error model, communications are often (in 61% of the cases) the final barrier before the incident occurred.

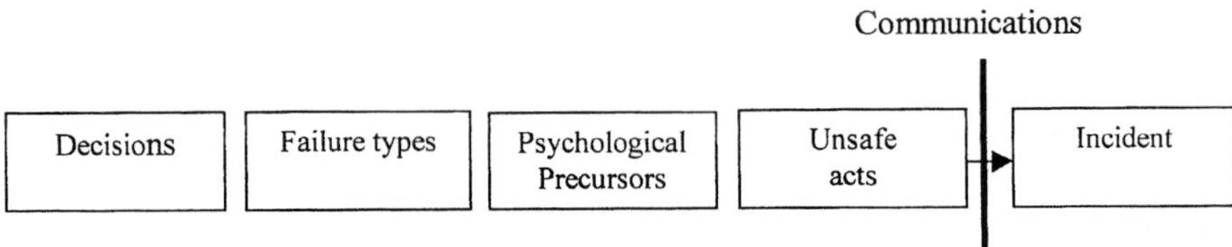

The implication of this is that, if communications had been better structured and accurate, then the organisational errors that had been promulgated through the system (e.g. The violations or planning errors) up to that point would have been noticed and rectified by communications. However, as the failures were most often based on the three failure types mentioned previously, this was not the case. The importance of communications to engineering work cannot be stressed enough, as the normal safety interlocking of signals is not in use, so all work is conducted by means of verbal and written communications.

Content Analysis

The results of the RCAT showed that poor communication played a prominent part in many of the incidents analysed. It became apparent through the reading of inquiry reports and the transcripts, that there might be something in the communications themselves which can be indicative of a failure. Therefore, it emerged that content analysis could be an appropriate technique to apply. This is a technique much utilised in the aviation industry (Cardosi, 1994, Burki-Cohen, 1995 amongst others) but which had never before been undertaken on the railways.

From previous work on assessment of communications and a literature search, a tool was developed to analysis communication content from three different perspectives: the quality of communications as determined by the rule book, the complexity of the communications and the rule which was violated. Both routine and incident engineering work transcripts were used, of the communications before the incident, so that comparisons could be drawn.

The results (Table 1) showed that there were no major differences in the quality of the communications, measured by the percentage of readbacks given, numbers spoken singly, phonetic alphabet usage, identification of self and location and the lead being taken. This is unsurprising given the results of the RCAT which showed that failures were grounded in more complex issues than those stated in the rule book.

Table 1. Percentage of rule book terms used

Rule book term	Incident transcript	Routine transcript
Messages repeated back	52	49
Rule book terms used	0	0
Numbers spoken singly	82	86
Phonetic alphabet used	16	0
Identification of self	69	63
Identification of location	38	25
Lead taken	0	0

However, statistically significant differences were found with regard to complexity of information. Firstly, a difference was found between the numbers of chunks of information in the routine versus incident transcripts. It was thought that the communication pre-incident would differ from the incident to the routine transcript. As measuring this was not easy, a simple way was to determine the 'chunks' of information in each communication, as a way of measuring complexity of each communication. Chunks were defined as being pieces of critical information such as signal numbers or mileages. The results showed that there was a relationship between chunks of information and incidents. Secondly, significant differences were found for levels of implicit readbacks. Implicit readbacks are where the critical details are not readback verbatim, but are merely inferred. Higher levels of these were associated with the incident transcripts. The likelihood of error is increased by the lack of clarity, lack of opportunity to notice any erroneous details and the raised propensity to make assumptions on the content.

Thirdly a correlation was found between higher numbers of implicit readbacks, and higher levels of complexity. Therefore, as the complexity of the communications gets greater, so the propensity for implicit readbacks becomes greater. It may be inferred that the more complex the information becomes, the less easy it is to repeat the details back explicitly, possible due to mental retention and storage problems and because rail communications can be rather verbose. Fourthly, a very significant difference was found in the levels of dysfluencies between the two transcript types. Dysfluencies are 'ummms' and 'errrrs', hesitations etc. which can be indicative of uncertainty of details. Far greater numbers of these were found in the incident transcripts, indicating that these can be an aural indication to operational staff that the other person may be hesitant or unsure of details they are providing. Finally a higher ratio of readbacks given to those expected was found in the incident compared to the routine transcripts. This result demonstrates the importance of readbacks in the comprehension and understanding of communications. Therefore, significant differences were found between transcript types and the implications of these are discussed in the next section.

With regard to the rule that was violated, there were distinct parts across the two types of engineering work (an engineering possession e.g.Tii and Tiii, two different ways of protecting work when normal train movements have been cancelled), as to where violations were more common, termed the Critical Failure Point (CFP). Although the sample size was small (n=11), the most common CFP was when communicating the signal protection. At this point, the signaller and operational staff should decide which signal will protect the work, how long for and any on-track protection which will be used. It is in the transferal of this information tat errors are most common. As a result of this section of the research, and the evident impact this could have on the rule book, a third piece of research was formed.

Critical Failure Point

It became evident that the same piece of work could be conducted using archive analysis of formal inquiry documentation. It was not so necessary for this piece of research to have transcripts, as formal inquiries provide enough information to ascertain where the CFP occurred in terms of the rule book. Analysis showed that again there were certain, commonly occurring rules where CFPs were found. With regard to Tiis possessions, the common CFPs were within protecting the line and communicating information before the work starts (which includes signal protection, as above). For Tiiis possessions the common CFPs occurred in communicating before a movement (of a train or an engineering machine referred to as an On Track Machine) takes place, the details regarding level crossing and the making of movements within a possession.

The results show that there are particular rules in the rule book which are more frequently not followed than others, with regard to communication. There is an aspect of these rules which makes them more susceptible to violations and the reasons for these needs to be established. It was not possible within the constraints of this research to answer this questions, but in light of the two pieces of work described above, the complexity of information to be transferred could be a cause, as could a lack of appreciation or understanding of the criticality of these details.

Implications for the industry

The implications of the outputs of these three pieces of research for the industry are considerable. Firstly, the RCAT found that understandings of communication failures that are common to the industry are incomplete and do not explore the failures at a sufficient level of detail to make changes. Moving away from the somewhat unstructured manner in which some parts of the railways communicate will require a significant cultural change and this can only be started by having a complete understanding of the failure types and the causes of these. The results also point to a lack of appreciation of the relationship between good communications and safe engineering works, given the number of incidents which have communication failures as a part and the number of occasions where better communications could have prevented an incident from occurring.

The Content Analysis result showed that the current quality of communications is poor for a safety critical industry and also that there are issues with the complexity of communications and readbacks. The implication for the industry is that consideration needs to be given to the way in which the rule book is structured with regard to the amount of information which is expected to be transferred. The rule book and awareness raising need to make explicit that information should be given in manageable chunks, and that after these are sent they could be explicitly read back to the sender. The sender then needs to make sure they are paying attention to this readback so any errors can be noted. Staff should also be made aware of the need to listen for any dysfluencies, as these may be an indication that an erroneous transmission is taking place.

Also, the violations analysis work has implications for the rule book. As there is a tendency for violations to take place with respect to specific points in the rule book, research needs to be undertaken to understand why this is occurring and whether any re-writing of the rules needs to take place.

Conclusions

This research has highlighted specific parts of safety communications on the railways which need to be improved. Until recently, there has been a distinct lack of research into rail communications. This research has raised awareness of both the complexity of the issues associated with communication failures and the way in which improvements can be made. Safety communications are of paramount importance during engineering works, due to the lack of standard safety interlocking, yet this criticality is often not reflected in the standard of communications. The results of this research are currently being implemented in the industry and it is hoped that they can be used to raise the standard and lower the incidences of incidents during engineering work.

References

Cardosi, K. M. (1994) 'An analysis of Tower Controller-Pilot voice communications.' *DOT/FAA/RD-94/15 Office of research and development, Washington, DC 20591*

Burki-Cohen, J. (1995) 'An analysis of Tower (ground) Controller – Pilot voice communications.' *DOT/FAA/AR-96/19 Office of research and development, Washington, DC 2059*

Murphy, P. 2001, The role of communications in accidents and incidents during rail possessions. In D. Harris (ed.), *Engineering Psychology and Cognitive Ergonomics Volume Five: Aerospace and Transportation Systems*, (Ashgate: Aldershot), 447-454.

Reason, J. (1990) 'Human Error'. *Cambridge*

MOTORCYCLE ERGONOMICS

MOTORCYCLING AND CONGESTION: BEHAVIOURAL IMPACTS OF A HIGH PROPORTION OF MOTORCYCLES

Sandy Robertson

Centre for Transport Studies, University College London,
Gower Street, London WC1E 6BT, UK
(020) 7679 1589

Following a study of the behaviour of motorcycles in congested conditions in the urban context, questions were raised with regard to the impact on behaviour of higher proportions of TWMVs in conditions where these vehicle types were the majority of road users. As under normal circumstances in the U.K. the proportion of TWMVs rarely exceeds 15 per cent, the location for filming the behaviour had to be at a place where there are special circumstances that generate both a high proportion of TWMVs and also a suitably high density of traffic. Qualitative descriptions of the behaviours were produced and quantitative measures of flows and of the density of TWMVs in queues were undertaken. Whilst, due to the nature of the sample, these measures are first approximations they do give some indication of the possible behavioural and other impacts of a move toward larger numbers of TWMVs on the road system.

Introduction

Motorcycling is an important mode of transport, which possibly has sustainable characteristics that until recently have been overlooked. To address this the Government has set-up an advisory group that has regular meetings with representatives of motorcyclists. DTLR commissioned study to look at the impacts of Two Wheeled Motor Vehicles (TWMVs) on congestion. Halcrow Fox & the Centre for Transport Studies at UCL have undertaken this work which is reported in Martin, Phull and Robertson (2001). This paper describes some of the behavioural work undertaken as part of an extension to that study. The objectives of the extension study were to obtain:

- A qualitative description of the ways in which TWMV behave when in high concentrations to identify how TWMV interact with each other and with other road users.
- Flows of TWMV in free-flow or near free flow conditions.
- The density of TWMV traffic in queues where they form.

The earlier behavioural work (Robertson 2002a and 2002b) focused primarily on typical urban commuting situations. It identified a number of behaviours that were exhibited by TWMVs.

Whilst the work reported here had the dual aims of investigating traffic flows and investigating behaviour, this paper focuses on the behavioural aspects. To investigate behaviours of high concentrations of TWMVs, a location with such concentrations was required. Potential sites included motorcycle shows and race meetings. Race meetings were rejected on the grounds that behaviour on the roads after being a spectator at a racing even might be modified. In order to get a reasonable sample a large show where many attendees arrived by TWMV was required. The BMF show at the East of England Showground is the largest motorcycle show in Europe that attracts enthusiasts and other motorcycle riders. There are a large number of TWMV on the roads around the East of England Showground during the weekend of the show. Anecdotal evidence suggested that there could be up to 90% TWMV on some roads around the show. In 2001 on it was held on Saturday 19th and Sunday 20th May. According to the BMF it attracted in the order of 30,000 participants on 19th May and 50,000 participants on the 20th May. According to organisers and the police, the traffic management system for the BMF show has evolved over the years and now provides a relatively congestion free approach to/exit from the show. TWMVs were segregated from other traffic and sent on different routes into the Showground. This gave the opportunity to film traffic with a high proportion of TWMVs in the traffic stream. Where traffic became clogged, other vehicles were sometimes directed along the routes primarily used by TWMVs. This allowed some filming of mixed traffic.

As far as it is known, no other studies have been undertaken in these conditions of high concentrations of TWMVs and it was uncertain what types of behaviours might be seen. While the use of the show provided a large concentration of TWMVs, there were some caveats with regard to the generalisability of the results. As the show attracts enthusiasts to a leisure venue, it was clear that the sample taken might not reflect the overall population of TWMV users. The very small proportion of scooters, (compared to the commuting situation), that were observed confirmed this. The attendees might also not reflect the general population of TWMV users in terms of their use of the road space. Additionally, as the show was a leisure activity, there might be less time pressure on attendees (compared to say a commuting journey). Hence, the behaviours of the road users might be different from that observed in rush hours in the city. The times and locations for filming were quite critical and should this exercise be repeated, the use of multiple cameras and filming locations would be of great benefit.

When considering the behaviours reported here, one must remain mindful that the sample of riders in this part of the study are not representative of the overall population of TWMV users. This is neither in terms of the observed vehicle mix nor journey purpose. We cannot, therefore, validly generalise the findings to the entire population of TWMV users, but some of the behaviours may well reflect the behaviours in the general population and provide guidance as to what is possible. The analysis reported here is qualitative.

Method

Observations were taken by roadside video at the BMF show 19th and Sunday 20th May 2001. A total of about 15 hours of video was recorded over the two-day period. Due to the tidal nature of the traffic not all of the locations chosen for filming yielded useful data. A qualitative analysis of the behaviours was undertaken as descriptions of the behaviours were required. 5 minute traffic counts were taken where appropriate and queue density was also investigated. Sites observed included dual carriageway and single carriageway roads. The most dense traffic was recorded on at a site on a single carrageway road on the approach to a pelican on the immediate approach to one of the entrances to the show. On the morning of the 20 May during a three hour observation period, an estimated 7000 vehicles passed. This had a proportion of

TWMVs of between 94 and 98 per cent with flows of between 1000 and 3000 vehicles per hour.

Behaviours in flowing traffic.

There were a number of ways in which TWMVs were observed to use the roadway. These are described below.
The major forms, shown in Figure 1 are:

- Single file- where the TWMVs ride one behind the other.
- Echelon- where the TWMVs are staggered so that the effective distance between TWMVs (i.e. front wheel to back wheel of the TWMV directly in front) is greater than the distance between the nearest TWMVs
- Side by side- where the TWMVs ride parallel to each other

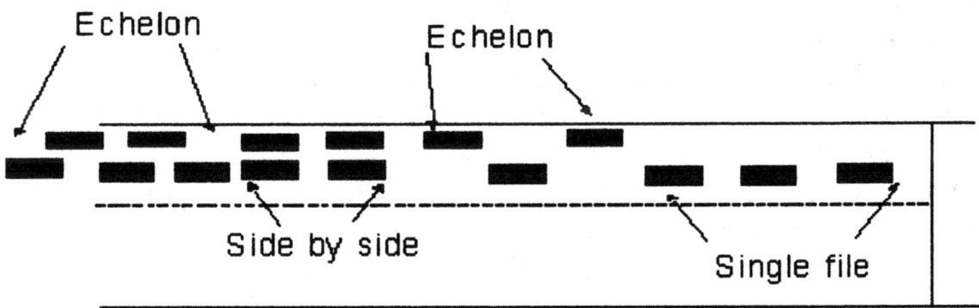

Figure 1. TWMV riding formations in TWMV only traffic.

Figure 1, above, shows the forms of roadspace utilisation for two lines of TWMVs The same principles apply to three or more lines of TWMVs. At low flows TWMVs were generally observed to be riding in single file or in echelon. As traffic flow increased so did the tendency to ride in echelon and/or side by side. There were changes in the ways in which TWMV users make use of the road space.

There appeared to be a switch between the conventional use of a single lane by a single line of vehicles under light traffic conditions to a more efficient use of road space under increasingly heavy traffic flows. With the limited data available it was not possible to determine the point at which this switch occurred and further work would be required to determine this. TWMV users appear to adapt their behaviours to match the road and traffic conditions. If it is assumed that the behavioural characteristics of the TWMV road users are similar to those of other road users

Lane discipline and other interactions between road users

In general, it was clear that lane discipline was very good when the traffic stream consisted entirely of TWMVs and all TWMVs remained on the nearside of the centreline with very few instances of overtaking even though there was virtually no traffic on the other carriageway. On some occasions when cars were present in the traffic stream, a proportion of TWMVs were observed to use conventional overtaking to pass the cars, then revert to strict lane discipline. This was particularly interesting in that there appeared to be at least two approaches to interaction with other road users dependent on the circumstances. Clearly there are implications for the modelling of flows of traffic and implies that TWMV users' behaviours attempt to maximise the capacity of a section of road. This also suggests that any models of TWMV user behaviours need to have their boundary conditions clearly defined.

Behaviours in queuing

Most of the observations of queuing were undertaken on the approaches to a pelican crossing. Queues formed both as a result of the use of the crossing by pedestrians and also due to the queue backing up from the entrance to the showground. Riders tended to respond to the change in signal state by slowing down some distance upstream from the signal so that relatively few actually stopped at the lights. Where they did so, in many cases a gap was left upstream from the stopline to allow other road users coming out of the adjacent side roads. While this caused difficulty for measuring queue length and density, the behaviours exhibited appeared to be generally oriented toward a smooth, courteous and safe riding style. The issue of responding in advance to the signals has implication for emissions in that by reducing the amount of braking and acceleration the levels of emissions may be reduced. It is not clear how different these riding styles are compared to the general population engaged in non-leisure activities.

One issue that may be related to the tendency to avoid coming to a standstill where possible is that TWMVs are easier to handle when they are moving. When stopping, the rider must change from a mode of riding in which the dynamics of the machine ensure stability to one in which stability is maintained by placing the feet on the ground and actively balancing the vehicle. In mixed traffic TWMVs would, on some occasions, queue on the offside of a car (when sufficient space was available).

The observed forms of queuing included side by side and echelon forms as shown in Figure 2. From the observations it was not clear when the different forms were used, other than side by side layouts being used when the riders wished to talk with each other while waiting.

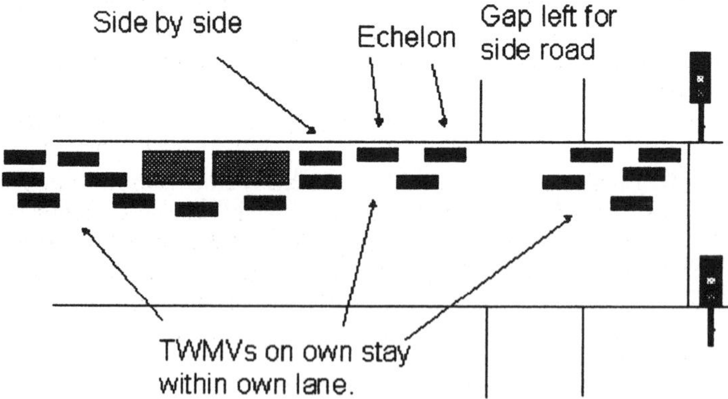

Figure 2. Types of queuing.

Discussion

The sample of the types of TWMVs obtained in this survey was not the same as the earlier urban surveys (Robertson 2002a,b, Martin, Phull and Robertson 2001) during peak hours. The distribution of vehicle types was clearly different as identified by the very small proportion of scooters. Observation of the type of vehicle at the roadside suggested that there were a considerably higher proportion of large (both physically and engine capacity) motorcycles at the BMF show than in rush hour traffic. This clearly has implications for queuing and possibly for observed flows of traffic. The impression given from roadside observation (for which quantitative data is not available to confirm) was that there were a high proportion of the more expensive type of vehicle. It may be suggested that the riders of such vehicles (which at the higher end of the market may be well in excess of £10,000 each) may be more cautious of getting into close proximity with other vehicles. Due to the limited number of days sampled and the specific time of year that this study was undertaken, neither seasonal variations nor effects of weather could be taken into account.

TWMV users appear to have a range of ways in which they utilise roadspace. They appear to modify the way they use the available roadspace on the basis of the traffic conditions. In light flows the TWMVs adopt single file or widely spaced echelon formations. As flows increased, the TWMVs increasingly used echelon and side by side formations. The echelon and side by side formation allows the TWMVs to retain a reasonable headway to the vehicle immediately ahead of them, but the capacity of the road (for TWMVs) is increased. On a few occasions the way in which the TWMVs used the road space were observed to change depending on the mix of traffic. Where TWMVs were present on their own, virtually no passing was observed. When some cars entered the traffic stream, TWMVs were observed to overtake the cars then continue in their place in the stream of TWMVs.

This change in the use of the road system and roadspace and the interactions with other road users suggests that road users do modify their behaviour (in terms of their use of roadspace) as traffic conditions change. This implies that development of traffic models that include TWMVs may need to treat TWMVs differently with different traffic levels, and that models which have been calibrated using low TWMV flows may not be reliable as the flow (and proportion) of TWMVs increases. The relatively small period of filming used in this study has precluded a detailed quantitative analysis. Such a quantitative analysis will be needed in future to allow a better understanding of the impact of the interaction between road users on network performance.

While the sample used in this study may not reflect the general population of TWMV users, it has provided a number of insights into the potential capacity of roads where TWMVs are the dominant mode. TWMV users appear to be able to make highly efficient use of roadspace and to respond to changes in traffic conditions by modifying their behaviours. Further work is needed to look at queuing in more detail and in particular the relationship between the behaviours of TWMV users and the roadspace taken up by them in queues. From observations of this sample, the presence of a small number of a different type of road user within a stream of traffic consisting of primarily one type of road user appears to elicit changes in behaviour. Further work is needed to investigate this.

Given the increasing numbers of TWMVs on the roads it is clear that knowledge of the ways in which TWMVs react with the road system and with other road users in the UK context will be important for the development of the road system in the coming decade

The views expressed in this paper are those of the author alone.

Acknowledgements

The author wishes to thank S. Phull of DfT for permission to publish this paper. The author wishes to acknowledge the assistance of the Simon Wilkinson of the BMF, and Inspector Nick Knight for facilitating this work at the BMF show.

References

Martin, B, Phull, S, and Robertson, S 2001, Motorcycles and Congestion. P*roceedings of the European Transport Conference* 10-13 September 2001 Homerton College Cambridge. Proceedings published as a CD, PTRC, London

Robertson, S, 2002a Motorcycling and congestion: Definition of behaviours. In P.T. McCabe(ed.) *Contemporary Ergonomics 2002*, (Taylor and Francis, London), 273-277

Robertson, S, 2002b Motorcycling and congestion: Quantification of behaviours. In P.T. McCabe(ed.) *Contemporary Ergonomics 2002*, (Taylor and Francis, London), 278-282

R^3 – A RIDING STRATEGY FORMULATION MODEL FOR THE RISK AVERSE MOTORCYCLIST

Steve McInally

Defence Engineering Group, University College London, Gower Street, London WC1E 6BT, UK

Motorcycle riding is a socially and economically important activity, but it is complex and demanding with many attendant risks. Depending on their motivation and level of skill, riders balance risk with progress to differing degrees. Skills are developed through a process of combining theory and training with the rider's own experiences. This paper proposes models for the relationships between rider, motorcycle, environment and journey; and goes on to describe the R^3 model for individual motorcyclist skill development.

Introduction

The motorcycle, with its distinctive features and abilities, has many roles to play in modern society. In the UK alone there are currently over 700,000 licensed motorcycles, scooters and mopeds. Motorcycles are compact, agile, consume less fuel, and are cheaper to buy and maintain than automobiles. Commuters, couriers and the emergency services use motorcycles to make progress through increasingly congested towns and cities where automobiles would struggle (RoSPA, 2001, 2.8. p3).

Despite these advantages, motorcycling is a relatively complex and risky activity. Recent reports from the UK Government Department of Transport's Driver Standards Agency (DSA) (DSA, 2000) and the Royal Society for the Prevention of Accidents (RoSPA) (RoSPA, 2001) show that a disproportionately high number of motorcyclists die or are seriously injured every year compared with car drivers. Both reports suggest that training plays an important role in reducing the numbers of accidents.

Training and Skill Development

There is recognition by government and motorcycling interest groups of the need for improved rider training, particularly for those new to riding, and those returning to riding after a long absence (DSA, 1999), (RoSPA, 2001). For more experienced riders, there are a number of charitable organisations like the Institute of Advanced Motoring

(IAM) and RoSPA, and commercial groups such as the British Motorcycling Foundation (BMF), that offer advanced training.

The UK motorcycling press also plays an active role in promoting skills development. Aside from the many advertisements for advanced training, track days, and 'off-road' experiences, the motorcycling press also regularly publish articles on how to improve riding skills. For instance, the UK's leading weekly motorcycling newspaper Motorcycle News (MCN) recently published a supplement with the title "Common riding mistakes and how to avoid them: 50 ways to improve your riding" (MCN, 2002).

Although training clearly plays a role in skill development and safety, most riders hone their skills through the day-to-day experience of riding.

Balancing Safety with Progress

It is assumed that most riders in most circumstances wish to make adequate progress whilst maintaining a reasonable level of safety, and that the balance between risk and progress depends on the type or urgency of the journey. For instance, the motorcyclist out for a leisurely ride on a sunny Sunday morning is less concerned about completing his or her journey quickly than a paramedic or police rider heading to the scene of an accident.

Not all motorcyclists will have the same degree of skill in balancing risk with progress. Consider figure 2. At point (*a*), the novice rider is making modest progress (*p1*) with a certain level of safety (*s1*), but the expert motorcycle rider, such as a trained paramedic or police motorcyclist, will be able to make either better progress (*p2*) for the same level of risk as a novice (point (*c*)) or ride at the same level of progress while improving levels of safety (point *b*, *s2*).

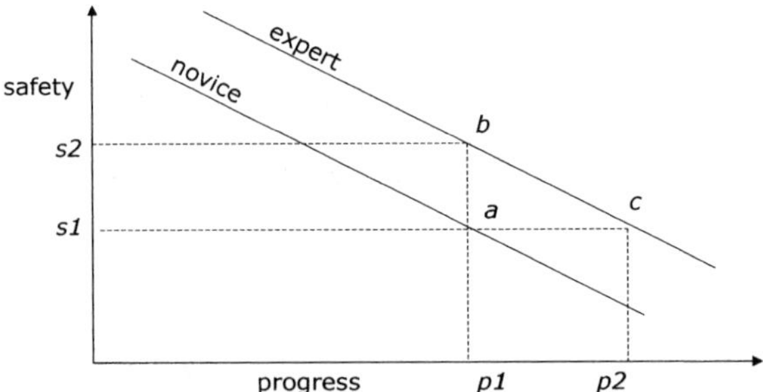

Figure 1 Experts can go faster and/or safer than novices

Note that the term 'progress' does not refer simply to vehicle speed. Progress refers to the elapsed time over a particular journey. The more complex the journey, i.e. the more hazards encountered, the greater the opportunity to smoothly and efficiently navigate those hazards, effectively making good progress.

Relationship between Rider, Bike and Environment

There are many complex interactions between rider, machine, and environment by way of sensory, physical and mechanical interfaces (figure 2).

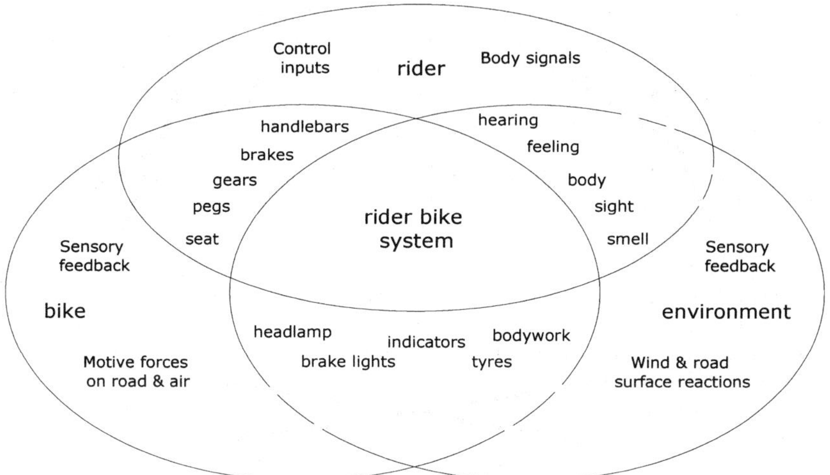

Figure 2 Rider/bike/environment interfaces and interactions

Given the sheer numbers of actions and reactions required of the rider, even the simplest manoeuvre is demanding. The challenge is further increased by the fact that the rider must carry out these complex actions and reactions over the whole journey. The more hazards encountered, the more decisions and manoeuvres to be made, the greater the demands on the rider's skills to ride safely and progressively.

As illustrated in figure 3, the novice or inattentive rider (a) may be unaware of the exact nature of a particular hazard, and ride through without paying much attention. The more expert rider (b) will perceive the hazard and pay more attention.

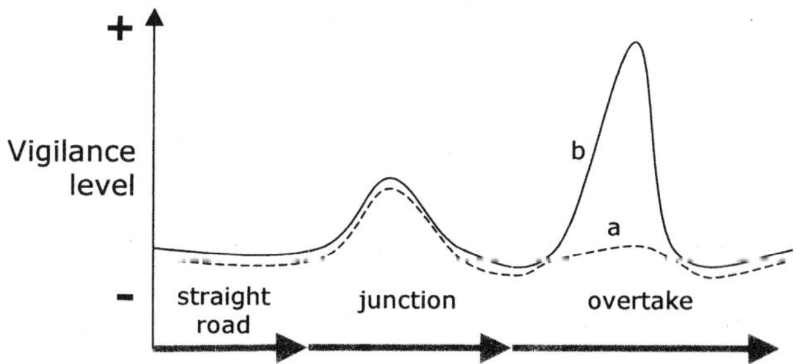

Figure 3 Vigilance levels vary for novice and expert

Although important, vigilance itself is not enough to avoid an accident. The rider must also choose an appropriate risk-reducing course of action in response to the perceived hazard and apply it at the right time.

R^3 – Risk, Reaction, Review

Riding skills can be developed in a number of ways, for example by learning from more experience riders; through experimenting with ideas from the motorcycling press; by generalising from car and bicycling experience; through formal training; and through personal experience.

During formal training, the rider is provided with theories, examples and demonstrations of safe, progressive riding. The trainee is then encouraged to emulate these best practices under supervision. The role of supervision then is to verify that the trainee rider's behaviour is consistent with the taught theory. However, it is possible for the novice to emulate instances of 'good' riding without fully understanding its importance and value. This is learning by *rote* or *parrot fashion*. The risk with rote learning is that the trainee does not understand the general applicability of examples and theory.

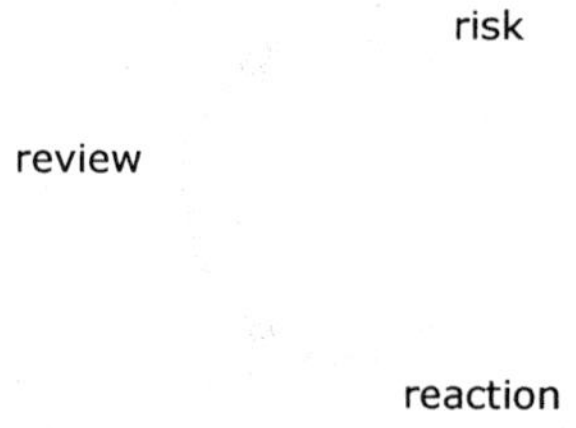

Figure 4 R^3 model

The proposed R^3 model in figure 4 is a structured way of exploring the meaning and value of theory during events in the course of the rider's own journey. The basic idea is that the rider explores three concepts continuously throughout a journey. The name for each concept is spoken aloud as a question or order to the self, cyclically and in a particular sequence.

"**Risk?**" This is the first concept and is always the starting point. The rider scans the environment, looking particularly at the forward field of view, exploring the changing evolving features of the environment for any hazards, that is 'threats to my health'. While scanning, the rider imagines what might happen at a particular feature; for instance, two rows of slow moving cars appear in the carriageway ahead. The question arises 'what do I know about this feature?' Depending on the rider's level of experience, a typical response might be 'cars change lanes unexpectedly'. The imagination may then provide a vision of the rider colliding with the car. There is an anxiety response, so the situation is deemed to be risky.

"**Reaction?**" Having 'seen' a vision of danger, the next question is what to do in response to this risk. The rider selects a strategy that maximises progress while minimising risk. In the case of the slow moving lines of cars, the rider may choose to cover the horn and the front brake lever, increase his or her vigilance of car movements, place the bike centrally between the rows of cars, and decrease speed.

"**Review!**" The review element is not as immediately important as the Risk/Reaction aspects of the R^3 model, but it can certainly help with the learning process. The idea is that the rider glances back to determine whether the chosen course of action was justified. In the case of filtering through two rows of cars, a glance back at a 'suspicious' car might reveal that it did in fact quickly change lanes, thus confirming the need for heightened vigilance and caution.

A feature of the R^3 model is the opportunity to introduce theory at Risk and Reaction phases.

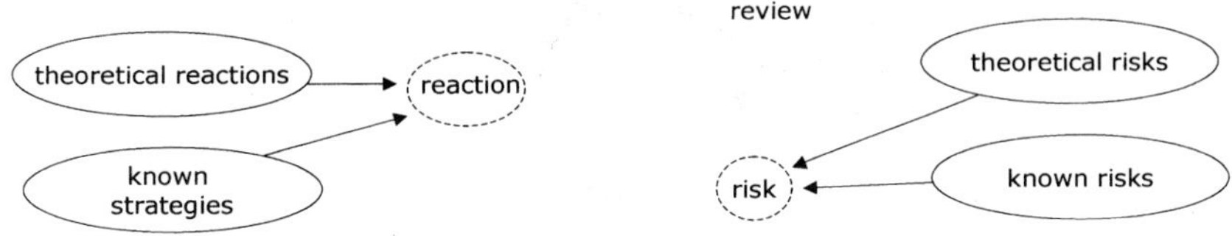

Figure 5 R^3 model adapts to theory

When scanning the environment for potential hazards in the Risk phase, two types of situations may come to mind, those which the rider has personally experienced (known), and those he or she has read or heard about (theoretical). For instance, whilst filtering through stationary traffic the rider may have encountered a car pulling out in front of them, but only heard about car doors opening unexpectedly.

Similarly in the Reaction phase, when deciding a suitable course of action, there are those manoeuvres that the rider has tried and tested, and those he or she may be aware of but not yet explored. For instance, in order to change direction quickly the rider might typically shift weight on the bike, but might have heard that pushing the handlebars in the opposite direction of travel is significantly quicker at higher speeds in an emergency.

Conclusions and further work

The approach is useful in that it helps riders to be conscious of hazards, to formulate avoiding manoeuvres, and to review the effectiveness of analysis and response, but it is not enough in itself to guarantee the safe progressive riding. It is best applied in conjunction with other riding systems, such as the police rider's system of motorcycle control (HMSO, 1978), the IAM's advanced driving method the RoSPA advanced driving method, or the Keith Code method (Code, 1997).

The ideas and method proposed in this paper are based on the author's own experiences as a motorcyclist (who has the occasional accident!) and researcher with an interest in systems engineering and experiential learning models. The ideas are strongly influenced by Lewin's concept of Action Research (Lewin, 1946), systems engineering principles (Hambleton, 2000), the UK army's OODA loop (observation, orientation, decision, action) (JDCC, 2001) and Kolb's four stage learning cycle (Kolb, 1984).

When researching this paper, it seemed that there was little published on motorcycle skill development and how training is designed, despite its social and economic importance, and of the obvious safety critical nature of motorcycling. Given these issues, more research into how motorcyclists behave and learn would be worthwhile, particularly in the areas of:

- Applying systems engineering and systems thinking principles to the development of behavioural models and training techniques.
- The application of educational and developmental models to the problem of learning to ride a motorcycle.
- A detailed study of how the motorcycle brings social and economic benefit.
- The role of cognitive kinaesthetic learning in motorcycle riding.
- The role of the motorcycling press in developing a culture of continuous skill development.

References

Code, K. (1997), '*A Twist of the Wrist*', Code Break Incorporated,

Driving Standards Agency, (1999) '*Safer Motorcycling*', UK Department of Transport consultation paper, http://www.roads.dft.gov.uk/roadsafety/dsa/consult/motorcycle/index.htm

Driving Standards Agency, (2000) '*Safer Motorcycle and Moped Riding*', UK Department of Transport consultation response report, p11, 20th December http://www.dsa.gov.uk/dsa/consult/safer_mc_moped_riding.htm

Hambleton, KG, (2000) '*Systems Engineering - an Educational Challenge*', Ingenia, November 2000, Issue 4, pp 61-67, The Royal Academy of Engineering, London.

Joint Doctrine and Concepts Centre (2001), '*British Defence Doctrine*', UK Ministry of Defence, Swindon.

Kolb, DA. (1984) *Experiential Learning*, Englewood Cliffs, NJ.

Lewin, K. (1946) 'Action Research and Minority Problems' Journal of Social Issues 2.

MCN, (2002), "C*ommon riding mistakes and how to avoid them: 50 ways to improve your riding*." 25th Sept. 2002, EMAP Automotive Ltd, Peterboroough.

Royal Society for the Prevention o f Accidents, (2001), 'Motorcycling Safety Position Paper', February 201, RoSPA, Bristol, http://www.rospa.org.uk/pdfs/road/motorcycle.pdf

UK Gov. Home Office, (1978), '*Motorcycle Roadcraft*', 14th impression 1992, HMSO, London

Bibliography

Gross, R.D. (1992), '*Psychology: The Science of Mind and Behaviour*', Hodder & Stoughton, London

DESIGN

EVALUATING MEDICAL EQUIPMENT FROM THE OUTSIDE-IN AND FROM THE INSIDE-OUT: RESCUE STRETCHERS

RS Bridger, E Bilzon, A Green, K Munnoch, R Chamberlain and J Pickering

Institute of Naval Medicine, Crescent Rd, Alverstoke, Hants PO12 2 DL, UK
hhfd@inm.mod.uk, 02392 768220

The paper reviews 2 studies carried out to investigate alternatives to the "Neil Robertson" stretcher (NRS) which has been in service in the Royal Navy for nearly 100 years. Having been selected using structured methods, four stretchers were evaluated, firstly in a ship board usability trial and secondly in a laboratory-based investigation of their capacity for cervical stabilisation. The ship-based trial involved carrying an 80 kg mannikin in each stretcher around a circuit simulating a route to a sickbay. Two of the candidate stretchers were considered superior to the NRS on a number of attributes. In the second trial, it was found that one stretcher provided the best stabilisation of the head, offering good usability while, at the same time, the greatest restriction of head movements.

Introduction

Ergonomic evaluations of medical equipment may require that the needs of two groups of "users" be taken into account - the medical or paramedical staff using the equipment to treat patients and the patients themselves. The issues pertaining to these "users" will differ. Ease of use and clinical effectiveness may not always be compatible. This paper presents the results of two evaluations of rescue stretchers that were carried out to identify a suitable replacement for the Neil Robertson Stretcher (NRS), which has been used by the Royal Navy for almost 100 years.

The NRS is fabricated from wooden slats and canvas and is believed to be a modification of a Japanese stretcher, made from bamboo (Figure 1). The NRS was designed to enable the casualty to be packaged into the smallest possible volume. It enables casualties to be lifted vertically, with minimal slippage, and is ideal for transporting patients through small hatches when traversing the decks of ships. However, because the NRS lacks rigidity, there is some concern about its suitability for transporting casualties with spinal injuries. Bilzon and Brace (1999) concluded that any replacement stretcher must have all the advantages of the NRS while providing better protection for the spine of the casualty. Further investigation revealed that there are already products on the market that claim to fulfill this requirement.

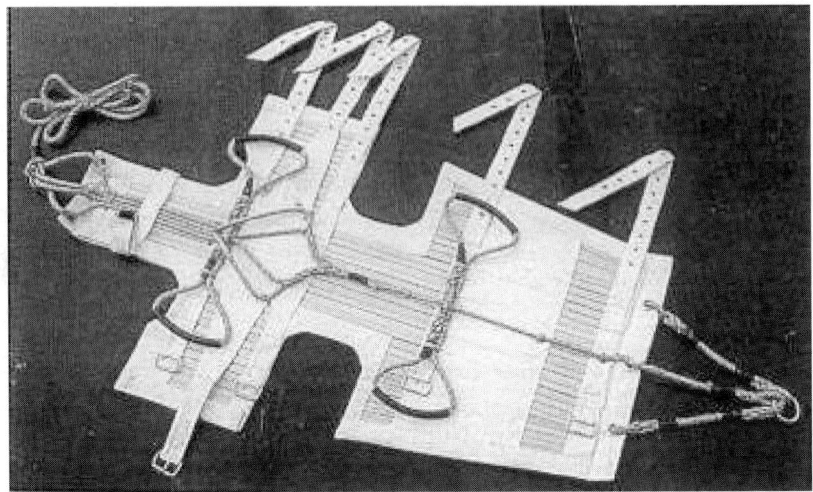

Figure 1. The Neil Robertson Stretcher

Three candidate replacement stretchers were identified, Stretcher 1 was made of plastic composite and unrolled to form a semi-rigid half-cylinder in which the casualty was secured using colour-coded straps. Stretcher 2 was based on a similar design principle but was made of ballistic corduroy. Stretcher 3 was fundamentally the same as the NRS in its design but was made of 18oz vinyl laminated nylon. An important feature was its "sleeve" design, which enabled a spinal board to be fitted, thus providing a firm base for head blocks to immobilise the cervical spine.

Ship-Based Usability Trial

A usability trial was carried out on HMS BRISTOL (Bilzon et al. 2002). Fifteen medical assistants participated. In stage 1, the trained subjects secured a 80kg mannikin into each stretcher. In stage 2 , the subjects were split into 2 groups (of 7 and 8). Six of the subjects acted as stretcher bearers, one as medical assistant and the remaining subject acted as an observer. The 80 kg mannikin was carried in each stretcher around a circuit to simulate a typical route to a sickbay. Time did not permit the groups to carry all the stretchers, in turn. Instead, the NRS and stretcher 3 were carried by one group and the remaining stretchers by the other group.

Securing the Mannikin.

In general, the time to secure the mannikin was shortest for Stretcher 3. It was found to be very simple to use and the groups quickly learnt how to secure the mannikin, asking very few technical questions. Stretchers 1 and 2 were found to be more complicated and took longer to secure.

Simulated Sick Bay Carry

Stretcher 3 was well-liked for its spinal board and ease of use. Stretcher 2 was regarded as being the most secure during casualty transfer. Table 1 summarises users' ratings on a number of stretcher attributes. The Freidman test was used to establish the significance of any differences between the stretchers. Highly significant differences were found between stretchers for ratings of durability, safety, quality, size and ease of use. Users rated stretcher 2 the most highly, but stretcher 3 was also highly respected and was quicker and simpler to use.

Table 1. Number of Subjects with Unfavourable Ratings for Each Stretcher Attribute (n=15)

Primary Attributes	Stretcher 1	Stretcher 2	Stretcher 3	NRS
Durability	6	1	8	15
Safety	11	3	11	15
Quality	11	10	7	15
Size	12	5	12	15
Manoeuverability	4	1	2	10
Ease of Use	14	8	7	15

Laboratory Study: Neck Posture and Head Movements

Discussions with domain experts revealed that one of the critical features of a rescue stretcher was its ability to stabilise the neck of patients with suspected cervical spine injury during rescue. Cervical collars are normally applied in these cases. Spinal boards are also applied at the earliest opportunity. Evidence from trials carried out on healthy volunteers indicates that cervical collars, used alone, do not immobilise the neck. Rosen et al. (1992) used goniometers to compare cervical range of movement (ROM) of volunteers wearing a variety of commercially available cervical collars. Two measurement devices were used; a head goniometer with scale marks in increments of 5 degrees and a hand held goniometer, consisting of a stationary arm and a moveable arm fitted onto a central scale, with increments of 1 degree. Measurements were read off the scales of the goniometers when the head reached the limit of its range in flexion, extension, lateral flexion and rotation. The measurements were made with the subjects sitting erect, to simulate extraction from a vehicle after a road accident. All of the collars permitted considerable neck movement in some cases, this was little difference from the free range of movement (from 21-39 degrees of neck flexion, 26-47 degrees of extension, 28-62 degrees of lateral flexion and 31-106 degrees of rotation).

In the present study, head movements were measured using a manual method and a visual method. In the manual method, a gravity goniometer (MIE Medical Research Ltd.) scaled to 1 degree was used to measure flexion and extension and a protractor (scaled to 5 degrees) was used to measure lateral flexion and rotation. In the visual method, head movements were captured on video and then digitised using the Peak Motus ® system. In both cases, measurements were made of head movements in order to obtain indices of flexion, extension, lateral flexion and rotation of the cervical spine.

Subjects

Eleven subjects (6 females and 5 males) participated in the trial. Ethical approval was obtained from MOD(N)PREC and all subjects gave their informed consent. All were trainee medical assistants in the RN with minimum P2 fitness.

Calibration of Neck Postures

For lateral flexion and rotation, the neutral posture of the neck is extant, due to body symmetry. For flexion and extension, *the neutral posture of the neck occurs when a standing person looks straight ahead in a relaxed posture*. The angle of the Frankfort plane with respect to the horizontal is used as the reference point for neutral in flexion and extension (the Frankfort plane is defined by straight lines extending from the upper edge of the external auditory canal to the lower rim of the ipsilateral orbit). In the present investigation, the ears were often covered when the subject was supported in the stretcher, so it was not possible to use the Frankfort plane as a reference for head posture. Instead, head position was calibrated using the goniometer. Subjects adopted a relaxed, standing posture and looked straight ahead so that the neck was in the neutral position. The goniometer was placed with one of its feet on the forehead and the other on the bridge of the nose. The moveable scale was then set to zero degrees. Subjects then flexed and extended the cervical spine 3 times with the goniometer held in place. The ROM of flexion and extension was measured manually and visually each time. When subjects were secured in the stretchers and were wearing the collar, their head position was measured again (Figure 2). Since the subject's body had been rotated by 90 degrees (the stretcher was horizontal) the position of stabilisation of the head was given by the difference between the actual measurement and 90 degrees. Angles greater than 90 degrees indicated *stabilisation in cervical extension.*

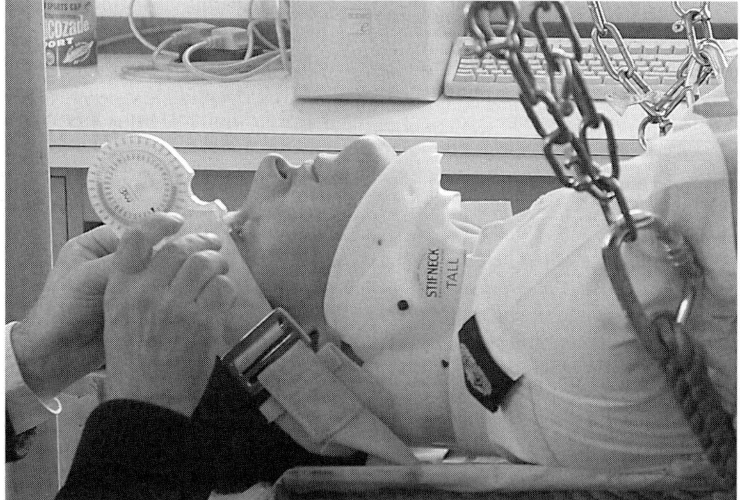

Figure 2. Measuring the position of stabilisation of the head in the Neil Robertson stretcher

Procedure

All measurements were taken in the same session for each subject. After the experimenters had explained the procedure and received informed consent, subjects were calibrated in the standing position. Next, the free ROMs of flexion/extension, lateral flexion and rotation were measured in the standing position. The head movements were videotaped at the same time as the manual measurements were made. The protractor was placed around the subject's neck with the nose pointing to 0 degrees. Rotations to the left and right were read-off the protractor using the nose as a pointer. Lateral flexion was estimated by placing the protractor behind the subjects head with the crown of the head pointing to zero. Lateral flexion was estimated by visual reference of the head with respect to the protractor. *All measurements were repeated three times.* Subjects were then fitted with a cervical collar and secured in each stretcher in a horizontal position. The position of stabilisation of the head was recorded each time. Measurements of flexion/extension, lateral flexion and rotation were then made, both manually and visually (Figure 3).

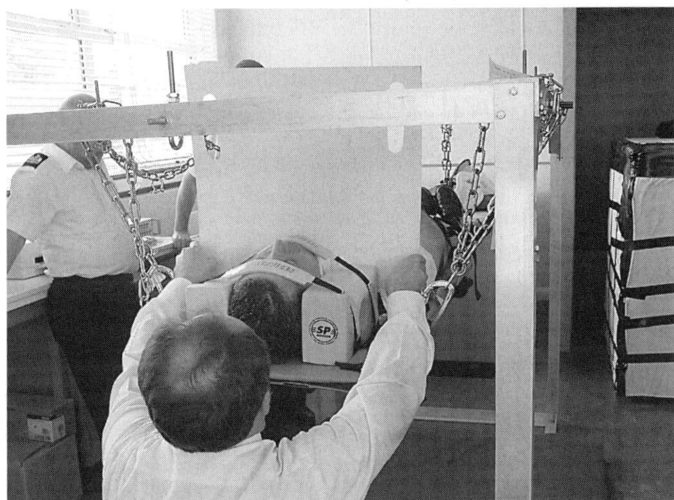

Figure 3. Measuring rotation using the protractor. The subject lies in Stretcher 3 and the head blocks are clearly visible. The video camera for visual measurement of rotation is behind and slightly to the left of the experimenter

Results

The methods were found to provide estimates of cervical ROM that were sufficiently repeatable to permit further statistical analysis. In general, the visual and manual methods gave similar results but only the manual methods are shown here due to space limitations.

Resting Position of the Head

In all cases, the neck was in extension when the subjects reclined in the horizontal stretchers. The mean angle of extension using the manual method was 18 degrees in stretcher 3 compared to 28 degrees in the NRS with intermediate values for Stretchers 1 and 2.

Cervical ROM in the Stretchers

There were statistically significant differences ($P<0.01$) between the stretchers for all head movements. In general, Stretcher 3 provided the most limitation in movement and was particularly effective in limiting lateral flexion and rotation. This is almost certainly due to the provision of head blocks, integrated into the basic design (Figure 3).

Discussion

None of the stretcher/collar combinations can be said to have "immobilised" the cervical spine in any of the anatomical planes of motion. Only Stretcher 3 came close. It had the lowest average ROM values of all the stretchers. This is because head blocks are integrated into the stretcher design. This stretcher/collar combination appeared to be particularly effective in reducing lateral flexion and rotation. According to McSwain (1989), the ideal is to immobilise the neck to within 11 degrees of the neutral position in cases of suspected cervical spine injury. In all stretchers the subjects' necks went into extension because none of the stretchers made allowance for the thoracic kyphosis. All of the stretcher bases were approximately flat and, with the thoracic spine resting on the base, the cervical spine had to extend for the occiput to come into contact with the base. The head blocks fitted to Stretcher 3 were effective in limiting some of the head movements but they were not effective in preventing the initial extension of the neck as the subject was placed in the stretcher. It is not surprising then, that none of the stretchers were able to stabilise the head such that the neck was within 11 degrees of the neutral position. On ships, stretchers have to be carried through tight spaces and it may not always be possible for the paramedics to stabilise the head manually. Of the stretchers evaluated, Stretcher 3, with its combined back board and head blocks provided the most stabilisation and would be the most suitable. Usability trials indicated that Stretchers 2 and 3 were the better stretchers. Fitting of the head blocks on Stretcher 3 was not an onerous task - patients could be secured in it more quickly than in Stretchers 1 and 2.

Conclusions

Stretcher 3 limited head movements more than the other stretchers. When used only with a stretcher, the cervical collars restricted head movements but they did not immobilise the head and neck. The head blocks on Stretcher 3 were effective in limiting lateral flexion and rotation of the neck. In all stretchers, the head was stabilised with the neck in an extended posture (beyond 11 degrees of extension). Neck posture in stretchers may require further investigation. Stretcher 3 combines a backboard with much of the simplicity and functionality of the NRS so there would seem to be little to be gained by redesigning or updating the NRS itself. If maximum immobilisation as close to the site of injury as possible is required, Stretcher 3 seems to be better than the alternatives tested here. Ease of use and clinical effectiveness do seem to be compatible in the case of Stretcher 3, which should be available on RN ships for cases of suspected neck injury.

References

Bilzon, E, Brace, C. 1999. A Gap Analysis of the Requirements and Provision of RN Rescue Stretchers. Unpubliushed MoD Report.

Bilzon, E., Munnoch, K., Chamberlain, R., Pickering, J. 2002. Ship-Based Usability Trial of Rescue Stretchers. Unpublished MoD Report.

Rosen, P.B, McSwain, N.E., Arata, M., Stahl, S., Mercer, 1992. Comparison of two new immobilisation collars. Annals of Emergency Medicine, 21, 1189-1195.

McSwain, N.E.Jr. 1989. Acute management. In McSwain, N.E.Jr., Martinez, J.A., Timberlake, G.A. (eds). Cervical Spine Trauma: Evaluation and Acute Management, New York, Thames Medical Publishers, 105-118.

A PAIN IN THE REAR: NEW UNDERGRADUATES' UNDERSTANDING OF ERGONOMICS IN DESIGN

Ian Storer, Deana McDonagh and Howard Denton

Department of Design and Technology
Loughborough University, Loughborough, Leicestershire, UK. LE11 3TU

Abstract

In a university context it is important to understand and appreciate prior learning and attitudes when planning teaching and learning experiences. This paper presents a study undertaken with new industrial design undergraduates with particular reference to their knowledge and application of ergonomic principles in a design task. Design teams of 3 students were established and given the task of *mind mapping* their approach to the design of a new bicycle seat. The mind maps were analysed in relation to the general fields emerging (e.g. initial ideas, research, materials and ergonomics). Follow up interviews were held to explore issues raised in more depth.

Findings indicate only an extremely basic level of understanding of ergonomics was demonstrated, which primarily focused on anthropometrics. Teachers appear to focus only at this level and students who have a deeper understanding of ergonomics are achieving this via personal interest and resources outside the classroom. Conclusions are drawn relating to teaching and learning ergonomics within undergraduate industrial design courses.

Introduction

Trainee industrial designers, at an undergraduate level, need a basic working knowledge of ergonomics if their design work is going to be effective. The authors' experience, as lecturers in industrial design, is that students arriving on undergraduate courses have a widely variable understanding of ergonomics. In order to be able to plan the most effective teaching and learning experiences for these students, it is necessary to learn as much as is possible about the range of ergonomics learning students bring with them.

This paper reports a survey of undergraduate students' learning of ergonomics when they were in schools (n = 60). The aim was to gain a picture of the students' level of ergonomic understanding and explore their school-based learning experiences.

Method

Firstly a benchmark was established. Three industrial designers (not ergonomists) completed a mind map on the ergonomic factors involved in the design of a bicycle seat. A total of 32 factors were identified and then grouped into three levels. Level 1 consisted of 4 general factors: use, users, comfort and fit. Level 2 consisted of 11 factors, sub-sets of level 1, for example type of user, adjustability and anthropometry. Level 3 consisted of 32 further sub-sets such as load, load direction and duration of use. Secondly, 60 undergraduates, in the first 3 weeks of their course, were given two minutes to individually mind map factors relating to the design of bicycle seats. This *warmed-up* the group prior to forming random groups of 4 and repeating the exercise as a group mind map over 4 minutes. The aim was to use group synergy to elicit the broadest range of factors that each group could reveal. Finally, the groups were asked to repeat the mind map focusing specifically on ergonomic issues.

The group results were then analysed against the benchmark and allocated a grade indicative of the number of factors identified. For example, a group may identify all the issues staff raised at level 1, but only 50% at level 2, and 10% at level 3.

Finally, a group of the six strongest students, identified from the individual mind map, were interviewed in two groups, three students in each group. The aim was to explore how their ergonomic understanding had been introduced and developed at school, prior to joining the degree programme. These particular students were chosen rather than random individuals as the authors wished to benchmark best practice at a school level. Of course the strongest students may not have come from the best practice in schools.

An interview schedule was developed after analysing the mind maps. It was apparent that a large majority did not know the principles of mind mapping. A question on prior experience of mind mapping and when teachers introduced it was included to gain feedback on this aspect. Other questions focused on the subjects in which students had used mind mapping, how this was done and any examples the students could remember. Questions were then asked on how student were taught ergonomics. These included when, how, areas covered, examples and whether subsequently used in their Advanced (‘A’) level design projects. Finally, questions examined their experience of bicycles, to discover whether the students had particular expertise in ergonomics or issues of bicycling.

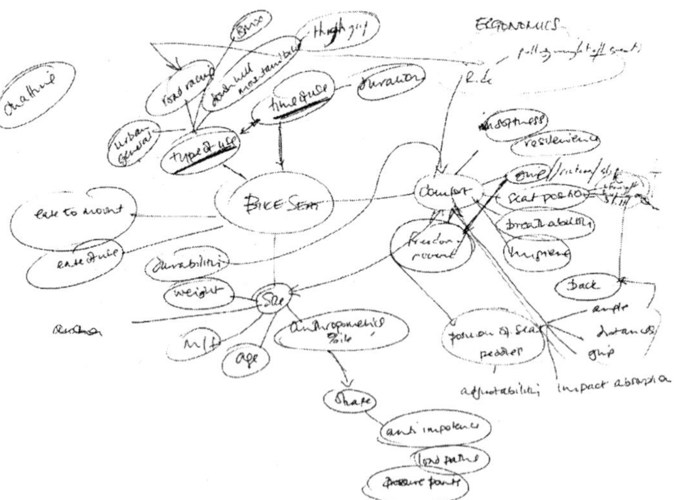

Figure 1. Staff benchmark mind-map

Results

Student group mind mapping

Table 1: The two student group mind maps against staff benchmark

Level Assigned	Benchmark generated by staff	a. Group mind map: factors in bike seat design		b. Group mind map: factors specific to ergonomics		Difference between a and b
		Mean	% benchmark identified	Mean	% benchmark identified	
One	4	3.0	75%	3.06	76.6%	+ 1.6%
Two	11	4.4	40%	4.1	37.3%	- 2.7%
Three	32	3.4	10.7%	5.125	16.0%	+5.3%

At level 1 the groups have a relatively high *hit rate*, but drop off considerably when addressing detail at levels 2 and 3. This is disappointing for students with a high proportion of good grades at 'A' level Design. Denton and Woodcock (1999) showed that only in Design and Technology do all 'A' level syllabi specifically mention ergonomics/human factors. In this survey students reported that approximately 70% of their 'A' level Design and Technology involved a 'medium to high' level of ergonomics.

Table 2: Group 'hits' against issues mind mapped by staff (32 possible)

rank	freq	issue	level	rank	freq	issue	level
1	**29**	'Fit'	1	24	**4**	Duration of use	3
2	**26**	Anthropometry	2	25	**4**	Seat width/length	3
3	**25**	Users	1	26	**4**	Breathability	3
4	**24**	Comfort	1	27	**4**	Pressure points	3
5	**21**	Adjustability	2	28	**3**	Use – general urban	2
6	**19**	Use	1	29	**3**	Professional use	3
7	**18**	Gender	2	30	**3**	Ease of mount/dismount	3
8	**13**	Materials	2	31	**3**	Load distribution	3
9	**13**	Cushioning	2	32	**3**	Softness	3
10	**13**	Specific adjust	3	33	**3**	Formability	3
11	**12**	Weight of rider	3	34	**2**	Use – bmx	2
12	**12**	Contour of rider	3	35	**2**	Flexibility of range	3
13	**11**	Shock absorption	3	36	**2**	Chaffing	3
14	**9**	Age of rider	2	37	**2**	Anti-impotence	3
15	**8**	Use – racing	2	38	**2**	Air pads	3
16	**7**	Riding position	3	39	**1**	Amateur use	3
17	**7**	Cycling action	3	40	**1**	Hygene	3
18	**7**	Grip on seat	3	41	**1**	Mechanical springing	3
19	**7**	Gel pads	3	42	**0**	Disabled user	3
20	**6**	Spinal position	3	43	**0**	Load direction	3
21	**5**	Use – mountain bike	2	44	**0**	Profile in 3 axis	3
22	**5**	Percentiles	3	45	**0**	Resilience	3
23	**5**	Inclusivity of range	3				

Student Interviews

All students reported having been taught some ergonomics. The norm for teachers to introduce the area appeared to be age 15 in this sample, though one student reported not having covered ergonomics until over 16 years of age. All but one was taught ergonomics using general schools design textbooks. These, of course, try to cover a broad range of design related knowledge of which ergonomics is but one. The depth of coverage, therefore, will tend to be basic. Only in one case was a specific ergonomics text used. Three students reported teachers using off-air video recordings of the *Better by Design* series (Design Council 2002). This series showed the process of product development including ergonomic factors, but did not focus specifically on ergonomics.

The interviews showed that the ergonomics taught focused on basic anthropometric data. Students reported that examples used to teach ergonomics were typically chairs and tables. Only one student appeared to be aware that ergonomics is far broader, than simply *fit*; for example light levels and noise. It is interesting to note that the individual student with the strongest grasp of ergonomics gained most of his understanding outside the classroom. All students reported applying ergonomics in their final projects at school. In some cases students simply took measurements directly from their own bodies and existing products. Only in one case did a student use conventional ergonomic data (Diffrient *et al* 1974). This was not the student who was actually introduced to this source of data at age 15. There was relatively little incremental development of ergonomic understanding by teachers in the students design experience at school.

All the interviewees had experience of riding a bicycle. One was an enthusiast and was aware of issues such as impotence due to pressure points (Orphan 2001) and was able to analyse ergonomic detail in some depth. One female student was aware that seats for males and females needed to differ. The others had basic awareness of comfort issues.

Discussion

Limitations

The method used has its limitations. Group mind mapping was intended to use the synergy effect (Hackman 1983) to support students in identifying factors. However, student ability to mind map was evidently limited and they operated at a relatively slow pace. This might be one reason why they were less able to explore in depth; rather than assisting students in uncovering prior learning, the technique may have been acting as a limiting filter. As the exercise was conducted out at the end of a long working day, they may have also been tired. The task did not carry marks and, therefore, motivation to tackle the exercise in depth may have been limited. Nevertheless, it is interesting to observe the simplistic level of mind mapping: no multiple branching or referencing between sections. The interviews showed that mind mapping is a technique also used in other subjects: art, science, history and religious studies, for example. As mind mapping is such a central tool for design this issue merits further investigation.

The exercise was also limited by the chosen topic: the bicycle seat. Whilst the industrial designers identified 32 issues relating to ergonomics other issues such as ambient heat, light and other environmental factors would not be identified by this topic. Nevertheless, it was surprising, considering the type of *lifestyle* and *techie* magazines commonly read by industrial design students, how few deeper issues were identified.

Ergonomic content and level of teaching

Both the analysis of the group mind maps and the interviews confirmed that most students appear to only appreciate ergonomics at the 'level 1' stage as described above. Whilst there may be a memory problem acting as a filter, the interviews support the mind map analysis by indicating that teachers primarily teach only a basic appreciation of ergonomics and focus, particularly, basic anthropometrics. The first 7 points in table 2 frequency were all anthropometrically orientated. These findings are similar to those in a survey by Denton and Woodcock (1999), of 353 undergraduates in both industrial design and engineering in their first week of their degree programmes.

Within the 1999 survey, 110 were engineering undergraduates. In this case whilst the industrial design students performed at a similar level to the survey reported in this paper, the engineers (again in their first week of university) were significantly weaker in their understanding of ergonomics. This is probably because the students taking engineering would have an 'A' level profile typically, of double maths and physics. Ninety-seven percent of the industrial design students had 'A' level design. All 'A' level design syllabii include a requirement to teach ergonomics/human factors (*idib*). Nevertheless even amongst these industrial design students only just over 40% scored at a 'high' or 'medium' level of understanding as described in the 1999 survey.

Teaching techniques used relating to ergonomics

The interviews attempted to identify when teachers first introduced ergonomics to pupils, how it was done and how this understanding was extended over time. The small sample indicated that these teachers did not introduce ergonomics until a mean of 15 years. The Denton and Woodcock survey showed that, at General Certificate of Secondary Education (GCSE) level (age 16), few syllabi mention ergonomics overtly. There is less pressure on teachers to teach ergonomics and, it may be hypothesised, few will unless they have a strong understanding of the subject and enthusiasm for the subject.

Both surveys show that teachers primarily use 'project work' as a medium for teaching subjects such as ergonomics rather than by formal class teaching. In practice this usually means that pupils are given a design problem to work through. With younger pupils this is often the same problem; with older pupils it may be self-chosen and, therefore, vary. The teacher will introduce materials, as they are required. Ergonomics would be raised as an issue relating to research where the pupil needs to gain data on target user-groups. It may be raised again in making design decisions and when evaluating final project outputs. Generally such approaches are considered good teaching practice: the subjects are taught in a practical context, when they are relevant to the stage of the project. However, it may be that teachers are not clearly 'signalling' the parts played by elements such as ergonomics when they are introduced. If this was the case students may forget they have been introduced to such work when at school.

A second point relating to the teaching of ergonomics within design projects is the question of iterative development. Good teaching would introduce basic principles to younger pupils and then incrementally increase them within projects, as they get older. This means that children are first given a simple, more easy to grasp, picture of an area such as ergonomics. This should, subsequently be developed as the pupil moves forward until, at 'A' level they can handle concepts such as ergonomics in a more sophisticated way. In practice what appears to be happening, if the self-report of these two surveys is accurate, is that many, though not all, teachers are leaving the introduction of ergonomics until as late as 15 years, but, more importantly students do not remember their understanding being developed incrementally. They are tending to stick at a very basic level of ergonomics. This may be because the majority of design teachers have little personal training in ergonomics themselves and so tend to avoid getting into too much depth. This is supported by the report that only one teacher in the student interviews took students beyond basic ergonomics within standard design text books and looked at specific ergonomics texts. Similarly whilst teachers did refer to ergonomics features in videos such as Better by Design (2002), they were not using specific materials such as the ergonomics training materials produced by The Design Council (1987).

Conclusions

The sample used represents a cohort with strong grades at 'A' level design, yet knowledge of ergonomics within the sample was limited to a basic level despite the fact that ergonomics/human factors are in all 'A' level design syllabi. This should cause concern, though the methodology used may not have been as effective as it was intended in uncovering students' understanding. Similarly the peripheral finding that these students were not able to use mind-mapping effectively as a technique requires further work to develop this highly useful skill in schools.

The survey indicates that students' limited understanding of ergonomics may be due to teachers rarely expanding the concept beyond basic anthropometrics. Teachers own training being limited in this area and this suggests that there is scope for in-service materials for teachers to be developed. If this were done it might also encourage teachers to bring forward the reported mean introduction age of 15 years and give more time for a logical development of the concept over a series of planned exercises. The survey also indicates that it may be necessary to review the efficiency of design projects as a means of introducing new concepts if this approach is used as the primary method of teaching and learning.

References

Denton, H. G. and Woodcock, A. 1999, Investigation of Ergonomics Teaching in Secondary Schools – Preliminary Results. In N.P.Juster (ed.) *The Continuum of Design Education* (Professional Engineering Publishing Ltd, London), 129 – 138.

Diffrient, N., Tilley, A. R., Bardagjy J C and Dreyfuss Associates. 1974, *Humanscale 1/2/3: a portfolio of information.* MIT Press: Cambridge, Mass.

Hackman J.R. 1983, A *Normative Model of Work Team Effectiveness.* Technical Report No 2. Research Project on Group Effectiveness. Office of Naval Research. Code 442. Yale School of Organizational Management.

Orphan, J. W. 2001, *Biking and impotence.* http://www.baystatetri.com/Dr.%20Orphan.htm (18 December 2002).

The Design Council. 2002, *Better by Design*. (The Design Council, London). Video.

The Design Council. 1987, Katherine Wykes: ergonomist, in M. Bennett (ed.) *Living by Design*, (The Design Council, London).

HARDER HARDWARE?
A PILOT STUDY INTO GAINING SOLDIERS PERCEPTIONS OF THEIR WEAPONS APPEARANCE

I Storer, GE Torrens, A Woodcock and GL Williams

Design Research Group,
Department of Design and Technology, Loughborough University, Loughborough
Leicestershire, LE11 3TU
i.j.storer@lboro.ac.uk

Making the soldier look more aggressive, powerful and threatening has long been seen as giving a psychological edge in combat. Modern warfare and equipment development has given this aspect low priority. The changing nature of the role of soldiers within society, has highlighted the need for the semantics of soldiers equipment to be addressed. This paper presents a pilot study undertaken with Royal Marines where alternative weapon designs were presented, based on existing SA-80 assault rifle architecture. The designs were presented to eight marines The outcomes were that soldiers prioritised functionality and initially were dismissive of the change in appearance between proposed concept designs. However, further discussion highlighted that a more aggressively styled rifle would help deter potential aggressors within peacekeeping activities. Although others might consider a less aggressively styled weapon could diffuse some situations. The outcomes of this pilot study warrant further investigation.

Introduction

The appearance of soldiers and in particular foot soldiers to their adversaries has always been of importance. Making the soldier look more aggressive, powerful and threatening has given a psychological edge within combat. The innate behaviour of animals demonstrates a period of potential combatants "sizing up" their opponent. In humans appearances have traditionally been altered by using; war paint, tribal tattoos, armour and uniforms to convey a more frightening image to potential aggressors.(see figure1 and 2) Traditionally threatening behaviour ranging from banging weapons on shields to bayonet practice has been used to intimidate the opposition and raise soldiers confidence.

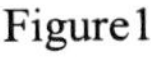

Figure1

Figure2

Modern warfare and equipment development has not given aesthetics a high priority. The changing nature of the role of soldiers within society, has highlighted the need for the semantics of soldiers equipment to be addressed. In disaster relief operations for example soldiers may need to appear less aggressive. This paper presents a pilot study undertaken with Royal Marines where alternative weapon designs were presented, based on existing SA-80 assault rifle architecture. The sketch concepts presented reflected different aesthetic styles ranging from highly aggressive to friendly. The designs were presented to twelve marines, with qualitative and quantitative data gathered via discussion.

Method

A researcher conducted semi structured interviews with twelve Royal Marines. The twelve participants interviewed were from one and a half "sections", with two non commissioned officers (NCO) present. The NCOs were interviewed separately to marines, in an attempt to elicit unbiased responses from the Marines. Officers were interviewed by the other researchers in a nearby location for the same reason. The marines were interviewed as a group. The Marines were all on exercise within the training complex at Sennybridge, South Wales, United Kingdom. All the participants were between 18 and 35 years old and had between one and fifteen years professional experience. The exercise upon which the Marines were training is known as the "All-Arms" course. The course involves practise and assessment of their capability to perform set military routines. The marines in this study had just finished taking part in a mock battle and were awaiting further orders. The interviews were undertaken in the autumn of 2001 in a clearing within woodland on the Sennybridge complex. The light level was low due the overcast and misty weather conditions. The Marines were also involved in sustainability activities such as eating, drinking and maintenance of equipment during the interview.

Based on previous experience of using soldiers in this context the operators were aware they would have a limited time with Marines. The operators introduced themselves and the purpose of their meeting with the participants. There was a period of introductory discussion about their perceptions of their equipment in general, followed by more focused discussion in relation to the proposed research question. The Marines were presented with alternative weapon designs, based on existing SA-80 assault rifle architecture. The concept designs were presented in sketch format. See Figure 3. The sketch concepts had been designed to reflect different aesthetic styles ranging from highly aggressive to friendly. The Marines were asked to comment on the relationship between function and styling. They were also asked for their opinion of the different styles and to consider them in relation to the changing roles of military personnel, specifically the role of peacekeeper.

The styling cues were taken from a range of semantic iconography. The sketches were drawn over the basic architecture of the SA80, so the function would remain largely as before. Only 15 minutes were spent sketching each specific weapon, and styling cues derived from household appliances and science fiction films were applied. The SA-80s two major styling features are the parallel linear elements and repetitive slots. The linear elements derive from the layout of the mechanical components of the weapon, and are not a result of conscious styling. The "friendly" styles were created by creating curved feature lines, covering the barrel and more random arrangements of holes. (see figures4-6) The "aggressive" styles used more converging and angled lines with bold features and sharp lines between forms.(see Figures 7-10) The sketches were put into context with a marine, and a SA-80 was also presented in the same format to act as a "control." (see figure 3)

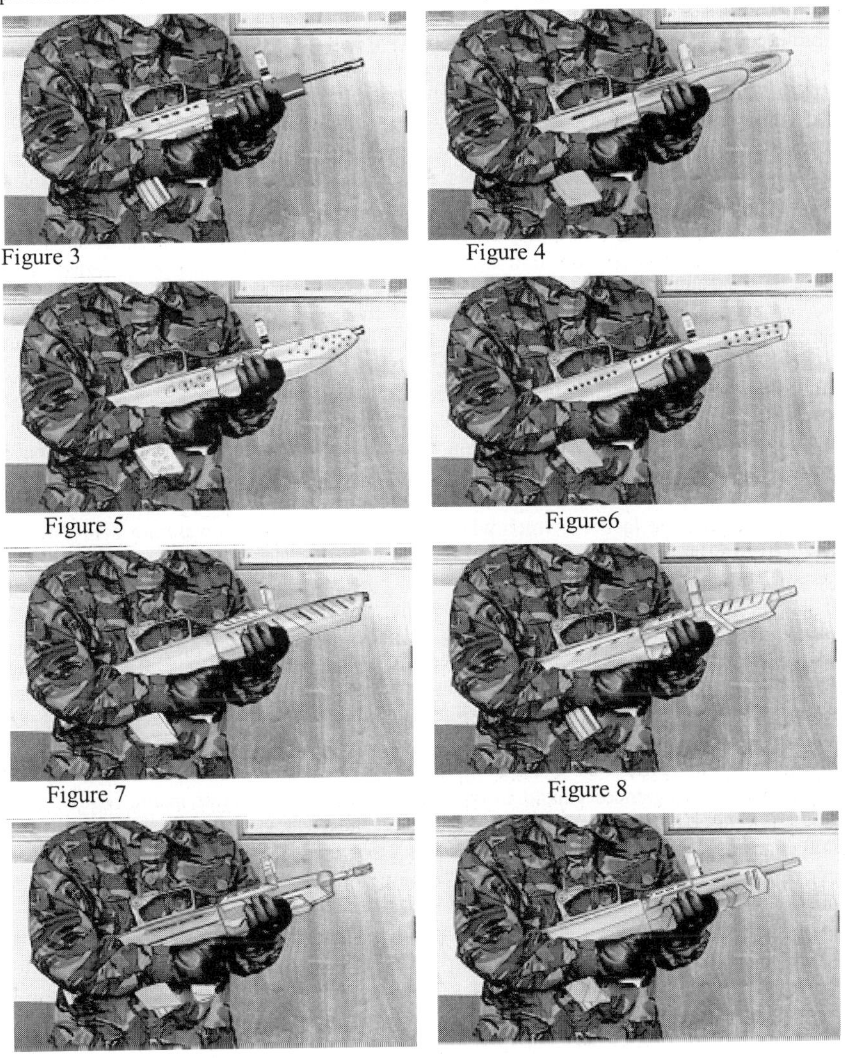

Figure 3

Figure 4

Figure 5

Figure6

Figure 7

Figure 8

Figure 9

Figure 10

Results and discussion

The outcomes were that soldiers prioritised functionality and initially were dismissive of the change in appearance between proposed concept designs. Accurate responses from marines are particularly difficult to elicit as they are unlikely to make any comments on or show preferences for a design that may make them appear "soft" in front of their colleagues. When initially asked the Marines made strongly worded derisory comments. At that time they perceived that there were still issues of functionality to be addressed within the SA80 rifle before any aesthetic refinement. Which showed that the question needed rephrasing to allow them to consider purely aesthetic criteria. Once the soldiers considered the functionality to be constant across the designs, more considered comments were made. Marines are unused to describing aesthetic terms, therefore some warm-up exercises where they classify product images into "friendly" or "aggressive" categories could have been useful.

The issue of environment and military role produced more positive comment. Soldiers were divided in their opinion as to whether a less or more aggressive rifle would help in a peacekeeping role, such as in Bosnia. From the more experienced members of the group they had found being less aggressive during dealings with local people enabled them to elicit more cooperation in their duties. The other viewpoint was that a less aggressive image would open them to be perceived as weak by locals who would then attempt to "take them down". Further discussion highlighted that a more aggressively styled rifle would help deter potential aggressors within peacekeeping activities. One example of soldiers using visual language to deter aggressors, is the Parachute Regiment in Northern Ireland wearing flak jackets under their normal camouflage jackets, which when combined with a wide webbing belt produces an illusion of a bulky, powerful upper body.

The group also acknowledged the 'Gucci kit' syndrome, where soldiers wanted the most up-to-date equipment that provided a level of social status. Trends and fashions may be considered as important within military society as in the general public.

The discussion had been led by one of the NCO's and it was noted that all others present from the two sections did not try to speak when this individual was making comment. There was a clear hierarchy within the group interviewed. This highlighted the need to divide the group up to get individual opinions. It was also highlighted by one of the soldiers that the American Armourlite rifle was considered the most desirable weapon by Asian military groups, due to past references gained from American war films shown in the region. This comment could not be easily corroborated and was considered speculation.

The total contact time was twenty to thirty minutes. However, focused discussion on the subject matter was only five to ten minutes. If running field-based trials, researchers should consider the following:

- There is likely to be limited contact time available to operators undertaking the interview or focus group during a military exercise, between five and thirty minutes;
- Research questions to be answered must focused and limited to one or two issues as they will be easily distracted;

- Military personnel may not have the necessary vocabulary to discuss design related issues. Hence warm-up exercises building their confidence and giving examples may elicit more useful data;
- The influence of the hierarchy of command must be managed even at a section level;
- A different Regiment, such as the Parachute Regiment, could produce different opinions from the group selected;
- The time of day and year of interviews must be factored into research planning when using visual aids; and,
- Any equipment to be used in the interviews or assessments must be lightweight, portable waterproof and wipe clean.

The outcomes of this initial pilot study are favourable in that some comment was provided by the participants. There were many constraints on the data gathering. Further interviews or focus group work would be more suited to a barracks-based activity. The semantics of military products may be considered of future importance for military personnel who are likely to be involved in changing roles and environments.

References

Bresman. J, 1999, The Art of Star Wars, Ebury Press, London
Wisher.W, De Souza. S.E, Phelps.N, 1995, The Art of Judge Dredd, Cambus Litho Ltd

RISKS AND HAZARDS

REFRAMING RISK USING SYSTEMS THINKING

Mark Andrew

Scenario
283 Glenhuntly Road
Elsternwick VIC 3185
Australia
scenario@bigfoot.com
www.scenario.fw.nu

Risk is an important factor when public issues are at stake. This paper develops a systems model of risk, aimed at public stakeholder domains. The model uses a narrative that recognises that decisions are subject to risk appetite, and so risk is not always undesirable. The model reframes traditional risk components in terms of systems thinking and in doing so critiques quantitative risk measures, which may have reached a natural 'limit of growth' when current public interests are considered. By applying the systems model of risk to cases such as transport, food and health, risk is seen as an emergent property resulting from *patterns of forces,* and is therefore governed by resilience. As users of public services currently have limited access to policy making, the model aims to increase access to risk judgements when risk quantification alone is inadequate.

Introduction

How do people influence social systems such as transportation, food production and health care? Access to decision processes requires that systems are 'fitted' to stakeholders' needs, which is a matter relevant to macro-ergonomics. This paper addresses one aspect of public/policy decision-making fit, namely the representation of risk. Risk is a feature of most systems as a direct result of uncertainty, and offers benefits and also potential loss. This duality of risk (reward and loss) has been neglected in the risk paradigms of recent decades (which emphasis loss) and a *reframing* of the concept of risk is overdue. This model therefore frames risk as a systemic commodity with value.

An argument for reframing risk

Reframing is essentially a process of relaxing an existing view and *re*-presenting it in an alternate manner to gain insight. Reframing is not a defined or 'invented' process with a designer or advocate but rather a natural facility that is constantly reinventing itself through original expression, although cognitive frames (Minsky 1974) are a helpful prompt. One may just as freely 'represent' a loss as a gain to experience, because the limits of sensible reframing soon become evident. A fatal accident could hardly be reframed realistically as a learning experience (except, perhaps in evolutionary terms).

The argument outlined in this paper proceeds as follows. Traditional views of risk are considered (stage 1) suggesting a stakeholder view of risk as a narrative (stage 2) to supplement quantitative measures. The key elements of systems thinking (stage 3) are then explored to reframe risk (stage 4) as an emergent property of systems (stage 5).

Stage 1 - current definitions of risk

Risk can be defined as the potential for hazard to cause loss. One definition of risk frames risk as the product of probability and consequence. For example, if probability is specified as an event frequency and consequence is specified as the level of loss, then risk can be computed for a particular transport mode. So risk analysts speak of jet air travel being safer than rail travel, which in turn is safer than car travel (based on fatalities per passenger mile). Safety is just one type of risk and the basic formula approach is applied to many other potential losses (such as business interruption or asset damage).

However life rarely keeps still long enough to measure all important things and it is difficult to see beyond figures to embrace complex reality. The formula approach betrays a feature of human cognition - our innate biases when comparing positive and negative outcomes. The risk literature is quite comprehensive regarding the meaning of risk when information is presented in different ways. A Dutch proverb for example, suggests that '*an ounce of illness is felt more than a hundredweight of health*' (Boyle 2000).

So different types of outcomes (such as loss and benefit) may *not* be considered as cognitive opposites. They are dealt with differently by our minds and can not be compared using traditional scales. Systematic biases are governed to some extent by people's concept of *control* and *dread* for example. For example, people generally feel more vulnerable when being driven as opposed to driving even when they rate the *other* driver as more competent. As a quantitative decision this is 'scientifically irrational', whilst feeling intuitively perfectly natural. So although the quantification of risk merits a place in policy decisions, the rote assumption of numbers as the sole expression of risk means that opportunities are lost for other views to be incorporated. For this reason, a more holistic view of risk is warranted.

Stage 2 – a richer picture of risk

If the engineering view of risk is considered as a 'linear' event-driven approach, then an alternate means of capturing risk complexity may be based on scenarios, where uncertainty is not so neatly compartmentalised into the front-end of risk descriptions. A scenario narrative should allow for interdependencies between risk components so that probability and consequence are seen to influence and affect each other. A more realistic view of risk then, recognises *multiple events leading to interacting exposures which may lead to consequences that impact in different ways depending on vulnerability*. The impacts should be both positive and negative because reward is also linked to risk. This notion is widely recognised in the finance sector where risk is illustrated as volatility. Volatility has different meanings under different contexts, such as impending parenthood in the case of life insurance, or retirement in the case of pension fund performance.

These factors suggest the need for a more systems-oriented view of risk. To progress this argument then, 'systems thinking' offers an opportunity to reframe our richer risk picture using the language of systems.

Stage 3 – What is systems thinking?

Systems thinking is a response to the technical focus of systems dynamics (Forrester 1969) and has provided a language suitable for organisations. It was motivated by a weakness of scientific analysis, which breaks a problem into parts and then studies the parts in isolation to draw conclusions about the whole. This approach is ineffective for issues that are interrelated and defy linear causation. Circular causation, where a variable is both the cause and effect of another, has become the norm rather than the exception in social systems. True exogenous forces are rare. Recognition that the components of complex systems are fundamentally interconnected has emphasised the role of endogenous feedback loops, as illustrated in figure 1.

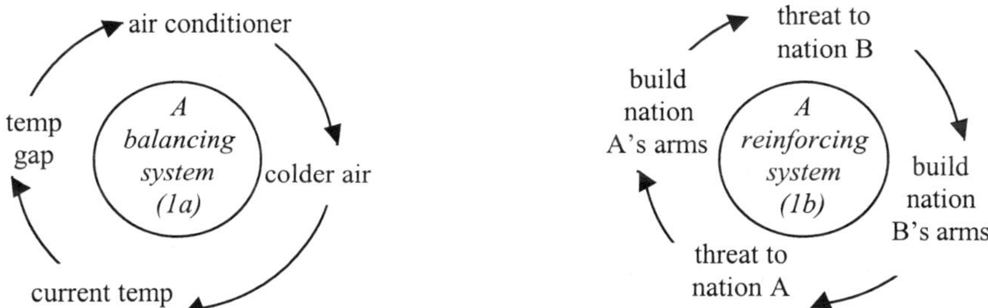

Figure 1. Two simple systems

A key tool of systems thinking is the use of arrows that denote *influence*. When an arrow is used in systems thinking, it does not denote linear cause but rather a circle of influence that may be both cause and effect (Senge 1990). A key feature of the arms race system (1b above) is that of time delay. Problem fragmentation can lead to an unexpected property that emerges from the system because lack of recognition that any given element may be both a cause and an effect contributes to continual growth. Reclassifying hospital trolleys as temporary beds to meet performance targets is an example.

So sustainable results should be understood to be a function of the structure and relationship between elements in the system. For example, a doctor's attitude to hierarchy amongst ward staff may erode effective communication, which in turn reinforces the attitude to the division of tasks and in turn justifies and increases hierarchy. In this instance, the risk of poor health outcomes grows due to a lack of balancing communication that would otherwise identify errors, and encourage learning.

This 'emergent property' construct is a powerful feature of systems thinking. Emergent properties in figure 1 are different because the systems have different structures *given the behaviour of the elements when acting together*. Although time delay is a feature of both systems, they are structurally different because the patterns of interaction result in different emergent properties - balance versus reinforcement. Although a reinforcing system is illustrated here by an arms race, it may also describe a property of a well-designed sales system (balanced hopefully by a production system).

Stage 4 – Systems views of probability and consequence

A systems perspective of probability is necessarily very different to a historical view. A systems view of probability also recognises future scenarios and possibilities. This is different to data-centric models of statistics, which rely on defining reality tightly to allow measurement. System variations present scenarios (possible futures) which have *varying* likelihood. These expose the system to different outcomes, and so basing probability purely on history is flawed. In systems thinking terms probability actually emerges from a causal loop because *patterns of events* are better descriptions of systems behaviour than discrete events. Complex systems make these limits appear quite problematic, encouraging us to differentiate between *detail* complexity and *dynamic* complexity as they relate to probability judgements.

Detail complexity describes a problem with many variables or events. By contrast, dynamic complexity more accurately describes most organisation systems where cause and effect are subtle and the effects over time of interventions are not obvious. Dynamic complexity explains why quantification fails to forecast future probability; it is not that the mathematics is flawed but rather that the goal of quantification is misdirected. In what way for example, would a number be useful when it describes only one instance amongst myriad others that can continually emerge? Dynamic complexity is evidenced for example, by the change in reliability of a rail network when infrastructure is commercially compartmentalised from rolling stock and maintenance, as occurred in the UK and Australian rail industries during the 1990s. Perhaps a more politically neutral description recognises *vulnerability* as an emergent property of compartmentalisation when the system does not provide balancing feedback. Having a number to define this type of rail transport risk is surely not as useful as being able to recognise structural influences that can identify and treat such risks.

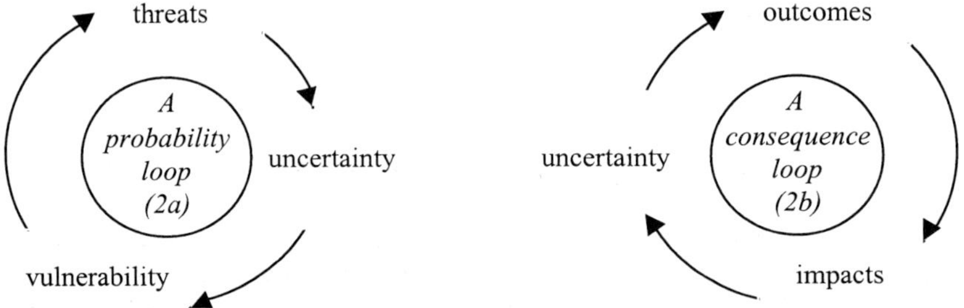

Figure 2. Systems models of probability (2a), and of consequence (2b)

Perhaps the biggest hurdle to overcome prior to presenting a systems model of consequence is to uncouple consequence as a property that necessarily occurs as a *result* of probability. It is just as valid to say that consequences pull probability, as it is to say that probabilities push consequence. Definitions of probability or consequence being leading parts of risk are fundamentally a matter of perspective and are not intrinsic to the meaning of risk. One illustration of this effect is provided by prospect theory (Kahneman and Tversy 1979), which describes our tendency to make different choices under different conditions and to literally redefine the relationship between probability and consequence. Prospect theory suggests that when people are in a position of gain they become increasingly risk averse (wishing to maintain their gain); when people are in a position of loss and losses increase they become more risk seeking (seeking to reverse their loss). In systems thinking terms, people change the way they deal with *elements as patterns* depending on context.

So every influence is both cause and effect, and consequence interacts with probability. More than any other features of systems thinking these insights suggest a systems model where risk is an emergent property of uncertainty. Figure 2 shows a systems view of both probability (2a) and of consequence (2b) comprising loops of elements that are both cause and effect. The models of probability and consequence shown in figure 2 both involve uncertainty as one of their elements and so the ground is prepared for a basic systems risk model which integrates these factors into a cause-effect-cause framework.

Stage 5 – A systems model of risk

This argument for a systemic approach to risk has developed a generic probability pattern and a generic consequence pattern. Combining these leads to a basic model of risk as an emergent property of probability and consequence, as shown in figure 3.

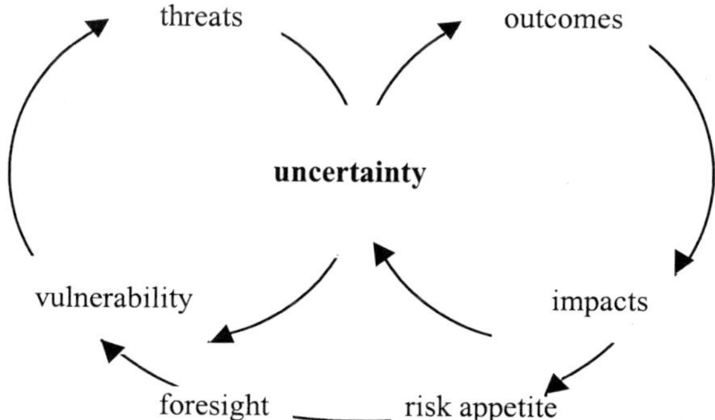

Figure 3. A systems model of risk with feedback

Figure 3 also provides feedback and feedforward that attempt to show how the management (not elimination) of risk can occur through foresight. In this model risk acceptability or tolerance is termed risk appetite, because this term does not cast risk as a commodity necessarily to be avoided. Risk appetite describes the way people recognise and judge future scenarios and so provides terms of reference to make judgements. Making risk appetite explicit is one of the key aims of the model, because public access to decisions can be improved by recognising and valuing risk as an *uncertain investment in potential benefit*.

So how can this systems model of risk be used to provide insight to a high profile public risk? By way of example, the disease variant Creutzfeldt-Jakob disease (vCJD) is now thought to be connected to a very rare combination of events. In the 1970s and 1980s feed supplements were given to UK calves from an earlier age when they were more vulnerable. Changes to the animal carcass rendering process may also be a contributory factor. Currently 101 people have died from vCJD but more future victims are still alive. The linear model of risk suggests that various events led to the public being exposed to hazard, which resulted in disease and fatalities. A systems thinking view recognises the interaction between events and the *patterns of events*. Public consumption pressure, farm economics and a particular regulatory environment contributed to changed production methods. These changes exposed a population vulnerable to outcomes that have a time delay and impact public health in unpredictable ways. Outcomes such as the public dependency on food production efficiency and the time delay between processing methods and disease incubation, may be as much a driver of probability as any specific event (such as the age of cattle that were fed supplements). Also, the *meaning* ascribed to food risks is different to say car travel, perhaps because dread plays a part in judgements of risks that people feel are not within their control.

Conclusions

Successful interventions occur by identifying leverage in systems. This systems model of risk attempts to model such leverage and resulted from three key strategies. First, the episodic view of probability and consequence was uncoupled, so that the cause-effect-cause properties of risk can be appreciated in systems thinking terms. Second, risk tolerance and foresight were made explicit by linking them to uncertainty. Third, the model was used to examine public risk issues (health, transport and food safety).

These steps frame risk as an emergent property of systems that if well managed, can contribute value. Having seen the systems model of risk reflect principles of resilience and vulnerability (such as clinical error management and rail network performance), and describe a disease episode (vCJD), it seems reasonable to suggest that 'well managed' refers to both feedback and feedforward. For risk judgements, these can be encouraged by identifying risk appetite. Science tends to highlight the *credibility* of probability judgements; public stakeholders can balance this by focusing on risk appetite (and the *credibility* of risk appetite judgements, as shown in figure 3). Although understanding appetite can not guarantee optimum risk/benefit outcomes, it does help with foresight and learning (which are essential elements in setting equitable policy). A systems model of risk illustrating the implications of choices in public policy is therefore proposed as an appropriate means of *supplementing* (not replacing) traditional quantified risk metrics. Using such a systems view particularly for macro-ergonomics decisions may enhance public access to policy issues, especially when risk is an important factor, or when numbers alone are inadequate for understanding risk appetite.

References

Boyle D. 2000, *The Tyranny of Numbers*, (Harper Collins).

Forrester J. 1969, *Urban Dynamics*, (MIT Press).

Kahneman D. and Tversky A. 1979, *Prospect theory: an analysis of decision making under risk*, Econometrica, 47.

Minsky M. 1974, *A Framework for Representing Knowledge*, memo #306 (MIT-AI Lab).

Senge P. 1990, *The Fifth Discipline*, (Doubleday).

HUMAN FACTORS AND MAJOR HAZARDS: A REGULATOR'S PERSPECTIVE

Martin Anderson MErgS, EurErg, MIOSH

HM Specialist Inspector Of Health & Safety (Human Factors)
Health And Safety Executive
Daniel House, Trinity Road
Bootle, L20 3TW

This paper outlines recent experience of the HSE's Human Factors Team in assessing human factors issues on major hazard installations. Our involvement on major hazard sites reveals that most duty holders do not adequately address human factors. Primary weaknesses include an imbalance between hardware and human issues, and focussing on the human contribution to personal safety, rather than to the initiation and control of major accident hazards.

However, following targeted inspection and awareness-raising in the industry, the profile of human factors and effective consideration of these issues is steadily increasing. We are beginning to see the results of these efforts reflected in contact with sites. We will continue to work with major accident sites and industry bodies to develop and share emerging best practice.

Introduction

Human failures are implicated in the majority of serious accidents in hazardous industries. Some recent examples include Bhopal, Texaco Milford Haven, Chernobyl, Piper Alpha and Flixborough to name a few. As technical safety measures improve, we can expect the significance of human factors in major accidents to increase. To help address these issues, the UK Health & Safety Executive (HSE) set up a new Human Factors Team in the Hazardous Installations Directorate in 1999. This Team provides site inspectors with specialist advice and support during inspections, investigations and enforcement; as well as preparing industry guidance on human factors issues.

The focus of the Team's activities is on those sites that fall within the scope of the Control of Major Accident Hazards Regulations 1999 (COMAH), although we are also active in the railway industry. Over the last three years, the team has become involved at numerous major hazard sites across the whole of the UK, including most oil refineries.

In addition to our inspection and assessment activities, we develop guidance and standards, train field inspectors, set policies for the field, manage applied research and promote human factors to industry (either directly or through intermediaries such as the

Institute of Petroleum, Institution of Chemical Engineers and the Chemical Industries Association). We are also involved in the European Commission PRISM network co-ordinated by the European Process Safety Centre (EPSC), the aims of which are to develop and disseminate best practice guidance on human factor topics.

What *we* mean by 'human factors'

The HSE document HS(G)48 presents a simple introduction to generic industry guidance on human factors. This guidance provides a useful definition:

> *'Human factors refer to environmental, organisational and job factors, and human and individual characteristics, which influence behaviour at work in a way which can affect health and safety'*

This definition includes three interrelated aspects that must be considered – the job, the individual and the organisation. In other words, human factors is concerned with what people are being asked to do, who is doing it and where they are working. Human factors interventions will not be effective if they consider these aspects in isolation. It is deficiencies in either of these three areas, or in the interactions between them, that lead to human performance problems. There are three types of human failures that may lead to major accidents:

- **Errors** are physical actions that were not performed as intended;
- **Mistakes** are also errors, but errors of judgement or decision-making;
- **Violations** differ from the above in that they are intentional (but usually well-meaning) failures, such as taking a short-cut or non-compliance with procedures.

The likelihood of these human failures is determined by the condition of a finite number of 'performing influencing factors', such as time pressure, workload, competence, morale, noise levels and communication systems. Given that these factors influencing human performance can be identified, assessed and managed; potential human failures can also be predicted and managed. In short, human failures are not random events.

Human factors is often seen as a rather nebulous concept and so it is convenient to break the subject down into a series of discrete topics. As a result of our site visits and assessment of COMAH safety reports, a small group of topics has emerged and we promote these as our 'top ten':

1. Organisational change and transition management
2. Staffing levels and workload
3. Training and competence
4. Organisational culture
5. Integration of human factors in risk assessment/investigations
6. Fatigue from shiftwork/overtime
7. Compliance with safety critical procedures
8. Safety critical communications (e.g. shift handover)
9. Human factors in design
10. Maintenance error.

Experience of regulating major hazard sites - positive issues

The Human Factors Team has a distinct advantage of having visited a broad sample of Major Accident Hazard (MAH) sites over the past three years, enabling us to construct a picture of best practice in human factors in the process industry. We are therefore able to facilitate the sharing of what works and what doesn't across the industry, through published guidance, seminars and individual site contact.

Clearly, the efforts of the Team in promoting these issues are beginning to reveal themselves in our contact with MAH sites. For example, some sites have addressed issues that we have raised at regional one-day events held in conjunction with industry bodies and associations (including consideration of the top-ten topics listed above). This has been reflected in the structure and content of their COMAH safety report submissions and in information available on site inspections.

As our capabilities are increasingly recognised within HSE, we are now finding that we are becoming involved at an earlier stage of the design lifecycle. For example, we are being consulted prior to site modifications (including proposals for organisational change) and are also involved in specifications for human factors in the design of new process plants. This opportunity to be involved at such an early stage will increase the impact of our involvement.

Although our approach has been new to many sites, we have received positive feedback following our interventions. For example, some sites apply the lessons learnt from an intervention to other installations in the organisation. Other sites have commented that we have provided them with a different perspective on their organisation, not obtained from previous 'independent' audits.

Once their awareness has been raised, some sites have clearly embraced the issues and are developing their in-house capability in human factors. We are seeing an increasing number of companies having a human factors champion on site, who acts as an 'intelligent customer' in dealings with the competent authority and external consultants.

Over the past couple of years or so, we have recognised that the major hazards industry is readdressing the balance between hardware and human factors. Given that the regulation of these issues has developed rapidly since the formation of a dedicated team of specialists, we expect that 'emerging best practice' will continue to develop across the industry.

Experience of regulating major hazard sites - negative issues

Although many MAH sites are managed by multi-national, blue-chip companies, the experience of the Team is that their consideration of human factors issues could be significantly improved. The main failings apparent in relation to human factors are discussed in detail below. These weaknesses have all been observed at numerous installations and are common threads rather than isolated occurrences.

Focus on engineering issues
Despite the growing awareness of the significance of human factors in safety, particularly major accident safety, many sites do not address these issues in any detail. Their focus is almost exclusively on engineering and hardware aspects, at the expense of 'people' issues. From reading many safety reports it would appear that these sites are unmanned, such is the lack of reference to human performance aspects.

For example, a site may describe alarm systems as being safety-critical and describe the assurance of their electro-mechanical reliability, but fail to address the reliability of the operator in the control room who must respond to the alarm. If the operator does not respond in a timely and effective manner then this safety critical system will fail and therefore it is essential that the site addresses and manages this operator performance.

Due to the 'ironies of automation' (Bainbridge, 1987), it is not possible to engineer out human performance issues. All automated systems are still designed, built and maintained by human beings. For example, an increased reliance on automation may reduce day-to-day human involvement in operations, but lead to greater maintenance activities, where human performance problems have been shown to be a significant contributor to major accidents (HSE Books, 2000).

Furthermore, where the operator moves from direct involvement to a monitoring and supervisory role in a complex process control system, they will be less prepared to take timely and correct action in the event of a process abnormality. In these infrequent events the operator, often under stress, may not have 'situational awareness' or an accurate mental model of the system state and the corrective actions required.

Focus on occupational safety
The majority of MAH sites tend to focus on occupational safety rather than on process safety. Those sites that consider human factors issues rarely focus on those aspects that are relevant to the control of major hazards. For example, sites generally consider the personal safety of those carrying out maintenance, rather than how human errors in maintenance operations could be an initiator of major accidents. This imbalance runs throughout the safety management system, as displayed in priorities, goals, the allocation of resources and safety indicators.

For example, 'safety' is often measured by Lost-Time Injuries, or LTIs. However, the causes of personal injuries and ill-health are not the same as the precursors to major accidents. Therefore, measures such as LTIs are not an accurate predictor of major accident hazards and sites may thus be unduly complacent in this respect. Notably, several sites that have recently suffered major accidents demonstrated good management of personal safety, based on measures such as LTIs. Therefore, the management of human factors issues in major accidents is quite different to traditional safety management.

In his analysis of the explosion at the Esso Longford gas plant, Hopkins (2000) makes this point very clearly:

'An airline would not make the mistake of measuring air safety by looking at the number of routine injuries occurring to its staff'.

Clearly, a safety management system that is not managing the right aspects is as effective in controlling major accidents as no system at all.

Performance indicators more closely related to major accidents may include the movement of a critical operating parameter out of the normal operating envelope. The definition of a parameter could be quite wide and include process parameters, manning levels, maintenance activities, accidental releases, alarm response times, procedural compliance, inspection activities or the availability of control/mitigation systems.

It is critical that the performance indicators should relate to the control measures outlined by the site risk assessment and/or detailed in the COMAH safety report. Furthermore, they should measure not only the performance of the control measures, but also how well the management system is monitoring and managing them.

Focus on the short-term

Where sites do consider human factors aspects in relation to major hazards this is usually in response to a major incident, an inspection by the HSE or both. In these cases, companies tend to view the initiative as having a short-term benefit, such as satisfying the requirements of the regulator, rather than as improving the long-term safety performance of the site. For example, human factors issues may be addressed in the COMAH safety report in order to facilitate acceptance of the report by the competent authority, rather than to make a real difference in major hazard safety.

Lack of ownership

This issue is related to the short-term outlook discussed above. Sites often consider human factors issues in relation to an immediate need to address a discrete topic. External consultants may be engaged to facilitate the intervention and too frequently, the expertise remains outside of the company reducing ownership of these issues by the site. In these cases, we propose that a senior member of site management adopts the role of a human factors champion.

Lack of realism

We are often informed by a site that operators are well-trained and experienced, partly as justification for relying on human actions. However, it cannot be stressed enough that highly skilled, motivated and experienced people do make errors, whether unintentional or not. It is human nature to take short cuts or break rules, for example when the pressure or inconvenience is high enough. It is also the case that unintentional errors occur, for example when workload is high, a task is complicated, or the situation is abnormal.

MAH installations frequently assume that an operator will perform certain actions in the event of a process upset. However, this assumption often fails to take account of the fact that human behaviour in an emergency situation is different to that in normal operations. Where there is reliance on operator actions in controlling or mitigating a MAH, this should be demonstrated to be realistic. In many cases, manual interaction or intervention could be replaced with reasonably practical physical measures. MAH control should not rely on the heroic actions of operators (Lucas, 2002, termed this the 'superman approach' to risk control).

On occasion, quantitative data is quoted by the operator without justification, for example: 'the probability of the operator failing to respond to the alarm is 0.0001' (the 'magic number' approach - Lucas, 2002). Such data is to be treated with caution, and we will require the site to make the assumptions explicit and demonstrate that the data is specific to the site.

Failure to identify safety critical aspects

Our experience is that MAH sites fail to produce an inventory of safety-critical tasks, roles, responsibilities and procedures. Without the identification of areas where human intervention is safety-critical, any consideration of human factors will be unfocussed. Human factors analyses, although productive, can be resource-intensive and in order for resources to have maximum impact, they should be targeted where their impact will be the greatest. Again, reference should be made to the MAH risk assessment; and the role of human intervention in the initiation and control of MAH scenarios reviewed.

The way forward

At its inception, the Human Factors Team outlined a strategy as follows:

- Increase awareness of the importance of human factors among MAH sites;
- Improve the integration of human factors in design, risk assessment and COMAH safety reports;
- Encourage continuous improvement and sharing of good practice;
- Codify knowledge in a useful way for HSE and transfer to field inspectors.

Over the past three years we have made significant progress towards achieving these objectives, including interventions on a large number of major hazard installations across the UK.

However, there remain considerable weaknesses in the approaches taken to human factors on many of the most hazardous sites in the country, operated by some of the world's largest oil and chemical companies. If further major accidents are to be prevented, duty holders are urged to examine whether any of the failings discussed in this paper apply to their organisation.

References

Bainbridge, L. (1987). Ironies of automation. In *New Technology and Human Error*. Edited by Rasmussen, J., Duncan, K. & Leplat, J. John Wiley and Sons Ltd

Hopkins, A. (2000). *Lessons from Longford: The Esso Gas Plant Explosion*. CCH Australia Ltd, ISBN 1 86468 422 4

HSE Books (1999). Reducing error and influencing behaviour, HSG48

HSE Books (2000). Improving maintenance - a guide to reducing human error

Lucas, D. (2002). We are all human: How human factors impact health and safety performance. IOSH Annual Conference, Manchester, 15-16 April 2002

THE EFFECTS OF FAMILIARITY ON WORKPLACE RISK ASSESSMENT - CONTEXTUAL INSIGHT OR COGNITIVE AVAILABILITY

Andrew K Weyman[1] and David D Clarke[2]

[1]*Health & Safety Laboratory, Broad Lane, Sheffield S3 7HQ, UK*
[2]*School of Psychology, University of Nottingham, Nottingham NG7 3RD, UK*

This study provides an insight into the effects of familiarity on magnitude of perceived risk amongst employees ($N = 104$) in large mechanised coal mines. The method of paired comparisons is used to establish the degree of consensus present between a range of grades of personnel for a set of commonly exhibited risk behaviours. Statistical testing of between group differences (ANOVA) reveals high levels of consensus between personnel groups at the level of rank order, but significant differences ($P < 0.001$) for relative perceived risk magnitudes. Magnitudes of perceived risk are higher amongst respondents with greatest exposure to / personal experience of the depicted risks. Results are interpreted with reference to congitive availability, specifically, habituation and familiarity effects. Implications of findings for staff and organisations engaged in risk assessments are discussed.

Introduction

The requirement for enterprises to "conduct a suitable and sufficient assessment of risk" for all major areas of activity was formalised in the Management of Health and Safety at Work regulations (1992). Employers are obliged to identify: hazards associated with their work activity; the range of exposed individuals / groups; assess the risks to exposed persons and introduce suitable and sufficient risk control measures.

It is widely accepted that developing an effective system of risk assessment and associated safety management infrastructure is significantly more onerous for those employers whose activity is typified by dynamic, changeable work environments. In sectors such as construction, mining and offshore industries, employees have need to routinely engage in active problem solving and decision making over systems and methods of working. Within all workplaces, effective risk management requires significant active input on the part of those responsible for planning work activity. Moreover, where environments are changeable, and levels of associated risk are similarly variable, there is increased need for senior staff to involve others, in possession of valuable point-of-work insight, in the risk management process (see HSE, 1991).

Insights from a range of social science perspectives, however, suggest that notable differences can be predicted with regard to the ways in which risk is perceived and reacted to by members of different social (Walker et al, 1998) and occupational groups (Ostberg, 1980; Rundmo, 1995; Rundmo & Sjoberg, 1998). A central issue here relates to the magnitude of perceived risk, and the potential for cognitive bias, attributable to experiential influences, in particular effects associated with familiarity and habituation.

Familiarity with hazards has been found to have both amplificatory and attenuative effects on magnitude of perceived risk. In an immediate sense, familiarity operates at the level of personal experience, but can equally encompass shared understandings (Walker et al, 1998). Attenuative effects have typically been attributed to the absence of negative outcomes (Weinstein, 1984). Conversely, familiarity has been found to lead to heightened perceptions of risk, dread, worry and concern, in instances where hazards are, easily visualised, imagined, recalled or otherwise brought to mind (Slovic, 1987). Such effects are commonly attributed to 'availability bias' (Tversky & Kahneman, 1974).

In view of the established evidence, clear implications seem apparent with regard to the selection of staff for conducting workplace risk assessments. A number of authors have cited organisational role as a potentially salient experiential effect (see, for example, Rundmo, 1995). However, rather than being based upon a systematic assessment of perceived risk magnitude, the majority of reported findings are based upon extrapolations from attitudinal data. A notable exception, in this respect, is apparent in the approach adopted by Ostberg (1980) in a study of forestry personnel.

The study reported on constitutes a replication of the method outlined by Ostberg (1980). In common with the original study, it sought to compare relative perceived risk magnitudes for a set of commonly encountered hazards, for a range of personnel groups employed in a high hazard industrial context, namely large mechanised deep coal mines.

Hypotheses

- A high degree of consensus exists amongst mine personnel for the set of depicted risks at the level of rank order.
- Organisational role (degree of familiarity) will impact on magnitude of perceived risk.

Method

In common with the method advanced by Ostberg (1980) respondents were required to perform a risk raking exercise using a development of Thurstone's (1927) *'Case V Method of paired comparisons'*, for a set of pictorially depicted hazards. The technique has been empirically demonstrated to provide a more reliable method of risk ranking than direct ranking, while permitting comparisons to be made between different respondent groups (Sjoberg, 1967), in the current context referenced to employment status / organisational role. Essentially, the technique produces a subjective scale for the set of depicted risks, for each personnel group. Between group comparisons are made using the ANOVA statistic, using data transformation techniques detailed in Bock & Jones (1968).

Procedure

In common with Ostberg's study a set of volitional risk behaviours were presented in the form of pictorial vignettes, supplemented by a brief caption (See Table 1 & Figure 1).

Table 1. Set of Depicted Risks

A	Standing on a belt conveyor to hang cables
B	Lifting with out of date tackle
C	Illegal man-riding on a belt conveyor
D	Going under unsupported ground to trim a roof
E	Riding on an Armoured Face Conveyor.
F	Lifting using a load binder
G	Standing in line of a chain under tension.
H	Changing cutter picks without isolating power

Tick the most dangerous ☐ ☐

Figure 1 - Example risk pairing

Data collection.

A stratified opportunity sample of respondents, drawn from nine large mechanised mines (see Table 2), performed a paired comparisons exercise, indicating for each of 28 pairing, which depicted risk they considered the *'most dangerous'*.Paired risk items were presented in the form of a booklet. To reduce the potential for order effects, four different versions (randomised pairings) of the booklet were produced.

Table 2. Sample (*N* = 104)

Job title	*n*
Senior manager	8
Safety manager	18
Supervisor	35
Craftsman	18
Workman	25

Analysis of results

Prior to commencing the analysis, tests of within respondent consistency (Kendall's Coefficient *'K'*) identified five response sets (4.8%) with unacceptable number of intransitive triadic relationships, e.g. A > B > C > A. These response sets were excluded from the analysis. Tests of within group consistency (Kendal's coefficient *'W'*) revealed levels of item order agreement, for all five groups, within acceptable limits (range 0.71 to 0.75 (see Kendall 1970). Having satisfied the necessary tests of consistency it was possible to proceed with the scale generation process for each respondent group.

Minor differences were apparent between the groups with regard to item ranking. Correlational analyses, performed to formally test the degree of congruence, revealed strong relationships between all groups ($r = 0.91$ to 0.96; r^2 0.83 to 0.94; $P < 0.001$, in all cases). In view of the apparent strength of revealed relationships, it was concluded that differences between employment groups, in this respect, were minor, reflecting a high level of consensus, to the extent that risk rankings were fundamentally homogenous.

The number of times each item was judged as more dangerous than every other item, and related judement proportions, were calcuated for each resppndent group. In common with the procedure adopted by Ostberg (1980), the derived proportions were transformed into arcsine deviates. These values were summed, the mean for each item becoming the *'risk value'* for that item on the resultant subjective scale. A final transformation was to reference each derived value to an anchor item, 'C', this item being set to zero.

From Table 3, it can be seen that there are few items which are not separated by more than the 95% confidence interval. Hence, it is reasonable to conclude that respondents within each group made consistent and differentiated judgements. Furthermore, the presence of differences in the sums of the assessment values, between groups, indicates that the groups assigned different over-all risk values to the items.

Table 3 - Calculation of risk item scale values

Item	Workmen	Managers	Craftsmen	Supervisors	Safety officers	Overall mean	Rank
A	-0.1576	-0.1354	-0.2463	-0.0752	-0.0822	-0.1394	7
B	-0.2425	-0.47	-0.3109	-0.3415	-0.3796	-0.3489	8
D	0.5489	0.5085	0.5429	0.7061	0.572	0.5757	1
E	0.6085	0.3988	0.5341	0.6215	0.6122	0.555	2
F	0.0281	-0.1939	-0.0405	-0.0011	-0.168	-0.0751	6
G	0.536	0.139	0.3876	0.3343	0.2676	0.3329	4
H	0.5938	0.2369	0.3749	0.4452	0.331	0.3963	3
C	0	0	0	0	0	0	5
rank	5	1	3	4	2		
mean	0.2394	0.0605	0.1552	0.2111	0.1441	0.1621	
SD	0.3812	0.3503	0.3673	0.3893	0.3813	0.3683	
95% CI	0.0319	0.0293	0.0307	0.0326	0.0319	0.0308	

Interpretation of this output permits the observation that, in relative terms, workmen and supervisors overestimated the risks (above mean ratings) whereas, managers, safety officers and craftsmen underestimated the risks (see Figure 2). In order to further test the reliability of findings, in common with the method outlined by Ostberg (1980), an analysis of variance was performed on the data set (see Table 4).

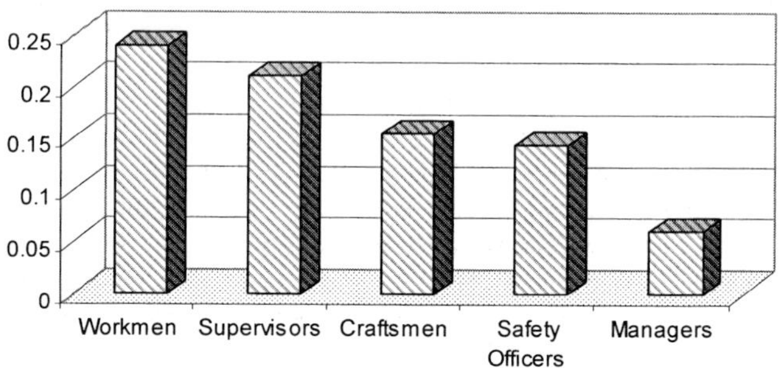

Figure 2. Mean scale values by workgroup

Table 4 - Results of two-way AVNOVA

	df	SS	MS	Variance ratio	
Between situations	6	4.07	0.68	r_6= 1-b/a 0.8624	
Within situations	28	4.2	0.09		
Between groups	4	0.17	0.04	F = BG/R 8.0723	*P* <0.001
Residual	24	0.13	0.01		
Total	34	4.37			

The high r_6 value in Table 4 indicates high inter-group reliability, i.e. agreement between the groups over the relative values ascribed to items (see Sjoberg, 1967). The significant *F* value suggests that absolute values of the items differed between the groups.

Principally, results indicate a main effect for items and an interaction between items and respondent group, and appear to have confirmed the hypotheses that there exists a notable consensus between the various occupational groups regarding the scaling of the risks, from the least to the most *dangerous* practices, i.e. differences in perspective are apparent, in terms of magnitude of ascribed risk, between the different personnel groups.

Conclusions

The study revealed two core findings. Firstly, that high levels of consensus exist between a range of personnel groups regarding their subjective assessments of the depicted risks. Secondly, notable differences exist between these personnel groups with regard to the relative magnitude of perceived risk. In essence, results reflect an inversion of the organisational hierarchy, with workmen rating the depicted risks as highest, and senior managers lowest. More specifically, results indicate that levels of perceived risk are highest amongst those personnel groups who can be characterised as having greater direct experience (familiarity) and hence potential exposure to the depicted risks, than members of more geographically and experientially distal groups.

In contrast to Ostberg's (1980) findings, where the effects of familiarity were found to attenuate level of perceived risk, findings from the current study appear to echo insights from psychometric risk research, where amplificatory effects have been mapped onto the broader constructs of 'dread' and 'unknown' risk (see Slovic, 1987). Within this context, the effects of familiarity are interpreted with reference to 'availability' and 'simulation' heuristics (Tversky & Kahneman, 1974), specifically the ease with which risk outcomes can be imagined and brought to mind. Beyond the psychometric framework, broadly complementary findings are provided by authors such as Rundmo (1995 and Greening (1997), who cite instances where personal experience of accidents tends to heighten 'victims' mental imagery and memory for such events, leading to elevated perceptions of the likelihood of comparable future events. In close knit communities it is perhaps also foreseeable that personal experience effects may generalise to include *close* secondary sources (see Lantz, 1958; Slovic, 1987). While interpretable in terns of cognitive availability, associated influences relating to differential levels of anxiety and concern might be also hypothesised (see Rundmo & Sjoberg, 1998).

Whether the relatively amplified perceptions of operational staff in the current study reflect influences associated with cognitive availability or affect ultimately cannot be ascertained on the basis of the available data. However, both interpretations highlight the salience of differential levels of exposure to / familiarity with sources of harm.

Returning to the issue of conducting risk assessments, findings from both the current study, and that reported by Ostberg 1980 indicate that differences in perspective, transparently attributable to workplace experience and job role, have potential to impact upon magnitude of perceived risk. It is further apparent that the degree of consensus in findings, regarding the effects of job role in this respect, may be variable between workplace contexts. Findings appear to reinforce HSE guidance (HSE 1991) which recommends that workplace risk assessments should be undertaken by groups of relevant personnel offering a range perspectives, rather than lone individuals.

References

Bock, R.D. & Jones, L.V. (1968) The measurement and prediction of judgement and choice. Holden-Day.

Greening, L. (1997) Risk perception following exposure to job-related electrocution accident: The mediating roles of perceived control. Acta Psychologica, 95; 267-277.

HSE (1991) Successful health and safety management; HS(G)65, HSE Books.

Kendall, MG (1970) Rank correlation methods, 4th ed Charles Griffin & Co Ltd, London.

Lantz, H.R. (1958) People of coal town. Columbia University Press, New York.

Ostberg, O. (1980) Risk perception and work behaviour in forestry - Implications for accident prevention policy. Accident analysis & Prevention, 12; 189-200.

Rundmo, T. (1995) Perceived risk, safety status, and job stress amongst injured and non-injured employees on offshore installations. J.of Safety Research, 26(2); 87-97.

Rundmo, T. & Sjoberg, L. (1998) Risk perception by offshore oil personnel during bad weather conditions. Risk Analysis, 18 (1).

Sjoberg, L (1967) Successive interval scaling & paired comparisons.Psychometrika,32(3).

Slovic, P. (1987) Perceptions of risk Science, 236; 280-285.

Tversky, A. & Kahneman, D. (1974) Judgement under uncertainty. Science 185; 261-275.

Walker, G.; Simmons, P.; Wynne, B. & Urwin, A. (1998) Public perception of risks associated with major accident hazards. Health & Safety Executive, HSE Books.

Weinstein, N.D. (1984) Why it won't happen to me: perceptions of risk factors and succeptibility. Health Psychology, 3; 431-457.

HUMAN ERROR

MEDICATION ERRORS AND HUMAN FACTORS

Ruth Filik[1], Alastair Gale[1], Kevin Purdy[1] & David Gerrett[2]

[1]*Applied Vision Research Unit*
[2]*Pharmacy Academic Practice Unit*
University of Derby
Derby DE22 3HL

Medication errors occur as a result of a breakdown in the overall system of prescribing, dispensing, and administration of a drug. There are numerous reasons for their occurrence, including: failed communication; poor drug distribution practices; dose miscalculations; drug- and drug-device related problems; incorrect drug administration, and lack of patient education. This paper reviews a number of human factors issues surrounding the occurrence of medication errors and describes preliminary research investigating aspects of drug package design.

Introduction

A medication error can be defined as, 'any event that could cause or lead to a patient's receiving inappropriate drug therapy or failing to receive appropriate drug therapy. The event could occur at any point from the decision to initiate therapy to the point at which the patient received the medication' (Edgar *et al.*, 1994, p. 1336). Errors may be potential (i.e., detected and corrected prior to the administration of the medication to the patient) or actual. In 1962, Barker and McConnell became the first to show that medication errors occur much more frequently than anyone had suspected – at a rate of 16 errors per 100 doses. More recently, one review article estimates that medication errors occur at a rate of one error per patient per day in most hospitals (Allan and Barker, 1990).

Medication errors range from the relatively minor, such as patients on a busy ward receiving their medication late, to death or paralysis resulting from a chemotherapy drug being wrongly administered by spinal injection. At least 13 cases of this more serious error have occurred since 1985 (Department of Health, 2000, p.13). As well as potentially causing injury or death to patients, and distress to families and healthcare professionals, medication errors can have considerable economic consequences. These include extended hospital stays, additional treatment, and malpractice litigation. The Department of Health (2000, p.60) report data from the Medical Defence Union stating that 25% of litigation claims in General Medical Practice are due to medication errors.

Only very recently has the medical profession made a systematic effort to reduce or eliminate the many preventable deaths and injuries that occur in hospitals each year. The serious public health problem posed by medication errors has been acknowledged by the Department of Health (2000) in a series of recent publications, including *An Organisation With a Memory*. One of the recommendations of this report is that serious errors in the use of prescribed drugs should be reduced by 40% by 2005 (p.86).

This increasing concern is also evident in the establishment of the National Patient Safety Agency (NPSA) in the UK in 2001. The NPSA is a Special Health Authority created to co-ordinate the efforts of the entire country to report and learn from adverse events occurring in the NHS, such as recent concerns over errors occurring during intravenous administration of potassium chloride solutions.

A similar system exists in the U. S., involving The United States Pharmacopeia (USP) and the Institute for Safe Medication Practices (ISMP). These institutions collect medication error reports (through the Medication Error Reporting Program (MERP)) and study them in an effort to provide feedback to practitioner, the Food and Drug Administration (FDA), and product manufacturers.

The occurrence of medication errors is currently under-reported. Methods of reporting medication errors include anonymous self-report, and incident reporting. One common reason for failing to report errors is fear of punishment. Therefore there is a need to overcome the mindset of a 'blame and punishment' oriented society. Though people actually commit errors, it is well known in the non-medical industry that the design of systems themselves allows people to commit errors (e.g., Bates, 1996). Therefore, rather than focusing on the persons involved, preventing errors requires a complete look at all the circumstances that allowed an error to occur before future errors can be prevented.

Cohen (1999, chap. 1) identified six categories of the causes of medication errors commonly seen by the ISMP. These are; failed communication, poor drug distribution practices, dose miscalculations, drug- and drug-device related problems, incorrect drug administration, and lack of patient education. The categories relevant to the current study are failed communication and drug- and drug-device related problems.

Failed communication

A number of factors can result in failed communication, including:

Handwriting – Poor handwriting can blur the distinction between two medications that have similar names, and is a widely recognised cause of medication errors (e.g., Brodell *et al.*, 1997).

Zeroes and decimal points – 'Trailing zeroes' (e.g., writing 2.0mg instead of 2mg) are a frequent cause of 10-fold overdoses (with 2.0mg being read as 20mg because the decimal point was not very clear or fell on a line). Lack of a zero before a decimal point can also lead to substantial dosage errors (for example .1mg being read as 1mg).

Abbreviations – For example the abbreviation 'U' for 'units' can be mistaken for a '0' or a '4' (Cohen, 1999, chap. 1).

Drugs with similar names – Name mix-ups (for example, confusion caused by drugs with similar looking names such as chlorpromazine and chlorpropamide) account for more than one-third of the medication errors reported to the USP MERP (http://www.usp.org/frameset.htm?http://www.usp.org/body.htm). A list of 645 pairs of look-alike and sound-alike drug names can be found in Davis *et al.* (1992). The list was derived by both subjective reviews of drug names and by actual mix-ups that have been published and reported to the authors.

Davis et al. report that confusion over look-alike names can be compounded by other factors, such as clinical dose overlaps (both drugs can be given in the same dose), availability of the same commercial strength or concentration, same dosage form available (e.g., both ointments), or same route of administration. Problems with confusable drug names can also be compounded by; illegible handwriting, incomplete knowledge of drug names, newly available products, similar packaging or labelling, and drugs being stored next to each other alphabetically by generic name.

Drug- and drug-device related problems
Similar looking commercial labelling and packaging is a common cause of medication error, often from the use of nearly identical packaging for two separate items. Edgar et al. (1994) report that the incidents most commonly reported to the USP MERP between August 1991 and April 1993 involved problems with the product (e.g., similar packaging or incomplete labelling).

Possible solutions to look-alike names/packaging
Cohen (1999, chap. 13) notes that manufacturers can use a variety of techniques to improve the readability of drug labelling and packaging. One method is reducing label clutter by having only necessary information (including brand and generic names, strength or concentration, and warning if any) on the front panel, and less important information on a side panel or package insert. Clarity can also be enhanced by the appropriate use of typeface elements such as serif or sans serif and upper- or lowercase letters (Nunn, 1992).

Another possible method is to use colour coding. Colour coding is the systematic application of a colour system to identify specific products. Although this system is already in use to some extent it is scientifically untested (regarding its use in drug labelling), and has a number of potential problems associated with its use. These include: the potential for people being less likely to read the label (relying on the colour to identify the product); the finite number of colours; difficulty in printing colour consistently, and the incidence of colour blindness.

One possible solution to the problem of look-alike names is to change the appearance of the names on computer screens, pharmacy and nursing unit shelf labels and bins, pharmacy product labels, and medication administration labels to emphasize different portions of a drug name. For instance, Cohen (1999, chap. 5) suggests that it is much easier to differentiate 'DOBUTamine' and DOPamine' than 'dobutamine' and 'dopamine'. The FDA Name Differentiation Project implemented this idea.

FDA Name Differentiation Project
Following FDA recommendations, The Office of Generic Drugs (in the U.S) requested that manufacturers of sixteen look-alike name pairs voluntarily revise the appearance of their established names in order to minimise medication errors resulting from look-alike confusion. The project spanned a two-month period beginning in March 2001 and ending in May 2001. In total, 142 letters were issued for 159 applications. The letters encouraged manufacturers to supplement their applications with revised labels and labelling that visually differentiated the established names with the use of 'Tall Man' letters (see e.g., http://www.fda.gov/cder/drug/MedErrors/nameDiff.htm for a list of the names involved and the recommended revisions).

Following these recommendations, a manufacturer would capitalise the letters in the drug name that are different from the name's stem. For example, two generic drugs that could become mixed up are chlorpropamide and chlorpromazine. Both names have 'chlorpro' at the beginning, so the remaining letters of each name would be capitalised and perhaps printed in a different colour, e.g., chlorproMAZINE and chlorproPAMIDE.

Experiment 1

The aim of the experiments reported in this paper is to begin to investigate methods by which medication errors can be reduced by improving drug package design. Specifically, investigating the use of 'tall man' letters to make look-alike drug names less confusable. It was hypothesised that if highlighting the different parts of similar looking word names on drug packaging (e.g., chlorproMAZINE and chlorproPAMIDE) makes them easier to distinguish from each other, then participants who are shown pairs of drug names should be faster to indicate that a pair of words are different from each other when these words contain 'tall man' letters than when the words are presented normally (all in lowercase letters).

Method

Participants
20 staff and students from the University of Derby participated (14 females, 6 males, mean age 34.7 years s.d. 9.78).

Materials and Design
Eighty pairs of generic drug names were selected as materials. Sixteen of these were pairs of drug names that had been recommended for the FDA Name Differentiation Project. The remaining 64 pairs were selected from Davis *et al.* (1992), who published a list of 645 similar looking and similar sounding drug names. These 64 pairs were selected on the grounds that they were similar to each other in length (i.e., did not vary in length from each other by more than one letter) so that people could not use word length alone in order to discriminate between them. Word pairs were presented with either both pairs in lowercase letters, or with both words having sections in 'tall man' (capital) letters. Word pairs could consist of two different words, or the same word presented twice. Participants did not see any word pair more than once. Therefore four stimulus files were created so that each word pair could be presented in each of the four conditions that are shown below:

chlorpromazine	chlorpromazine	(lowercase – same)
chlorpromazine	chlorpropamide	(lowercase – different)
chlorproMAZINE	chlorproMAZINE	(tall man – same)
chlorproMAZINE	chlorproPAMIDE	(tall man – different)

The program to present the stimuli was written in Psychology Software Tools E-Prime version 1.1. The words were presented on a 17 inch CRT monitor in Arial font as black text on a white background at a screen resolution of 640 x 480. The shortest word used (digoxin) subtended 3.28° of visual angle, and the longest word (medroxyprogesterone) subtended 10.29° of visual angle to the participant. The height of the uppercase letters was 0.76° and the words were separated by 2.14° of visual angle.

Procedure

Participants were informed of the nature of the study and asked to sign a consent form. They then read a set of instructions on the computer screen. Their task was to decide whether the pairs of words that were presented to them were the same word, or two different words. They had to make this decision as quickly and as accurately as possible and indicate their response via two keys on a serial response box. Participants fixated a cross at the centre of the screen that disappeared after one second and was replaced by the word pair. The word pair remained on the screen until the participant pressed one of two keys on the button box. Four practice trials preceded the eighty experimental trials. The experiment took approximately five minutes in total.

Results

Reaction time was taken as a measure of how easy the word pairs were to distinguish. Mean reaction time for lowercase word pairs that were the same was 2213.63ms (standard deviation = 979.16). Mean reaction time for lowercase word pairs that were different was 1752.25ms (standard deviation = 649.93). Mean reaction time for 'tall man' word pairs that were the same was 2222.65ms (standard deviation = 893.83). Mean reaction time for 'tall man' word pairs that were different was 1713.11ms (standard deviation = 577.69). A 2(similarity) x 2(letter style) ANOVA was conducted on the results. There was a significant effect of similarity, $F_{(1, 19)} = 22.83$, $p<.0005$, with longer reaction times when the two words were the same than when they were different. There was no effect of letter style and no interaction (both Fs < 1) so participants could discriminate between the 'tall man' words no faster than they could discriminate between the lower case words.

Discussion

Reaction times were longer when word pairs were the same then when they were different. This could simply be because participants could judge word pairs as being different on detection of a single difference between the words. In order to judge them as being the same, however, participants had to scan the entire word. Results from Experiment 1 indicate that word pairs containing 'tall man' letters are no easier to distinguish from each other than words that are presented normally (in lowercase). However, it is possible that in a real life situation, people would be aware that the uppercase sections of the words were particularly informative, and therefore adopt a different strategy based on this knowledge. This possibility was investigated in Experiment 2.

Experiment 2 - Method

20 staff and students from the University of Derby (8 males, 12 females, mean age 22.9, s.d. 4.39) participated. The materials and design were identical to those in Experiment 1. The procedure was also identical to Experiment 1 except that participants were given the additional instruction that in 'tall man' word pairs, the capital letters were informative.

Results

Mean reaction time for lowercase word pairs that were the same was 1571.20ms (standard deviation = 594.28). Mean reaction time for lowercase word pairs that were different was 1451.38ms (standard deviation = 448.23). Mean reaction time for 'tall man' word pairs that were the same was 1520.54ms (standard deviation = 540.27). Mean reaction time for 'tall man' word pairs that were different was 1319.53ms (standard deviation = 364.42). Once again there was a significant effect of similarity, $F_{(1,19)} = 13.14$, $p<.005$, with longer reaction times for words that were the same than words that were different. There was also a significant effect of letter style, $F_{(1,19)} = 4.71$, $p<.05$, with shorter reaction times for the 'tall man' word pairs than for the lowercase words. There was no significant interaction, $F_{(1,19)} = 1.30$, $p = .269$.

Discussion

Results from Experiment 2 show that if participants are told that the 'tall man' lettering is informative, then they can discriminate between word pairs with 'tall man' lettering more easily than word pairs that are presented in lowercase. Taking Experiment 1 into account, this would suggest that 'tall man' lettering does not intrinsically make look-alike word pairs easier to differentiate, but when participants are informed of it's purpose, it can be effective. It is necessary to view results tentatively for a number of reasons, for example, some of the word pairs might not have been particularly confusable, therefore not hard to discriminate in lowercase (so the benefits of including the 'tall man' lettering would be limited). It is also necessary to use the results from laboratory studies of this nature in conjunction with evidence from more 'real life' studies in order to obtain a clearer picture of the effectiveness of various methods of reducing medication errors.

References

Allan, E.L., & Barker, K.N. (1990). Fundamentals of medication error research. *American Journal of Hospital Pharmacy*, **47**, 555-571.

Barker, K.N., & McConnell, W.E. (1962). The problems of detecting medication errors in hospitals. *American Journal of Hospital Pharmacy*, **19**, 360-369.

Bates, D. (1996). Medication errors: how common are they and what can be done to prevent them? *Drug Safety*, **15**, 303-310.

Brodell, R.T., Helms, S.E., KrishnaRao, I, Bredle, D.L. (1997) Prescription errors: legibility and drug name confusion. *Archives of Family Medicine*, **6**, 296-298.

Cohen, M.R., (ed.) (1999). *Medication Errors*. Washington: American Pharmaceutical Association.

Davis, N.M., Cohen, M.R., and Teplitsky, B. (1992). Look-alike and sound-alike drug names: the problem and the solution. *Lippincott's Hospital Pharmacy*, **27**, 95-110.

Department of Health (2000). *An organisation with a Memory*. London: HMSO.

Edgar, T.A., Lee, D.S., Cousins, D.D. (1994). Experience with a national medication error reporting program. *American Journal of Hospital Pharmacy*, **51**, 1335-1338.

Nunn, D.S. (1992). Ampoule labelling – the way forward. *Pharmaceutical Journal*, March 14.

REPETITIVE INSPECTION WITH CHECKLISTS: DESIGN AND PERFORMANCE

Blaine Larock and Colin G Drury

University at Buffalo, State University of New York, Department of Industrial Engineering, Buffalo, NY 14260, USA

Tasks in a number of industries require long, complex procedural inspections. Inspectors are typically given checklists, but after gaining familiarity with the checklist they will often perform the task from memory, e.g. in civil aviation (Pearl and Drury, 1995). Our questions were whether such behaviour would occur under controlled conditions, and whether the design of the checklist contributed to the behaviour. This study used a 108-step procedure for an overnight check on a common airliner, taught to 24 students and then repeated on eight different days. After training to perform the task in the order given on their checklist, with signoffs where specified, participants returned eight times to perform the task on simulated aircraft systems. Checklists were either arranged by function or by spatial location. There were either individual signoffs for each of the 108 items, or 37 signoffs for logical subsets of items. There was no difference in probability of defect detection between conditions, but the performance times and rates of both sequence and signoff errors changed significantly. All participants tended to follow a spatial sequence whatever their checklist, so that users of the functional checklist made more sequence errors (and were also slower) than those using the spatial checklist. Signoff errors did not differ between groups, although participants were quicker with fewer signoffs, and preferred that condition in post experimental ratings.

Introduction

Aircraft inspection tasks play a vital role in ensuring that aircraft systems are operational and are functioning properly. The use of a job-aid checklist or workcard during an inspection task is quite common in the aviation industry and is used as the primary tool to assist in the inspection process. Presently, aviation maintenance technicians (AMT) use a job-aid checklist or workcard to assist them in identifying the locations and tasks required to be performed during the inspection. Aircraft pilots also use a checklist while in the cockpit to assist them in performing flight deck procedures.

The importance of the inspection checklist and its effect on the quality of the inspection has lead to significant human factors research into the design and improvement of aircraft inspection documentation. Several studies have investigated the design and layout of the job-aid checklist to improve the efficiency of the inspection and to reduce the risk of human error.

Degani and Wiener (1990) performed a field study to investigate several normal flight deck checklists. This study revealed that design weaknesses existed with the traditional paper checklist and the humans that interact with it. In response, a list of design guidelines for normal flight deck checklists was developed. Specifically, Degani and Wiener (1990) found that the sequencing of checklist items or tasks should follow the geographical organization of the items in the cockpit and should be performed in a logical flow.

Patel, Prabhu and Drury (1993) found that during the initial inspection of an aircraft on arrival at maintenance, the sequence of tasks in the checklist or workcard did not match the sequence of tasks that aircraft maintenance technicians typically followed. In a related study Pearl and Drury (1995), questionnaires and videotapes showed that mechanics tended to sequence their tasks using spatial cues on the airplane rather than the order specified on their workcard. The study also revealed that aviation maintenance technicians who performed low level inspections used spatial locations of tasks to sequence them. In addition, many aircraft mechanics rarely used the checklist and viewed it as a guide only for inexperienced mechanics. Experienced inspectors felt they had acquired sufficient skill to perform the inspection task using their memory and referred to the checklist occasionally.

A more recent series of investigations (Ockerman and Pritchett 1998a, 2000a, 2000b) have examined the relationship between the medium (paper and wearable computers) on which the procedure was displayed, the presentations of procedure context, over-reliance and inspection performance for a preflight inspection task. The studies found that inspection performance could be influenced by the presence of procedure context information presented with procedures. Their 1998 study also observed that one third of the participants used their memory and not the task guidance system to perform the preflight inspection. They observed that in some sessions the subjects performed the task from memory and only consulted the checklist to see if anything was forgotten.

All the studies quoted involved the use of a job-aid checklist or workcard while performing a task. They differ however in content and in the task itself, ranging from checklists in cockpits, to pre-flight inspection to detailed maintenance inspection. They all raised the issue of task sequencing: while checklists can be organized in either a functional or spatial manner, the spatial organization appears more natural. One issue not addressed, but of great concern to maintenance workcard designers was that of the number of sign-offs. Typically, every task step has to be signed by the inspector or mechanic, although as observed by Pearl and Drury (1995), in many cases a logical group of task steps is performed as a unit, then all are signed off at the same time. This is a breach of procedure, but it raises the concern that too many signoffs will be treated as groups in practice.

Methodology

The main objectives of this study were to investigate the effects that the sequencing of tasks and checklist signoffs has on a frequently performed aircraft maintenance task. The aim was to determine whether the effects observed in the industrial studies above also appeared in a simulation with non-aviation participants, and how such behaviours developed over time. The study consisted of four study conditions all combinations of functional task sequencing with 37 or 108 signoffs for the 108 tasks. The study is modeled on the aviation maintenance technician's typical work in that the same inspection task is often performed every night for extended periods. This study was made more rapid, as it was not necessary to walk a great distance to inspect the aircraft locations in the simulation. In addition, to reduce the skill level requirements, all tasks were given easily recognizable outcomes. For example, a low oil level was represented by a picture of a reservoir with the oil level below a critical mark.

Three boards, 2.4m wide and 1.2m high were placed on adjacent walls of the laboratory, two boards on opposite walls and one on the end wall. The three boards represented the left and right exteriors and the interior of a Boeing 737-300 twinjet aircraft. Affixed to each board were a number of 215mm x 280 white panels with hinged covers. The white panels represent the parts of the Boeing 737-300 (i.e. Left Main Landing Gear, Radome, etc.) that required checking. The thirty-seven panels representing the parts of the Boeing 737-300 were placed in the appropriate locations simulating the actual aircraft. Inside each panel, one or more paper sheets contained a graphical representation (picture, graph, pictogram) of each of the aircraft system requiring inspection. There were a total of 108 subtasks distributed among the thirty-seven panels that required inspection. Participants were trained to perform all tasks in sequence by repeated task performance with accuracy feedback before the main study.

Four groups representing the four checklist conditions each containing six subjects participated in the main study. Each subject was randomly assigned to a group representing the type of checklist that was to be used in their inspection task. Following the training session, subjects were required to perform their inspection task on eight separate occasions over a span of one to two weeks. The 24 participants consisted of 17 male and 7 female students aged from 18-48 years (mean = 24 years). During the eight trials, subjects were introduced to six (6) discrepancies for detection. Discrepancies were distributed to give 2 trials each with 2 and 1 discrepancies, and four trials with no discrepancies.

Task performance was video-recorded in order that errors could be identified, classified and counted after task completion. After trials 1, 4, and 8, participants rated workload on the NASA TLX scale, and after study completion used the Patel et al (1993) ratings of workcard suitability. Performance measures were error type, frequency and time to complete the task.

Results

Total time was defined as the time between when checklist was handed to subject and when the subject handed the checklist back to researcher. An analysis of variance for total time indicated significant differences exist between checklist layout ($F = 16.66$, $P = 0.01$), number of signoffs ($F = 11.59$, $P = 0.003$), trials ($F = 89.25$, $P < 0.001$) and the interaction of layout and the trials ($F = 8.74$, $P < 0.001$). Subjects using a spatial checklist took less time (655.8 s) to complete the task than did subjects using a functional checklist (976.0 s). The same was true when subjects signed off only thirty-seven times on a checklist (682.4 s) rather than one hundred and eight times (949.4 s). Using either task sequence layout (spatial or functional) the total time to complete the task significantly decreased from trial one to trial eight but at different rates (Figure 1).

Errors were divided into Outcome errors, i.e. where a wrong outcome would be realized, and Procedural errors, where the correct procedure was not followed, irrespective of whether the outcome was or was not correct.

Table 1 summarizes the Outcome error distribution across the four conditions. There were very few errors, so that the only test showing significance was the difference between layouts for total errors ($\chi^2(1) = 4.00$, $p < 0.05$).

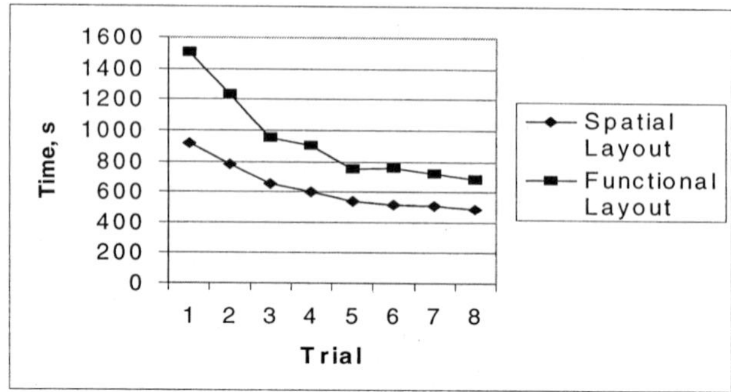

Figure 1. Total Time Means-Significant Interaction between Layout & Trial

Table 1. Outcome Error Distribution Across Conditions

		Spatial	Functional
37 Signoffs	**Miss**	1	3
	False Alarm	0	0
	Failure to record task step	2	3
108 Signoffs	**Miss**	1	1
	False Alarm	0	4
	Failure to record task step	0	1

Sequencing errors were defined as deviations from checklist sequencing protocols. Participants were required to follow the sequence of tasks exactly as it was arranged on the checklist. Total errors under each of the four conditions were compared using Chi-Square tests, which showed significant effects of layout ($\chi^2(1) = 193$, $p < 0.001$) and a layout X signoff interaction ($\chi^2(1) = 41.1$, $p < 0.001$). The mean sequencing error rates for the four conditions are shown in Figure 2.

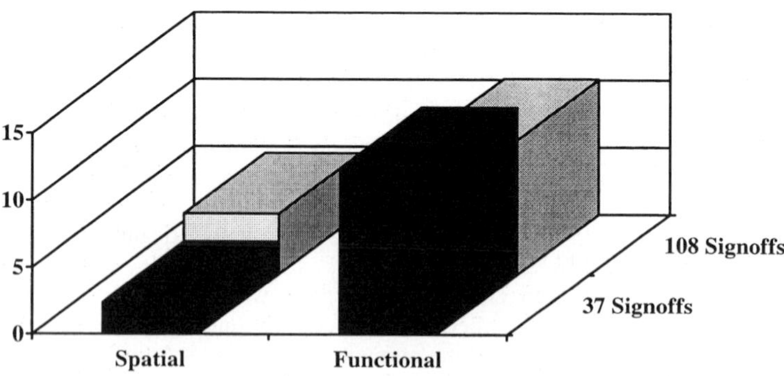

Figure 2. Sequencing Error Means by Study Condition

Signoff errors were defined as deviations in signoff inspection protocol that occurred after the subject has made a decision on the status of the aircraft system. During the task, subjects are required to sign off (a check mark is placed in a blank box on the checklist) or record a discrepancy for every task or group of tasks, depending on the checklist condition. Total signoff errors were tested with a Chi-square test, giving significant main effects and interaction: Layout ($\chi^2(1) = 260$, $p < 0.001$), Signoff ($\chi^2(1) = 511$, $p < 0.001$), interaction ($\chi^2(1) = 686$, $p < 0.001$). The interaction is shown for mean errors in Figure 3. No significant effects were found for scores on NASA-TLX or the Patel et al (1993) ratings.

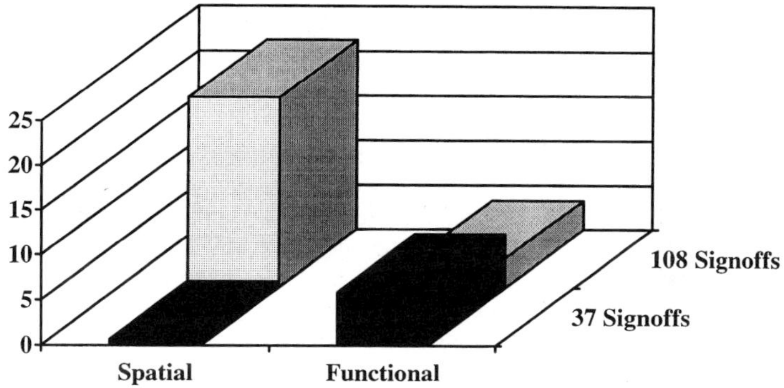

Figure 3. Signoff Error Means by Study Condition

Discussion

The major results of this study were that outcome error rates were low, and that participants moved towards using a spatial layout whatever the layout they were given. The rate of missing discrepancies was only about 4% overall, while for false alarm rates it was 0.02% and for failures to record 0.04%. The results indicated a higher level of outcome accuracy as compared to the Ockerman et al. (2000a) study where 51% of discrepancies were missed. Our task was simplified in that no judgment was required in detecting errors, but equally we used non-aviation participants.

In terms of Layout effects, we found results under our controlled conditions very similar to those observed in the field studies quoted earlier. The tendency to work spatially appears not just to be a function of expertise in piloting (Degani and Wiener, 1990; Ockerman and Pritchett 1998) or maintenance / inspection (Pearl and Drury, 1995; Patel et al, 1993). We have shown that this behaviour is characteristic of the task itself by training naïve participants to perform correctly, then having them perform the same task repeatedly with just the instruction to follow the rules given in training. For maintenance mechanics and inspectors, workcards are still being produced with a functionally layout (Drury, Kritkayski and Wenner, 1999) despite improved designs being available. We have produced a Documentation Design Aid (DDA) that gives guidance on the production of improved documents and have proved its usability and error-reduction potential in a number of studies. For example, Drury et al (1999) showed that comprehension error rates for the most extreme work card were 27% and 51% for existing workcards but only 4% for the DDA-designed workcard.

There was an overall difference in outcome error rates between the two layouts with spatial giving a total of 4 errors and functional giving 12 errors, but the differences were not significant for the individual error categories in Table 1. Also, from Figure 1 we can see that the times were much lower for the spatial layout, about two-thirds of the times for the functional layout. The spatial layout had far fewer sequencing errors than the functional, in a ratio of about 1:3, as would be expected when the layout of the job aid matched the way the participants performed the task in practice. The signoff errors give a totally different picture (Figure 3) with about two-thirds of the total signoff errors occurring in the spatial layout/108 signoff group. A spatial layout appears to encourage signing off a block of task steps together. When the spatial layout was combined with signing off by logical groups, the error rate was much the lowest of all four groups, showing that all aspects of the job aid must be matched to the task itself if errors are to be avoided. Indeed, the combination of spatial layout and 37 signoffs had the lowest error rate in all of the error measures, and the shortest task time.

While we could not find other studies in the literature of the effects of number of signoffs, this has been a contentious issue in maintenance. Any missed signoff is of great economic consequence to the airline, with large fines per flight completed with improper paperwork. It is also of great job security consequence to the inspector or mechanic, whose job is in jeopardy if he/she makes such an error. In human factors terms, the design of the whole task should be integrated, with error-prone situations being avoided. Our research suggests strongly that a combination of a spatial layout and signing off of logical groups of tasks produces the highest performance on all measures. Interestingly, this finding was not matched by either workload scores or ratings of design elements of the workcards, even though the latter had been a sensitive measure in previous research (Patel et al, 1993).

Acknowledgement

This work was supported by a grant from the FAA AFS-300, contract monitor Ms. Jean Watson.

References

Pearl, A. and Drury, C. G. 1995, Improving the Reliability of Maintenance Checklists. *Human Factors in Aviation Maintenance* (Office of Aviation Medicine, Wash, D.C.).

Degani, A. and Wiener, E. L. 1990, The Human factors of flight deck checklists: The normal checklist. (*Contractor Report 177549*) (NASA Ames Res Ctr, Moffett Field).

Patel, S., Pearl, A., Koli, S. and Drury, C. G. 1993, Design of Portable Computer-based Workcards for Aircraft Inspection. *Human Factors in Aviation Maintenance - Phase Four Report* (Office of Aviation Medicine, Washington, D.C.)

Ockerman, J. J. and Pritchett, A. R. 1998, Preliminary investigation of wearable computers for task guidance in aviation inspection. Paper presented at the *Second International Symposium on Wearable Computers*, Pittsburgh, PA.

Ockerman, J. and Pritchett, A. 2000a, Reducing Over-Reliance on Task-Guidance Systems. *Proc HFES 44th Annual Meeting,* San Diego.

Ockerman, J. and Pritchett, A. 2000b, Wearable Computers for Aiding Workers in Appropriate Procedure Following. Unpublished paper submitted to the *International Symposium on Wearable Computers.*

Drury, C. G., Kritkausky, K. and Wenner, C. 1999, Outsourcing Aviation Maintenance: Human Factors Implications, *Proc HFES 43th Annual Meeting, Houston, TX*, 762-766.

DO JOB AIDS HELP INCIDENT INVESTIGATION?

CG Drury[1], Jiao Ma[1] and K Woodcock[2]

[1] *University at Buffalo, State University of New York, Department of Industrial Engineering, Buffalo, NY 14260, USA*
[2] *University of Toronto, Canada*

A previous study established that investigators only collect a fraction of the available facts, and further select facts for their reports. The current study was designed to measure the effectiveness of job aids in improving the thoroughness of investigations of incidents in aviation maintenance. The methodology involved having participants investigate a known incident scenario by asking the experimenter for facts, as they would in their normal investigation routine. The two job aids used were the Maintenance Error Decision Aid (MEDA) developed by Boeing and the Five Principles of Causation (Marx and Watson, 2001). Both are used extensively in aviation maintenance. We tested a total of 15 experienced users of the two job aids, where the investigators were provided with the job aid they had been trained to use. Eleven of the 15 participants used their job aids during the investigation but four did not. The results showed a significant improvement in investigation performance when the job aids were actually used.

Introduction

Aviation Accident Investigation has been recognized by most countries as a necessary component of aviation safety. Many countries and many military services having an equivalent of the National Transportation Safety Board (NTSB) charged with determining the causes of accidents and incidents so that preventive measures can be implemented.

The genesis of the current project lies in the work of Marx (1998a) who studied the causation of accidents using classical attribution theory. He found that people in aviation maintenance have certain consistencies in attribution of incidents, and proposed a set of causation conditions based on these consistencies. However, our point of departure from his work was our assertion that the investigation process itself is an active rather than a passive task, and depends intimately on human cognition. Thus, an investigator must actively choose what lines of investigation to pursue, and when to stop following each causal chain.

These decisions are likely to be influenced in a dynamic manner by the number and sequence of facts discovered, as well as by any biases or prejudices of the investigator. Hence, a study of attribution of causes and blame needs to be paralleled by a study of what set of facts an investigator discovers, and what sequence is used to discover them.

Earlier we (Drury, Wenner and Kritkausky, 1999) developed an incident investigation methodology for understanding how aviation personnel investigate maintenance incidents. The methodology has professional participants investigate incident scenarios. This methodology was originally developed by Woodcock and Smiley (1999) for analyzing how industrial accident investigators performed their task. Each scenario consists of a relatively exhaustive listing of facts pertaining to the incident. The facts are initially unknown to participants whose task is to elicit facts from the experimenter until they are satisfied that they have satisfactorily investigated the incident. At that point they provide the experimenter with a synopsis of the incident in their own words. Their success is judged primarily by their depth, i.e. the number and type of facts they elicit and the number and type of facts they choose to include in the synopsis. Earlier, we found that overall only about 32 of the available set of facts (out of a maximum of 40-115 facts) were requested by participants. Of this total requested, only about 9 appeared in the participants' synopsis of the incident. There were differences in total facts requested between personnel job types, mainly as a result of including a sample on non-airline professional accident investigators, who found about 20% more facts than AMTs, managers or Quality Assurance (QA) investigators.

In the current study we were specifically concerned with the use of investigative tools to determine how they affect (hopefully improve) the depth of the investigation. Within the aviation maintenance domain, a number of incident investigation methodologies are currently in use. Perhaps the earliest was Boeing's Maintenance Error Decision Aid (MEDA) described more fully below. One of MEDA's developers (D. Marx) went on to produce the Aurora Mishap Management System (Marx, 1998b) that expands the concepts introduced in MEDA. Marx then produced a tool that is more an aid to logical reasoning and analysis than a methodology for investigation, the Five Rules of Causation, again described below.

The literature on incident investigation (Ferry, 1981; Rasmussen, 1990) (typically sees it as a four-phase process. An initial Trigger starts the process, which then has a Data Collection phase, followed by a Data Analysis phase, and the process is completed with a Reporting phase. On the basis of our earlier results we have developed a more realistic descriptive model (Figure 1) of how people actually investigate incidents. The Data Collection and Analysis phases could not be separated in our study, and indeed it is doubtful whether they ever can be in practice. Initial hypotheses are formed, data is collected to test these hypotheses and new analyses performed based on the outcome, in an iterative process. After the trigger stage is the exploration of the boundaries of the system under study. This is primarily a temporal exploration, as the spatial boundaries are largely implicit, e.g. the hangar or the departure gate. In this Boundary Stage the investigator extends the information from the Trigger to help structure the rest of the data collection and analysis, so that in one sense this stage provides a logical bridge to the Sequence Stage.

Investigation Job Aids

***MEDA*:** The MEDA investigation consists of an interview with the mechanic(s), who made the error, to understand the contributing factors. A decision is then made by management as to which contributing factors will be improved to reduce future errors. Central to the MEDA process are the MEDA Results Form, and the MEDA Users' Guide (Boeing, 1997). The MEDA Results Form has six sections, moving the investigator from the background information on the incident in a logical manner towards error prevention strategies. Note that a single-incident may trigger more than one MEDA Results Form if more than one error contributed to the incident. MEDA's sections are:

Section I. General Information. Background data such as date, time and aircraft details.

Section II. Event. A classification of the event outcome (e.g. operations process event, aircraft damage event, personal injury) plus a short narrative event description.

Section III. Maintenance Error. A classification of the error (e.g. Installation error, servicing error) plus a short narrative description of the error.

Section IV. Contributing Factors Checklist. Here, a large number of contributing factors under 11 categories (e.g. Information, Job/Task, Individual Factors, Environment) are listed exhaustingly. The investigator checks each factor and provides a short narrative description pertinent to the factor.

Section V. Error Prevention Strategies. This section examines the barriers that were breached for the error to have propagated (e.g. Maintenance Policies, Inspection). From these, a list of recommended error prevention strategies is generated, with each keyed to specific contributing factors from Section IV.

Section VI. Summary. A narrative summary of the event, error and contributing factors is required.

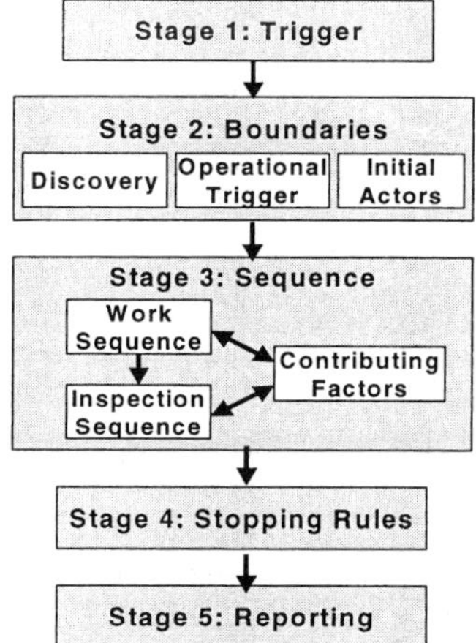

Figure 1. Model of Aviation Incident Investigation

MEDA was developed by Boeing in conjunction with several airlines, labor unions and the FAA. It is the most widely used aviation maintenance incident investigation tool, with Rankin (2000) reporting implementation in over 120 organizations, and active use by two-thirds of these. One airline reported decreasing flight departure delays due to mechanical problems by 16%, while another reduced operationally significant events by 48% over two years after implementing MEDA.

***Five Rules of Causation*:** The causation system pioneered by Marx (e.g. Marx and Watson, 2001) was developed to fill a need in incident reporting systems, and particularly the recommendations coming from existing systems. This system is intended to increase the rigor of deriving recommendations from incident data. Note that the Five Rules of

Causation were never intended as an investigative job aid, only to assist with making recommendations based on the investigation.

Based on attribution theory (Fiske and Taylor, 1984) and a data collection involving participants who derived attributions from scenario data, Marx originally developed seven causation rules which have since been truncated to five rules and taught extensively to airlines, the armed forces and medical practitioners:

1. Causal statements must clearly show the "cause and effect" relationship.
2. Negative descriptors (such as poorly or inadequate) may not be used in causal statements.
3. Each human error must have a preceding cause.
4. Each procedural deviation must have a preceding cause.
5. Failure to act is only causal when there is a pre-existing duty to act.

Methodology

Incident Scenarios: The three scenarios from previous studies were used. Each incorporated 50-120 facts, classified by Fact Type into Task, Operator, Machine, Environment and Social (see Drury and Brill, 1983).

Participants: We recruited fifteen participants who actually conduct maintenance accidents/incidents investigations at work. Their average experience as an investigator was about 4 years, and they had investigated on average about 16 cases in the previous year.

Experimental Design: Each participant was tested on a single scenario, with the order randomized across participants. The participants made quite different uses of the job aids we provided. At the first two sites, participants used the job aids extensively, while at the third site they did not refer to the job aids.

Interview Protocol: The data collection was in interview format, where the participants asked questions which were answered by the experimenter. The job aids were laid out in front of the participants. In addition, participants were given a pad and pencil to record facts if they desired. The incident trigger paragraph was given to the participant. At this point, the participant was prompted to ask questions of the experimenter, just as they would ask the same questions of personnel in the incident. The experimenter answered the participant's questions from the data sheets developed for each scenario. When the participants declared that they would stop the investigation, they were asked to provide a verbal synopsis of the incident, as they would in writing a report. They were asked to list the contributing factors in their synopsis.

Analysis Methods: Analysis of the audiotape allowed a separation of the two parts of each interview: the data collection stage and report stage where a synopsis was given. The number of facts requested for each scenario was the primary measure of data collected. From a transcript of the participant's report, the total number of synopsis facts was measured. The ANOVA model used was a 3 (Sites) X 3 (scenarios) X 5 (Fact Type) fixed effects model with participants nested under groups. Subsidiary variables such as years of experience, organization, human factors training etc. could be treated as covariates.

Results

Three different styles of using the job aids were observed. These observations were not analyzed further, but are used later as a partial explanation of some of the results we found.

Style 1. Job Aid as Checklist. Two participants in the MEDA group relied on the job aids extensively. For example, they went through each item on the MEDA Results Form in more or less the given order, and asked relevant questions based on those items. They quoted the phrases or read aloud the contents in the form.

Style 2. Job Aid as Back Up. Three MEDA investigators and three Five Rules investigators were observed to first conduct their own investigation independently of the job aid, taking extensive notes. At a certain point of the investigation, when they had apparently asked all the questions they could, these investigators started referring to the job aids.

Style 3. Job Aid Rarely Used. One MEDA and three Five Rules investigators conducted the investigations completely independent of using the job aids. They structured their own investigation while taking extensive notes.

Analyses of Variance (ANOVAs) of Number of Facts Requested and Number of Facts Requested in Synopsis were performed. First, a correlation analysis was performed of all of the demographic and performance variables for each participant. The correlation between Number of Incidents Investigated in the previous year and Number of Facts Requested was 0.645 ($p = 0.009$). Number of Incidents Investigated was thus used as a covariate in our analyses, making these Analyses of Covariance (ANCOVAs).

For Number of Facts Requested, the covariate was marginally insignificant ($F(1, 27) = 4.10$, $p = 0.052$) but was significant for Percentage of Facts Requested ($F(1, 29) = 4.27$, $p = 0.047$) and so is included in these results. There was a significant effect of Fact Type ($F(4, 29) = 15.91$, $p < 0.001$) and a significant interaction between Fact Type and Scenario ($F(8, 29) = 3.40$, $p = 0.007$). Such effects and interactions have been found in previous years, and indicate that not all fact types were investigated equally. For the Synopsis, the number of facts was significantly different by Fact Type ($F(4, 29) = 5.10$, $p = 0.003$), but not significant in its interaction with Scenario ($F(8, 29) = 1.97$, $p = 0.087$).

Task facts were still the major contributor, with Operator and Social facts close behind. Fewer Machine facts were requested, although a greater fraction of these appeared in the synopsis. Finally, Environment facts were requested and reported rarely, especially for Scenario 2.

It is also of interest to check the times taken to complete the investigation, although accuracy rather than speed is our primary concern. ANOVAs were run of the data using Stop Time as dependent variable and Site and Scenario as crossed factors. The only significant result was that for Stop Time, Site had a significant effect ($F(2, 4) = 9.31$, $p = 0.031$). The MEDA site averaged 51 minutes for the investigation task, the Five Principles site averaged 35 min, while the third site averaged only 19 min. Thus, use of the job aids led to different times, with MEDA taking longest and the site not using either job aid taking the least time.

Because the same type of participants were used in this study as had been used in the baseline study, a direct comparison of the results between these two years can provide direct evidence of the efficacy of job aids for incident investigation. Baseline data were selected for only those participants who were tested with the same three scenarios used in the current study, giving 20 participants. Analyses of Variance were performed for the combined data set with the factors of Job Aid used, Scenario and Job Type. The Job Aid used in baseline was in fact no job aid. In the current data it could either be MEDA (Site 1), Five Principles

(Site 2) or no job aid for Site 3. The only significant results were for Number of Facts Requested and Stop Time, where Job Aid was a significant factor. The significance was $F(3, 19) = 5.28$, $p = 0.008$, for Number of Facts Requested and $F(3, 16) = 3.32$, $p = 0.047$ for Stop Time. These two measures correlated highly ($r = 0.912$) but failed to reach significance ($p = 0.088$).

The overall picture is that the depth of investigation increased with the use of job aids compared with baseline, and with Site 3 in the current study, although only the MEDA job aid gave a statistically significant improvement.

Discussion

The job aids would be expected to improve performance, even though one (the Five Principles) was never intended as a job aid during the investigation process itself. Indeed, they did make such an improvement overall. We found a significant effect of Site, where different sites used the different job aids, with the whole of the baseline data being classified as a single site. Clearly, job aids are effective, but only if they are actually used during the investigation. When we classified the Number of Facts Requested by Job aid use Style, we also got highly significant results. The effect of Style was highly significant ($F(2,30) = 7.68$, $p = 0.002$). Our Style 1 participants, who worked systematically through the job aid requested 61.5 facts on average. Style 2 participants who used the job aid as a back up requested 54.0, while the rest who were in Style 3 requested only 33.4.

Acknowledgement

This work was supported by a grant from the FAA AFS-300, contract monitor Ms. Jean Watson.

References

Marx, D. and Watson, J. 2001, *Maintenance Error Causation.* FAA, Office of Aviation Medicine.

Marx, D. 1998a, Discipline and the blame-free culture. *Proceedings of the 12th Symposium on Human Factors in Aviation Maintenance,* (CAA, London, England) 31-36.

Drury, C. G., Wenner, C. and Kritkausky, K. 1999, *Development of process to improve work documentation of repair stations.* FAA/ Office of Aviation Medicine (AAM-240) (National Technical Information Service, Springfield, VA).

Woodcock, K. and Smiley, A.1999, Developing simulated investigations for occupational accident investigation studies, prepublication draft.

Marx, D. A. 1998b, *Learning from our Mistakes: A Review of Maintenance Error Investigation and Analysis Systems* (with recommendations to the FAA), (Galaxy Scientific Corporation, Atlanta).

Ferry, T. S. 1981, *Modern Accident Investigation and Analysis* (John Wiley and Sons, NY).

Rasmussen, J. 1990, The role of error in organizing behaviour.*Ergonomics*.**33.10**,1185-1199.

Boeing Commercial Aircraft Division 1997, *MEDA User's Guide.*

Rankin, W. L. 2000, The maintenance error decision aid (MEDA) process. *Proceedings of the IEA 2000/HFES 2000 Congress*, 3-795 – 3-798.

Fiske and Taylor 1984, *Social Cognition* (Addison-Wesley, Reading, MA).

Drury, C. G. and Brill, M. 1983, Human factors in consumer product accident investigation, *Human Factors*, ***25.3***, 329-342.

TRAINING

THE RELATIONSHIP BETWEEN RECRUIT TEST 4 SCORES AND MECHANICAL COMPREHENSION DEPENDENT TASKS

Kathy Munnoch, Roger Pethybridge and Bob Bridger

Kathy Munnoch
Institute of Naval Medicine
Crescent Road, Alverstoke
PO12 2DL
E Mail: hsohf2@inm.mod.uk

The aim of the present study was to establish whether a relationship exists between recruit RT4 (mechanical comprehension) scores and two mechanical comprehension dependent tasks within the Royal Marines (RM) recruit training courses. This aim was achieved through the collection of subjective and objective performance data of RM recruits carrying out the tasks. Statistical analyses revealed that performance in map reading was not related to RT4 score. However, performance at weapons training was significantly related to RT4 score. On the basis of the findings it was recommended that a cut-off point for potential recruits should be introduced at the Armed Forces Careers Office (AFCO) stage. Candidates who achieve a score of less than 10 on their RT4 should be rejected, as they will not be demonstrating the level of mechanical comprehension required for safe and efficient weapon handling during training.

Introduction

The Royal Marines (RM) recruit training course is considered to be one of the most arduous military training regimes in the world. This results in a high level of training wastage, despite the extensive modification to the training course in recent years. These modifications have resulted in a reduction of injury-related discharge. However wastage remains high at 46%. A previous study found attrition to be predicted by a combination of two sources of information: an individual's performance data from tests undertaken at the Armed Forces Careers Office (AFCO) and physical performance data obtained from tests taken at the Potential Royal Marines Course (PRMC). Tests undertaken at the AFCO include the "Recruiting Tests". These multiple choice paper and pen tests were developed to supplement information obtained from the selection interview, for career guidance. The 4

Naval Recruiting Tests (RTs) measure reasoning (RT1), numeracy (RT2) literacy (RT3) and mechanical comprehension (RT4). Pethybridge *et al* (2000) found that the four main variables that were found to relate to voluntary withdrawal from training are:

a. Recruiting Test 4 (RT4) score – lower pass rate with scores of less than 15.
b. Assault course time – lower pass rate with "long" completion times (exceeding 4 minutes).
c. Bleep test performance – lower pass rate with "low" scores (under 11).
d. Age – lower pass rates for those aged 16 and 17.

The term "mechanical comprehension" was found to reflect multiple aptitudes relating to the understanding of the working and maintenance of machinery. Perceptual, spatial and problem-solving skills are assessed within the RT4 test. It was also established that the RT4 score obtained by an individual was an artefact of the test items, rather than a definitive assessment of potential mechanical ability (Anastasi, 1990). In other words, the score obtained on a test of mechanical comprehension will reflect test items, and the cognitive skills examined will differ if the test items differ. Other well-validated tests of mechanical comprehension were also identified and studied, such as the Australian Council for Educational Research (ACER) Mechanical Comprehension (Australian Council for Educational Research, 1977) and the Bennet (Bennet, 1994) Mechanical Comprehension Test (BMCT). A content analysis revealed the BMCT to be the most similar to the RT4. The RT4 was developed by Vernon (1949) to measure an "*....understanding of basic mechanical principles using both mechanical elements (e.g. gears and pulleys) and domestic leisure applications (e.g. cars)*" (Directorate of Naval Recruiting, 2001). RT4 was intended to assess an individual's potential to learn, and different elements of the RT tests have been found to be predictive of performance in different jobs within the Armed Forces (e.g. Cawkill, 1991., Duffy, 1992). However, only RT4 was found to be predictive of attrition from Royal Marine recruit training. Observation of the tasks undertaken during Royal Marine recruit training revealed map reading and weapons training as both requiring a number of the aptitudes measured by RT4. The aim of this study was to investigate the relationship between RT4 scores and performance on the identified mechanical comprehension dependent tasks.

Method

Data

Performance on two map reading tests and instructors' assessments for 4 sessions of weapons training were obtained from 1 troop of RM recruits on the 30 week training course. The RT4 scores for 54 of the 56 recruits were extracted from records on a weekly basis. It is important to note that methods of assessment intrinsic to the course were used. In other words, the test scores were from RM tests undertaken by each recruit as part of the course. The instructors' assessments were the same as those recorded on the Training Record Cards (TRCs) by the instructors. Performance data for map reading and weapons training was collected for Troop 823.

Subjects
Troop 823 began their foundation course as the study was being set-up, and so could be followed from the beginning, through foundation and into the later stages of training. The troop consisted of 56 male recruits, with ages ranging from 16-30.

Procedure
Instructors of Troop 823 were asked to complete a form to assess the performance of each recruit on certain tasks. Each instructor would report on their section (10 recruits). The measures taken were on a 5 point Likert scale, with space for additional comments from the instructors. The scale (range 1-5) was a measure of help given by an instructor to a recruit with a weapons training task, where 1 = None, 3 = Some and 5 = A lot. Data from test scores were also collected. The map reading test scores consisted of standardised measures of a combination of speed and accuracy of a recruit carrying out particular map reading tasks appropriate to the level of training.

Hypotheses
Null Hypothesis 1: Extra help given by an instructor to a recruit performing a weapon handling task was not related to his RT4 score.
Experimental Hypothesis 1: Extra help given by an instructor to a recruit performing a weapon handling task was related to his RT4 scores.

Null Hypothesis 2: Test scores of a recruit performing a map reading task were not related to RT4 score.
Experimental Hypothesis 2: Test scores of a recruit performing a map reading task were related to lower RT4 score with 'low' task performance associated with 'low' RT4.

Statistical Tests
The Kolmogorov-Smirnov (KS) 2 sample test was used for hypothesis testing. The test allows the RT4 score distributions for two groups to be compared. The groups were composed of recruits demonstrating a 'poor performance on task' and those recruits demonstrating 'other performances'.

Results

Map Reading (2 tests)
The analyses for map reading tests 1 and 2 were undertaken separately. There was no relationship between the map reading scores and RT4 scores for either test 1 (45 recruits) or test 2 (37 recruits). The KS test yielded no significant differences in the RT4 score distributions for two groups of recruits. Group one consisted of recruits with map reading test scores below 90 (8 recruits on test 1; 20 recruits on test 2) and Group two consisted of recruits with map reading scores of at least 90 (37 recruits on test 1; 17 recruits on test 2).

Weapons Training Assessments
The 54 recruits (with known RT4 scores) were assigned to 2 subgroups - Group 1 included those recruits with any of the four instructor's assessments indicating "below standard" or

"extra tuition was required"; whilst Group 2 included all the others. The KS test yielded a significant difference ($P<0.05$, 1-sided test) between the RT4 distributions for the 2 groups, with Group 1 having lower RT4 scores. Fourteen recruits of the 56 had RT4 score below 15. Summary RT4 scores for the 2 groups can be seen in Table 1.

Table 1. Distribution of RT4 Scores for Recruits assessed as being Poor Performers (Group 1) and other RM Recruits (Group 2) during Weapons training Sessions

	Group 1	Group2
Number of Recruits	7	47
Minimum RT4 Score	9	11
Median RT4 Score	14	18
Maximum RT4 Score	17	28
% RT4 Score Under 15	57	21

Discussion

This study of a relatively small sample of recruits has shown a statistical association between instructors' ratings of recruits' "performance" and ratings of "perceived extra help required" during weapons training and RT4 score; those with "low" RT4 scores generally experienced more problems and performed "below standard" during weapons training. Historical evidence on the success rate on the 30 week training course relative to AFCO/PRMC selection tests/performance include RT4, multi-fitness test and assault course time. The key indicators are used in the "risk zone model"; a computer program developed by INM and used by OC PRMC in recent years. The current study provides evidence of a statistical association between RT4 and performance during weapons training.

A methodological consideration of the study is that the performance data from the map reading tasks was from a test. Although most recruits appeared to do extremely well on these tasks, the data do not indicate what level of extra tuition was needed by each recruit to reach the standard they were at when they sat the test. The data should be interpreted with some caution, as two recruits could perform equally well on a map reading test, where one has received much additional tuition and the other has not. Repeating the study could rectify this problem by collecting performance data as indicated by the instructors for the recruits during map reading lessons, rather than performance data as measured by the test.

Approximately 63% of recruits with RT4 < 15 do not complete training. The main concerns regarding this are twofold. Supporting those with low RT4 scores through training is a financial burden on the MoD. A recruit who prematurely opts-out, takes any training they have received with them. This incurs a loss of the investment of time and money that has been put into that individual's training. As well as the drain on material resources, there is a drain on instructors. Recruits with low RT4 scores may have low mechanical comprehension ability. However, the task performance difficulties exhibited by recruits with low RT4 scores not only have an effect on their own ability to learn tasks involved in weapon handling, but also have an effect on the instructor. For each recruit that struggles with their weapon handling tasks, extra time may be required with the instructor during the lesson, or additional time may be required with the instructor after class in order to enable that recruit to perform to the same standard as the others in his troop. In the case of extra help during the lesson, the rest of the class suffers due to lack of time from the instructor. In the case of extra help after the lesson, the instructor suffers due to the extra workload and administrative burden. This is unnecessary effort and expenditure, and does not guarantee a higher number of recruits passing out of training. Even with all the additional effort that is necessitated by struggling recruits, only approximately one third of recruits with RT4 scores of less than 15 pass out of training.

So far, it has been suggested that the RT4 test is a combined measure of such characteristics as mechanical comprehension, experience (hence its correlation with age) and worldliness. Another suggestion is that it is a measure of what is known as "cognitive rigidity". This is considered largely to do with a person's approach toward problem–solving, rather than their ability *per se*. They are less able to consider alternative approaches toward a problem, and hence become rapidly frustrated, disillusioned and de-motivated despite their good intentions and perseverance at first (Levenson, 1974). They demonstrate a limited ability to develop new ideas and find difficulty in thinking flexibly (Patsiokas *et al*, 1979). Whereas a person who does not exhibit cognitive rigidity would change tack and generate new strategies toward solving a problem, a person with high rigidity would continue to be ineffective and see no way out other than to quit, despite the solutions being gradually presented to them (Levenson and Neuringer, 1971). It may be appropriate to study cognitive rigidity's relationship with performance and whether it is inherent, learned or develops with age and experience. The answers to these questions could then contribute to a development programme for those potential RM recruits who demonstrate good mental and physical ability, but lack the problem solving skills to be effective at tasks such as weapon handling.

Finally, the tasks (weapon handling and map reading) themselves need to be examined in detail. The conditions in which the first few lessons are taught could make a real difference to the recruit's ability to learn and retain that information. For example, it was observed that map reading lessons are taught in a brightly lit classroom. However, weapons training lessons are taught outside in a weapons stance. These are cold and dark, particularly in winter, and could have a detrimental effect on an individual's ability to learn at the important beginning stages of training. Interestingly, weapons training is taught in a classroom in the Royal Navy (RN).

References

Anastasi, A. 1990, *Psychological Testing,* 6th Edition, (Macmillan Publishing, New York)

Australian Council for Educational Research (n.d.). 1977, *Manual for ACER Mechanical Comprehension Test,* (ACER: Victoria)

Bennet, G.K. 1994, *Bennet Mechanical Comprehension Test Manual,* 2nd Edition, (The Psychological Corporation, Harcourt, Brace and Company, San Antonio)

Cawkill, P. 1991, Validation of rating selection/allocation test performance, NAMET and GCE/GCSE type education against weapon engineering mechanic part III ordnance training performance, *Unpublished MoD Report*

Directorate of Naval Recruiting. 2001, *Selection for the Royal Navy and Royal Marines:* Procedures and practice test booklet

Duffy, B. 1992, Assessment of the feasibility and utility of introducing minimum RT3 (numeracy) score for radio operator entry, *Unpublished MoD Report*

Levenson, M. 1974, Cognitive characteristics of suicide risk. In C. Neuringer, I.L. Springfield & C.C Thomas (eds.) *Psychological Assessment of Suicide,* 150-163

Levenson, M. and Neuringer, C. 1971, Problem-solving behaviour in suicidal adolescents, *Journal of Consulting Clinical Psychology,* **37,** 433-436

Pethybridge, R.J., Shariff, A., Pullinger, N.C. and Allsopp, A.J. 2000, Statistical evaluation of Royal Marine recruit selection criteria. *Unpublished MoD Report*

Patsiokas, A.T., Clum, G.A. and Luscomb, R.L. 1979, Cognitive characteristics of suicide attempters, *Journal of Consulting Clinical Psychology,* **47,** 478-484

Vernon, P.E. and Parry, J.B. 1949, *Personnel Selection in the British Force,* (University Press, London)

THE VALUE OF AN ONLINE, DISTANCE EDUCATION PROGRAM IN HUMAN FACTORS

Marcus Watson[1] and Tim Horberry[1, 2]

[1]Key Centre for Human Factors & Applied Cognitive Psychology, University of Queensland, Brisbane, Qld 4072, Australia.

[2]Accident Research Centre, Monash University, PO Box 70A, Victoria 3800, Australia.

Distance education programs have had a long history in many countries; the growth of the Internet as a mass communication tool is now an invaluable aid in delivering such programs. Australia, which has a large geographical area, a comparatively sparse population and a high standard of education, has been one of the leaders in developing education to remotely based students. This paper focuses on the postgraduate Human Factors distance education program coordinated by the University of Queensland, Australia. Two novel features of this program are: First, it involves the collaboration of experts from several different universities around Australia, each offering one or more courses in their areas of specialisation. Second, as previously mentioned, it is by distance learning, supported by on-line tutorials and assessments. The overall strengths and weaknesses of online distance education courses in Human Factor will be discussed, focusing on the lessons learnt from the University of Queensland program and how they may be applied to other distance education programs. Conclusions are made in regard to advancing the development of the Human Factors profession and related disciplines by the wider use of distance education courses.

Keywords: Human Factors, distance learning, education, teaching methods.

Introduction

Human Factors (HF) is one of many new multi disciplinary areas of employment expanding due to the information revolution. Major growth areas are usability, complex systems design, consumer/pleasure based human factors, and organisational human factors (Helander, 1997). In light of this many universities around the world have introduced dedicated HF programs or included HF subject as part of the corniculum of existing courses. In Australia there has been a large commercial demand for qualified HF professionals; however, there are no critical mass students or academics in any one location to meet this need (Horberry & Watson, submitted for publication). The demand has come from employed tertiary qualified professionals rather than school leavers looking for a career. The majority are interested in part time education while they continue their full time employment. This combination of factors makes the development of a tradition campus based HF course impractical. The solution adopted by the Key Centre of Human Factors and Applied Cognitive Psychology is to offer an online HF postgraduate suite with subjects provide by a number of Universities throughout Australia.

Postgraduate Distance Education in Human Factors

As most Australian universities already have well developed infrastructure for distance education, a HF program provided by many universities is possible. Distance learning technology has developed far beyond paper and pencil programs posted to students in the mail, or lessons given over a radio. The growth of the Internet as a mass communication medium and specialised distance learning software has increased the interactivity with fellow students and lecturers to a level similar to class based postgraduate courses. Distance education has the ability to bring together the critical mass of students and experts required to develop first class postgraduate HF courses. It is, however, important to recognise the costs involved in cross institutional courses. In this paper we will outline the pros and cons of the cross institutional course for the five major stakeholders: (1) the student, (2) the universities, (3) the employers, (4) the course providers and (5) the program coordinator.

As with other course based postgraduate studies, a large proportion of HF students will undertake the study to facilitate their full time employment. In general they want a course that can be undertake part time with a large proportion of the course directly relevant to their immediate employment requirements (Horberry and Watson submitted for publication). To define and meet this demand we have spent many hours talking with current as well as prospective students and large HF employer groups including aviation, road, maritime transport and defence organisations. Similarly a large amount of liaising has occurred both within The University of Queensland and inter university. For each of the five stakeholders we have been able to identify the advantages (Table 1) and the disadvantages (Table 2) of a distance educational course.

We have avoided providing courses that focus on fulfilling employer groups' demands to provide electives that solve the company's immediate problems rather than providing something that will provide a skill base beyond the current year. The course does, however, offer the students the ability to focus on their employers' problems during the research component of the Diploma and Masters course. As some electives may only be available every second year, such an approach would prove costly in terms of yearly course rewrites. To get around these problems we have focused on provided electives that are compatible with existing course taught at the different node university, that offer long term educational benefits. Any highly specialised techniques or topic area can be better dealt with in the research projects where a supervisor can provide more appropriate guidance.

Table 1. Stakeholder Advantages in participating in a distance education HF program

Stakeholder	Advantage
Students	• The student can tailor the course to suit them due to the wide range of electives and research topics on offer that could not be offered at a single Australian institute, • Structure allows reasonable flexibility as to when students take each unit, • Greater flexibility on when students conduct their studies, • The student population benefits from cross discipline and work experience pollination, • The students are able to combine work and study, • Overseas students can obtain the qualification without incurring the cost associated with travel and accommodation for a full time class-based course
Universities	• Increased teaching income without increasing classroom pressure, • Ability to attract overseas student enrolments without having to provide accommodation, social and cultural support, • No critical mass of HF academic at one university required to provide a program. • Ability to offer HF electives to other postgraduate courses
Employers	• Employees can continue full time work while studying, • Most research projects can focus on the employers HF problems, • Some of the electives will be directly related employers needs.
Program coordinator	• Less restricted by the faculty staff areas of interest; therefore, the ability to offer a wide range of HF related courses.
Course provider	• The online course can be based on and run in conjunction with existing classroom courses • Courses can easy to teach with much of the day to day discussion groups and assignment marking run by tutors, • No lecturing time only online discussion monitoring, • Potential to identify appropriate PhD candidates • The course need not be taught from the office.

In preparing the program we have focused on developing material that will both educate the student for current and future requirements as well as holding their interest. The course material does more than follow 'traditional' lectures methods, where the student as passive recipient of information; the material incorporates a mixture of tutorial style problems, reading material and short research projects. As many of the students already enrolled are employed in HF jobs, they bringing their own experiences and needs to the program; therefore, they are able to share these experiences with students from other sub-disciplines who would benefit from the cross-fertilisation of approaches and knowledge. The online discussion groups have focused on sharing the students' own experience as it relates to the course assignments. The assignment in turn has helped to focus students' interest on both the core material and peripheral readings in a broad range of HF. The results of the design of the two subjects are reflected in both the retention of students (100% in 2002) and the few extension required by students (3 out of 126 assessment submissions across the courses). The retention rate and the number of assignments completed on time reflected the effort put into developing the course.

Table 2. Stakeholder Disadvantages in participating in a distance education HF program

Stakeholder	Disadvantages
Students	• Little or no face contact with the teaching staff and other students leading to students feeling isolated (Rossman, 1997) • Delays in receiving feedback to questions about course material • Subjects that require controlled laboratory practical work will require travel and accommodation costs • Electives may not always be available at the most appropriate time • Distance education methods may inhibit some practical techniques from being learnt
Universities	• Differences in between university course workloads • Difficulties in cross institutional enrolment • Financial management of course development • Financial management of student fees • Student union requirements • Promotion of the course – who pays the bills?
Employers	• Electives may not always be available at the most appropriate time to support work activities
Program coordinator	• Workload administrating the students with the University is higher than for an on campus course • Workload administrating of inter university payments for students • Time spent liaising with other universities to achieve electives of equal workload and course availability • Liaising with major employment groups to determine how best to structure course availability • Work in designing the course to meet international HF requirements • Making sure that elective material provided by other universities do not overlap each other or the introductory courses
Course provider	• High workload in developing the course material that focus learning activities and research projects directly on students' own workplaces or experiences • Assignment marking takes longer as there is no face to face discussion between student and supervisor

The design of the program has avoided 'gimmicky' interactive computer-based presentation of the material, as it is easy to create amusing interactive diagrams and similar activities that have at best questionable educational value. We have made use of the many free Internet sites that already exist, that have interactive diagrams useful in conveying appropriate information, which the learning package can be linked. Such sources of information are have proven appropriate for supplementing lecture notes and problem solving activities rather than as the sole method of teaching a concept. The HF programs should rely on simplicity and a high degree of interactivity; which also helps to demonstrate some basic HF principles for conveying information.

Developing material that does not rely on interactive diagrams did not reduce the cost substantially as a large amount of effort went into the design of the course to balance the students' interest and workload over the semester. Since the course is also aimed at overseas students and equipping Australian students for worldwide HF employment, cultural and international recognition were taken into account when designing the program. Careful attention was given to the design and review of the material in relation to the objectives of the course, not only by the developer but also by education experts. The reliance on simplicity has reduced the costs in placing the material in a web environment; however, there are cost associated with the administration and distribution of the material.

Another method that has increase the amount of time students had available for study, involved sending students the relevant lecture materials, providing the students with high quality resource for each module (Horberry & Watson submitted for publication). This helps to ensure that when students work through their material a minimal amount of time is spent waiting for information to be downloaded.

The range of distance learning media used in the introductory subjects includes:

- Print — Learning Guides containing the basic lessons and activities plus a Course Readers containing published articles, book chapters that the student needs to read as part of their course. Plus the course textbook??
- Video/ Voice — on websites, including material made freely available by other websites
- Computer —website which contains features such as an asynchronous discussion board, email, glossary, calendar, assignment submission zone and exam hosting area.

The high proportion of employed students in the HF program (and likely to be found in other HF distance education program) allows for supervised human factors research projects to be undertaken in the students' regular working environment. Flexible projects allow students to put their newly acquired human factors education into practice, both benefiting the student and their organisation. This has already led to one large organisation nominating the University of Queensland as the preferred provider. In offering appropriate supervision for research project many companies are proving ready to pay or at least supplement their employee's student fees. As the projects are more relevance to the student needs we argue that this provides a more appropriate delivery method of Human Factors techniques than can be achieved in classroom interactions.

A well designed program has the ability to meet both the student and employers needs while not over taxing the universities and the academic team. Interactivity is the key to successful distance education; therefore, as much as possible, students should interact with their fellow students and course lecturers via discussion boards and emails. Small team projects also help to foster interactions and increase the sharing of HF experiences. Team project combined with a variety of assessment methods encourages the students to go beyond the material delivered throughout the course so long as appropriate reference material is made available as a starting point. The level of interaction with the students by the course coordinator and/or tutors may actually be less than those required by conventual instructional lectures and tutorials (Table 3). The possible saving in both time and money in running the course is offset by the long developmental time.

Table 3. Development and course administration time

Activity	Minimum weekly time	Maximum weekly time	Average weekly time	Total time (hours)
Course development time				
Average for the first three course	NA	NA	NA	117:00*
Program coordination				
First year (13 week semester)	2:00	20:30	11:30	149:30†
Estimate time after the first year				*104:00*
Discussion forum and online course administration				
HUFA7001 Introduction to Human Factors 1	0:50	4:35	1:51	24:00
HUFA7002 Introduction to Human Factors 2	1:20	5:00	2:00	26:00
Assessments				
HUFA7001 Four written assignments	NA	7:40	NA	19:00
HUFA7001 One exam	NA	5:00	NA	5:00
HUFA7002 Four written assignments	NA	8:00	NA	19:30

* Developmental time does not include the amount of time contributed by the online providers at The University of Queensland.

† Coordination time does not include time commitments of other universities in liaising with the program coordinator.

A key cost associated with a cross university distance education program is the high cost in administrating the program. Factors such as ownership of course material retention of staff will differ between the institutes involved but need to be consolidated within the program. Simply developing course of equal unit loads is very difficult and time consuming. There are also problem associate with the confirmation of degrees. There is a need to balance the administration, institute recognition and international acknowledgment for any postgraduate qualification. The University of Queensland will award the current program qualification; however, in future other universities involved in the program may wish to offer the same degrees. Institutes choosing not to offer the program, but instead teach into the program, may be motivated by the teaching income, access to potential PhD candidates and contacts with industry.

Conclusions

Many of the obstacles involved in learning the profession of HF can be overcome by careful program planning by the lecturers, support by major employers and motivation by the students. In Australian due to the absence of a 'core mass' of potential students and the vast geographical distances involved, perhaps distance learning is the only feasible way of expanding the teaching of HF in Australia. Distance learning can be a viable way of teaching HF and developing strong industry relations; however, any HF distance educational course will involve a high developmental workload especially if they are cross institutional. The advantage of cross institutional HF distance education is that it meets the growing needs of students and industry.

Acknowledgements

We would like to acknowledge the institutes that are contributing to the Postgraduate Suite in Human Factors: The University of Queensland, University of South Australia, University of Adelaide, University of Sydney and the University of Newcastle. We also acknowledge the Australian Research Council for providing the seed funding.

References

Helander, M. (1997). The Human Factors Profession. In G. Salvendy (Ed), *Handbook of Human Factors and Ergonomics* (3-16). USA: J. Wiley.

Horberry, T.J., and Watson, M. (submitted for publication). Technical Note: Human Factors and Distance Learning. *Australian Journal of Psychology*

Rossman, J. (1997). *The Internet Traveler's Guide to Distance Education*. Retrieved 2nd April 2002 from http://www.albemarle.cc.nc.us/~jrossman/archives/dist.html

GENERAL ERGONOMICS

GM INFORMATIONAL LABELS – DOES SHAPE MATTER?

A. J. Costello, E. J. Hellier & J. Edworthy

Department of Psychology,
University of Plymouth,
Plymouth
PL4 8AA

How information is presented on labels indicating GM content may affect how hazardous the product is perceived. Previously, Costello, Hellier, Edworthy and Coulson, (2002) demonstrated that colour, wording and attributed information source, influenced hazard perception and purchase intention. Here we extend these findings to examine the effect of shape, colour and font size for labels indicating GM and non-GM content on measures of purchase intention and hazard perception. Analysis of variance revealed significant main effects for label shape, colour and content, but not font size on both measures. The control (no label) condition resulted in higher hazard perception and lower purchase intention than all GM labels and most non-GM labels tested. Although no overall effect of gender was found, for purchase intention, a significant interaction between gender and label shape was found.

Introduction

Many studies and surveys in this country have indicated that the majority of Europeans are not in favour of Genetically Modified (GM) foods (e.g. EuroBarometer, 2001). This has resulted in campaigns against GM foods from major environmental groups such as Greenpeace and Friends of the Earth. Previous telephone survey research suggests that an individual's acceptance of GM food products is related to their perception of the risks and benefits associated with GM technology (Hoban and Kendall, 1992). Consequently those perceiving greater risk from the use of biotechnology in food production are more likely to reject GM food products on principle. However little research is available which tests the link between the perception of risk of GM food products and actual behaviour. Social cognition models such as the Theory of Planned Behaviour (Ajzen, 1991) have been suggested to form a basis for predicting purchasing behaviour with regard to GM foods (Bredahl, *et al*, 1998). The Theory of Planned Behaviour (TPB) would take into account attitudes to a behaviour (in this case, the intention to purchase GM foods), the perceived views of others and control beliefs which may facilitate or inhibit performing the behaviour. Behavioural intention would then predict actual behaviour. However, findings from warnings research would indicate that the manner in which an individual is informed about a risk, might also predict purchase intention (Edworthy and Adams, 1996).

Campaigns exist to lobby for compulsory labelling of all GM foods products to allow consumer choice (e.g. The Consumer Association). In the UK, GM products where the end products are significantly different in terms of DNA or protein content, need to be labelled (Food Standards Agency, 2002). However the effect of such labelling on public perception is not known. Previous research into the effect of variation in the characteristics of warning labels on hazard perception would suggest that the type of label used to inform consumers may affect how much hazard is perceived and consequently purchase intention (Costello, *et al*, 2002).

Costello *et al*, (2002) examined the effect of variation in label colour, information source, GM content and probabilistic Vs definite wording on self reported measures of hazard perception and purchase intention for a food product. They found a strong inverse relationship between hazard perception and purchase intention, as hazard perception increased, purchase intention decreased. Using analysis of variance, a main effect of colour and content was found for measures of purchase intention and hazard perception. Red labels, as opposed to blue, green and white labels increased hazard perception and reduced purchase intention, as did labels indicating GM as opposed to non-GM content. In addition a main effect of wording was found on hazard perception whereby labels with the wording 'contains GM' as opposed to 'may contain GM' increased hazard perception. A main effect of information source on purchase intention was also found where labels attributed to a 'Consumer Association' as opposed to the 'Department of Health' or 'Manufacturer' increased purchase intention. These results suggest that public perceptions of specific GM food products are subject to variation dependant on the way in which information indicating GM content is presented. This research suggests that the findings from warning label research generalise across to GM food labels, specifically the effect of colour, information source, and wording.

This experiment furthers this work by examining the effect of other label design characteristics which have been shown to influence perceived hazard in warnings research. Here, the effect of GM label shape, colour, font size and GM content on self-reported measures of hazard perception and purchase intention is investigated.

Warning label research suggests that label colour will affect how much hazard is conveyed by a warning label with red labels indicating higher levels of hazard than green and blue labels (Braun and Silver, 1995). Riley *et al,* (1982) investigated the effect of label shape variation on hazard perception. Their results, from asking participants to rank warning label shapes for hazardousness, suggest a label shape scale where more angular shapes are viewed as indicating a higher level of hazard and more rounded softer shapes a lower level of hazard. Adams and Edworthy (1995) investigated the effect of font size, border width, white space and colour on hazard perception. Their results indicate that colour and font size variations had the greatest effect on the perceived urgency of the warning labels.

It was expected that a main effect for label colour, shape, font size and content would be found, that red GM labels with larger font size and a more angular shape would result in the highest perceived hazard and lowest purchase intention ranking. Thereby generalising the findings from warnings research into the area of GM informational labels and adding validity and reliability to the previous findings by Costello *et al*, (2002).

Method

Design

There were four within subject variables, colour (red, blue, green, white), shape* (diamond, hexagon, oval), font size (8pt, 12pt) and content (contains GM ingredients, contains no GM ingredients). *The shapes chosen were ranked high, medium and low hazard indicators by participants of the experiment by Riley et al, (1982), (diamond, 2.5, hexagon, 6, oval, 16.5).

Participants

40 participants aged between 18 and 54 took part in the experiment, 23 were female and 17 male. All were literate and English speaking. The participants were recruited through posters displayed at the University of Plymouth and paid £2.50.

Stimuli and Materials

Factorial combinations of all levels of colour, shape, font size and content resulted in 48 different labels. The labels were scanned onto identical pictures of a food product. A picture of the food product with no label acted as the control. Each label had the word Notice as a header. The wording for the labels indicating that the product was GM was 'Contains genetically modified ingredients'. The wording for the labels indicating that the product was GM free was 'Contains no genetically modified ingredients'. The labels although different shapes, were approximately equivalent in area.

Procedure

Participants were presented with folders containing 49 photographs of the food product each with a different label, in one of 4 different randomised orders. When viewing the pictures, participants were told that each product was the same price and quality and they should assume in principle that they would buy the product. The participants were asked 'How likely would you be to buy the product?' and directed to indicate the number that best represented their answer on a scale which ran from 0 to 8 where 0 represented the answer 'Not at all' and 8 represented 'Extremely likely'. Participants were also asked 'How much hazard is indicated by the label?' and indicated their answer on a scale which ran from 0 to 8, where the number 0 represented the answer 'No hazard' and the number 8 represented 'Extreme Hazard'.

Results

Repeated measures analysis of variance was used to investigate main effects and interactions. Post hoc tests were not carried out due to the within subjects nature of the experimental design. A strong inverse relationship ($r = -.979$, $p<0.001$) was found between hazard perception and purchase intention across all label types.

Hazard Perception Main Effects

The mean score for the control condition where the food product was shown with no label was 1.436. A main effect of colour was found $F(3, 117) = 25.154$ $p<0.05$, with red labels (3.782) scoring higher on hazard perception than blue (2.739), green (2.574) and white (2.297) labels respectively. A main effect of label shape was found $F(2, 78) =$

5.647 p<0.05, with diamond shaped (2.959) labels scoring higher on hazard perception than hexagon (2.818) and oval (2.767) labels respectively. A main effect of label informational content was found F (1, 39) = 58.279 p<0.05), with GM labels (4.017) scoring higher on hazard perception than non GM (1.679) labels. No significant main effect of gender was found.

Hazard Perception Interactions

A significant interaction was found between colour and informational content F (3, 117) = 10.049 p<0.05. For labels indicating GM content there was little difference between blue and green label scores. However, for GM free labels, blue labels scored higher (1.748) for hazard perception than green labels (1.419). The colour red also appeared to have a greater effect for GM (5.23) as opposed to non-GM (2.33) labels. A significant interaction was also found between font and content F (1, 39) = 5.089 p<0.05, as shown in figure 1.

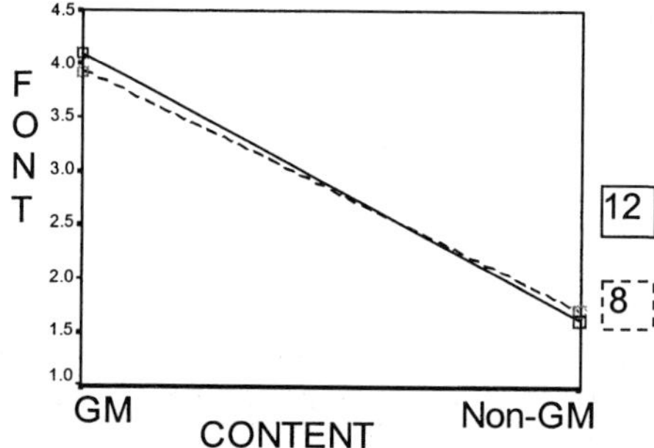

Figure 1. The interaction between font size and GM content

For GM labels, the larger size font, 12pt, (4.101) resulted in higher scores for hazard perception than the smaller font, 8pt, (3.932). Whilst for non-GM labels, the smaller font (1.723) resulted in higher scores for hazard perception than the larger font (1.635)

Purchase Intention Main Effects

The mean score for the control condition where the food product was shown with no label was 4.825. A significant main effect of label colour was found F (3, 117) = 14.620 p<0.05. Red labels scored lowest (3.095) for purchase intention followed by blue (3.775), green (3.865) and white (4.227) labels respectively. A significant main effect of label shape was found F (2, 78) = 10.321 p<0.05. Diamond shaped labels scored lowest for purchase intention (3.633) followed by hexagon (3.748) and oval (3.840) labels. A significant main effect of label informational content was found F (1,39) = 58.448 p<0.05. GM labels scored a great deal lower (2.374) than non-GM labels (5.107) for purchase intention. No significant main effect of gender was found.

Purchase Intention Interactions

A significant interaction was found between font and informational content F (1, 39) = 12.202 p<0.05. For GM labels, the smaller font (8pt) scored higher for purchase intention, whilst for non-GM labels, the larger font (12pt) scored higher for purchase intention. A significant interaction was found between gender and label shape F= 4.204

(p<0.019). The female participant scores for purchase intention were highest for the hexagon, then oval shaped labels followed by the diamond shape, with little difference between the hexagon and oval scores. In contrast, the male participants scored the oval shape highest, then the hexagon, followed by the diamond. For non-GM labels, label shape had a greater effect on male rather than female participants as shown in figure 2.

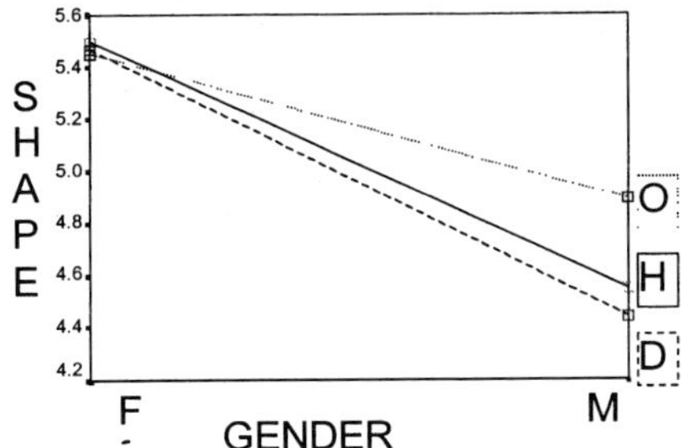

Figure 2. The interaction between shape and colour for non-GM labels.

Discussion

These results confirm the findings of Costello *et al*., (2002) that GM informational label colour has a main effect on both purchase intention and hazard perception. In addition, these findings are extended to include a main effect of label shape for both measures. The effect of shape variation on hazard perception was in the order predicted by the scale of Riley *et al*, (1982) with the diamond shaped labels yielding the highest hazard perception scores, followed by the hexagon and then oval labels. For purchase intention, the results were reversed with the highest purchase intention mean score for the oval labels followed by the hexagon and diamond. Hence again demonstrating the inverse relationship between measures of hazard perception and purchase intention. However, an interesting interaction was found between label shape and gender. These results suggest that men are more sensitive to label shape variation than women, especially for labels indicating non-GM content.

Whilst no main effect of font size was found, significant interactions were found between font and GM content for both measures. For GM labels, the smaller font scored higher for purchase intention, whilst for non-GM labels, the larger font scored higher for purchase intention, whilst the reverse was found for hazard perception scores. This suggests that it is only best to use a larger font with information that is likely to be perceived as non-hazardous in order to increase purchase intention.

The control condition (with no label indicating GM content) yielded higher hazard perception than all GM labels and most non-GM labels in white and green. This contrasts with Costello *et al*, (2002) where the control condition yielded lower hazard perception and higher purchase intention than *all* other labels regardless of content or design. These results suggest that particularly red and blue GM free labels are still associated with hazard regardless of the informational content. A similar pattern was indicated in the purchase intention mean scores where the control condition resulted in a score that was higher than all GM labels and most non-GM labels except red hexagon and diamond shaped labels. It appears, therefore that red and angular shaped labels can override the increase in purchase intention and decrease in hazard perception related to this example of GM free wording.

A significant effect of label content was shown with GM labels scoring significantly higher for hazard perception than non-GM labels. In addition to this finding a very strong inverse relationship was found across all label types between hazard perception and purchase intention. This finding has important implications for those wishing to market GM produce in that it is likely that any efforts to make the products appear less hazardous will increase purchase intention. This finding also suggests that the predictive power of social cognition models such as the Theory of Planned Behaviour may be improved by the inclusion of label design characteristics in the model when predicting the purchase of GM foods.

In summary, these results support the findings of Costello *et al*, (2002) of a main effect of colour in informational labels indicating GM and non-GM content. These results have been expanded to include a significant main effect of label shape. It is suggested that a future repetition of this experiment using a wider selection of shapes from the ranking developed by Riley *et al,* (1982) will further calibrate the effect of label shape on perception. In particular the suggested difference in gender effects of label shape for GM and non-GM labels.

References

Adams A. S., and Edworthy J., 1995, Quantifying and predicting the effects of basic text display parameters on the perceived urgency of warning labels: trade offs involving font size, border weight and colour. *Ergonomics*, **38** (11), 2221-37

Ajzen I., 1991, The Theory of Planned Behaviour, *Organisational Behaviour and Human Decision Processes*, **50**, p179-211

Bredahl L., Grunert K. G., and Frewer L. J., 1998 Consumer Attitudes and Decision-Making with regard to Genetically Engineered Food Products – A Review of the Literature and a Presentation of Models for the Future. *Journal of Consumer Policy*, **21**, 251-277

Braun, C. C., and Silver N. S., 1995, Interaction of signal word and colour on warning labels: differences in perceived hazard and behavioural compliance. *Ergonomics*, **38**, 2207-20

Costello A. J., Hellier E. J., Edworthy J. and Coulson N., 2002, Can food label design characteristics affect perceptions of Genetically modified food? In McCabe P. T. (ed.) *Contemporary Ergonomics 2002*, (Taylor and Francis, London), 442-446

Edworthy J., and Adams A. S., 1996, *Warning design: a research perspective*, (Taylor and Francis, London)

EuroBarometer, 2001, European Commission, http://europa.eu.int/comm/dg10/epo

Food Standards Agency, 2002, http://food.gov.uk/science/sciencetopics/gmfoods/gm_labelling

Hoban T. J., and Kendall P.A., 1992, *Consumer attitudes about the use of biotechnology in agriculture and food production*, (Raleigh, NC: North Carolina State University)

Riley M. W., Cochran D. J., and Ballard J. L., 1982, An investigation of preferred shapes for warning labels, *Human Factors*, **24** (6), 737-742

IDENTIFYING THE TASK VARIABLES THAT INFLUENCE ASSEMBLY COMPLEXITY

Miles Richardson[1], Gary Jones[1], Mark Torrance[2] & Guy Thorne[1]

[1]*Centre for Psychological Research in Health and Cognition, University of Derby, Western Road, Mickleover, Derby DE3 9GX*
[2]*Department of Psychology, Staffordshire University, College Road, Stoke-on-Trent, Staffordshire ST4 2DE*

Little is known about the variables that affect performance in assembly tasks. Seven assembly task variables were identified based on previous literature and a task analysis of self-assembly products. As real world assemblies could not be controlled sufficiently, the values of the assembly task variables were systematically varied in 16 abstract assemblies. Sixteen sets of assembly instructions were presented to 38 participants who were instructed to compare them to a final model and indicate whether following the instructions would result in the production of the final model (the time taken to accomplish this is a reflection of assembly complexity). The five task variables studied were found to be significant predictors of assembly task complexity. The results show the need for continued investigation and are a step towards the development of methodologies for the prediction of assembly difficulty.

Introduction

Self-assembly products have become increasingly common as they offer good value by reducing transport and labour costs. Illustration-only instructions have also become increasingly common as they avoid translation costs. However, there is a general lack of understanding as to what factors affect assembly task performance and the use of diagrammatic instructions. It has been noted that there has been insufficient research regarding information processing in assembly (Prabhu *et al,* 1995), human cognition in assembly performance (Shalin *et al,* 1996) and diagrams in assembly (Novick and Morse, 2000).

The present research identifies the physical characteristics or features of an assembly task, referred to as task variables, that impact on assembly complexity. The task variables are tested while holding the mode of task presentation (single step diagrammatic instructions) constant.

Assembly task variables

There has been a surprising lack of research into object assembly and much of the research conducted has considered the format of the assembly instructions. It is essential to go beyond comparing instruction formats and to study the structure of the object to be assembled, as it is this that ultimately impacts on assembly performance. Understanding the effect of task variables on complexity and performance allows the development of methodologies for the prediction of assembly complexity that will provide practical recommendations regarding what levels of complexity can sensibly be managed in a single assembly step. The methodologies for the prediction of assembly complexity provide tools for the development of assembly instructions.

There has been little research on the task variables affecting difficulty in the assembly process. The assembly research identified generally studied external factors. Morrell and Park (1993), whilst investigating cognitive load and age–related performance differences in assembly tasks, concluded that other unknown variables play extremely important roles in the performance of procedural assembly tasks with illustration only instructions.

It is proposed that performance is related to task complexity and that many of the unknown variables identified by Morrell and Park must be certain physical characteristics of the assembly tasks. The present research asks how the fundamental task variables of assembly tasks, when taken together, impact on assembly complexity.

Assembly task variable identification and definition

A hierarchical task analysis of a range of self-assembly products was used to identify the task variables. The task analysis identified four secondary operations of an assembly procedure, (selection of the components required for next assembly procedure; orientation or rotation of component to allow positioning; positioning of component to allow fastening; fastening of component to current assembly) that map onto four task variables: Selections, Symmetrical Planes, Fastening Points and Fastenings. Further, the four secondary operations are performed a number of times during an object assembly depending upon the number of components, creating a fifth task variable: Components. Also, the ease by which these secondary operations can be carried out during the assembly depends upon both the physical characteristics of the component parts and whether or not the assembly procedure has been performed before, creating a sixth task variable: Novel Assemblies. A seventh task variable was derived from the literature: Component Groups.

Operational definitions with which to score individual assembly steps or entire assembly tasks for each task variable were written. As the study did not involve full assembly, two of the task variables that relate specifically to the assembly process were not considered (Selections and Fastening Points).

The operational definitions for the remaining five task variables are listed below:

- Symmetrical Planes: The mean number of symmetrical planes measured in three planes, X, Y and Z per component in the assembly step or task being evaluated.
- Fastenings: The total number of fastenings required in the assembly step or task being evaluated.
- Components: Number of components added in the assembly step or task being evaluated (excluding fastening devices such as nuts and bolts).

- Novel Assemblies: The number of unique assemblies in the assembly step or task being evaluated.
- Component Groups: The number of component groups in the assembly step or task being evaluated. Theoretical justification such as Baggett and Ehrenfeucht's (1988) finding that when building an object people conceptualise it as a hierarchy of sub-assemblies. Component Groups was defined as a combination of components that are clearly separated from other groups of components in the assembly. To be clearly separated, the parts connecting the component groups must be able to be cut in a single plane that physically separates the component groups. (The definition was tested in a short pilot study to ensure it produced reliable results.)

Definition of abstract assemblies

As real world assemblies could not be controlled sufficiently, a set of assemblies that systematically varied the levels of each task variable was required so that the influence of each can be examined independently of the others. To create these assemblies, each task variable was assigned a value, either high or low. Although only five of the seven task variables are examined here, all seven task variables were considered in the assembly definition as future experiments would involve full assembly.

The seven task variables, each with two levels (high or low), were used to generate an orthogonal design that produced a random sample of 16 of the 128 possible combinations of task variable levels. The values of each task variable for each assembly were then used to develop 16 abstract assemblies (Figure 1).

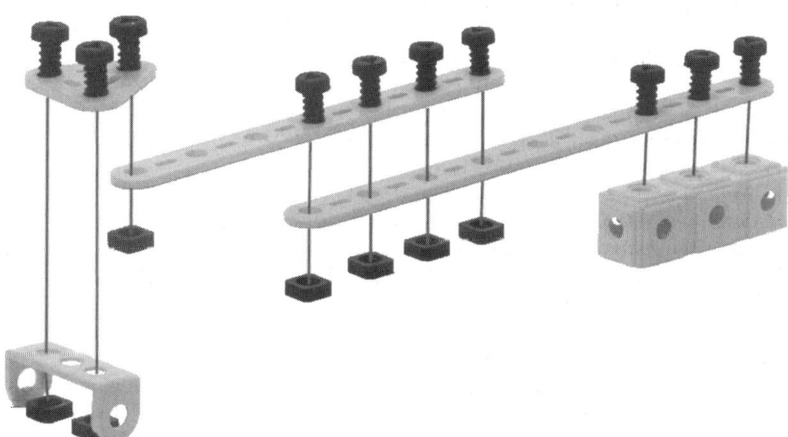

Figure 1. Example of the instructions for an abstract assembly.

Junior Meccano was used for assembly creation because it offered the flexibility in design required and would be easy to handle. The task variable levels for each assembly were then used to design each assembly. For example, based on the high/low levels for each task variable, Assembly Four had: a low number of component parts; a low variety of components; a high level of symmetrical planes; a high level of fastening points; a low number of bolt fastenings; a high number of distinct component groups. Selections can be controlled by adding unneeded items to the components in studies involving assembly.

On completion the assemblies were photographed and a graphics package used to create exploded isometric views of each of the assemblies, in order to create basic one-step assembly instructions.

Method

The study sought to validate the identified task variables influence on assembly task complexity. The measure of complexity used was the encoding time of the instructions. The assembly instructions were shown to participants, who viewed them for as long as they required. The instructions were then removed to reveal a picture of a completed assembly (half of the completed assemblies were subtly changed so that they differed from the instructions), participants deciding if the instructions would lead to the construction of the assembly displayed.

A program to present the instructions and completed assemblies was written in Macromedia Authorware. 38 Psychology undergraduates participated in the study. Participants were told that they should follow the directions for the experiment on the screen. Participants were presented with the single step instructions for four practice assemblies one at a time. They could view the assembly instructions for as long as they wished. When the participant pressed a key the assembly instructions disappeared and they were presented with a photograph of the completed assembly. They then had to indicate if the instructions matched the completed assembly or not. To ensure that participants attended to the task fully they were informed if they were correct after each assembly. After completing the four practice assemblies participants were presented with the single assembly step instructions for the 16 assemblies in a random order.

Results

Encoding time was used as a measure of complexity. The mean encoding time from 608 cases was 12.3 seconds, (sd = 9.0). As the encoding times were skewed towards zero, a log transformation was performed.

Encoding time correlated significantly with all of the included task variables at the $p=0.05$ level. The correlations between task variables were also studied and no indication of collinearity was found. To predict encoding time based on the assembly task variables, hierarchical repeated measures multiple regression analysis was used.

The 608 observations of encoding time were included in the analysis. As there was multiple data for each participant, dummy between subject variables for each participant were entered first to control for between subjects' variance. The five suitable assembly task variables (Components, Symmetrical Planes, Novel Assemblies, Fastenings and Component Groups) were then entered.

The between participant variables in model 1 of the multiple regression analysis gave R=0.63 and Adjusted R^2=0.36, suggesting that 36% of the variance in perceived difficulty was accounted for by between subjects effects. The final model including the five task variables gave R=0.77 and Adjusted R^2=0.56 ($F(42,565)=19.42, p<0.001$), with the R Square Change figure suggesting that 20% of the variance in encoding time was related to some combination of the task variables. Overall, 56% of the variance in encoding time was accounted for.

All of the entered task variables were significant predictors of encoding time at $p<0.05$. Standardised regression coefficients for each task variable suggest that an increase in encoding time was associated with increased Novel Assemblies (0.395), an additional number of Fastenings (0.209), higher Component Groups (0.169), a reduced number of Components (-0.137), and fewer Symmetrical Planes (-0.094).

Discussion

This study was designed to identify and validate task variables that influence object assembly complexity. Significant correlations between the task variables and encoding time were found and the results of the multiple regression analysis have shown how the five task variables were significant predictors of encoding time. This provides evidence that the task analysis and task variable definition process was valid and that the task variables relate to assembly complexity. It has also been shown that abstract assemblies can be used to systematically control task variable levels.

The standardised regression coefficients of the task variables indicate how the attributes of an assembly task relate to assembly complexity. An increase in novel assemblies, or the variety of components, complicates the assembly, while the number of components does not increase assembly complexity. This suggests that it is the nature of the components that is important rather than the simple number of component parts. A variety of components requires the completion of a variety of novel assembly procedures, whereas a high level of identical components may simply require the same procedures to be repeated several times. The standardised regression coefficient suggests that an increased number of components reduces assembly complexity, this is a weak and unexplained effect. Replication in future studies would be of interest.

A further task variable found to relate to assembly complexity is the number of fastenings. This simple measure relates to the number of connections between components and the intrinsic complexity of the assembly procedure (Marcus *et al,* 1996). Any additional information processing will impact on cognitive load, and therefore assembly complexity.

Of particular interest is the significant relationship between the number of component groups and assembly complexity. The definition and measurement of component groups was problematic, yet despite this it has been found to be significant, showing that there is psychological justification underpinning its inclusion. It suggests that the manner in which people conceptualise an object and how it breaks into sub-assemblies affects assembly complexity. Further investigations of this task variable, and its definition and measurement, are required as it has received little attention beyond Baggett and Ehrenfeucht (1988).

Finally, the number of symmetrical planes was found to be a weak, but significant predictor of assembly complexity. As this task variable was derived from the need to orient components before positioning and fastening and as the study was two-dimensional and did not involve assembly, the weak standardised regression coefficient is not unexpected. Future studies involving full assembly should continue to investigate this.

To conclude, the task variables correlated with assembly complexity as measured by encoding time and the multiple regression analysis has shown the five task variables to be significant predictors of complexity. At least some of the unknown variables that play important roles in the performance of procedural assembly tasks with illustration only instructions suggested by Morrell and Park (1993) have been identified. It is possible to use the regression equation to predict the complexity of assembly tasks based upon the task variable levels that are inherent within the assembly. Such a method for evaluating the likely complexity of assemblies based upon relevant physical characteristics has great potential and value, both applied and theoretical, and should be developed. Further, the interaction between the task variables and human cognition allows cognitive models of the assembly process to be explored. It is clear that the direction taken in this research is of value and further experiments involving full construction of assemblies should be conducted.

References

Baggett, P. and Ehrenfeucht, A. 1988, Conceptualizing in assembly tasks. *Human Factors,* **30**, 269-284

Marcus, N., Cooper, M., and Sweller, J. 1996, Understanding instructions. *Journal of Educational Psychology,* **88**, 49-63

Morrell, R.W. and Park, D.C. 1993, The effects of age, illustrations and task variables on the performance of procedural assembly tasks. *Psychology and Ageing,* **8**, 389-399

Novick, L.R. and Morse, D.L. 2000, Folding a fish, making a mushroom: The role of diagrams in executing assembly procedures. *Memory and Cognition* **28**, 1242-1256

Prabhu, G.V., Helander, M.G., and Shalin, V.L. 1995, Effect of product structure on manual assembly performance. *International Journal of Human Factors in Manufacturing,* **5**, 149-161

Shalin, V.L., Prabhu, V. and Helander, M.G. 1996, A cognitive perspective on manual assembly. *Ergonomics,* **39**, 108-127

TEACHING ERGONOMICS TO 4 TO 6 YEAR OLDS

Andrée Woodcock, Emma Clarke, Joanne Cole, James Cooke and Charlotte Desborough

School of Art and Design, Coventry University, Gosford Street, Coventry, UK.
email: A.Woodcock@coventry.ac.uk

Previous research by Woodcock (1998) has indicated that primary school children empathise with others and take a user centred approach to their design and make activities. She suggested that this could form a basis on which to introduce ergonomics into the curriculum more formally. This paper considers the development of an activity for 4 to 6 year olds to introduce them to the concept of user centred design. The activities took the children from looking at differences in the size and environment of animals, through to a consideration of how these might influence where animals lived, and the design of a house to accommodate them. The next part of the activity would be looking at the differences and similarities between children in the group. The project is used as a case study for considering the type of issues which need to be considered when designing ergonomics educational material for schools

Introduction –4 to 6 year olds

The design brief given to information design students was to design an activity pack for children to introduce them to the concept of user centred design. A group of four students undertook the project for one term, and selected 4 to 6 year olds as their target audience.

Obviously the age of the group and their cognitive ability imposed restrictions on the type of activity that could be developed. The final learning resources consisted of a set of activities which could be followed by a group of children over a period of weeks. Animal metaphors were used to introduce concepts of similarity and difference and the way in which differences have to be accommodated in design (in this case design of homes for animals).

A starting point for the development of any educational material has to be the cognitive capabilities of the target audience. In this case, this was especially important as we needed to make sure that the children understood the concepts and more importantly whether they would be engaged in and enjoy the activities.

Science is taught at primary school successfully if children are encouraged to take an investigative approach to their work in terms of design, collecting and analyzing data and following through their problems to the end (Goot and Duggan, 1996). By Key Stage 1 (5-7 year olds and the upper end of the age group we were considering), children should be able to plan experimental work, obtain and consider evidence (DfE, 1995), contribute to group activities, take part in the reporting and discussion of ideas (NCC, 1991).

This means that an approach to teaching ergonomics where the children find information and try things out themselves might be appropriate for this age group; also it could be expected that they might be able to work co-operatively and share ideas. Unfortunately, even at this age there is a wide variation in ability in areas of learning (e.g. mathematics), where some Key Stage 1 children may perform as ably as a Key Stage 3 child (an 11-14 year old), e.g. Chawla-Duggan and Pole (1996). We certainly found a very great variation in ability and experience in the children observed as part of this study. This affected the design of the activity in so far as it had to accommodate and not disenfranchise children who may have different backgrounds e.g. who had only seen animals on television, or did not have a pet. Additionally activities should be developed which gives all children an equal sense of worth and opportunities for success.

In using animal metaphors it was important that children could understand the use of metaphor and 'relate' to the animals in question e.g. a giraffe is tall, so it would need a room with high walls. Leekam (1993) has shown that children as young as 2 show empathic reactions to another's distress and have the ability to engage in hypothetical thinking by imagining themselves in another person's situation. By the age of 4 there is a quite good consensus among children about the type of situations that will provoke emotions like happiness, sadness and anger (Barden et al, 1980)

Also at this age there is evidence for analogical reasoning. In research using animals as a subject , Carey (1985) studied the attribution of biological property in 4 to 7 year olds. When asked whether 'strange' animals had biological properties such as 'can eat, have bones, have babies and have a heart', the attribution of biological properties fell off as the animals became less similar to people, with the fly and worm faring worse. Goswami (1992) showed that children can identify biological properties to animals on the basis of analogy. For example a 4 year old will say that 'a dog breathes because I do' and a 5 year old will explain that baby rabbits must inevitably grow 'because like me, if I were a rabbit, I would be 5 years old and become bigger and bigger'. Additionally, Brown, working with 3 year olds, showed that children of this age can use biological analogies to solve problems e.g. the tactics an animal might use to avoid predation – through colour or shape change or mimicry. These and similar results have led Goswami (op cit, p95) to conclude that 'research….has established that when children understand the relations of an analogy , then they can reason about relational similarity at ages as young as 3 years'.

It therefore seems likely that if a child centred approach is undertaken and activities presented to children in a way that makes sense to them, that they can appreciate complicated concepts (such as predation/camouflage) at a young age and use analogy to solve problems. Given that they also learn through observation and empathy, they might also start to understand, at a very simple level, the need for design to accommodate individual differences. Additionally animals would seem to be a good choice because they are easily accessible to the age group and are not subject to gender bias.

Developing a set of requirements for the activity packs

Prior to commencing the development of the teaching activities the designers observed children both in and outside the classroom (for example at the zoo), the type of activities they engaged in, what they enjoyed and the educational resources available to them. The observation showed:

- The wide variation in children's ability and experiences
- Short attention spans, and the fact that children have no qualms in not continuing with an activity which they find uninteresting
- The apparent inability of the age group to sit still
- Simplicity of current learning material (e.g. puzzles with 5 to 50 pieces) and games such as Lotto, matching games, simple Snakes and Ladders and books such as 'The Fish who could Wish' and 'Max's Book of Friends and Strangers' which compares people you know in your life and uses similarities and differences

In terms of activities 'people play' allowed imaginative play which may be based on real life experiences; for example, introducing children and adults into games and re-enacting their home lives, or engaging in a full range of imaginative play from making houses, to aircraft games with intricate rules of engagement, to games based just on pretending to use noisy guns. Simple scenes were created in much of the play activity, and this was bounded by a set of rules developed out of real life experiences. For example, in traffic jams, cars did not bump into each other; and when 'people' were put in the middle of the road, it was because the road was free of cars. When brick 'houses' were constructed even though they did not look like actual houses the children knew what each part was and what its function was. Children played co-operatively in groups (e.g. in making a traffic jam) and shared experiences whilst playing. For example, when playing with a helicopter one child indicated that the helicopter was on fire, and when another touched it, she was told she would get burnt and hurt.

Observation, interviews with teachers and a literature review of child development enabled an understanding of what children could achieve physically and mentally, e.g.

- they learn best through doing
- their attention span is small so learning has to be flexible to achieve goals
- 'Watching adults interacting and copying is essential to children's learning.' The idea of putting oneself in another's shoes is an idea which could be taken forward e.g. what does it feel like to be a fireman?
- There is a need to reward children for their smallest efforts
- Because of the range of abilities and the need for positive reinforcement, a 'progressive' pack is required, which can challenge the more advanced child.

Additionally, the design of current teaching material was reviewed: teaching packs typically had very plain fonts with bold titles, were black and white with a clean and simple layout with bold, simple pictures with a set style, and some words are inverted (white on a black background). They were generally assumed to be very easy to use for people with little time.

Deconstructing ergonomics for the age group

User centred design concerns designing to meet the needs of the target users. This requires an understanding of relevant characteristics and a wish to tailor the design to accommodate those needs and characteristics. Children can understand similarities and differences, although to what extent they can do this, and whether they can appreciate the effects of these on design, has not been determined. Animals were used as a starting point for the exploration of this topic because their similarities and differences are very obvious, and these can be easily translated into the sort of home they might need. Additionally children are interested in animals, and making safe/play areas in their play activities. Once children understood this relationship it was hypothesized that they could go on to look at the design of their own homes and furniture, and how they were designed to accommodate them.

The activities

Unfortunately it was not possible to design, test and evaluate the whole activity pack in the time frame available. Material was developed and tested for stages 1 to 5 (below) and tested on a group of 5-6 children.

Table 1: Overview of the activities

Stage	Topic	Activity
1	**Pets** Look at pets and animals you see everyday	Make histogram of the pets. Ask children to talk about them, draw them, write about them. Talk about the differences between household pets. Pet diary, matching game.
2	**Everyday animals** Common animals they might see eg cows, horses, birds	Discuss how these are different from pets, why you couldn't have a cow as a pet. 'Old Macdonald' sing along. Redecorate the home corner as a home farm.
3	**Exotic animals** Animals from different countries	Trip to the zoo, to show them real animals and where they have come from. Activity sheets (See Figure 2).
4	**Choose an animal** Consider more detail about the favourite animal	Show the way the animal moves, what it sounds like. Bring in examples of the type of food it may eat. Link to PE, drama and food technology.
5	**Making a house** Make a house for the favourite animal	All the information they have learnt about the animal should inform the design of the house eg size. Make model of the animal, the house.
6	**Comparing themselves** How is the house that they have made for their animal different from the house they live in	Imaginative play – what would it be like for them to live in the house, what other things would they need and why.
7	**Comparing each other** How are the children different from each other? How are these differences manifest in their own houses?	Look at anthropometry, other ways in which children are different from each other eg in what they like to do. What would their ideal room look like. Also focus on similarities.

Figures 1 and 2 show examples of the final worksheet developed for the trip to the zoo and a group activity board used to explore where animals would live in the jungle. The children selected a favourite animal, coloured it in, and placed it in the appropriate location on the board, explaining why the animal would be in that position e.g. the lion would hide in the bushes so it could eat the other animals. A crocodile could not be kept as a pet because 'it would bite your legs off', 'break all your toys', 'be smelly', and 'it would bite and kill you'.

Figure 1: Activity sheet for zoo

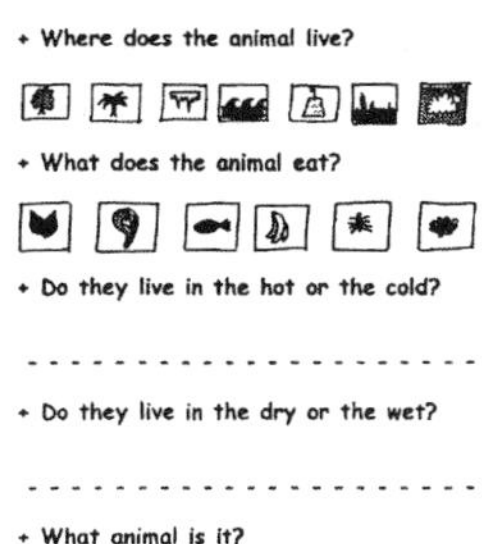

Figure 2: Picture board for animals

Evaluation

Few measures of evaluation have been identified for this age group so we relied primarily on input from teachers and observations with children using the material. The designers adopted an action learning approach to their work with the children, reflecting on what had happened, and the extent to which children had been engaged, enjoyed and understood the material which was presented.

The videos showed the intensive input required to get all the children engaged in the session. This is not a reflection of the material but of the age group! Also children varied enormously in their experiences. Many children had not seen wild animals before and thought that their natural habitat was the zoo. All children interacted during sessions and were not inhibited in shouting out answers about why animals would be in certain places.

Some of the activities were not accessible to all children; for example, a week 1 activity was to keep a pet diary, but not all children have pets. The failure of this exercise might have disenfranchised children and made them feel disadvantaged. The solution was that if children did not have a pet they could keep an imaginary one. Additionally the results of the initial trials showed that some of the earlier activities were too simple for the age range. Teaching material, even for this age group, has to be linked to the National Curriculum. This means that the programme of activities has to provide opportunities to employ numeracy and literary skills, to design and make, engage in creative play, co-ordinate activities, engage in drama and PE, look at food issues.

Discussion

Although we were not able to see this work through to completion, it has raised several issues concerning the design and testing of ergonomics educational material for this age group.

Firstly, in terms of the design of material to teach children about ergonomics:

- Very young children form naïve hypotheses about the world, and learn very quickly. Teaching of ergonomics concepts which fits into their interests and beliefs may be possible.
- Material has to address the wider aims of the National Curriculum
- Investigative approaches to the collection and analysis of data work well with primary school children. An activity pack, where ergonomics is learnt through discovery, may be most appropriate.

Secondly, in relation to the design of the material

- There appears to be little information for designers wishing to produce material for this age group
- Very few instruments exist for measuring learning, pleasure and engagement in young children
- The adoption of an action learning approach to the development of the learning activities can work well as part of the design process.

Acknowledgements

The authors would like to thank Lee Gunn for her support during this work.

References

Brown, A.L.(1989), Analogical learning and transfer: what develops, in S. Vosniadou and A. Ortony, (eds), Similarity and analogical reasoning, p369-412. Cambridge: Cambridge University Press.

Barden, R., Zelko, F., Duncan. S. and Masters, J. (1980), Children's consensual knowledge about the experimental determinants of emotion, *Journal of Personality and Social Psychology*, 39, p968-978

Carey, S. (1985), *Conceptual Change in Childhood*, Cambridge Mass;MIT Press

Chawla –Duggan, R. and Pole, C.J. (eds) 1996, *Reshaping education in the 1990's: Perspectives on Primary Schooling*, Falmer Press, London

Department for Education and the Welsh Office (1995), *Science in the Nationsl Curriculum,* London, HMSO.

Goswami, U. (1992), *Analogical reasoning in children*, Lawrence Erlbaum, London,

Gott, R. and Duggan, C., 1996, Primary Science in the Mid-1990's: Grounds for Optimism, in Chawla-Duggan and Pole, p33-49

Leekam, S. (1993), Children's understanding of mind, in Bennett, M (ed.), *The Child as a Psychologist.* New York:Harvester, Chapter 2, p 26-61

NCC (1991), *National Curriculum Council Consultation Report: Science in the National Curriculum*, NCC, York

Woodcock, A. and Galer Flyte, M.D. 1998, Ergonomics: It's never too soon to start, *Product Design Education Conference*, University of Glamorgan, 6th-7th July

Woodcock, A., Gunn, L. and Georgiou, D. 2002, Usability and evaluation issues in student design projects, Poster paper at *International Workshop on Interaction Design for Children*, September 2002, Eindhoven. Available on request from authors

HUMAN-CENTRED DESIGN IN A SYSTEM ENGINEERING CONTEXT

Brian Sherwood Jones[1], Jonathan Earthy[2], Stuart Arnold[3]

[1]*Process Contracting Limited*
6 Burgh Rd, Prestwick, Ayrshire, Scotland KA9 1QU

[2]*Lloyd's Register of Shipping*
71 Fenchurch Street, London England EC3Y 4BS

[3]*Qinetiq*
Fort Halstead, Sevenoaks, Kent England TN14 7BP

Ergonomics has traditionally taken a 'systems' approach, and has considered the role of people within a system and stakeholders external to it. Recent developments in systems engineering, notably the development of ISO/IEC 15288, have provided a project environment that sets out to support the application of ergonomics. The ergonomics contribution to system development and operation has been defined by the Human-Systems (HS) model, currently with ISO to be published as a Publicly Available Specification. The HS model is a significant expansion of the successful standard ISO TR 18529, moving from an interactive product to a full system. The model has a number of differences from more traditional descriptions of how ergonomists go about their work. The ability to produce such a model indicates a maturing of the profession, with all the opportunities that standardisation brings with it. It also offers opportunities to define practitioner competence within accepted frameworks.

The form of process standards and their use

A process standard is essentially a hierarchical task analysis (HTA) of a specified scope of activity. ISO TR 15504 has defined the format for such standards. Each process has a purpose, an outcome (a result, not an output), work products (inputs and outputs) and a description. The system engineering standard ISO/IEC 15288 and the HS model have three levels: process category, process, and practice. ISO TR 18529 has two levels: processes and practices. The performance of a process is assessed using a capability level, ranging from 0 (no achievement of results) through to 5 (optimisation to meet business needs) via 1(ad hoc), 2 (monitoring of time and quality) and so on. Most software is developed by processes at level 1 (using appropriate models), and it is expected that most Human-centred design (HCD) assessments will be at levels 0 and 1.

The 'plan' information normally found in an HTA is intentionally omitted. It has been found that the top levels of HTA comprise continuing goals and plans are not appropriate.

This is the level of specification in the standards. The methodologies of the 1980's were characterised by large manuals and long flowcharts. They proved hard to use and could not accommodate project-specific requirements and constraints. The scheduling of processes in a process model is left to the project team, but with the expectation that all processes need at least a watching brief most of the time.

HCD processes are characterised by multi-disciplinary teams (ISO 13407). Socio-technical systems theory advocates semi-autonomous teams. Accordingly, the allocation of function in the HS process standard describes the tasks required for a team to deliver usability[1] rather than the tasks that might be expected of an ergonomics specialist. The processes therefore extend beyond ergonomist-centred descriptions of ergonomics. A process owner is likely to have more responsibility (and accountability) than a human science technical advisor. As a minimum, a process owner would be responsible for providing assurance as to how well the process was achieving its outcomes. At most, responsibility could extend to authorisation of a system for safe operation.

Process standards are principally concerned with two sorts of use, process improvement (PI), and capability evaluation (CE). However, they have proved to have other uses e.g. planning, competence definition. PI is typically conducted by an organisation for its own benefit (but can be mandated by a customer) and comprises a cycle of self-assessment, structured improvement and re-assessment. CE can vary in formality, but at its most formal is done by an accredited third party for the purposes of pre-contract award vendor assessment. The systems community has developed COTS tools and methods for both PI and CE, one of the many benefits from adopting the process standard approach.

Content of process standards in system engineering and HCD

Table 1. ISO/IEC 15288 (2002) Systems engineering - systems life cycle processes

Agreement processes						
Acquisition			Supply			
Enterprise processes						
Enterprise environment management	Investment management	System life cycle processes management	Resource management	Quality management		
Project processes						
Project planning	Risk manage -ment	Project assesment	Configuration management	Project control	Information management	Decision making
Technical Processes						
Stakeholder requirements definition	Requirements analysis	Architectural design	Implementation	Integration		
Verification	Transition	Validation	Operation	Maintenance	Disposal	
Special processes						
Tailoring						

[1] Defined as 'The capability of a system to enable specified users to achieve specified goals with effectiveness, productivity and satisfaction in specified contexts of use.'

ISO/IEC 15288 has four process categories and a tailoring process (See Table 1). People can be part of a system and/or a stakeholder in its design and operation (Arnold *et al* 2002). A life cycle is also specified. An 'enabling system' is associated with each stage of the life cycle e.g. a system to develop a factory for the implementation stage of a product system. Of particular interest to the delivery of usability is the 'project process' category. This is crucial to the system engineering objective of developing a 'common picture' to harmonise diverse implementation technologies (e.g. software, electronics, biology, hydraulics, mechanics) and the diversity of engineering specialities (e.g. safety, operability, supportability). To contribute to this integration, each technology, and each specialist discipline needs to portray its activities in a form that others can understand, translating its specialist toolbox of techniques into statements about project achievement. The HS model does this for HCD, as ISO 12207 does for software.

Table 2. The Human-Systems (HS) model processes

<table>
<tr><th colspan="5">HS.1 Life cycle involvement</th></tr>
<tr><td colspan="5">Human-system issues in:</td></tr>
<tr><td>conception</td><td>development</td><td>production and utilization</td><td>utilization and support</td><td>retirement</td></tr>
</table>

HS.3 Human-centred design	HS.4 Human resources
Context of use	Human resources strategy
User requirements	Define standard competencies and identify gaps
Produce design solutions	Design staffing solution and delivery plan
Evaluation	Evaluate product system solutions and obtain feedback

HS.2 Integrate human factors	
Human-system issues in business strategy	HF data in trade-off and risk mitigation
Human-system issues in quality management	User involvement
Human-system issues in authorisation and control	Human system integration
Management of human-system issues	Develop and re-use HF data

The HS model has four process categories, see Table 2. These processes address:

HS.1. Human Factors in the Life Cycle – issues related to people in the stages and enabling systems for the system. The processes here are related to anticipating and tracking the particular needs of each stage in the life cycle. The intent is to improve the efficiency of the life cycle by timely resolution of human-system issues.

HS.2. Human Factors Integration – issues related to people in the system of interest. The processes here are concerned with strategy, procurement, planning. This process ensures that human-system issues are addressed by the appropriate stakeholders. It reduces life cycle costs by ensuring that "design for people" is used within the organisation.

HS.3. Human-centred design – design for people using the system of interest in the context of use. The processes here are technical, concerned with requirements, design and evaluation. This process enables user-centred technical activity to be focused appropriately. It contributes to a 'better' system by the inclusion of processes that deliver usability. It is a development of ISO TR 18529.

HS.4. Human Resources process – the provision of the correct number of competent staff for the system of interest. The processes here are concerned with selection and training, and harmonising organisational development and technological change. This process provides the means to resolve issues concerned with the 'implementation' of the human part of the system, rather than the equipment-centred part. It ensures continued, timely delivery of the correct number of competent people required to use the most suitable equipment.

The first three processes have been the subject of extensive international review and form part of the body of the proposed standard. The last process is considered to have received less review, and so forms an Annex. Bringing together the processes concerned with 'putting the machine in front of the user' with the processes concerned with 'putting the user in front of the machine' can be considered the essence of Manpower and Personnel Integration (MANPRINT) and Human Factors Integration (HFI).

The HS model can be considered as four views of system and enterprise that complement ISO/IEC 15288. The interfaces between the HS model and ISO/IEC 15288 are shown in Figure 1.

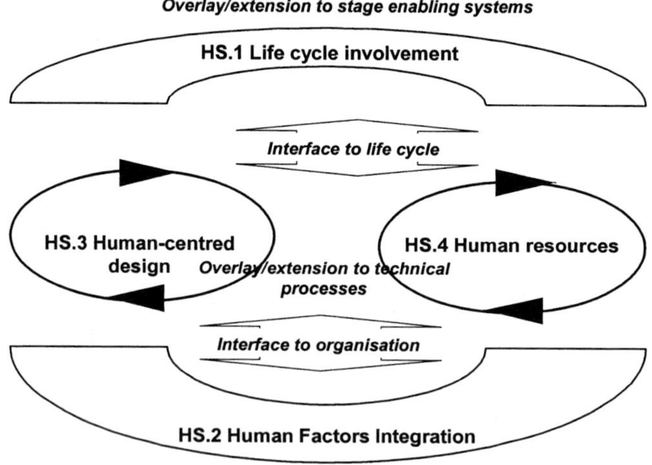

Figure 1 Interfaces from the HS model to the project and to ISO/IEC 15288

Implications of process standards for HCD

There is the risk that ISO TR 18529 and the HS model are seen as a constraint on the individual genius of the ergonomist. However, the view taken by their proponents is that such clarity and standardisation of process is essential for the widespread uptake of HCD. One of the most important problems for ergonomics is its own usability by the systems development community. This lack of usability is considered detrimental to uptake and successful application. Based on Earthy *et al* (1998), the HS model offers the following benefits when assessed against ISO 9241, part 10 Dialogue principles:

Suitability for the task: Ergonomics can now explicitly support project managers in the effective and efficient development of a usable system, rather than be seen as an isolated specialism. The standard offers assurance of delivering outcomes.

Self-descriptiveness: Feedback from ergonomics activities and any explanations of results should be immediately comprehensible to developers rather than long obscure texts. Ergonomics is presented as an integrated system engineering discipline.

Controllability: The HS model does not propose methods and techniques that require standard, expensive and time-consuming fixed steps. Effort can be assigned on the basis of risk to achieving outcomes.

Conformity with user expectations: The standard identifies work products that are compatible with software and system engineering. Software and system engineers are familiar with the concepts of process models, CE, PI.

Error tolerance: The provision of assurance metrics supports corrective action. The integration process (HS.2) maximises error recognition and recovery opportunities.

Suitability for individualisation: Process standards support structured tailoring.

Suitability for learning: Skilled behaviour is marked by economy of effort and the subtle adaptation of well learned routines. Ergonomics practitioners, both individually and collectively need International Standards to develop such skilled behaviour.

The recent HF NAC report (HF NAC 2002) put 'Design Process – tools, techniques and integration' as a top priority for research. It is unclear from the report whether this is to take advantage of recent HCD and system engineering standards, or whether its community still views methodology as a research topic. If the latter, there is the real risk of re-inventing the wheel and delaying the maturing of professional practice. The constructive way ahead for the discipline is to build on the recent process standards. Any proposed advances in specific methods and techniques can take advantage of the context provided by HCD standards, rather than have to re-invent it (cf. Older *et al*, 1997).

Ergonomics is claimed to be vital; human error, inclusive design, usability are all touted as silver bullets for system effectiveness. However, ergonomists typically see themselves as performing obscure, risk-free specialist services to 'the designer' who carries all responsibility, authority and liability. HCD process models are concerned with providing assurance of usability and potentially open the door to supporting Authorisation for use. They offer the ergonomist a different role and the possibility (risk?) of greater responsibility in addition to the current specialist role.

A process standard provides a structured statement of an area of activity that can be used for defining competence. ISO/IEC 15288 is being considered by INCOSE and IEE as the basis for defining system engineer competence. The HS model has the potential as a COTS product to be given similar consideration by the Ergonomics Society and IEE activity to define competence in HCD or ergonomics practice. Such a basis would further help to integrate ergonomics as a systems approach.

Conclusion

The work by the US Software Engineering Institute on the software Capability Maturity Model (Paulk *et al* 1993) noted that as technologies developed they moved away from a 'bespoke' approach to one in which methods are standardised and subject to quality assurance. ISO TR 18529 and the HS model represent a maturing of the discipline. They can be considered an application of 'the ergonomics of ergonomics' in a number of ways. They are in essence a task analysis of ergonomics as applied to equipment and systems. They were developed iteratively with user involvement and are written in the language of the user i.e. system engineers and other project engineers. They can be used for assessment, so enabling task performance (including the performance of the ergonomist) to be measured. The HS model describes the user needs for ergonomics rather than the view of the ergonomist as supplier. These models, and the new context provided by ISO/IEC 15288 represent both an opportunity and a challenge to the practitioner and to the research community. There is the hope that they provide much of the incorporation of Human Factors into system development called for by Holman (1988).

Acknowledgements

Part of the authors' work was carried out under projects for the European Commission and the UK Ministry of Defence. The support of these bodies is gratefully acknowledged. The paper has benefitted considerably from correspondence with David Beevis. The opinions expressed in this paper are the authors' own and not those of Lloyd's Register or Qinetiq.

References

Arnold, S., Sherwood Jones, B.M, and Earthy J.V. 2002, 'Addressing the People Problem - ISO/IEC 15288 and the Human-System Life Cycle' *12th Annual International INCOSE Symposium*, Las Vegas, July 2002

Earthy, J., Shackel, B., Courteney, H., Hakiel, S., Sherwood Jones, B., and Taylor, B. 1998, Human-Centred Processes, in MAY, J. et al, *HCI'98 Conference Companion, adjunct proceedings of the 13th BCS annual conference on HCI, HCI'98*, Held at Sheffield Hallam University, Sheffield, Sept. 1998, ISBN 0-86339-795-6

HF NAC 2002, Industrial Priorities for Human Factors Research in UK Defence and Aerospace *2nd Report of the Human Factors National Advisory Committee*, DTI

Holman, R. 1988, Integrating Human Factors into the systematic development process. In E.D. Megaw (ed.) *Contemporary Ergonomics 1988*, (Taylor and Francis, London), 183-188

ISO 9241:1996-10 Ergonomics of office work with VDTs - Dialogue principles

ISO 9126:2000 Software product quality - quality model

ISO 12207:1995 Software process – Software lifecycle processes

ISO 13407:1999 Human-centred design processes for interactive systems

ISO/IEC 15288:2002 System engineering – System lifecycle processes

ISO PAS tba:2003 A specification for the process assessment of human-system issues, 2003 *in press*

ISO TR 15504:1998-2, Software process assessment - A reference model for processes and process capability

ISO TR 18529:2000 Ergonomics of human system interaction - Human-centred lifecycle process descriptions.

Older, M.T., Waterson, P.E. and Clegg, C.W. 1997, A critical assessment of task allocation methods and their applicability. *Ergonomics*, **40**: 151-171

Paulk, M.C., Weber, C.V., Garcia, S.M., Chrissis, M.B. and Bush, M. (1993). Key Practices of the Capability Maturity Model, Version 1.1. SEI-93-TR-025, Software Engineering Institute, Pittsburg, PA.

Further information on the HS model, and details of a late draft (the HSL model) can be found at http://www.processforusability.co.uk/HFIPRA/

STANDARD SYMBOLS - USING THE ISO TEST METHOD

Anne Ferguson and Wendy Brown

Anne Ferguson and Wendy Brown
ITS-Research and Testing Centre
Davy Avenue, Knowlhill
Milton Keynes, MK5 8NL UK

Work is underway to update a CEN standard for the design of Automated Teller Machines (ATMs) and other card-reading machines such as car-park barriers and train ticket dispensers. Part of this work is the approval of symbols to supplement user information on the machines. The CEN working group (TC224 WG6) set up a project team (PT05) to evaluate symbols using the ISO 9186 test method. This paper reports the work and considers the relevance of ISO 9186 when compared to another method (ETSI Multiple Index Approach). Advantages and limitations of ISO 9186, in particular in relation to the needs of people with disabilities, and suggestions for its revision are also reported.

Introduction

Card-reading machines, such as cash machines (ATMs) and train ticket dispensers are becoming increasingly common. The European Standard, EN 1332: *Identification card systems – Man-machine interface,* is currently being revised and extended to include provisions for physical accessibility, such as wheelchair user access, and symbols related to the key actions needed to use a card-reading machine. Funding was obtained from the European Commission to research, develop and test symbols, taking into account the needs of people with disabilities. The work was carried out by a project team, *PT05: Icons, symbols and pictograms for card-reading devices*. Symbols were developed using criteria from various ISO standards (see references) and tested to ISO 9186, after consideration of an alternative ETSI method (ETR 070), and with users with a visual or hearing impairment. Work reports (CEN PT05 2001 and 2002) were submitted to the relevant CEN working group, TC224, WG6. This paper concentrates on the relative merits of the test methods with proposals for amendment to ISO 9186.

Terminology

In this paper, the terms icon, symbol or pictogram have been used, without distinction, in whatever context, to indicate 'a figure which transmits information independently of language'. The term **referent** indicates the function a symbol is intended to convey and **variant** the different versions of symbols related to a particular function.

ISO 9186 test method for understanding of symbols

The ISO 9186 method was originally created specifically for public information symbols but, in its current form (2001), is proposed as suitable for all kinds of graphical symbols. ISO 9186 contains two test methods, although it is not necessary to carry out both:

ISO 9186: comprehensibility judgement (Test A)
This is used to assess which version of a symbol is most appropriate for a particular function. Subjects are shown all the symbol variations (variants) for a particular meaning (referent) displayed around a circle (with a radius of 80mm) on a sheet of paper or on a computer screen. Participants are told what a symbol means and must assess what percentage of the population they think would understand that symbol. This test can be used however many variants there are (within the physical limitation of how many symbols could fit around the circle and still be distinguishable on the page). The Standard requires that the exercise is carried out in at least two countries with a minimum of 50 subjects in each, selected to resemble the eventual user population, in terms of age, sex, educational level, occupation, cultural background and (where relevant) physical ability. Symbols must pass the **criterion of acceptability**, a figure available on application to the relevant ISO Technical Committee, TC145.

ISO 9186: comprehension test (Test B)
In this 'naming test', each symbol is presented in the centre of a page (sheet of paper or computer screen). Subjects are told where they might expect to see the symbol (in this work: on ATMs or other card-reading devices) and asked what they think the symbol means and what action they would take in response to it. Test sets of no more than 20 pages are made up, for each subject to work through, with the different variants for each referent in different test sets and only one variant for any given referent in any test set.

This test is normally used where there are three or fewer variants, although four have been used, successfully (Foster 2001). It may be the second phase, if testing of a greater number of variants using the comprehensibility judgement method has not resulted in any variant reaching the relevant criterion of acceptability. The test is carried out in a minimum of three countries, again with at least 50 subjects in each, selected in the same manner as Test A. Three judges are required to interpret the scores for each symbol.

Alternative ETSI Multiple Index Approach to testing symbols

This method was originally developed by ETSI, the European Telecommunications Standardization Institute, for symbols to be used on videophones (ETSI 1993). It aims to select the most suitable **set** of symbols from a number of alternative sets. The questionnaire is organised in three parts. The first uses a matching method to assess **pictogram associativenes.** Functions are explained and then presented randomly. Subjects must select the most appropriate symbol, for that function, from a set of symbols, in a setting as close as possible to that in which the symbol will be used. Subjects say, on a five-point scale, how certain they are of their choice - **subjective certainty** and their impression of how well the symbol represents the referent - **subjective suitability**. Data is also collected on **hit rate** (score of correct associations between the referent and variant) **false alarm rate** (number of associations with the wrong referent) and **missing values** (number of times a respondent did not answer).

In the second test, of **pictogram preference**, subjects choose which symbol best represents the referent. In the third, **pictogram set preference**, subjects choose their preferred **set** of symbols. Data is collected through paper or computer presentation. The method does not specify the number of users, nor countries for testing. Significant analysis is required to interpret the data.

Comparing test methods: ISO 9186 and ETSI Multiple Index Approach

In many respects, the ETSI method more closely accords with the form of symbol presentation to be used in ATMs, where the user approaches the machine asking 'which button (or symbol) should I press for the function I want'. However it is complex to administer and requires presentation of a set of symbols. ISO 9186 was the method the project team was directed to use. Both Foster (2001) and Brugger (1999) had found the Comprehensibility Judgement test a good predictor of the outcome of the Comprehension test. Brugger concluded that it was the most valid measure of comprehensibility when compared with an **Appropriateness ranking** method (subjects order symbols from best to least appropriate) and **Appropriateness class assignment** method (subjects categorise symbols in one of three levels of appropriateness). It is relatively simple to administer and may be used regardless of the number of variants. It was selected to be used for the first stage of current work. A decision between computer and paper-based trials was made after consideration of the cost benefits, see Table 1.

Table 1. Comparison of cost benefits of computer and paper-based trials

Computer via internet	Networked computer in lab	Paper-based forms in lab
no control over subjects	need suite of computers for effective trials	location not dependent on new technology
potentially higher set-up costs: need to write program in reproducible form (pdf or html)		no special programming for setting up
presentation quality may vary depending on computer system		guaranteed consistency of presentation of materials
easy to randomise presentation of symbols		presentation not randomised (impractical)
no additional materials costs if number of subjects significantly increased		cost of materials increases with number of subjects,
cheap to use large numbers of subjects	additional subjects require additional supervision	
computer literate subjects required	need computer literate subjects or assistance.	not influenced by level of computer skill
allows variation in presentation of symbols		less flexible presentation
analysis can be incorporated in program, may reduce costs		manual tabulation/scanning required, prior to analysis

Using the ISO 9186 method

Work by Easterby and Hakiel (1981), Cairney and Sless (1982), and Akerboom (1993) has all pointed to differences in symbol comprehension by different age, ability or cultural groups. The requirements of the ISO method (50 subjects in each of two countries) were considered insufficient to reflect the variations across Europe. Thus a total of 342 subjects across three countries: the UK (Germanic based language/North European country), Spain (Latin based/Southern European) and Romania (different economic and cultural base to others) were recruited. Incorporating disabled people in the sample in proportion to their incidence in the national population, as proposed by the ISO method, was also thought to give insufficient data to be meaningful for critical groups such as those with a visual impairment. Given the importance of ensuring that users with disabilities had equal access to card-reading machines a second qualitative phase of testing was undertaken, targeted on special needs.

The ISO comprehensibility judgement method requires that subjects estimate how many other people would understand a symbol, rather than indicating if they, themselves, understand it, to avoid any implication that the respondent was being stupid if the symbol had no meaning for them. Foster (2001) found that, whilst this might not seem an easy concept to grasp, very few people queried the task. Those who asked for clarification were satisfied by the explanation that it was 'the percentage of the general population who would understand' and not of some specialised population. Current work also found that the concept could be understood but required careful explanation and it was deemed necessary to ensure that the first forms completed by subjects were checked by supervisors to ensure full understanding. Also guidance on percentage equivalents (e.g. that half the population equated to one in two or 50 per cent) was made available to subjects, to aid any who were unfamiliar with the use of percentages. Pilot work established that subjects could be batched in groups of around 15 with one supervisor and that the test could be completed within one hour, allowing several batches per day.

The importance of having written information or **keywords** to accompany symbols was recognised, drawing on sources such as Reddy (1996) and Akerboom (1993). The ISO method is limited only to symbols, but work by ETSI in the field of telecommunications (2002) was used in the development of a questionnaire to be completed by subjects after the comprehensibility judgements of symbols.

Additional testing with people with disabilities

The first stage in symbol testing is typically to select the most appropriate symbol from a set of variants and the second is to rate symbols for ease of understanding (Poppinga, 2001). Qualitative work on symbols and keywords was undertaken with people with a visual or hearing impairment to ensure that their comprehension was the same as those without these disabilities. Some work was also undertaken in this phase on clarity of symbols with people with a visual impairment. Three separate tests were carried out: Test 1, a lab-based test with 21 hearing-impaired and 22 visually-impaired subjects, in which subjects were asked to identify, from the set of symbols found most acceptable in the main tests, the symbol which related to a particular function or keyword; Test 2 a web-based questionnaire, for which 25 visually-impaired and 15 hearing-impaired respondents rated the acceptability of the same set of keywords tested in the main tests. Users would have used whatever adaptations they had on their own computers to cope with websites, for example computer-spoken words and Test 3, lab test of symbol clarity, using 15 visually-impaired subjects, in which a limited set of symbols were displayed, one at a time, on a computer screen and subjects were asked to adjust the symbol size to the minimum and optimum reading size and rate the clarity of the symbols. A similar exercise was undertaken with different colour combinations.

Principal results from symbol testing

For the symbols tested, a very similar pattern in the ranking of variants was observed across cultures, although scores were generally highest in Spain and lowest in Romania. Similar rankings were also obtained from groups of different ages, although actual scores differed in magnitude. Despite (or perhaps because of) the large number of subjects in the trials, few variants reached the criterion of acceptability (85 per cent) for judged comprehensibility, required for approval to the Standard. However many referents had variants with scores of 65 to 80 per cent. Resource limitations prevented use of the comprehension test from the ISO Standard but the qualitative work with visually-impaired and hearing-impaired users provided good support for the results from the main tests, for both symbols and keywords. Some differences occurred, for example in the preference for keywords 'Take' or 'Remove'. 'Remove' as in 'remove money' was preferred in the UK main trials, whereas 'take' money was preferred by subjects with a visual or hearing impairment. Findings were sufficiently robust to enable proposals for symbols and accompanying keywords for 12 referents in all countries.

Conclusions regarding possible revisions to the ISO method

The work undertaken confirmed the value of ISO 9186 as a method for assessing symbols but resulted in proposals for a number of possible revisions which have been communicated to the relevant technical committee. These include:

Reviewing the criterion for acceptability

It is extremely difficult to devise new symbols which are instantly intelligible across cultures and Davies *et al* (1997) have pointed to the possible advantages of selecting symbols on the basis of how easily they may be learned rather than on how well they are understood when first seen. The current score of 85% for the criterion of acceptability for informative symbols is considered too high. Anecdotal evidence suggests that even some well-established safety symbols may score less than this. It is proposed that 70%, 75% or at most 80% would be more appropriate, especially where testing is undertaken in three countries (as in our tests). Including definitions for 'minimal score' and 'criterion of acceptability', together with associated scores, would simplify the stages of preparing for testing.

Increasing the amount of guidance

A pilot exercise is invaluable to smooth the process for full trials but additional information in the Standard could also aid this process. For example, providing data on the time tests may take, what test forms should look like, useful supplementary information (such as a guide to percentage equivalents) and how to brief supervisors. Guidance could also be included on procedure such as who to contact within ISO to check whether a symbol has been standardised, what to do with the submission for test forms and how long the original questionnaires should be held by the Test House. This could all be included in the form of an informative annex.

Reviewing sample size and structure

Fifty subjects is not sufficient to adequately reflect the population (using the categories as defined in the current version of the standard) but reflecting the different cultural groups within the population should be retained. It may be that some 'socio-economic' scale could be used rather than educational level and occupation. Including 'physical ability' within the sample means that there are insufficient numbers of the critical samples (visual impairment etc) to have an impact whereas mobility impairment, for example, is not relevant and thus need not be recorded.

The addition of tests for keywords and people with disabilities
Tests for keywords, using samples representative of the national population, should be added. Separate qualitative tests targeted on special needs are also proposed, in particular tests for keywords (hearing and visually-impaired subjects) and tests of symbol clarity preferably in a likely context (visually-impaired people). Although not explored for the current work, tests with people with learning disabilities would also be beneficial.

References

Akerboom S.P. 1993, *The effects of pictogram and consumers' attributes on noticing, comprehension of and compliance with the pictogram,* Proceedings of the International Conference on Product Safety Research

Brugger C. 1999, *Public information symbols: a comparison of ISO testing procedures,* Visual information for everyday use, design and research perspectives, edited by Zwaga et al., (Taylor & Francis, London)

Cairney P. and Sless D. 1982, *Communication effectiveness of symbolic safety signs with different user groups,* Applied Ergonomics, **13**(2), 91-97

CEN (Comité européen de normalisation)

EN 1332-1 *Identification card systems – Man-machine interface, Part 1: Design principles for the user interface* , unpublished draft

TC 224 WG06, PT5, 2001, *Icons, symbols and pictograms for card reading devices. Work Report* A1-A9 and 2002, *Final Report: Phase B,* Volumes 1 – 4, unpublished

Davies S., Haines H.M. and Norris B.J. 1997, *The role of pictograms in the conveying of consumer safety information,* URN 97/751 (Department of Trade and Industry, Consumer Safety Research) UK

Easterby R.S. and Hakiel S.R. *Field testing of consumer safety signs: The comprehension of pictorially presented messages,* Applied Ergonomics, **12**(3), 143-152

European Telecommunications Standardization Institute

1993, *Human Factors (HF); The Multiple Index Approach (MIA) for the evaluation of pictograms,* ETSI Technical Report ETR 070

1998, *Human Factors (HF), Framework for the development, evaluation and selection of graphical symbols,* EG 201 379 V1.1.1 (1998-12) 6

2002, *Generic Spoken command vocabulary for ICT devices and services,* STF 182

Foster J. 2001, personal communication

International Organization for Standardization

ISO 9186:2001 *Graphical symbols - Test methods for judged comprehensibility and for comprehension,* Second edition 2001-04-01

ISO 3864 *Safety colours and safety signs*

ISO TR 7239 *Development and principles for application of public information symbols*

ISO 11581 *User system interfaces and symbols – Icon symbols and functions*

ISO 80416 *Basic principles for graphical symbols for use on equipment.*

Poppinga J. 2001 *Design and evaluation of icons - a literature review*, unpublished

Reddy M.T. 1996, *Adoption of symbols within standards of interest to consumers,* (Consumer Policy Unit, British Standards Institution, London)

WORK DOMAIN MODELS FOR COGNITIVE ERGONOMICS: AN ILLUSTRATION FROM MILITARY COMMAND AND CONTROL

John Long[1], Martin Colbert[2] and John Dowell[3]

Ergonomics and HCI Unit,
University College London,
26, Bedford Way,
London, WC1H 0AP, UK

([1] Now at: University College London Interaction Centre (UCLIC); [2] Computer Science, Kingston University; [3] Computer Science, University College London)

There is general agreement that 'work' is a primary concept for Cognitive Ergonomics (CE). However, there is little agreement how the domain of work might best be modelled. This paper assesses two contrasting approaches to such modelling. The first, and implicit approach, derives from domain experts. The second, and explicit approach, derives from domain research. The approaches are illustrated by an initial analysis of the domain of military command and control and specifically of models of the Vincennes incident. Implicit and explicit domain models are assessed in terms of the incident events. It is concluded that both models have potential to support design, but the explicit model also has potential to support research. The need for explicit domain modelling to support validation of CE design knowledge is underlined.

Introduction

There is general agreement that for Ergonomics, and in particular for Cognitive Ergonomics (CE) – the concern here – 'work' is the primary concept, expressing its scope. Ergonomics was originally derived from two Greek words: 'ergos' meaning work and 'nomos' meaning (natural) laws, and so 'laws of work' (Murrell, 1971). Work, domain, environment etc. reflect Ergonomics' and so CE's concern with work (Murrell, 1971; Card et al., 1983; Long 1987; Galer, 1989; Carroll, 1991; Pheasant, 1991; Dix et al., 1993; and Oborne, 1994). There is also agreement as to the other primary concepts of CE, for example users, technology and performance etc. (Flach, 1998; Vicente, 1998; and Long 2001). However, there is little or no agreement as to how the domain of work should be modelled for CE design and research. Two contrasting approaches can be discerned.

The first, and implicit, approach derives from domain experts (Murrell, 1971; Card et al., 1983; Galer, 1989; Pheasant, 1991; Dix et al., 1993; and Oborne, 1994). These experts are highly knowledgeable about individual domains of work and their knowledge informs CE design and research. The second, and explicit, approach derives from research into domain models (Dowell and Long, 1998; Flach, 1998; and Vicente, 1998). Domain models represent work explicitly.

The aim of this paper is to assess implicit and explicit domain models for CE design and research. The assessment is based on an illustration, taken from an initial analysis of the work domain of military command and control (C^2) and specifically as embodied in the Vincennes incident.

Illustration: the Vincennes Incident

The Vincennes incident was widely reported (see The London Times, 4 and 5 July, 1988; and Time Magazine, 18 July 1988) and was also the object of an official report (USDoD, 1988). The incident took place during the Iran/Iraq war, initially a land battle. In the late 1980s, Iraq attempted to disrupt Iran's oil trade. Iraq launched air attacks against Iranian oil installations. Iran's response was to disrupt oil transport in the Persian Gulf. The US response was to send naval forces to ensure oil supplies. The incident occurred on the morning of 3 July, 1988. The Iraqi Air Force had successfully attacked Iranian forces near the North Persian Gulf. Iranian retaliation of small boat attacks on commercial shipping was expected. The incident occurred when the USS Vincennes mistakenly shot down civilian Iran Air Flight 655, while simultaneously engaging a group of Iranian small boats.

Implicit Domain Model

The domain of work here is that of C^2. The main sources of information about C^2 are military domain experts using C^2 systems. Here, the C^2 model is implicit in the experts' descriptions of the Vincennes incident in the official report. A selection is summarised in the left column of Appendix 1.

The incident events are described in C^2 systems expert language. For example: "0649 Flight 655 adopts its flight path, which is toward the Vincennes. The Vincennes challenges its air contact (actually Flight 655), but receives no reply". The language comprises both ordinary expressions ("receives no reply"), as well as technical expressions ("Flight 655 adopts its flight path"). However, neither expresses C^2 explicitly, only in the specific terms of the events. Further, C^2 domain aspects ("The missiles destroy Flight 655") and C^2 system aspects ("Vincennes' air contact appears electronically to identify itself as a military aircraft") both appear, but are not distinguished. The domain of the Vincennes incident is undoubtedly that of C^2, derived from experts, but the model of C^2 remains implicit. The model informs the experts' explicit descriptions, but is not itself explicit.

Explicit domain model

The domain of a worksystem is the work it performs (Dowell and Long, 1998). The domain comprises objects, constituted of attributes, having values. A worksystem comprises interactive user and device behaviours, which, when executed, perform tasks effectively (transforming the domain, so to achieve desired goal states). Here, C^2 performs two types of work, and so is modelled as two domains: the domain of plans for armed conflict and the domain of armed conflict. The domain of plans comprises a single object – an **interest** – with a single attribute – **security** (model concepts are in bold initially). An interest specifies a use of resources by a nation state. Uses may be political, military etc. Resources include the land, sea, air, space and installations etc. Security is the potential to realise an interest.

Within the domain of armed-conflict (military operations), the domain objects are identified as: **friends**, **enemies** and **neutrals**. These objects are distinguished by the ends pursued. A friend supports the interests pursued by C^2. An enemy, in contrast, pursues incompatible interests. Neutrals pursue interests, compatible with those of both friends and enemies. The domain also comprises a **resource** object, whose use attribute is interest. The three classes of participant object exhibit between them attributes of **power**, **vulnerability** and **involvement**. Power, here, is the potential to secure interests by the display of force.

One participant's potential to damage another implies the latter's weakness. That is, power implies vulnerability. The power of one participant, and the reciprocal vulnerability of another, implies the involvement. Friends and enemies may exhibit power/threat, vulnerability and involvement. Neutrals, in contrast, only exhibit vulnerability and involvement (for a complete description, see Colbert and Long , 1995).

This explicit domain model of C^2 was used to model the Vincennes incident. A selection from the application is summarised on the right of Appendix 1. Domain concepts are in bold. The appendix shows how the concepts are able to describe the domain transformations of the left column. For example, "The small boats and the Vincennes continue to close" is expressed by the domain model as: "The **involvement**, **power** and **vulnerability** of the **friend** (Vincennes) and the **enemy** increase again". The appendix also demonstrates the explicitness of the concepts – by their identification. Finally, the model expresses only the transformations of the domain, not the behaviours of the C^2 worksystem.

Domain model assessment

Both CE domain models, implicit and explicit, constitute CE design knowledge, intended to support design and research. CE design can be conceived as the diagnosis (of design problems) and the prescription (of design solutions) (Long, 2001). CE research can be conceived as the acquisition and validation of design knowledge (Long, 2001). Implicit and explicit models are now assessed for their potential to support design and research.

The implicit model of C^2 would appear to have potential to support design as both diagnosis and prescription. For example: "Flight 655's altitude is misread. The air contact is perceived as diving towards the Vincennes; in fact, it is climbing away from it". Misreading the altitude is obviously an error and hence constitutes a design problem. Further details concerning the misreading of the flight altitude might be expected to inform a prescription, that is, a design solution. Altitude might be more clearly displayed, using colour or spatial coding.

In contrast, the implicit model of C^2 would appear to have little potential for informing research. Validation of design knowledge requires its conceptualisation, operationalisation, test and generalisation. Since the model is implicit, it is not (explicitly) conceptualised. Hence, it cannot be operationalised, tested or generalised, and so validated.

The explicit model of C^2 would also appear to have potential to support design. For example: "The **involvement** and **vulnerability** of the **neutral** and the **friend's power** with respect to it continue to increase rapidly". The greater involvement and vulnerability of a neutral is obviously an instance of ineffectiveness – hence a design problem – here resulting in "The **friend's power** with respect to the **neutral** is realised, with catastrophic results". The model also informs prescription, that is, a design solution, as it specifies the required domain state, that is, an 'uninvolved and non-vulnerable neutral'. Note the model does not inform design of the behaviours of the C^2 system itself, that is, how the neutral's uninvolvement and non-vulnerability might be brought about.

The explicit model would also appear to have the potential for informing research. As the C^2 concepts are explicit, it meets the requirement of research for conceptualisation. Since conceptualised, the model offers the potential for operationalisation, test and generalisation. Iran Flight 655 can be conceived and operationalised as a neutral, because its interest – civilian, commercial air transport – is compatible with both those of friends and enemies (see earlier). Once operationalised, the concepts can be tested and generalised. Note that the explicit model characterises C^2 as a class of domain, the Vincennes incident constituting an instance of the class.

Conclusion

This paper has assessed two contrasting approaches for their potential to model the domain of work for CE in support of design and research. It is concluded that both implicit, expert and explicit, research domain models have the potential to support design practice. However, the nature of the support is different. Implicit models support (re-) design of the instance, here embodied in the Vincennes incident, but not of the C^2 class (which remains implicit). In contrast, explicit domain models express the instance, embodied in the Vincennes incident, as a member of the C^2 class. Explicit model support for design is thus more general than implicit support. In contrast, explicit support is more limited, as it excludes the C^2 system itself. As concerns research, only explicit domain models have the potential to provide support, because implicit C^2 models cannot be validated in the absence of conceptualisation.

It is concluded, that implicit domain models of C^2 are needed, in the shorter term, to inform design of individual C^2 systems. Explicit modelling is currently in its infancy. However, explicit domain models offer the possibility of validation by research and so a better guarantee, in the longer term, of support for both design and research. The importance of explicit, domain models needs to be underlined and its further development encouraged. Only in this way can CE hope to progress from a craft to a more formal engineering discipline, so increasing the guarantee of its design knowledge.

References

Card, S.K., Moran, T.P., and Newell, A. 1983, *The Psychology of Human-Computer Interaction,* (Erlbaum, New Jersey)

Carroll. J.M. 1991, *Designing Interaction,* (Cambridge University Press, Cambridge)

Colbert, M. and Long, J. 1995, Towards the Development of Classes of Interaction: Initial Illustration with reference to Off-Load Planning, *Behaviour & Information*

Technology, **15** (3), 149-181

Dix, A., Finlay, J., Abowd, G. and Beale, R. 1993, *Human – Computer Interaction* (Prentice Hall)

Dowell, J. and Long, J. 1998, Target paper: Conception of the engineering design problem, *Ergonomics*, **41**, 126-139

Flach, J.M. 1998, Cognitive systems engineering: putting things in context, *Ergonomics*, **41**, 163-167

Galer, I. 1989, Ergonomics – Introduction and applications. In Galer (ed.) *Applied Ergonomics Handbook,* (Butterworths, London)

Long, J. 1987, Cognitive ergonomics and human-computer interaction. In P. Warr (ed.) *Psychology at Work*, (Penguin, Harmondsworth)

Long, J. 2001, Cognitive Ergonomics: some lessons learned some remaining. In M.A. Hanson (ed.) *Contemporary Ergonomics 2001,* (Taylor and Francis, London), 263-271

Murrell, K.F.H. 1971, *Ergonomics: Man in his Working Environment,* (Chapman and Hall, London)

Oborne, D.J. 1994, *Ergonomics At Work*, (John Wiley & Sons Ltd., Chichester)

Pheasant, S. 1991, *Ergonomics, Work and Health*, (MacMillan, Basingstoke)

USDoD 1988, *Formal Investigation into the Circumstances Surrounding the Downing of Iran Air Flight 655 on 3rd July, 1988*, US Department of Defence Investigation Report, Office of the Under Secretary of Defence, Public Affairs, Washington

Vicente, K. 1998, An evolutionary perspective on the growth of cognitive engineering: the Riso genotype, *Ergonomics*, **41**, 156-159

Appendix 1: An implicit, expert model (left column) and an explicit domain model (right column) of the Vincennes incident

0330 USS Montgomery reports that about seven Iranian small boats have challenged a tanker. Explosions are heard.	**0330** The **power** of the **enemy** (the small boats) with respect to the **resource** (tanker) increases. The **enemy** makes attempts to realise its **power**. The **security'** of the US **interest** is reduced.
0412 The Vincennes is ordered to the area to support Montgomery and investigate reports of small boats challenging a tanker.	**0412** The **friend's** (Vincennes') **power** and **involvement** increase, and the **vulnerability** of the **enemy** increases accordingly.
0610 The Vincennes' helicopter is diverted from a patrol to reconnoitre the small boats. By so doing, the helicopter and the small boats come within range of each other.	**0610** The **involvement, power** and **vulnerability** of the friend (Vincennes' helicopter) and the **enemy** increase rapidly.
0615 The helicopter is fired upon by the small boats, but no damage is inflicted. Having established the small boats' intentions, the helicopter returns to the Vincennes.	**0615** An attempt to realise the **friend's vulnerability** fails. Levels of **power, vulnerability** and **involvement** reduce.

0620 The small boats and the Vincennes continue to close.	**0620** The **involvement, power** and **vulnerability** of the **friend** (Vincennes) and the **enemy** increase again.
0643 The small boats and the Vincennes continue to close. Two small boats turn towards the Vincennes, while the other small boats manoeuvre erratically. (By so doing, the small boats have been drawn away from the tanker.)	**0643** The **involvement, power** and **vulnerability** of the **friend** and the **enemy** continue to rise. (The **friend's vulnerability**, however, is relatively low.) The US **interest** appears more secure.
0647 Flight 655 takes off from Bandar Abbas, and is detected by the Vincennes, as an 'unknown, presumed enemy'.	**0647** The **involvement** of the **neutral** (Flight 655) and the **friend** begins to increase.
0649 Flight 655 adopts its flight path, which is towards the Vincennes. The Vincennes challenges its air contact (actually Flight 655), but receives no reply. The period immediately after take-off is a busy time for flight crew, so they may not have been monitoring Air Distress frequencies. For a moment, Vincennes' air contact appears electronically to identify itself as a military aircraft (due to freak weather conditions?). Flight 655 is approaching the range of Vincennes' missiles.	**0649** The **involvement** and **vulnerability** of the **neutral**, and the **friend's power** with respect to it, increases rapidly.
0651 One of the Vincennes' guns jams when one of the small boats is about to adopt a dangerous position. A sharp change in course brings the remaining gun to bear on the small boat posing the greatest threat. Further challenges to the air contact receive no reply. Flight 655 is now within range of the Vincennes' missiles. Flight 655's altitude is mis-read. The air contact is perceived as diving towards the Vincennes; in fact, it is climbing away from it.	**0651** The **power** of the **friend** with respect to the **neutral** temporarily falls sharply, but is soon restored. The **vulnerability** of the **enemy** fluctuates accordingly. The **power** of one **enemy** also rises slightly before falling back. The **friend's vulnerability** fluctuates accordingly. The **involvement** and **vulnerability** of the **neutral** and the **friend's power** with respect to it continue to increase rapidly.
0654 Two surface to air missiles are launched by the Vincennes, in the belief that the ship is under attack from an enemy fighter. The missiles destroy Flight 655.	**0654** The **friend's power** with respect to the **neutral** is realised, with catastrophic results.
0703 The small boats leave the area. One has been destroyed and the Vincennes has incurred superficial damage from small arms fire or shrapnel. The Vincennes learns that is has mistakenly shot down a commercial aircraft.	**0703** The **power** of the **enemy** (with respect to the **friend**) reduces, as does the **involvement** of all parties. The **friend's power** with respect to the **enemy** has also been realised.

INFORMATION PROCESSING MODELS: BENEFITS AND LIMITATIONS

Brendan Wallace, Alastair Ross, John Davies

Brendan Wallace, Alastair Ross, John Davies
University of Strathclyde
141 St James Road
Glasgow G4 0LT

This paper looks at the three main information processing models from the point of view of researchers in confidential human factors databases. It explores conceptual problems with two of theseinformation processing models, and goes on to explore possible advantages of adopting a 'connectionist' paradigm. Links between connectionism and 'situated cognition' are demonstrated. Practical work carried out using a connectionist/situated cognition model is described, and the way in which the 'situatedness' of discourse can influence the kind of data that can be collected is discussed. Finally it is argued that more emphasis should be placed in ergonomics on sociation, sitautedness and embodiment, and that this might help to deal with problems faced in creation and interrogating databases: especially as regards the creation of coherent and reliable 'coding taxonomies'.

Introduction

The information processing approach to human cognition remains very popular in the field of ergonomics. However, in recent years, an alternative approach has arisen: 'situated cognition'. Situated cognition stresses the extent to which human beings related dynamically with their environment (Hollnagel, in press). However, there are few generalised discussions of the advantages of both approaches in a pragmatic human factors context.

The purpose of this paper is to show that there are generally agreed to be three main information-processing approaches, and that at least one seems to be compatible with situated cognition. We will then go on to discuss how useful the situated cognition approach has been in our own work. Finally there will be a brief discussion of the implications for ergonomics.

Information Processing

Our work at the University of Strathclyde has been with qualitative research in safety management and human factors. There are three major projects we have been involved with which are of importance here. Firstly there is CIRAS. CIRAS is the Confidential Incident Reporting and Analysis System, the national *confidential* reporting system for the UK Railways. The University of Strathclyde set up the system and established it as a national database (Davies, *et al*, in press). We have also run a number of focus group studies of train drivers within the railway industry (Wallace *et al*, 2002). We have also just begun to run CARA (Confidential Accident Reporting and Analysis), the nation-wide confidential reporting system for the railway system of the Republic of Ireland.

It was because we come from a psychology background that we looked to information processing theories for a theory of 'human error'. However (as is not noted frequently enough in discussions on this subject) there are not one but three main information processing theories.

The best known of these is the Atkinson and Shiffrin model (1968). This posits memory as having two main components. Firstly there is a short-term memory (STM) store, which has a limited amount of storage space. Data is inputted into this store, before being passed onto the long-term memory (LTM) store that has an infinite capacity. Because of this distinction, this is normally known as the two stage, or 'modal' memory model.

However, there are a number of problems with the concept of two distinct systems of memory. Here we must remember a number of correlates of Atkinson and Shiffrin's theories. For example, by definition, data must pass through the short-term store, before passing into the long-term store. But as Shallice and Warrington (1970) have showed, it is perfectly possible for some brain-damaged patients to have perfectly normal long-term memory but no short-term memory. This is impossible in the Atkinson and Shiffrin theory.

Of course, there are many experiments that purport to demonstrate that there are indeed two memory stores. However, they can usually be interpreted in other ways, ways that fit more closely with 'ecological' studies of how human beings actually function in their 'natural environments'. For example, Reisberg, Rappaport, and O'Shaughnessy (1984) demonstrated that it is easy, by using their fingers as *aides memoires*, to improve the capacity of the 'short term memory store' immensely. As Glenberg (1997) comments (in a very important paper to which we will return) 'This evidence might be interpreted as evidence for a new "finger-control" *module*, but it seems more sensible to view it as a newly-learned *skill*' (italics added). This is a very important point: memory behaviour is better seen as being a skill or activity that can improve through practice. And if this is the case, then talking about the 'absolute capacity' of short-term memory store, as though it can be 'filled up' with digital data seems to be meaningless. As Glenberg writes: '(in recent years) much of the evidential basis for a separate short- term (memory) store has been eroded' (Glenberg, 1998). Given that this is the case, then simply renaming STM 'working memory' will not solve the problem, although 'working memory' does solve, ironically enough, a far deeper problem with the whole concept: the idea of memory as a store (Baddeley, 1986).

It was as a result of such problems that an alternative Craik and Lockhart (1972) model of memory was proposed. It is generally agreed that it at least blurs, and possibly eliminates the STM/LTM distinction (Hayes, 2000), positing instead a single memory store. In this model, things which are processed more deeply are better remembered than those which are not. It therefore avoids the STM/LTM problem above.

But there is a deeper problem with both these theories, and one which cuts to the root of the matter: is memory a store at all? The problem here is conceptual and empirical. To take the conceptual problem first: The key point to remember is that in order to 'recognise' something, both the 'information processing' models described propose that the *operator must retrieve a stored memory from a memory store, and then match it with the information in front of him or her.* If it matches, then the object is 'remembered'.

However, there is a fundamental problem with this approach, which demonstrates its reliance on the now outmoded 'cognitivist' model of cognition. Cognitivism was discarded because it fell prey to the 'homunculus' fallacy: it posited an internal 'person' who provided the motor for cognition. However, this then led to infinite regress (who controlled the homunculi? Obviously, other homunculi). (Davies *et al*, in press. See also Dreyfus 1992).

The concept of a memory store seems to fail for the same reason. If the operator sees their car, for example, how do they know where it should be looked for in the memory store, in order to 'remember' it? S/he would have to remember that it was his/her car (and, in fact that it was a car at all) before the search through the memory store for 'my car' could begin. The alternative is that the entire memory store would have to be searched through for every act of recognition.

Even if the memory was retrieved, how would the operator know it was the 'correct' memory? The answer, never stated but implicit, is that a homunculus would compare the internal 'symbol' in the memory store with the symbol of the external object, because the homunculus would remember what a 'car' looked like and be able to compare it with the external symbol. In other words, they object would have to be recognised (and therefore remembered) before it can be compared in the store: what the model is attempting to explain (Wilcox and Katz, 1981).

Moreover the concept of memory 'stores' seems to be empirically false as well. If memory is a store, it is simply a passive receptacle of events. But Bekerian and Baddeley (1980) conducted a classic study which studied how new radio frequencies were remembered. Despite bombardment by the media, people still did not retain the information, which suggests that the theory that communication is simply a matter of repeating information often enough until it is 'put into' the store, is false.

The Third Information Processing Model

There is, however, a third information processing model. This is the parallel-distributed processing (PDP) or 'connectionist' model. There are two main differences between connectionist and the other two main information processing models. Firstly, in connectionist models, memories are not 'stored' but instead are distributed throughout the system (Bechtel and Abrahamson, 2002). Secondly, connectionist models posit one, not two, memory 'systems' (Sougné, 2000).

We have discussed connectionism extensively elsewhere (see Davies *et al*, in press). However, the key point to realise is that connectionist models *are* conceptually plausible in a way that the other two models of information processing are not. Moreover, as far as a connectionist system's processing is concerned, 'there is no difference between reconstructing a previous state, and constructing a completely new state (confabulating)' (Bechtel and Abrahamsen, 2002: 50). In other words, in the connectionist view memory is a dynamic act of reconstruction (Clancey, 1997). A number of experiments have demonstrated these points (for example, Loftus and Loftus, (1975), see also Zadny and Gerard (1974))

To return to Glenberg's paper, he argues, on the basis of the embodied cognition theory of Lakoff and Johnson (1999), that: 'memory evolved in service of perception and action…and that memory is embodied to facilitate interaction with the environment' (Glenberg, 1998). Memory is a tool which facilitate the organism in dealing with the environment, and we would argue that this is a good model for all behaviour, linguistic or otherwise.

As we have seen, connectionist models avoid the most obvious problems resulting from the other two 'information processing' approaches. However, *situated cognition* goes further by stressing the situated aspect of language and behaviour: the extent to which these are specific responses to specific situations. The main aspects of this model we have found useful in our own research is that of the situatedness of discourse.

Accident Investigation

It tends to be generally accepted within the field of safety management that discourse about past events (what one might term memory discourse) is veridical, that is, that discourse is an attempt to communicate the 'truth' from one subject to another. But this misses an important point about communication; that is, as stated above, the extent to which memory is an active, situated response to a situation.

Despite its expression in the language of situated cognition, this insight would strike most non-psychologists as being common sense, especially in a situation such as an accident investigation, where people still attempt to deny that they were 'to blame' when investigations try to reveal 'what caused the accident'. For example, in a study of a major train operator (Wright, Ross and Davies 2001), discussion groups were held with drivers who had passed a signal at danger and those who had not, in order to establish risk factors for Signals Passed at Danger (SPADs). Drivers with no SPAD on their driving record stated that they had been caused more by dispositional factors (i.e. things to do with the other drivers, such as inability to concentrate) than the drivers who had previous SPADs. By contrast, this latter group remembered system and environmental conditions (e.g. training, railhead conditions) for their own incidents, thus implying that SPADs were "caused" by external factors beyond their control. Thus memory discourse of SPADs seemed to be presented in a functional manner and showed a self-serving bias, *dependent on drivers' own personal histories.* This is not to suggest that drivers were 'lying'. Instead it must be remembered that all discourse exists in a specific situation, and that it was created to cope with the needs of that situation. Consciously or not, any subject will *tend* to remember things (to be precise, produce memory type discourse) which is situation specific and which facilitates their own goals (this of course applies to management as much as to workers at the man-machine interface.) This research suggests is that approaches to safety which suggest that the purpose of discourse in situations such as these as *primarily* veridical are flawed.

One way of dealing with this problem is to treat discourse (instead of being veridical or otherwise) as being 'stable' or 'unstable'. That is, to acknowledge that there are certain kinds of discourse that will be produced in *any* normal situation (for example, the subject's name): this is 'stable' data. Other kinds of discourse will only be produced in specific situations (for example, in this respect, the answer to the question 'why did you have a SPAD?' will be very different depending on whether the subject is in a courtroom during an inquiry, or in the pub with co-workers): this is 'unstable'. Therefore a process of 'triangulation' can be used, in which data are obtained from subjects in different situations. This does not of course have to be done for every fragment of data or with every subject. But even a small scale study might indicate the relationship between stable and unstable data, and help to give an insight regarding the extent to which the data obtained in research is stable or not (for details as to how this might be done, see Davies, 1997).

Safety Culture

Another way of using situated discourse is to see whether it can give help in terms of defining safety culture. Despite common usage, there is no agreed definition of 'safety culture'. Indeed Cummings and Worley (1997: 479) state that 'Despite the increased attention and research devoted to corporate culture, there is still some confusion about what the term *culture* really means when applied to organisations' *(emphasis as original).*

What we would suggest here is that safety culture could be interpreted as the social complex of intersubjective meanings of an organisation expressed through language. In other words, the reconstructions of 'memory', or rather the linguistic behaviours we call 'memory behaviours' are an indication of the safety culture itself. That is, if a reporter chooses to remember an event as being managerial in origin, rather than being 'his own fault', then this can be studied in its own terms (it must be remembered that CIRAS and CARA are confidential systems and so no 'check' on the veridical nature of the memories can be performed). Therefore, we can study texts of memory events as texts to be classified and coded.

We have described elsewhere our methodology for transcribing and coding texts (Davies *et al*, in press) and performing a consensus or reliability trial. However, the key point is that codings must be reliable (i.e. socially agreed upon). The matrix produced in figure 1 in order to code the transcriptions is a general ethnographic matrix: detailed codes are project specific. This, therefore, is the JCP (Job, Communications,Procedures) 'coding matrix' which produces data.

	Job/Task	**Communications**	**Procedures**
Managerial	Codes	Codes	Codes
Supervisory	Codes	Codes	Codes
Frontline	Codes	Codes	Codes

Figure 1. The coding matrix currently used in CARA

Once the discourse has been coded and entered into a database, then the data can be analysed. Analysis in this way is therefore an ethnographic (and hence, interpretative) classification taxonomy of discourse. In this discourse reporters write in and recreate events and situations which happened to them, or that they saw or experienced. These may be 'true' or 'false' but of course because of the confidential nature of the database, we can never know this.

Reliability and Discussion

We have found situated cognition (or rather, situated behaviour and discourse) a useful model for our work (not least in terms of the 'embodiment' of cognition as stressed by Lakoff and Johnson (1999) a topic which lies outside the scope of this paper. See however, Wallace *et al* 2002). But the key point in terms of discourse is that of consensus or reliability. It is taken for granted within most fields of the social sciences that in situations where behaviour, either in terms of physical behaviour or linguistic behaviour, has to be 'judged' or 'coded', that coders have to agree with each other. In other words, two coders coding the same text would code it the same way, all other things being equal. However we found consensus impossible to achieve when attempting to use 'information processing' concepts to create a matrix of 'human error'. Situated concepts on the other hand, proved a far better model as the situation they reporters were in (the workplace) could more easily be verified from the reports. It should be noted that the matrix in figure 1 is an organisational matrix which does not posit discrete 'internal states' which 'cause' 'human error'.

Another implication of this research is that discourse and behaviour are situated. One way of dealing with this has already been suggested: triangulation. However if this is not possible, the situatedness of behaviour should at least be controlled for. That is, if a database is being built up based on discourse (or from questionnaires, which are simply tabulations of discourse), the discourse producing situations should be as similar as possible.

This must be stressed, as, during research still ongoing, we have discovered that databases concerning the 'causes' of car accidents are frequently compared to each other, regardless of how the data was collated. For example, some police forces specifically ask drivers involved in accidents questions regarding whether certain 'risk factors' were present or absent (for example, if the driver was using a mobile phone at the time of the accident). Others do not, and only note this data *if the driver volunteers the information.* But it is clear that differences between these databases do not reflect 'real' differences in accident causes, but demand characteristics of the situations itself. How frequently will drivers *volunteer* information that may land them in court? However, data from these two sources is frequently compared as though one source is 'reliable' and the other falls victim to 'under-reporting', whereas clearly this bias is situational.

It is also not impossible that other features of 'objective' statistical databases are produced by the situatedness of interpretation as well as of discourse: and this is particularly the case with confusing or incoherent coding matrices. The writer J.L. Borges once created a fictitious classificatory system of animals thus: 'Animals are divided into those that belong to the Emperor: embalmed ones: those that are trained: suckling pigs: mermaids: fabulous ones: stray dogs: those that are included in this classification: those that tremble as if they were mad: innumerable ones: those drawn with a very fine camel's hair brush: others: those that have just broken a flower vase: those that resemble flies from a distance' (Borges, 1964, 108). This is clearly absurd, but many 'coding' matrices used in safety management are not much better, the key problem being most events could fit plausibly into any category. This will again confuse and produce incoherent data. Lakoff (1990) a proponent of situated embodied cognition argues that incoherent categories arise from failure to consider this view of cognition. In our own research we would stress that reliable taxonomies are best produced from coherent approaches to cognition and activity, and that the situated, sociated view of cognition seems to provide such a view.

Bibliography

Atkinson, R.C., and Shiffrin, R.M. 1971, The Control of Short Term Memory *Scientific American*, **225**, 2, 82-92

Baddeley, A. 1986, *Working memory*, (Oxford University Press, Oxford)

Bechtel, W., and Abrahamsen, A. 2002, *Connectionism and the Mind* (Blackwell, Oxford)

Bekerian, D.A., and Baddeley, A.D. 1980. Saturation advertising and the repetition effect. *Journal of Verbal Learning and Verbal Behaviour,* 19, 17-25.

Borges J.L. 1964, *Other Inquisitions* (University of Texas, Press Austin)

Clancey, W., 1997. *Situated Cognition,* (CUP, Cambridge)

Craik, F.I.M. & Lockhart, R.S. (1972) Levels of processing: A framework for memory research, *Journal of Verbal Learning and Verbal Behavior*, **11**, 671-684

Cummings, T.G., & Worley, C.G. (1997) *Organization development and change*, (South Western, Chicago)

Davies, J. B. 1992, *The Myth of Addiction*, (Harwood Academic Publishers, London)

Davies, J.B., Ross, A., Wallace, B., Wright, L. in press. *Safety Management: A Qualitative Safety Analysis* (Taylor and Francis, London)

Dreyfus, H. 1992 *What Computers Still Can't Do*, (MIT Press, NY)

Glenberg, A.M. 1997. What memory is for, *Behavioral and Brain Sciences* **20**, 1, 1-55. Online http://www.bbsonline.org/documents/a/00/00/05/54/bbs00000554-00/bbs.glenberg.html

Hayes, N. 2000, *Foundations of Psychology*, (Thomson, London)

Hollnagel, E. Cognition as control: A pragmatic approach to the modelling of joint Cognitive systems. *IEEE Journal of Systems, Man and Cybernetics* (in press). Online: http://www.ida.liu.se/~eriho/Publications_O.htm

Lakoff, G. 1990, *Women, Fire and Dangerous Things,* (UCP, Chicago)

Lakoff, G., and Johnson, M. 1999, *Philosophy in the Flesh.* (Basic Books, New York)

Loftus, G.R. and Loftus, E.F. 1975, *Human Memory: The Processing of Information.* (Halsted Press New York)

Morris, P.E. 1982, Research on Memory in Everyday Life. In P. Sanders, N. Hayes, R. Brody, and L. Jones (eds) *A Handbook for GCSE Psychology Students* (ATP Publications, Leicester)

Reisberg, D., Rappaport, I. & O'Shaugnessy, M. 1984, Limits of working memory: the digit digit-span. *Journal of Experimental Psychology: Learning, Memory, and Cognition,* **10**, 203-221

Shallice T. and Warrington, E.K. 1970, Independent functioning of verbal memory stores: a neurophysiological study. *Quarterly Journal of Experimental Psychology,* **22**, 261, 273

Sougné, J.P. (Submitted). Short Term Memory in a Network of Spiking Neurons. Neural Computation and Psychology Worshop NCPW7 Brighton, Online http://www.ulg.ac.be/cogsci/jsougne/ncpw7.pdf

Wallace, B., Ross, A., Davies, J.B., Wright, L. 2002, Information, Arousal and Control in the UK Railway Industry. In A. Thatcher, J. Fisher, K. Miller (eds) *Proceedings of CybErg 2002: The Third International Cyberspace Conference on Ergonomics*. (IAE Press, Johannesburg)

Wilcox S., and Katz S.(1981) A direct realist alternative to the traditional conception of memory. *Behaviorism* **9**: 227-239

Wright, L., Ross, A., Davies, J.B. 2001, *SPAD Risk Factors: Results of a Focus Group Study.* (Railway Safety, London). Available on 'Human Factors Research Catalogue CD-ROM, July 2001'

Zadny, J., and Gerard, H.B. 1974. Attributed Intentions and Informational Selectivity, *Journal of Experimental Social Psychology*, **10**, 34-52

SITTING STRESSES INSIDE THE BODY

Niels CCM Moes

Delft University of Technology, Dept OCP/DE, Section Integrated Concept Advancement, Landbergstraat 15, 2628 CE Delft, The Netherlands
C.C.M.Moes@IO.TUDelft.nl

Abstract: The internal stresses and distortions in the buttock area of a person seated on a flat, hard and horizontal seat was analyzed by Finite Element Analysis. The model was based on real geometry. The actual, non-linear material properties are being investigated, and the analysis was carried out for different assumed values of the elasticity. The results show stress concentrations inside the soft tissue and along the ischium surface. These stresses exceeded significantly the interface stresses. The maximum interface pressure was, due to the model specificities, not confined to one single point, but showed a region of distribution.

Introduction

Sitting upright in a chair implies the transmission of the upper body weight to the chair via the upper leg and the buttocks. The main part is transmitted via the ischial tuberosities, the soft tissue below the ischial tuberosities, and the skin (Moes, 2002). Since the gluteal muscle group moves laterally if the hip is flexed, the soft tissue contains mainly adipose tissue and no muscles (Zacharkow, 1988). Although it was argued by many authors, for instance (Levine et al., 1990; Sprigle et al., 1990), that the main criteria for shape design of a seat is based on avoiding the shear deformation of the soft tissue, much research has nevertheless been reported on the interface pressure distribution, mainly because the internal biomechanical processes can not be accessed for measurement in vivo, but only simulated by Finite Elements Analysis (FEA). Such analysis requires modelling of the geometry, the tissue material properties, etc, which are known te be difficult and extremely time consuming in terms of personal effort and cpu-time. As a consequence only few research has been done in the area of FE modelling for a sitting body. Also, the systems supporting it were missing for a long time; even now the system requirements do often exceed the possible hardware capabilities of current generation computers. Past research applied extreme simplifications, such as linear material properties, modelling the ischial tuberosities and the soft tissue part as half spheres (Chow and Odell, 1978; Dabnichki et al., 1994), or as applying 2D representations of the 'real' 3D buttock shape (Todd et al., 1990; Todd and Tacker, 1994).

Apparently there is a need for knowledge on the stresses and strains inside the body, and their relationship with (i) the shape of the support, and (ii) the personal characteristics. Because of a lack of such knowledge of the biomechanical effects of an external load (sitting) on the internal body regions (the continuum), it was the intention to analyse (i) the internal and the surface stress and strain distributions, and (ii) the effect of the soft tissue elasticity on these distributions. The search for the optimal material properties was described in (Moes and Horváth, 2002) and is still going on.

This paper presents the effects, in terms of the stresses and strains, of loading a 3D FE model of the buttock region by a flat seat and an upright posture for a range of material properties. No arm rests or backrest were included in the model. The model, see the next section, is based on in vivo obtained geometry and on assumed, varying material properties. In the discussion section, the main assumptions of the FE model will be critically reviewed. In the model the seat was represented by a flat and horizontal geometric surface.

The model

The development of the applied FE model has been described in detail elsewhere (Moes et al., 2001; Moes and Horváth, 2002; Moes and Horváth, 2002). It consists of two major parts: a bone part (femur and pelvis bones) and a soft tissue part, representing skin, adipose tissue, muscles, etc. The shape of the soft part was generated by vague discrete geometric modelling including the upper leg and the buttock regions. The bone part originated from the VHP data set (VHP, 1997) and was geometrically transformed to fit in the soft part. Initially the model included the upper leg and the buttock completely, but its size was reduced because of computational reasons. The current model contains the lower half of the ischium and the soft tissue of its immediate neighbourhood. Figure 1a shows a top view of the complete and of the reduced model. The bone parts are represented as spatially fixed nodes to simulate infinite stiffness and to enable force balance during the FE analysis. In figure b the ischium is therefore shown as a hole in the continuum. The thin-lined rectangle is the contour of the seat, that was pushed against the lower side of the model until a force of 300 N was exerted (about half the upper body weight). Since the material is defined non-linear, the loading was done in four loadcases to enable small time steps.

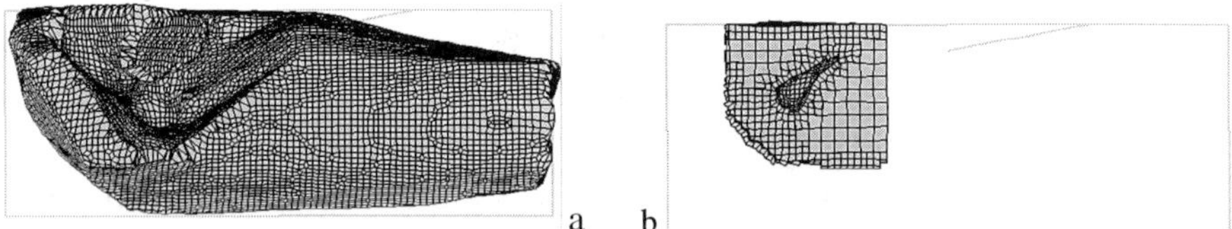

Figure 1. Top view of the complete and the reduced model. The thin rectangle is the support.

Although the model is a first trial representation of the living pelvis, containing many limitations, it enables us to draw conclusions from the analyses for different material constants.

The elements were 8-node hexahedral elements with Hermann definition. The definition of the material properties was based on isotropy, incompressibility, but no creep or viscosity were taken into account. The contact was frictionless, and no element separation was allowed. Since the expected deformations are extermely large, nonlinear behaviour has to be considered. The constitutive equation for the elastic behaviour was based on the generalized Mooney-Rivlin strain energy function of elastomeric materials (MARC, 2001):

$$W = \sum_{m=0}^{N} \sum_{n=0}^{N} C_{mn}(I_1 - 3)^m (I_2 - 3)^n$$

were I_1 and I_2 are the first and the second strain invariants. Following (Vannah and Childress, 1996) only the coefficients C_{10}, C_{01} and C_{11} were used in the constant ratio $C_{10} = 4C_{01} = 0.5C_{11}$, so that only one independent variable had to be considered.

Analysis

The FE calculations were done with the Mentat/MARC2000 software package running on a Linux 2.4.10 OS. Ten values of C_{10}, varying from 500 Pa to 20 kPa were applied. The analyses included, among other things (not discussed in this paper), three quantities that have ergonomics relevance: the hydraulic pressure, the tissue distortion, and the maximum interface pressure.

Because of the large amount of water in the human soft tissues, the *hydraulic pressure* is usually considered an irrelevant quantity since human tissue is practically incompressible. But a change of this pressure accounts for the drainage of intercellular tissue fluid, lymph and blood.

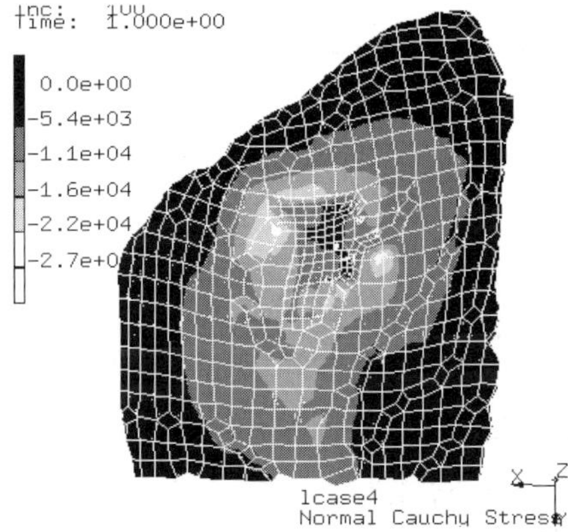

Figure 2. Typical pattern of the hydraulic pressure at a level of 1 mm above the seat.

In figure 2 a typical pattern of this pressure is shown graphically in a transversal cross section located 1 mm above the seat. The Mentat/MARC equivalent of the hydraulic pressure is the normal Cauchy stress. The brightness in the figure increases with the magnitude of the pressure. In the centre the effect of adaptive remeshing can be observed as a reduction of the element size. Since the average distance of the not remeshed edges in this figure is about 10 mm, it can be concluded that the gradient has maximum values of the order of 100 to 200kPa/m, but it is small in the peripheral regions.

The shear deformation is considered as a main cause of decubitus of bed ridden or wheel chair bound people. This deformation is expressed as the *Shear Total Strain*; the corresponding stress is the *Equivalent von Mises Stress*.

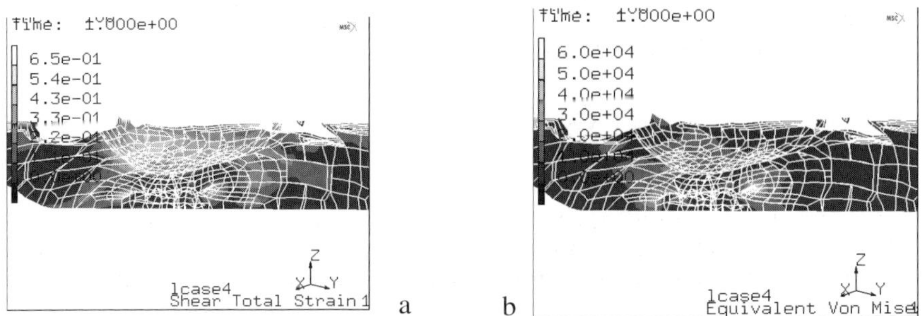

Figure 3. The deformation in the soft tissue (left) and on the ischium (right).

Figure 3a shows the distortion and the figure b the von Mises Stress distribution in a vertical section along the ischium (called the *cutting plane*). It can be seen that the distortion, and thus also the shear stress, is concentrated in two vertical bands or regions, with a minimum distortion below the lowest part of the ischium. This is in agreement with the findings of (Zhang and Roberts, 1993).

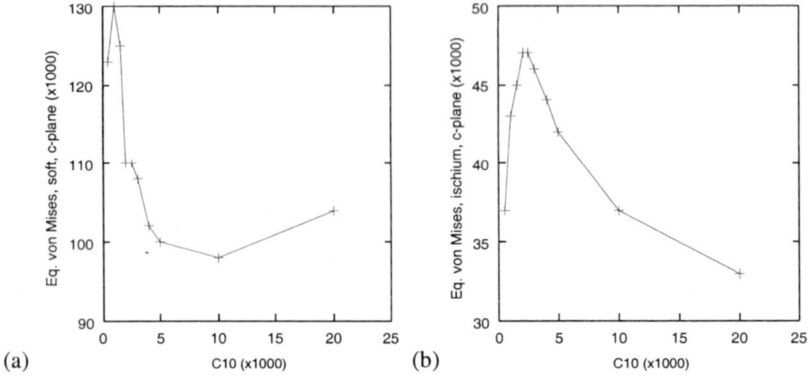

Figure 4. The von Mises stress in a vertical cutting plane along the ischium for the soft tissue (left) and for the surface of the ischium (right).

Figures 4a and b show the graphs of the maximum von Mises stress along the cutting plane for the investigated elasticity values. Figure a for the soft tissue between the bone and the skin, and figure b for the lower aspect of the ischium. It is clear that the maximum von Mises stress within the soft tissue is about three times the maximum von Mises stress on the bone surface for all values of C_{10}.

The *maximum interface pressure* between the seat and the skin, has been calculated since it is an often applied, important measure in the fields of anti-decubitus research and the reshaping of seat cushions and seat shells.

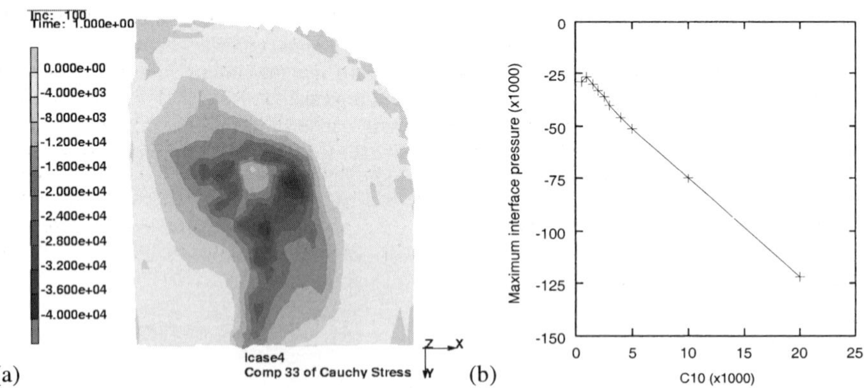

Figure 5. The normal interface pressure.

In figures 5a and b the component '33' of the Cauchy stress, which is the representative of the interface pressure, is shown for the seat contact region. Note that the blackness increases with the magnitude of the pressure. In figure a the maximum pressure is not concentrated in one small area, but distributed over the contact area. Future research must resolve this problem; it stems probably from specific mesh properties or from the adaptive remeshing that was applied during the analysis. In figure (b) the maximum pressure is shown as a function of the elasticity. This dependency is used to compute the optimum material properties by comparison with empirically obtained maximum interface pressure data (Moes and Horváth, 2002; Moes and Horváth, 2002). The lowest maximum pressure, ca 25 kPa, was found for $C_{10} = 1$ kPa.

Discussion

Personal observation is that the in vivo distance between the skin surface of the buttocks and the ischial tuberosity is much smaller than the distance that follows from the FE analysis. It is known to be ca 3 mm depending on individual characteristics. The initial thickness of about 5 cm is reduced to ca 2 cm. Two possible reasons for this discrepancy were found. First, in vivo circumstances a structural rearrangement of tissue takes place where particularly neighbouring regions of the subcutaneous adipose tissue is separated. Therefore a valid FE model should contain one or more layers of fat cells that are only loosely connected.

The second reason is the drainage of several fluid substances in the skin and subcutaneous tissues. Interstitial fluid is knows to migrate easily and quickly if the tissue is pressurized. Also blood and lymph are subject to these effects. A more correct constitutive material model should therefore consider hybrid modelling of a combination of elasticity, viscosity and drainage.

The presented model contains no separate skin layer. Such a layer could significantly modify the internal biomechanical tissue behaviour. Non published, premature results of further analysis have shown that the assignment of a stiffness to a skin layer equal to three times that of soft tissue, does not result in significant differences. Further research is needed.

The results show sharp pressure singularities within the tissue, especially in the central region below the ischial tuberosities. This is possibly a result of the mesh generation procedure. Future models must take into account geometric attachment surfaces that follow the expected stress trajectories. Moreover, the element definition must be reconsidered. Currently the 8 node brick element is used, but 20 node hexahedral elements show more adaptivity to regions of strong pressure and deformation gradients.

It was argued in (Moes and Horváth, 2002) that for decreasing soft tissue stiffness the maximum interface pressure should initially decrease, and then, after reaching a minimum show an increase again. This hypothesized effect was applied in the search for the best material properties in the sense of agreement with empirically obtained data (Moes, 2002). This minimum is probably seen in the very left part of the graph of figure 5. However, lower values of C_{10} resulted in extremely large deformation and therefore also in instability of the element matrix. The solution for this problem is possibly in the allowence of separation (for the adipose tissue) and in applying hybrid modelling so that elasticity and fluid flow can be incorporated simultaneously.

Conclusions

The distortion of the soft tissue between the ischial tuberosity and the skin surface and the corresponding stress was shown to exceed the values at the ischium. Should the model be sufficiently valid then it can be concluded that medical sitting problems, such as decubitus, are possibly initiated within the soft tissue.

It was shown that the hydraulic pressure changes rapidly within the soft issue. This is possibly related with the structural tissue rearrangement and the drainage of the viscous components of the soft tissue. The FE model must be updated by renewed meshing and modified material definitions to obtain better agreement with the empirical results of pressure distribution.

Nevertheless a good starting point has been made for furher development and improvement of the understanding of the biomechanical effects of sitting as far as the continuum of the buttock region is considered.

References

Chow WW and Odell EI (1978). Deformations and Stresses in Soft Body Tissues of a Sitting Person. *Journal of Biomedical Engineering*, 100(may):79–87.

Dabnichki PA, Crocombe AD, and Hughes SC (1994). Deformation and stress analysis of supported buttock contact. In *Proceeding of the Institution of Mechanocal Ebgineers*, volume 208, pages 9–17.

Levine SP, Kett RL, and Ferguson-Pell M (1990). Tissue Shape and Deformation Versus Pressure as a Characterization of the Seating Interface. *Assistive Technology*, 2(3):93–99.

MARC (2001). *Analysis of Visco-Elastic and Elastomeric Materials.* MARC Analysis Research Corporation, Palo Alto, CA 94306 USA.

Moes CCM (2000a). Distance Between the Points of Maximum Pressure for Sitting Subjects. In Marjanovic D, editor, *Proceedings of the 6th International Design Conference DESIGN2000*, pages 227–232, Zagreb. University of Zagreb.

Moes CCM (2000b). Pressure Distribution and Ergonomics Shape Conceptualization. In Marjanovic D, editor, *Proceedings of the 6th International Design Conference DESIGN2000*, pages 233–240, Zagreb. University of Zagreb.

Moes CCM (2002). Modelling the Sitting Pressure Distribution and the Location of the Points of Maximum Pressure for Body Characteristics and Rotation of the Pelvis. *Ergonomics*, ? Status: submitted aug. 2002.

Moes CCM and Horváth I (2002). Estimation of the non-linear material properties for a finite elements model of the human body parts involved in sitting. In Lee DE, editor, *ASME/DETC/CIE 2002 proceedings*, pages (CDROM:DETC2001/CIE–34484), Montreal, Canada. ASME 2002.

Moes CCM and Horváth I (2002). Optimizing the Product Shape for Ergonomics Goodness Index. Part II: Elaboration for material properties. In McCabe Paul T., editor, *Contemporary Ergonomics 2002*, pages 319–322. The Ergonomics society, Taylor & Francis.

Moes CCM, Rusák Z, and Horváth I (2001). Application of vague geometric representation for shape instance generation of the human body. In Mook DT and Balachandran B, editors, *Proceedings of DETC'01, Computers and Information in Engineering Conference*, pages (CDROM:DETC2001/CIE–21298), Pittsburgh, Pennsylvania. ASME 2001.

Sprigle S, Chung KC, and Brubaker CE (1990). Reduction of sitting pressures with custom contoured cushions. *J Rehabil Res*, 27(2):135–140.

Todd BA and Tacker JG (1994). Three-dimensional computer model of the human buttocks, in vivo. *Journal of Rehabilitation Research and Development*, 31(2):111–119.

Todd BA, Thacker JG, and Chung KC (1990). Finite Element Model of the Human Buttock. In *Proceedings of the Resna 13th annual conference*, pages 417–418, Washington DC.

Vannah WM and Childress DS (1996). Indentor tests and finite element modeling of bulk muscular tissue in vivo. *J. of Rehab. Res. and Devel.*, 33(3):239–252.

VHP (1997). The Visible Human Project. URL: http:// www.nlm.nih.gov/ research/ visible/ visible_human.html.

Zacharkow D (1988). *Posture: Sitting, Standing, Chair Design and exercise.* Charles C Thomas, Springfield, Ill.

Zhang M and Roberts VC (1993). The effect of shear forces externally applied to skin surface on underlying tissues. *J Biomed. Eng.*, 15:451–456.

WORK SMARTER NOT HARDER! ERGONOMICS IN A LEAN BUSINESS ENVIRONMENT

Joanne Smyth

Boots Manufacturing,
Motherwell Street, Airdrie,
North Lanarkshire, ML6 7HR

Boots Manufacturing are currently implementing lean manufacturing into their factory production operations and this paper will demonstrate how ergonomics and lean share the same common goals. By applying lean principles to ergonomics problem solving exercises, we have developed improvements that not only have a dramatic impact on reducing ergonomics risk exposures but also generate significant value benefits to the business. This paper will describe how we have used detailed process activity mapping and value stream management techniques to make changes to work processes which support the objective to achieve process excellence throughout the company.

Introduction

Boots Manufacturing (BM) develop and produce a vast range of cosmetics, toiletries and healthcare products for sale in the UK and overseas and are the main supplier for Boots stores. The stores rely on BM to supply the correct quantity of each product at the right time and in today's marketplace, the accelerating pace of product innovation and new product introduction has made it increasingly challenging to maintain these levels of on-shelf availability. The availability of products in Boots stores is crucial to meeting demanding sales targets and we must address any issues within the supply chain that prevent us from fulfilling this requirement, particularly when aggressive promotions programmes add further complexity and make demand more volatile.

Any business is a complex intermingling of people, technology, organisational structures and financial operations, the arrangement of which dictates the flow of the supply chain. To improve business performance, it is necessary to look at how workplace organisation and technology can best be linked with human abilities to create a slicker and more holistic view of the supply chain with end-to-end process flow. Fundamentally, this means that the work environment and processes must be carefully designed to remove any obstacles that prevent people from doing their job effectively and to the highest quality standard. To meet our company strategy to be the best supplier to our customers, Boots have chosen to use lean thinking as a technique to assess and improve our business processes since lean has proven to deliver the high standards of quality, cost, delivery, and people development that BM are looking for in many other businesses and industries.

Lean thinking

Lean thinking has evolved from well known business management disciplines such as the Toyota Production System (TPS), Just-in-Time (JIT) and Kaizen. The core principles of lean are fundamentally the same as these other disciplines, but lean thinking has developed this theory into a generic concept that can be more readily applied in a diverse range of industries using a more people focussed approach. Lean thinking is more than an initiative, it is an all-encompassing business ethic that every function throughout the company supply chain must be committed to if the company is to achieve an integrated approach to improving our products, processes, people and plant capability.

Lean thinking is based on creating value driven activity by defining the value stream of a product from the customers' perspective. The value stream is the entire collection of activities and information flows that are essential for producing and delivering a product or service. Waste is the opposite of value. It is everything we do that does not add value to the product and it's not just materials or components we discard, it includes unnecessary moving or handling of materials, reworking the product to remove defects, using inadequate processes that cannot produce a product of the desired quality, etc.

Taiichi Ohno, the founder of TPS and JIT, summarised this into seven key areas of waste (figure 1). These wastes must be challenged to find ways to reduce their impact or eliminate them from the process so that only essential operations are performed that will add value to the product or service being delivered.

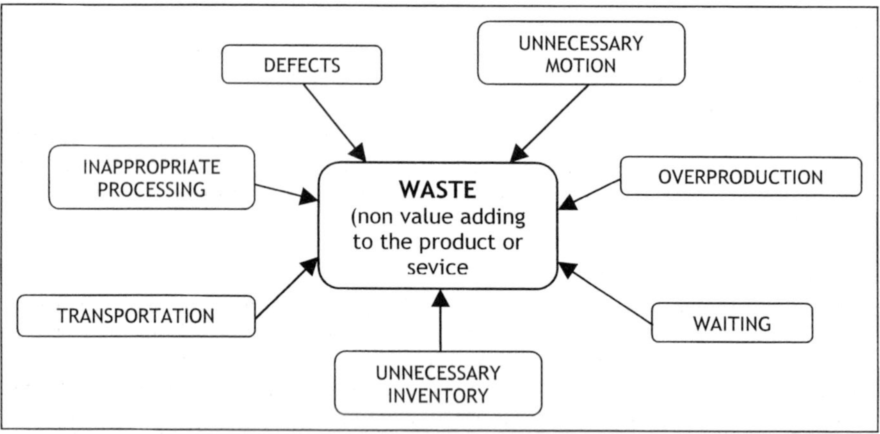

Figure 1. The seven wastes

Although lean thinking is based on creating value and value driven activity within business processes, it is not efficiency obsessed. Taiichi Ohno once said that the real objective of the Toyota Production System was to "create thinking people" (Bicheno, 2000) and he considered it to be a waste on the company's part to undervalue the creative brainpower and potential of their employees. 'Untapped human potential' is now often classified as an eighth type of waste. BM rely on everyone to develop, implement and sustain practical solutions to eliminate wasteful or potentially risky operations and this requires a culture of trust, support and mutual respect within the business. This culture is being developed via a programme called 'lean culture' and also through the issue of 'behavioural safety standards'.

Lean Culture

Before lean thinking was introduced on the Airdrie site, a team was formed to consider the reasons that have previously affected our success, or lack of success, in the introduction of change programmes. The team soon realised that the most fundamental issue was the lack of involvement and commitment of some employees to any programme of change and the strength of resistance against even the most trivial changes from a small but influential number of staff. The business recognises that the quality of our products can never exceed the quality of our people so we knew we needed to gain commitment so that everyone could direct their energies towards the same goals and be more positive about change within the business.

The theory behind the lean culture programme is to firstly use basic motivational psychology techniques to get the teams to the point of actually recognising the issues that may be affecting them. The next stage is to benchmark with the teams what they consider to be acceptable and unacceptable behaviours relating to the identified issues at all levels the business and to encourage people to challenge the things that they feel are unacceptable. This is particularly powerful in a shop-floor environment since the teams are encouraged to take on new challenges to resolve issues that they feel impede on their ability to perform their job in the safest or most effective manner.

Through the lean culture implementation, BM aims to develop a quality company where everybody makes important contributions to maintain high quality standards, works together to achieve them and above all, people are proud of their job, department and company. The lean culture programme has improved motivation in the areas it has so far been implemented and, together with the scope for everyone to be involved in non-production activity and make their own improvements on issues that they feel affect their work, this has done a great deal to improve psychosocial aspects of the working environment.

Behavioural safety standards

Behavioural safety standards have been designed to set minimum behavioural expectations of all staff within BM on specific issues (e.g. pallet handling, use of forklift trucks). Each one can be delivered in a fifteen minute briefing and they do not replace any other form of more detailed training (e.g. manual handling or truck license training), they simply act as a summary or reminder on the behaviours that should be demonstrated to complete the job in the safest possible manner. The behavioural safety standards are regularly briefed to all staff and also displayed in production and other areas so that people can reference them as required. Once issued, these standards can be used by anyone, regardless of their position, to challenge and correct anyone they observe to be demonstrating unsafe behaviours which may lead to an accident or injury. When followed, the standards promote a culture of safe behaviour which will result in the improved health and safety of staff and visitors whilst on our premises.

Integrating lean and ergonomics

The integrated approach of lean and ergonomics provides a powerful means of increasing process efficiency and improving the quality of work, and also dispels the myth that sacrifices in safety or comfort must be made by the individual for the company to achieve greater productivity. In a dynamic working environment, with a large product inventory and short packing runs, it is very tempting to go straight for the obvious 'quick fix' solutions (e.g. lifting equipment, adapted tools, etc) but such improvement measures do not necessarily deal with the root cause of the problem are often just a superficial cover for fundamental flaws and inefficiencies within the process. Many of the musculoskeletal risks and problems identified on the site are symptoms of waste within the processes, and almost all of the seven wastes (figure 1) relate to ergonomics in some way, in particular unnecessary motion and inappropriate processing.

Unnecessary motion is the waste that has the strongest connection with ergonomics. Whenever someone has to stretch, bend, reach or unduly exert themselves, not only is the physical risk to the operator increased, but ultimately productivity and quality of the product will also be affected. When implementing lean thinking, effective ergonomics design in the workplace is considered to be both ethically desirable and economically sound as good workplace organisation will reduce the risk of poor postural movement and in the broader scope will improve the "quality of worklife" (Bicheno, 2000). Inappropriate processing refers to people using machines and processes that are not fully capable or using the wrong set of tools, procedures or systems when a simpler or different approach would be more effective. The side effect of using inappropriate processes is that people may be expected to perform any of the operations that the machine can not and such activities can present a high risk of musculoskeletal injury to the operator.

Lean thinking uses a series of analytical techniques to detect areas of waste and measure production flows and we have found that using these as an alternative method of task analysis for ergonomics problem solving helps to make the value-adding solutions more obvious and makes it easier to capture the benefits. These solutions often require little capital investment but even if investment is required, the long-term benefits to the supply chain will far outweigh the cost of implementation. Two examples of the methods that can be used to achieve this are process activity mapping and line balancing, and the basic principles and benefits of using these methods will now be discussed.

Process activity mapping

Process activity mapping is a technique used extensively in lean problem solving projects to highlight areas of waste within the supply chain. For the areas of waste most closely associated with ergonomics risk factors, this is the most effective technique to use when generating process solutions that create value for the business. Process activity mapping was traditionally used in work study or industrial engineering so it is not a new method, the difference is in the method of application (Hines & Taylor, 2000). In previous change programmes (such as JIT) it was seen as a specialist role of the work study or industrial engineer to collect the data and suggest ways of re-engineering the process to gain efficiency improvements. However, it is the operators that work in the area who ultimately have the most detailed knowledge of the process, so a participative approach is now used where the operators champion collection of the data and act as key contacts when considering changes to the process or working environment.

Detailed process mapping is a simple technique to understand and apply, yet dramatic changes and results can be achieved. When completing a process activity map, each step in the process is recorded in a standard chart format to show what is actually happening, where this is taking place, and how much time is spent carrying out each activity. To aid further analysis, the observations are classified into one of six main types of activity to indicate the process flow and each of these are identified by a symbol (figure 2).

Type of activity	Symbol	Definition
OPERATION	○	Indicates the main steps in a process, method or procedure where the physical or chemical characteristics of a part, material or product is modified or changed.
HANDLING	⬭	Any task relating to the physical handling of an item.
TRANSPORT	⇨	The act of physical transport from one location to another
INSPECTION	□	Examination of an object for identification or to verify quality or quantity for any of it's characteristics.
STORAGE	▽	Any time where an item had been placed in a warehouse location up to the point where it is first removed for use.
DELAY	D	Waiting time between transactions that would ideally have continuity under normal circumstances.

Figure 2. Definition of categories for classification of process flow

To analyse the data, the value adding activities are separated from the non value adding activities to assess the proportion of activity that is actually adding value to the product. Typically the proportion of value adding activities is surprisingly low, and it is not unusual to find that less than 1% of total process activity is essential operation. The rest of the process steps are considered to be a waste within the process and the validity of any activities that do not create value must be questioned so that the process can be re-engineered to optimise the production flow.

Safety is one of the prime foundations for lean manufacturing and by using detailed process analysis techniques it is possible to eliminate both safety and ergonomics risk factors, or at least change the state in which the problem is presented so that it is easier to control, by developing more suitable working methods and restructuring the process. This approach generates solutions to ergonomics and safety risks that create value for the business rather than requiring capital investment and even if investment is required, the long-term benefits to the supply chain will far outweigh the cost of implementation. To support lean thinking principles, the focus of cost benefit analysis has shifted to value benefit analysis. Even if there is no direct or obvious immediate cost saving, we can justify change by quantifying the long-term potential value to the business in terms of removing unnecessary operations from the supply chain (e.g. saving eight hours labour per work order by eliminating double handling of materials).

Line balancing
With such a diverse variation in the types of products produced at BM it is difficult to achieve high levels of automation, particularly in the gift pack assembly area where all of the operations have to be performed manually. Due to the seasonal nature of the work, this department uses a high proportion of temporary staff, each with a different skill level which makes it difficult to arrange the lines so that everyone can cope with the rate of work. This problem has many implications for ULD risk and does not reflect the lean principles of work flow.

Lean thinking has helped BM to apply a different method of line balancing and this has so far had the biggest impact on the gift assembly area, primarily because it has an increased consideration of individual capability. When a line is balanced well, you should know exactly how many products should be packed every hour and be able to predict, based on this consistent flow rate, exactly how long the job will take to finish. The method used to achieve this involves breaking the tasks down to their smallest, individual components and recording how long it takes to complete each one. The work rate is determined by balancing out the time taken to complete each operation, putting extra people on those tasks which take longer to complete and only grouping together the shorter pace tasks once a consideration has been made for the complexity of task movement. This means that the work flow is arranged in a pattern where the task or tasks allocated to each workstation takes roughly the same time to complete and using a theory of one piece flow (Hines & Taylor, 2000), the pace of work is indicated to the person at the first stage of the operation using markers that are fixed onto the packing line. Using this arrangement, the tasks are simpler to perform and with the time pressure caused by the old working method eliminated, the risk of ULD is significantly reduced. With a reduction in the number of multi-tasking activities, there was a concern that apathy may result but the team has been kept enthused by training the operators in the line balancing process and encouraging them to set-up the lines for each product themselves. This has generated a good working atmosphere where the teams regularly monitor performance and congratulate each other when they manage to achieve the goals and targets set.

Conclusion

The lean thinking approach is designed to challenge the traditional business mindset and this paper demonstrates applying supply chain thinking to ergonomics problem solving generates business focussed solutions with quantifiable value benefits. Integrating ergonomics with lean thinking provides a consistent focus towards the same goals of improving production output and quality, meeting customer demands and adapting process flexibility whilst also reducing the risk of accidents and errors, undue stress and strain, absenteeism and turnover. All of these factors contribute to optimising the supply chain and helping BM achieve its strategy of being the best supplier to our customers.

References

Bicheno, J. 2000, *The Lean Toolbox*, Second Edition (PICSIE Books, Buckingham)
Hines, P., Taylor, D. 2000, *Going Lean – A Guide to Implementation,* First Edition (Lean Enterprise Research Centre, Cardiff)

ERGONOMICS – IS THERE A NEED FOR GUIDELINES ON ROLE COMPETENCIES?

Iain S MacLeod

Atkins Aviation and Defence Systems
650 Aztec West
Almondsbury
Bristol, BS32 4SD

Ergonomics education is normally of a high standard but associated and subsequent skill development is largely left to chance. Often reliance is place on the use of Ergonomics /Human Factors Standards as 'bibles' substituting for skill and expertise. Unfortunately, these standards are, for the most part, merely guidelines. In many professions Continuous Professional Development (CPD) regimes are striving to fill the gulf between education and progressive skill/expertise development. Levels of society membership are also an indication of skills and expertise. Competency implies a sufficiency of the skills and expertise necessary for an individual or team to fulfil a work role. Acknowledgement is needed of the importance to practitioners and their employers of an understanding of what constitutes the competence levels required for particular roles in the application of the profession.

Introduction

Competencies are an amalgam of skill and expertise. Accepting that an individual or team has the competence to perform given work tasks implies that some metric exists to assess what skills and expertise is sufficient to indicate a fitness or qualification to perform the work. This essay will argue that such metrics do exist but are scattered over diverse phases of skill and expertise development. Moreover, that their interrelationship and applicability to the assessment of the competencies required for any role is poorly considered. To assist in the matching of skills and expertise to competence requirements for a work role requires the use of guidelines on the competencies needed to perform actual work.

Skill is the appropriate application of practical knowledge as mediated by the ability of the practitioner(s). There are many forms of skills varying between skills that are manifestly physical to skills that are primarily cognitive. Inherent in the peer acknowledgement of skills is knowledge of the reliability of the application of the skill. Human reliability is considered more as a consistent ability to apply simple, complex, and/or open skills (Poulton, 1957) to the quality and goal related address of natural events and artefacts of the operating environment. Experience in the application of skills results in the creation of levels of the expertise owned by practitioner(s). Expertise can thus be defined as the reliable and knowledgeable application of high skills within a particular context.

Situational awareness (SA) is associated more with complex 'open skills' than with 'closed skills' (Poulton, op cit), in that its maintenance requires a good deal of appreciation of, and interaction with, external events. SA is a necessary and intrinsic property of applied skills. It is also a product of the quality and experience of skill application, a product enhancing such skills, but is variable under the influences of certain personal and organisational factors (MacLeod, Taylor, & Davies. 1995). SA exists in a symbiosis with complex 'open' skills.

The SA capabilities of each individual, or team of individuals, are different. These differences in capabilities are a part result of variations in the levels of basic ability between individuals. Therefore, it is important that individuals are carefully selected as suitable candidates for particular skill training. In some cases the ability levels of the individual will bar improvement in their skills or SA. However, the SA level of the individual can usually be raised through education, the training of associated skills, and SA should improve in parallel to work practice and a gain of experience with relevant work environments.

SA resides in cognition. Cognition is normally taken to refer to the part processes of the mind involved in human knowing such as conscious and unconscious reasoning processes, understanding and judgement. Hosting human cognition are mental or psychological processes. These processes are global and cover the total remit of mind functions and processes including such as emotion and long term memory. Particular cognitive functions of concern would encompass spatial reasoning, judgement and cognitive skills.

> "The critical thing about doing shared tasks is to keep everyone informed about the complete state of things. The technical term for this is situation awareness."
>
> (Norman, 1993)

All the above emphasises the difficulties in defining the competencies of individuals to apply their skills and levels of expertise on their own or within teams. Nevertheless, the quality completion of work is highly reliant on the competencies of the work participants.

The remainder of this essay will consider the properties and interrelationships of different contributing factors to the initiation, build, maintenance, and classification of competencies.

The Road to Competency

The road to competency passes through phases requiring formal acknowledgement of the levels of competency reached as is appropriate to each particular phase. The form of competency achievable at each phase can be considered as a step towards the competencies required to fulfil work roles. Examples of the themes of these stages include education, training, continuation training, work experience, qualification or certification, and role adoption. These phases are not necessarily addressed in order and may be repeated. However, the interaction or cross fertilisation between these developmental phases forms the 'glue' through which role competency achieves coherence. Some of these phases will be discussed to indicate their relevance to the determination of role competence.

Education and Training

The achievement of the basic education qualifications required of an individual for skill training is reliant on but different from their innate ability and intelligence. A person's latent ability and intelligence are, within narrow limits, assessable by the time that person seeks training for employment. However, within the bounds of the person's ability to learn, and available monetary budgets, they can be educated at any time to a required level prior to vocational or skill training. A lack of education may be more indicative of missing or inaccessible educational opportunities than of a person's ability.

Nevertheless, the training of any skill requires a basic standard of education in relevant subjects before the training can hope to be assimilated. Arguably it is not necessarily true that a high level of education will provide knowledge to aid the development of all forms of skills. The training of skills should always have an advised associated entry level of education. Training is about the pragmatic use and handling of knowledge within a context. Further, the training of skills requires a working knowledge of basic rules; for example car drivers need knowledge of the Highway Code and footballers need knowledge of the rules of the game and their teams approach to play. All education results in some certification of the resulting educational qualification, this authenticated through some formal means such as the supervision by an appointed authority competent to fulfil the role. This indication of competence is often used as an indication of competence to enter training.

Ab Initio Training

Ab initio or initial training teaches the first basic building blocks of skill required as the foundation for the development of higher level skills. The training environment must not only create a suitable environment for the training, that is also part representative of the intended work organisation as a whole, but it must make best use of training strategies and assessments of the results of the training programme. As SA depends on the level of development of skill, as well as experience, it is unlikely that SA will be well developed even at the end of this training phase. During this level of training the trainee learns how to assess their capabilities with relation to given tasks, to assess the main effects of the working environment on task performance, and to extract pertinent cues from the environment.

The relevance and importance of environmental cues are taught at this stage of training as are the principles of situation assessment, team working and expected roles within a work hierarchy. Ab initio training is typically a mixture of whole task and part task training. Basics are initially taught as whole task to train the fundamental interrelationships between the skilled tasks appropriate to a multi-task environment. The competencies achieved from such training are intended as a foundation for more specific role training.

Role Training

Role training takes the form of the conversion of initial skills into those required to satisfy a particular role or roles. It might also be necessary when an individual is changing roles, having gained some expertise in a previous role. This stage of training places emphasis on organisational rules and expectations from the role. It may place emphasis on the influences of particular environments or equipment states on the conduct of work. The competency achieved by role training is probably sufficient for the individual to work in the role with the guidance of rules and under supervision.

Forms of Continuation Training

Continuation training is required to promote skill development at all levels. It is also necessary to ensure that skilled individuals maintained their skills and expertise and that these can be developed through work experience. Continuation training is also used to learn new skills, if and when required, and to allow the level of skill of the operator to be assessed and, through standardisation procedures, to prevent habitual behaviours replacing and inhibiting skill development.

During this form of training, SA will be under continuous improvement depending on the effectiveness of the training, advice from peers, and the exposure it offers to the operator of diverse environments and situations. Currently assessment of an aviator's level of SA is primarily performed by peers (Macmillan et al, 1995). Indeed, the constructive criticism of peers is important feedback to the development of aviator's flying related skills and SA.

For instance, the training of team related SA of aviators is more problematical because of the number of dimensions involved in the workings of a highly skilled team. Occasional crew checks on operational and training missions, and the use of mission simulators, can provide feedback on crew teamwork over diverse situations. It should be noted that although simulators can be used to train many skills that present little opportunity for practice in the air, for example the handling of emergencies, they cannot simulate all the stresses and cues offered by the real flying environment or the practical awareness of their significance that is so important to the development of SA.

It should be noted that the enforcement of rule based behaviours related to recommended practices may inhibit the development of system operating skills and SA, and like some equipment operating procedures, only lead to the establishment of practices created to overcome deficiencies in equipment or tool design rather than promote improvements in tactical execution. Within safe limits, individuals and teams must to some extent learn by trial and error if they are to move onto levels of expertise related to higher skill levels (Rasmussen, 1986).

Therefore, continuation training of skills and the development of expertise and competence is a continuous process. However, the process is often indistinct or poorly traced. This is not assisted by the poor definition of what process or processes are most relevant to the sustained professional development and quality of work produced by an individual, team, or organisation. Much work has been addressed to this problem (see Earthy, 1999 & SEI Capability Maturity Model Integration). However, like all good initiatives their efficacy must be assessed considering the purposes of their inventors, the cost of their application, and the education needed to make participants adhere to new principles. For example, the SEI CMM was initiated by the USA Department of Defence as a possible means of assessing the competency levels of companies bidding for defence work. Though the scheme promotes of the professional development of all company employees, among many other things, it does not necessarily assist in a determination of the role competency of a particular individual or team in that company as might be required by an alternate employer.

Expertise and Breadth of Work Experience

Breadth of experience, like highly developed skills, is a background foundation for the achievement of expertise and competency, a foundation that only fades slowly with time and lack of practise. Because of their differing experience levels, experts and beginners recognise situations differently and then evoke analytical reasoning differently. Therefore, their level of skills and SA is different. The Situation Recognition and Analytic Reasoning Model (SRAR) gives one explanation for these differences. Table One below shows the basic tenets of the model. However, the quality of the work experience, or its influence for good or bad on individual competence, is hard to assess on its own without support from elsewhere. One form of support would be the use of competency guidelines as a template on which to assess the degree of relevance of particular experiences to work.

	SITUATION PATTERNS	ANALYTICAL KNOWLEDGE
Beginner	Small, well defined, and inflexible situation patterns.	A few large analytical knowledge chunks.
Expert	Larger, fuzzy and flexible / dynamic situation patterns compared to beginners.	Many analytical knowledge chunks compared to beginners. However, the chunks are small.

Table One - Derived from SRAR Model (Boy, 1995)

Requisite Practice

Considering the comments of the previous paragraphs it is important that designed practice is sufficient to maintain the level of skill over a selected number of important skills and procedures. The form of practice is also important to the development of skills and expertise. Pure procedural practice in good for the practice of emergency procedures, but is inappropriate to develop skills relate to cognition based multi task analysis and control, these as related to complex tasks such as may be found in sport.

Further, there is a difference between the operator tasks as devised and trained for a system and the operator work activities as applied in reality under the auspices of complex skills. Under the umbrella of training, contrived practice may unnecessarily concentrate on what is planned rather than on the essence of that required to convert plans to the reality of work. What should be exercised by training, to promote cognitive skills and expertise, are the Cognitive Functions that convert known and understood tasks to intended activity.

Cognitive Task > Cognitive 'Function(s) > Cognitive Activity > Feedback

Expanding on the above and on other theories of human behaviour, Neisser (1976) formulated his concept of the ever active 'Perceptual Cycle' where the interaction between the human and the environment shapes the human's perceptions and actions. Neisser argues those schemas are always activated but that the activation of particular schema is as an oriented response to the environment. This oriented response selects new information for attention which in turn activates appropriate schema and so on. Thus, Neisser presents a description of SA in that human attention and awareness of the environment or situation is actively modified by their changing appreciation of information gleaned from that environment. This illustrates one model on the means of developing expertise and competence through awareness brought about by the experience and interaction with diverse environments increasing knowledge and skill. Figure One below illustrates the basic perceptual cycle.

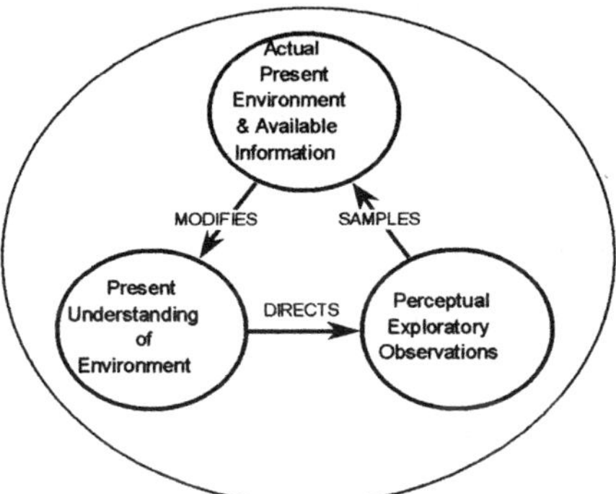

Figure One - The Perceptual Cycle

Cognitive skill and expertise cannot be developed by the slavish application of professional 'Standards' or 'Guidelines' as if they were Laws. Such practice is an antithesis of the development of role competence. It is rule bound and has a tendency to ignore the working situation and environment

Qualification and Certification

Qualification is initially achieved post the achievement of a recognised level of education and training. Certification of competency is awarded by trade or professional societies through a formal assessment of the individual, team, or organisation. This assessment is normally based on a combination of assessments that may include appraisal of areas such as educational qualifications, experience in the application of skill, references on capability, relevance of work environment, supervision, and any undertaken Continuous Professional Development (CPD) activities.

CPD is increasingly being promoted by Professional Societies as a means of guiding practitioners on methods of maintaining their expertise. Some societies such as the British Psychological Society anticipate that CPD evidence will need to be produced on a yearly basis to indicate that a sufficient level of work is being performed to support the continuation of a member's professional status within the society. However, the question must be asked as to whether an adherence to CPD guidelines or rules, depending on the society, will serve as a means of indicating the competence of the individual rather than a means of a society validating the professional worth of its members.

The above statement is not meant to decry the good intentions of the majority who support some form of CPD as one of many means of ensuring that professionalism is sustained within their discipline. Indeed, a well formulated CPD regime could promote and classify the forms of competence of its participants. However, the question would still have to be addressed as to the fit of that competence with role if not specifically addresses by the CPD. Further if role is specifically addressed by the CPD, how easily would the indicated competence be matched with the requirements of an alternate role – a role that may fit the capabilities of the person or persons in question.

Competency Guidelines

CPD on its own will not allow an employer to fit the man to an existing or future work role within their organisation. That fit is always problematical and arrived at by the evaluation of a collection of evidence on the individual or team's expertise and competence. What makes the fit all the more problematical is the frequent lack of explicitness in the definition of the work form or the competencies needed to satisfy that role. One method of assisting the fit would be to promote sets of competency guidelines with relation to particular work roles. Such guidelines must help to complete the criteria for assessing individual or team competencies. Competency guidelines may be primarily a guide for the employer but also can serve as a good source of advice on what the employer currently anticipates their role competency requirements to be. This advice would ideally be used as guidance to the particular competence aims of other phases of competence build.

An example of a guideline that proffers advice on the criteria to be used to assess expertise with relation to work is contained in the *IEE - Safety, Competency and Commitment: Competency Guidelines for Safety-Related Systems Practitioners*.

The usefulness of the above form of guideline can be strongly argued as highlighting the need for consideration of the collective *How, Why, When, Where* and the *Form* of any professional application as a means of understanding the required competencies for any work role. Such guidelines are needed to complete the understanding of the requirements applicable to the build of any role competency. Possibly the necessary co-ordination and interaction between the competency aims of the different phase of role competency build, and guidelines on the requirements of role competencies, is more an ideal than a practical suggestion!

References

Boy, G., (1995), Knowledge Elicitation for the Design of Software Agents, in *Handbook of Human Computer Interaction*, 2nd edition, Helander, M. & Landauer, T. (Eds), Elsevier Science Pub., North Holland.

Capability Maturity Model Integration, *CMMI-SE/SW/IPPD/SS, Version 1.1*, March 2002, Software Engineering Institute, Carnegie, Melon University, USA

Earthy, J. (1999), *Usability Maturity Model: Processes*, Version 2.2, Proceedings of INTERACT'99.

IEE, Safety, Competency and Commitment: Competency Guidelines for Safety-Related Systems Practitioners, IEE Publications, Stevenage, UK, 1999.

MacLeod, I.S., Taylor, R.M. and Davies, C.L. (1995), *Perspectives on the Appreciation of Team Situational Awareness*, Proceedings of International Conference on the Experimental Analysis and Measurement of Situational Awareness, Embry Riddle University Press, Fla.

McMillan, G.R., Bushman J., and Judge C.L.A. (1995). Evaluating Pilot Situational Awareness in an Operational Environment, in *Situation Awareness: Limitations and Enhancement in the Aviation Environment*, 79th AGARD AMP Symposium. AGARD Conference Proceedings, AGARD, Neuilly-sur-Seine. April.

Neisser, U. (1976), Cognition and reality: Principles and implications of cognitive psychology, W.H. Freeman, San Francisco.

Norman, D.A., (1993), *Things that make us smart*, Addison-Wesley, Reading, MA.

Poulton, E.C. (1957) 'On the stimulus and response in pursuit tracking', *Journal of Experimental Psychology*, 53, pps 57-65.

Rasmussen, J. (1986), *Information Processing and Human machine Interaction*, Elsevier Science Publishing Co. Inc., New York.

Opinions expressed are those of the author and not necessarily those of his employers.

IDENTIFICATION OF WORKING HEIGHT RANGES FOR WHEELCHAIR USERS IN WORKSHOP ENVIRONMENTS

Eoin O' Herlihy and William Gaughran

Department of Manufacturing and Operations Engineering,
University of Limerick,
Limerick,
Ireland.

Analysis of wheelchair user ergonomics in workshop environments has largely been ignored. Currently there is little evidence of any standards or data that has been collected on correct working heights for wheelchair users under the age of 18. This study investigates working height ranges for wheelchair users in engineering and technology environments in second level education. For experimental purposes it is assumed that an able-bodied person placed in a wheelchair will carry out tasks in the same manner as a wheelchair user (Goldsmith, 2000). Fifty able-bodied subjects were used as wheelchair users to take part in the study. The results indicate that an adjustable height range of between 600mm and 850mm would suit all operators for tasks such as writing, reading, using a vice and using a drill and the ideal height to place controls on a machine is between 750 and 950mm.

Introduction

Insurance companies, regulatory boards, and many employers are becoming increasingly aware of the potential for injury and illness as a result of inadequate working posture. O'Sullivan and Gallwey (2000) have identified that table height is an important design issue as it determines the gross body posture for the majority of tasks. Correct working heights will reduce injury and back pain, awkward and unnecessary working postures as well as increase safety and productivity. However incorrect working heights will lead to uncomfortable working postures and the consequences can result in both long-term and short-term effect on the operator and the company (Neumann et al., 2001 and Mital et al., 1999).

Several researchers have carried out research for adults into ideal working heights for various types of industrial work (Kroemer and Grandjean, 1997, Human Factors Society, 1988, Pheasant, 1996 and Konz, 2000). The most widely used and referenced for industrial workplaces are that of Kroemer and Grandjean (1997).

For wheelchair users correct working heights are critical, as the most common physical complaint from wheelchair users is back pain (Sargent, 1981). It is also critical that the wheelchair user can work in a neutral body position. The more the seated position at the workstation differs from neutral body positions; the greater are the stresses on the musculoskeletal system (Abdel-Moty and Khalil, 1991). Working height ranges for wheelchair users have been investigated with regards to table heights and counter heights for living and office environments (Goldsmith, 1984, 2000 and ADAAG 1997). However, little if any investigation into working heights for industrial workplaces and the wheelchair user has taken place. As students work in engineering and technology environments in schools, colleges and at part time work, there is a definite need to research this area to identify working heights for various manual tasks that take place.

Method

Approach

The purpose of this experiment is to identify working height ranges for various tasks in engineering and technology environments for students who use wheelchairs in post primary schools (12-18 years old). For experimental purposes it is assumed that an able-bodied person placed in a wheelchair will carry out tasks in the same manner as a wheelchair user. Collectively therefore, in the context of anthropometric data, it is admissible for able-bodied people to be surrogates for wheelchair users (Goldsmith, 2000). Therefore according to Goldsmith (2000), able-bodied people can justifiably be employed for the purpose of wheelchair testing.

The use of fitting trials is a common technique in ergonomics evaluation. "Fitting trials are an established technique in ergonomics where a product or workplace is evaluated by trials (perhaps on a mock up or prototype) using a carefully selected user group that is representative of the target population. Typically, subject selection would be based on age, gender, size etc. and total sample sizes limited to perhaps a few dozen" (Case et. al 2001). Each subject was chosen according to his or her stature. This would allow for tall, medium and small wheelchair users. The type of wheelchair that the subject would use was then decided upon by measuring their standing stature of the research subjects. This technique thus introduced "wheelchair users" of different height ranges.

Experimental Design

The experiment set out to examine the working height range for various tasks in a wheelchair. Table 1 below indicates the various tasks and the minimum and maximum heights that were used to identify the limiting factor. Height increments of 50mm were used in all the tasks.

Table 1. The tasks and limits used

Task	Minimum Height	Maximum Height
Vice work*(bench surface)	550mm	950mm
Manual work	550mm	950mm
Writing/Reading	550mm	950mm
Controls mounted on wall	300mm	1450mm
Machine controls	300mm	1450mm
Bench drill table height	400mm	900mm

*The upper edge of vice is 90mm higher than heights shown.

Apparatus
A specially designed rig was used to carry out the test. The rig was height adjustable using an automated lifting column. On the rig there was a work surface designed to allow maximum access for wheelchair users, a simulation of the drill mounted on a stand and a vice stand to hold the Record 90mm high, compact multi-purpose vice.

To account for the different heights of the subject's two wheelchairs (DMA self propelled 218-23WHD Heavy-duty wheelchair and DMA self propelled wheelchair for children) were used. In all there were three seat heights used (between two wheelchairs), 450mm, 460mm, and 540mm and 50% of the subjects completed the experiment with armrests fitted on the wheelchairs and the other 50% without. Standard bench tools were used for the manual task, a textbook, pen and paper were used for writing/reading task, metal files and aluminium test pieces were used for filing and a Holtain anthropometer was used to record the anthropometric data.

Experimental procedure
Each subject read and signed an informed consent form and received a verbal explanation of the experiment. The first step was to gather anthropometric data in a standing position and then in a seated position in the wheelchair. The next step was to carry out the fitting trial for the various tasks. Latin Square ordering was used for the various tasks and the fitting trials were carried out in ascending and descending orders (Pheasant, 1996) using the heights shown in Table 1 above. After completing each task the subjects indicated whether or not they felt the height was too low, too high or satisfactory. The subjects also indicated their preferred optimum and the results were recorded.

Results

Fifty subjects (twenty-five males and twenty-five females) from a post primary school participated in the study. The mean age was 15.4 years (SD=1.6) and the mean stature was 1662.2mm (SD=92.8mm). Anthropometric data for seated stature and knee height were also collected and these are shown in Table 2.

Table 2. Anthropometric Data

Percentile	Stature	Knee Height
5th	1118	585
50th	1226	671
95th	1354	757

The heights of controls are the most critical factor in the design of machines. Figure 1 below shows the results of the fitting trial that determined the height range for machine controls. From Figure 1 the subjects have identified the heights between 550 and 1350mm as the limiting position to place controls on a machine. From analysis of Figure 1 it is clear that there is no one position that suits everyone but the height range of 750 to 950mm would cater for the greatest range of operators.

The plot of the fitting trial on control heights mounted on walls is shown in Figure 2. This illustrates that the subjects have identified the height of between 350 and 1450mm as ideal positions for controls. Examination of Figure 2 would indicate that a suitable height to mount controls to be accessible by all would be between 700 and 900mm above floor level. Figure 2 also illustrates that over 90% of wheelchair users would find the heights of between 700 and 1000mm accessible.

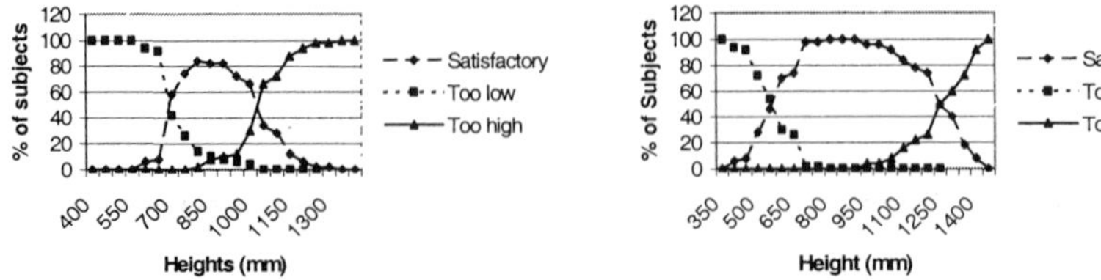

Figure 1. Machine Controls **Figure 2. Wall Controls**

Little research if any has been carried out to identify the correct working height range for wheelchair users working on a bench vice. Figure 3 below indicates the results of a fitting trial for filing tasks. The results in Figure 3 indicate the clearance (that is needed underneath the vice stand for working at a vice) should be adjustable between 550 and 900mm. The results also indicate that a height of above 600 and below 850mm was the optimum range that suited all users.

Figure 4 also indicates that if the clearance had to be fixed it should be between 700 and 750mm to cater for the majority of users. It is clear from Figure 4 that the working height for carrying out manual tasks should be adjustable with a range of 600 to 900mm. The results illustrate that a height of above 650 and below 900mm would suit all users. If the work surface had to be fixed it would have to be fixed somewhere between 700 and 800mm to cater for all wheelchair users.

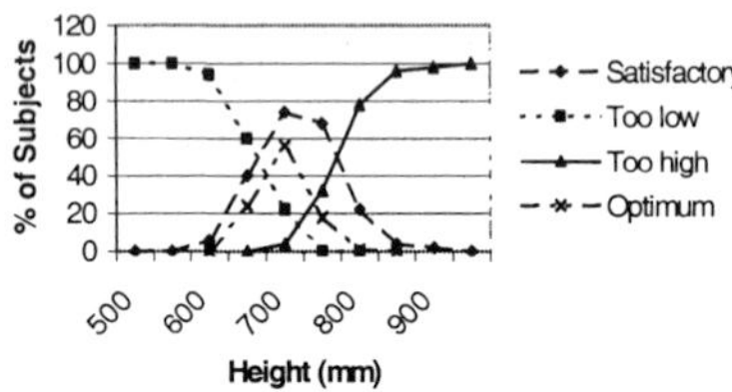

Figure 3. Vice Heights

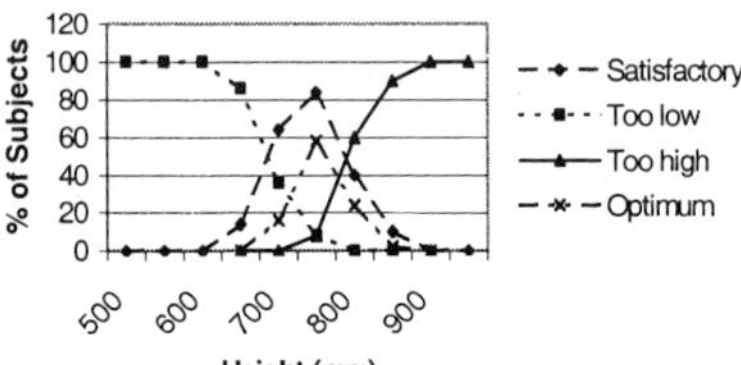

Figure 4. Manual Tasks

Writing and reading are carried out in engineering workshops while working on drawings and during theory classes. Figure 5 shows the results of the fitting trials for writing/reading tasks. The results in Figure 5 indicate that the working height for a writing/reading task for wheelchair users should be above 600 and below 900mm. The results also indicate that a height of above 600 and under 800mm would suit all users. If the work surface had to be fixed, it would have to be placed between 700 and 750mm to cater for all wheelchair users.

The results in Figure 6 indicate that the working height for the drill table should be above 500 and below 900mm. The results also indicate that a height of between 600 and 850mm would suit all wheelchair users. If the work surface had to be fixed, it would have to be placed between 700 and 750mm to cater for all students using wheelchairs.

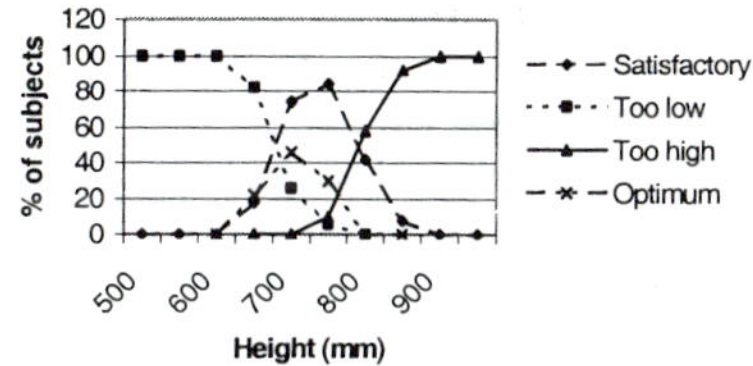

Figure 5. Write/read

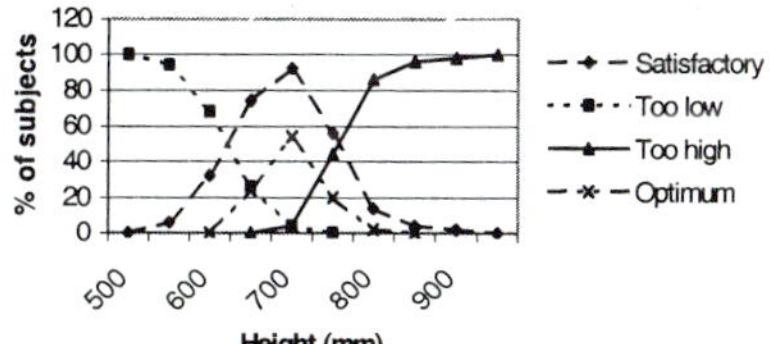

Figure 6. Drill Table

Discussion

For all the results (except Figure 3) the graphs for 'too high' and 'too low' intersect. This intersection emphasises that in the majority of situations there is no one height that is suitable for all operators thus the need for adjustable height. Kroemer and Grandjean (1997) , Saunders and Mc Cormick (1993) and Konz (2000) are just three of the many authors that have discussed working heights for workplace design. Despite the fact that these ergonomists promote and suggest that adjustable height is of great importance in the workplace there is little evidence that adjustable height has been implemented.

O'Herlihy and Gaughran (2002) noted that in educational engineering and technology environments the current work surface height ranged from 770mm to 800mm. Analysis of Figures 4 and 5 indicate that these heights are unsuitable for 34-60% of subjects while carrying out manual work and 34-58% of subjects while carrying out writing/reading tasks. On existing workbenches in educational establishments the working heights of a standard vices range from 950mm to 980mm above floor level. Figure 5 indicates that this height is unsuitable for 96% (at 950mm) and 98% (at 980mm) of subjects.

O'Herlihy and Gaughran (2002) also state that the average height for emergency switches mounted on walls was 1382mm. Figure 4 indicates that controls mounted at a height of 1382mm would only cater for 14% of wheelchair users. The results in Figure 4 also suggest that the average height of the emergency switches is 382mm too high to include all operators, using wheelchairs.

Conclusions

- To cater for all operators' adjustable height work surfaces are essential in engineering and technology environments.
- The clearance underneath the vice stand should be adjustable between 600 and 850mm above floor level.
- To carry out manual work the height of a work surface should be adjustable from 650 to 850mm.
- For writing and reading tasks the work surface should be adjustable between 600 and 850mm.
- The clearance underneath the drill table should be adjustable between 600 and 800mm to allow safe operation.
- The ideal height at which to place controls on a machine is between 750 and 950mm.

- The data suggests the height and position of the standard vice used in post primary schools is unsuitable for nearly all wheelchair-using operators.
- Finally the results have identified that a height of between 700 and 1000mm would be the best position to place controls, sockets and switches on a wall to cater for wheelchair users.

References

Abdel-Moty, E. and Khalil, T. M. 1991, Computer-aided Design and analysis of the sitting workplace for the disabled, *International Disability Studies*, **13** (4), 121-124

ADAAG. 1997, Americans with Disabilities Act Accessibility Guide, Access Board, (US Architectural and Transportation Barriers Compliance Board, Washington, DC)

Case, K., Porter, M., Gyi, D., Marshall, R. and Oliver, R. (2001), Virtual fitting trials in designing for all, *Materials Processing Technology*, **117**, 255-261

Goldsmith, S. 1984, *Designing for the Disabled*, Third Edition, Fully Revised, (RIBA Publications, London).

Goldsmith, S. 2000, *Universal Design*, (Architectural Press, Oxford)

Human Factors Society (1988) American national standard for human factors engineering of visual display terminal workstations, (ANSI/HFS 100-1988), (Santa Monica, CA: Human Factors Society)

Konz, S. A. 2000, *Work Design: Industrial Ergonomics,* Fifth Edition, (Holcomb Hathaway, Scottsdale, Arizona)

Kroemer, K.H.E. and Grandjean, E. 1997, *Fitting the Task to the Human: An Ergonomic Approach,* Fifth Edition, (Taylor and Francis, London)

Mital, A., Pennathur, A. and Kansal, A. 1999, Non-fatal occupational injuries in the United States; Part 1- overall trends and data summaries, *International Journal of Industrial Ergonomics*, **25**, 109-129

Neumann, W.,P., Wells, R.P., Norman, R.W., Frank, J., Shannon, H., Kerr, M.S., and the OUBPS group. 2001, A posture load sampling approach to determining low-back pain in occupational settings, *International Journal of Industrial Ergonomics*, **27**, 65-77

O'Herlihy, E. and Gaughran W. 2002, Universal access in engineering and technology environments for paraplegic Operators. Proc. 1st Cambridge Workshop Universal Access and Assistive Technology (CWUAAT), (Cambridge), March 25-27. pp 105-112

Pheasant, S. 1996, *Bodyspace: Anthropometry, Ergonomics and the Design of Work*, Second Edition, (Taylor and Francis, London)

Sargent, J.V. 1981, *An Easier Way.* Iowa: Iowa State University Press, Ames

Saunders, M.S. and Mc Cormick E.J.1993, *Human Factors in Engineering and Design*, Seventh Edition, (Mc Graw-Hill)

A PROCESS FOR CAPABILITY ACQUISITION

CE Siemieniuch[1] and MA Sinclair[2]

[1]*Department of Systems Engineering, Loughborough University, LE11 3TU, UK*
[2]*Department of Human Sciences, Loughborough University, LE11 3TU, UK*

The paper outlines lessons learnt from a case study of developing a generic 'Capability Development and Deployment Process' for an organisation in the aerospace industry. The paper commences by indicating the need for such a process, and then outlines what the process is. For reasons of commercial confidentiality, details of the process are omitted, though the structure corresponds to the usual 'new product introduction' process using a stage-gate philosophy, widely described in the literature. However, in generating the process, a number of lessons of ergonomics significanee became evident, and nine of these are reported below. Some of them are reiterations of what has been known for some time; others appear to have novelty value. The effects of staffing philosophies, and the difficulties of metrication of 'soft' aspects are two of the latter.

Introduction

Capability Acquisition is the upgrading of a company's ability to compete in its chosen markets, by collective improvements to its technology, its processes, and its people.

There are six basic processes in any company intent on survival into the future within a capitalist environment:

- Delivery of valued products to customers
- Capture and redistribution of money for these products
- Designing new products
- Knowledge lifecycle management
- Capability acquisition, both for design and for operations.
- Resourcing the organisation (Human resource Management, Estates, legal, etc.)

We concentrate on the penultimate of these, in recognition of the fact that the long-term survival of the company is critically dependent on this process. Its medium-term survival is also critically affected by this, too; Table 1 below illustrates this, for two large companies in the aerospace industry.

In this industry (and others like it, such as the automotive and IT industries), there has developed a consensus that costs must be reduced by about 2% per annum, in order to stay competitive with other companies. Using published figures and making the twin assumptions that (a) the average gain from an improvement project is £100,000, and that (b) the average duration of an improvement project is one year (feel free to substitute your own values), we arrive at Table 1.

Table 1. Example of two major aerospace companies, to demonstrate the 'pressure' of improvement projects required per annum on employees (see text)

	BAE SYSTEMS (2001)	Boeing (2000)
Sales (£million)	13,138.00	35,271.52
Costs	11,878.00	29,562.42
2% improvement p.a.	237.00	591.25
Employees	120,000	198,000
Employees/project	50	33

It is unlikely that this could be sustained year-on-year, and major re-engineering of business processes must be undertaken, as an ongoing strategy. In turn, this points to the need for a generic process for capability acquisition to enable a collection of big and small projects to be managed efficiently and concurrently. We discuss one particular company's approach to the development of a 'Capability Development and Deployment Process' (CDDP), and some of the lessons learnt.

Development of a Capability Development and Deployment Process

'Airco', in the aerospace industry wished to develop a generic process that would deliver capability (defined as any combination of people, technology, and processes) into the organisation with minimum pain and maximum benefit. In their opinion, their current performance in delivering new capability left much to be desired, with many projects running over time, failing to deliver expected benefits, and many being highly localised in their use. 'Enough', they said; 'let us be more organised'.

A rapid-prototyping, user-centred approach was adopted. A scoping study took place, followed by the development of an initial CDDP. A sequential process of case-study/revision-of-CDDP took place, culminating in Version 5 at the present time, shown as a concept diagram in Fig. 1. Each of the smaller boxes in Fig 1 comprises a set of activities that must be accomplished if the project is to progress through subsequent review gates, shown in Fig. 2. These activities include the following classes (though not all projects will include all activities):

- Management of a change project:
  - Execution of process and procedures for a change project
  - Execution of activities for leadership development
  - Monitoring of the change project for dilution of the vision
  - Maintaining the strategy of the change project
  - Evaluation of the performance measures for the change project
  - Management of the stakeholders
- Development of a new capability:
  - Development of new process capability
  - Execution of associated processes for organisational redevelopment
  - Execution of processes for redevelopment of the IT infrastructure
  - Execution of processes for location-specific aspects of the deployment.

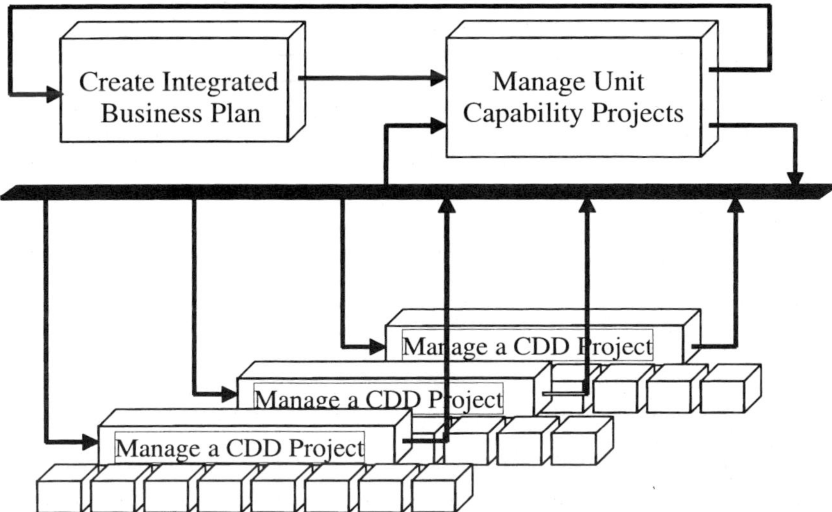

Fig. 1. Representation of structure of CDDP. At bottom of diagram, each CDDP project must be managed through a series of 'gates' until the project is completed or abandoned. Above this is a portfolio management function, co-ordinating the resourcing and operation of the individual projects. At top left is a project identification function, bringing order to the day-to-day chaos of ideas and suggestions for improvements.

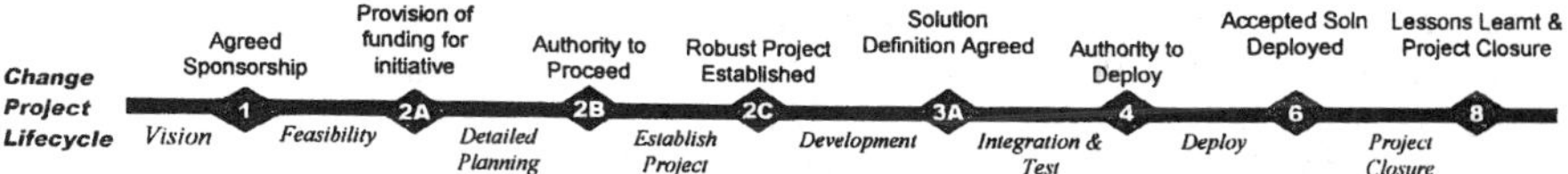

Fig. 2. Representation of 'stage-gate' development process

Lessons

A number of lessons are apparent form this investigation.

1. It may be possible to define a generic process, but instantiations of this will vary greatly from one CDDP project to another. CDDP projects can range from organisational revolutions to 'tweaks' to IT code; they can occur in different levels of environmental uncertainty (de Meyer, Loch et al. 2002). Consequently, there can be no single way to execute CDDP projects; a conclusion we reached was that a viable solution is a generic version from which the Project Manager, in conjunction with the Chair of the Reviews, could derogate as necessary. The success of this depends on appropriate answers to lessons 3 and 4 below. This generic process would need to consider the lifecycle of a CDDP project as a three-stage process, since it is apparent the projects could start and finish at different points in this lifecycle:
   - Proof of concept, concerned more with the underlying logic and inter-connectivity of the system
   - Pilot study/demonstrator, proving the viability of the system in some 'real' environment

- Full deployment, where the system is inserted into everyday operations and supported within a process

2. The level of detail that is appropriate for the generic process is unresolved, and is probably not able to be resolved. Everyone can agree on a one-page description of a process; almost no-one will agree on a 200-page version. The solution adopted is to specify the top four layers, to the point where individual projects might deviate markedly from the generic version. This is an unresolved issue within the ergonomics discipline (and many others, according to discussion forums), appearing in many process development contexts.
3. It is unwise to embark on CDDP projects unless the organisation has appropriate infrastructures and support processes for the conservation of knowledge, and for the management of a portfolio of CDDP projects. The essence of CDDP projects is to insert order where uncertainty has existed. This reduction in information entropy implies a growth of knowledge, from any CDDP project, likely to be useful in the future. Therefore, such projects should only occur within a well-established knowledge management environment equivalent to Level 5 of the Capability Maturity Models in the software industry (Crosby 1979; Humphrey 1989; (SPICE) 1996)
4. It is very unwise to embark on CDDP projects unless an appropriate culture and associated policies are in place. Widening Lesson 3 above, the readiness of the organisation to undertake CDDP projects is critical to their success. This may mean several years of preparation for a CDDP before instituting a formal CDDP approach.

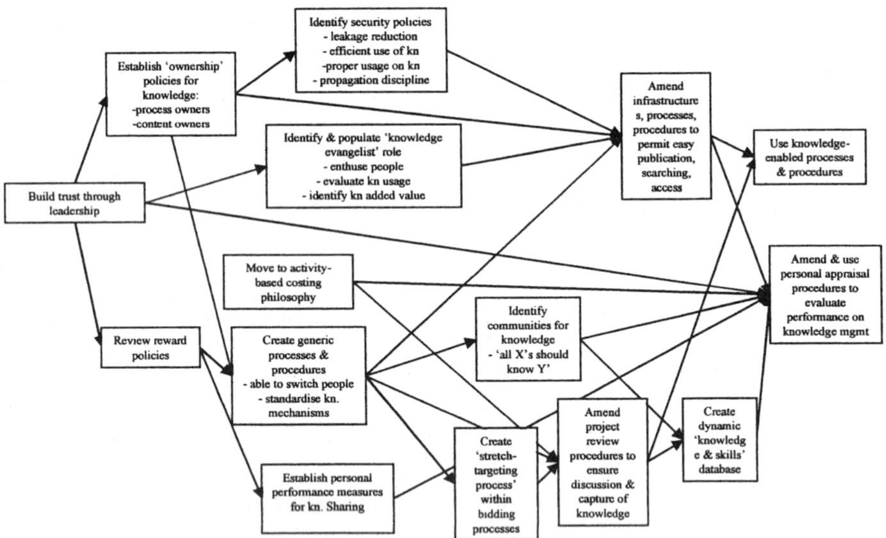

Fig 3. A generic sequence diagram illustrating the steps necessary to create a suitable knowledge management environment within which a CDDP could operate.

5. The staffing philosophy for a CDDP project can have a strong influence on both the nature of the delivered capability, and on its re-usability elsewhere in the organisation. For example, a preponderance of staff seconded from a given Business Unit will result in a system well-geared to that Business Unit, but of dubious relevance or use to others. Equally, staffing from a central function (e.g. IT) may result in a system of

real beauty, but inappropriate for day-to-day use in the Business Units. A simple tool to explore this issue was developed, to enable four classes of stakeholders (responsible person, dogsbodies, constraint providers, advice providers) to be plotted on a process chart, and to illuminate the relationships between them utilising a Role Relationships Matrix. This enables predictions to be made regarding team-working on a given CDDP, and the likely operational performance of that team.

6. The 'capturing' process to identify, choose, plan and fund CDDP projects is an exercise in creating administrative order within a happenstance environment, and will require considerable effort and time from senior management, both in understanding the inter-relationships between different CDDP projects, their relevance to company goals, and their utility in the light of changes in the operational environment. It is unlikely that a central process will be sufficient for this; consequently, Business Unit budgets should include a subvention for localised CDDP projects, not under central control. This is a recognition of the parallelism and complexity that exists within organisations. Expressed slightly differently; discrete annual budgets for CDDP activities do not sit well with the occurrence of need for projects, which are spread over the year. Some of these may be foreseen, others will emerge as a 'surprise'.
7. A 'living documents' approach is probably optimum for capturing the knowledge from a CDDP for re-use within a knowledge management process, and for auditing purposes. The implication here is that an editable document network could be presented to the Project Manager from within a workflow application, enabling an audit trail to be maintained, the project history to be described, and minimal extra work to be created at review time. Concomitant with this is the need for a good search engine.
8. Metrication of a CDDP project is difficult; especially in attempting to provide predictive metrics and methods (e.g. answers to questions such as 'If I put this group together for this project, what are the chances of success in 18 months?'). It is an unfortunate fact that the discipline of psychology has so far failed to provide a comprehensive, proven, generic model to underpin such a process, providing both the inter-relationships between relevant variables and easily-maintained metrics for their assessment. Currently-advocated approaches seem to be based on the Balanced Scorecard (Kaplan and Norton 1996), but this in turn still needs a human behavioural model; were such a model to be provided, it is possible that many of the issues that currently affect the metrication process (e.g. Crandall 2002) could be obviated.
9. The main value of a generic CDDP is for planning purposes, not for management purposes. For this reason, a concomitant activity for any attempt to introduce a capability acquisition process should be accompanied by the development (or acquisition) of a decision support system, to enable any of a CDDP project's stakeholders to explore options for the project.

References

Crandall, R. E. (2002). Keys to better performance measurement. *Industrial Management* **44**(1): 19-26.

Crosby, P. B. (1979). *Quality is free: the art of making quality certain*. New York, McGraw Hill.

deMeyer, A., C. H. Loch, et al. (2002). Managing project uncertainty: from variation to chaos. *Sloan Management Review* **43**(2): 60-67.

Humphrey, W. S. (1989). *Managing the software process*. Reading, MA, Addison-Wesley.

Kaplan, R. S. and D. P. Norton (1996). *The balanced scorecard*, Harvard Business School Press.

SPICE. (1996). Software process assessment, part 5: an assessment model and indication guidelines, ISO/IEC JTC1/SC7/WG10/OWG.

THE MARKS AND SPENCER BROOKLANDS PROJECT: INTEGRATING ERGONOMICS INTO THE DEVELOPMENT OF RETAIL EQUIPMENT

Tom Stewart

Tom Stewart
System Concepts Limited
2 Savoy Court, Strand
London, WC2R 0EZ

This case study describes the five year development programme for the Brooklands Refrigerated Display Case for Marks and Spencer and the role of ergonomics in the management of the project.

Marks and Spencer's top priority in developing a new case was ensuring that food would be presented to the customer attractively and in optimum condition – both in terms of temperature and hygiene. They wanted a design which met the unique requirements of their food business so they established a consortium of three manufacturers supplemented by consultants in product design, hygiene, structural engineering and ergonomics to develop the case. I was appointed as Project Director.

The end result was not just an innovative design of refrigerated case (selected as a Millenium Product) but a single design which gives Marks and Spencer enormous flexibility in planning and efficiency in operation in order to provide their customers with high quality merchandise in peak condition. We have also learned a great deal about the effective integration of ergonomics in design and development projects and I believe some of the lessons we learned can be transferred to other industries and contexts.

Introduction

This case study describes the five year development programme of the Brooklands Refrigerated Display Case for Marks and Spencer and the role of ergonomics in the management of the project.

Marks and Spencer started some years ago with a clean sheet of paper and an internal team from the Store Development Group drew up a design brief to meet the needs of the business well into the next millennium. Number one priority was ensuring that food would be presented to the customer attractively and in optimum condition – both in terms of temperature and hygiene. Basic dimensions of the shelves and case carcass were designed to ensure easy access for customers, including wheelchair users, as well as improving staff health and safety when merchandising and cleaning the cases. The brief also reflected the business needs for reliability, energy efficiency (minimising greenhouse emissions), sophisticated monitoring and control, efficient use of space and value for money.

Following a design competition in 1994, Marks and Spencer established a consortium of three manufacturers supplemented by consultants in product design, hygiene, structural engineering and ergonomics to develop the case.

Four years later, as the Brooklands Case (named after the first store where the final design was installed) started to roll out through the business, the design was selected as a Millennium Product for its innovation in energy efficiency, ergonomics and accessibility.

The Business Requirement

Marks and Spencer's top priority in developing a new case was ensuring that food would be presented to the customer attractively and in optimum condition – both in terms of temperature and hygiene. Basic dimensions of the shelves and case carcass were designed to ensure easy access for customers, including wheelchair users, as well as to improve staff health and safety when merchandising and cleaning the cases. The brief also reflected the business needs for reliability, energy efficiency (minimising greenhouse emissions), sophisticated monitoring and control, efficient use of space and value for money.

Marks and Spencer wanted a design which met the unique requirements of their food business so they established a consortium of three manufacturers supplemented by consultants in product design, hygiene, structural engineering and ergonomics to develop the case. I was appointed as Project Director. Acting both as Project Director and ergonomics consultant definitely helped ensure that customer ergonomics was taken very seriously, not simply because I had the authority to make decisions, but rather because these issues were addressed right at the beginning before other factors had clouded the picture. How many times have ergonomists been told 'if only we had realised this was important earlier, we could have done it differently'.

The Ergonomics of Display Cases

From the customer's point of view, a refrigerated display case is a very simple product. All it has to do is present food in a convenient accessible location for browsing and selecting and keep it at the right temperature. So ergonomics for the customer involves making the case structure as unobtrusive as possible and ensuring that most people can reach and pick up the merchandise without falling into the case or hitting their heads when they lean forward.

Other major considerations in setting basic dimensions obviously include access for cleaning and maintenance (a traditionally difficult area in supermarkets with space at a premium) as well as where to locate the surprising amount of machinery necessary to maintain an even temperature in an open fronted case (fans, temperature probes, evaporator coils, valves, electronic controllers and so on).

In addition to ensuring that most customers could reach most of the case, it was important to avoid the 'trapped in a canyon' effect which tall display cases often create in aisles. Cutting the top of the case back reduces this effect. The multi-disciplinary team including product design (Priestman Goode), ergonomics (System Concepts) and refrigeration specialists from Hussmann Manufacturing, Carter Retail and George Barker devised an aerofoil front canopy edge which was cut back, unobtrusive and delivered an appropriate air flow.

The other main 'customers' for the case are the staff in the stores who load merchandise, clean and maintain the cases. The team included a hygiene consultant from the Food Safety Consortium to ensure where possible that surfaces were easy to clean, with no sharp corners and accessible with a minimum of fixings (which get lost) and tools. Time trials conducted during the development process showed that it was quicker to clean the Brooklands case to Marks and Spencer's demanding standards than conventional cases.

However, one of the biggest ergonomics benefits for staff was the design of a 600mm shelf module with integral brackets. Traditionally, display cases have large metal shelves with separate brackets and removing them for cleaning is a major manual handling challenge. The smaller shelf, designed to suit the Eurotray standard merchandise unit, is easy for even the smallest shop assistant to move and adjust. The shelf frame also provides a base for different inserts – clear acrylic to improve visibility, baskets for loose packs, pin bars for hanging merchandise and so on. A major spin-off from the smaller shelf was that staff found it easier to adjust the shelving to suit the merchandise – giving them much more control over product display and making more efficient use of space.

Human-Centred Design

Although Brooklands Cases are monitored and controlled by sophisticated electronic controllers networked together, we would not normally think of them as 'interactive systems' of the kind covered by ISO 13407:2000 Human-Centred Design for Interactive Systems. The idea behind that International Standard is that by following the human-centred design process described, project managers can ensure that whatever computer based system they are developing will be effective, efficient and satisfying for its users. Yet there are significant similarities between the process which evolved during the Brooklands Case development and the standard.

The standard describes four key elements of human-centred design:

- a clear understanding of the 'context of use' ie users, tasks and environment
- the iteration of design solutions using prototypes
- the active involvement of real users
- multi-disciplinary design

So how do these match up to the Brooklands Case development?

A clear understanding of the 'context of use' ie users, tasks and environment

The standard points out that most products have a variety of users and the Brooklands Case was no exception – customers, shop assistants, cleaners and engineers all had to use or work on the case during its life. The initial brief was drawn up by a Marks and Spencer team including the Store Development Group (Design, Engineering and

Merchandising), Store Operations and the Food Group. They spent more than a year reviewing the requirements of all parts of the business and obtaining Board approval.

The iteration of design solutions using prototypes

One of the eye-openers working with refrigeration experts was that even in a mature engineering discipline like refrigeration, theory only goes so far and prototypes play an important part in the development process. Where the Brooklands project was different from normal engineering development was that prototyping was used in two parallel streams.

A **functional prototype** was tested in controlled laboratory conditions to refine the technical design to meet Marks and Spencer's demanding performance standards. A **merchandising and design prototype** was developed to test aesthetic and operational requirements in parallel. Of course, there had to be frequent interaction between the two streams to ensure that design solutions were compatible with technical performance constraints and vice versa. But developing in parallel saved time and allowed significant advances in both areas. One benefit of the design prototype was that it could be constructed from easy to work materials like foam board and card which allowed many different options to be tried in quick succession.

Marrying the two prototypes was a significant exercise as was 'productionising' the final design. This process was made more complex because there were actually three technical prototypes – one in each of the three manufacturers in the consortium representing different models of case (end cases, low height cases and tall cases).

The active involvement of real users

Eventually, working prototypes were ready which could be installed in store for real trials with staff and customers. Two stores were selected and trial gondolas (blocks of cases) installed in Glasgow Argyle Street and Richmond. Although these designs performed well, the real life trials resulted in a myriad of small but significant design enhancements. The final improved product was installed in Brooklands Store in late 1997 and the newly named Brooklands Case started to roll out throughout the business.

Feedback from other users – shop assistants, cleaners and maintenance engineers all helped to refine the design – and continued over a significant period with regular design review meetings.

Multi-disciplinary design

Most retailers have their own requirements and manufacturers are expected to modify their standard products to suit. But because these designs are based on other products, different suppliers' offerings are often incompatible in dimensions, styling and even performance characteristics. One of Marks and Spencer's requirements was to be able to mix and match case sections from different suppliers to meet operational requirements.

They therefore established a Core design team with representatives from the three manufacturers, a product designer, ergonomist and hygienist as well as the Marks and Spencer team (Merchandising, Engineering and Design). At different times during the development, the team was extended with other Marks and Spencer specialists and other consultants including structural engineering, fluid dynamics, quantity surveying and intellectual property specialists. A parallel project developed the electronic controller with Elm Electronics and Linton Electronics (part of George Barker).

Lessons from the experience

We have also learned a great deal about the effective integration of ergonomics in design and development projects and I believe some of the lessons we learned can be transferred to other industries and contexts.

Start with requirements, don't rush into solutions
It's easy to fall into the trap of jumping straight to solutions instead of starting with requirements. This is especially true in a relatively conservative industry like refrigeration where design engineers will start 'back of envelope' calculations on such issues as compressor output before even considering whether other technologies might be more appropriate. Having a multidisciplinary team helps to stimulate creativity.

Get senior management requirements early to avoid later show stoppers
Although we thought we understood the business requirements fully, we only really exposed our ideas to very senior management once we had a working prototype. One argument was that there was no point promising something until we were certain we could deliver. Unfortunately we learned the hard way what a 'show stopper' issue was and had to do some rapid redesign after the initial pilot.

Beware manufacturers only designing what's buildable for them
In manufacturing engineering, the design staff are required to ensure that their designs are buildable with the technology available (which is often a function of history). Hence manufacturers with extensive sheet metal working facilities tend to design in sheet metal. If you want more creative use of materials, this can be frustrating. We used three different manufacturers (with different manufacturing bases) and industrial designers with a wide range of experience of materials to avoid this trap.

Don't be fobbed off by experts claiming 'it can't be done'
Sometimes the Project Director has to ignore the initial 'it can't be done' from experts and hold the vision in the design. I found this was particularly true with issues concerning ease of use and the robustness of infrequently used components eg access flaps. With hindsight, I would have liked to have been even tougher.

Sort out Intellectual Property issues early
In any innovative design process involving multiple team members, the question of Intellectual Property (who 'owns' design solutions) can become an obstacle to collaboration. I believe the cost and effort of sorting this out early pays enormous dividends later (when people forget what was agreed informally and the stakes are higher).

Do not be embarrassed about ergonomics methods
As I mentioned earlier, I was surprised how reliant design engineers were on prototypes and trials (just like ergonomics). In fact our strategy of parallel prototypes seemed to work particularly well and allowed us to phase together activities which, if sequential, not only take longer but also reduce design options. Parallel prototypes allowed us to try out

the operational and the performance impact of ideas at the same time – for example the shape of the shelf edge (which had aesthetic as well as air flow implications)

Conclusion

One of the most satisfying aspects of the project was to see people from competing organisations and different disciplines work together to solve common problems. Some issues were contentious and difficult to resolve but the whole team worked well together. The end result was not just an innovative design of refrigerated case but a single design which gives Marks and Spencer enormous flexibility in planning and efficiencies in operation in order to provide their customers with high quality merchandise in peak condition.

Finally I would like to acknowledge the extensive team involved in the project and thank Marks and Spencer for trusting me with this significant project.

AN INTEGRATED APPROACH TO SUPPORT TEAMWORKING

[1]Andrée Woodcock, [2]Christoph Meier, [3] Rolf Reinema, [4]Marjolein van Bodegom

[1] VIDe Research Centre, School of Art and Design, Coventry University, Gosford Street, Coventry, UK. A.Woodcock@coventry.ac.uk
[2] Fraunhofer Institute for Industrial Engineering (FhG-IAO), Nobelstrasse 12, D-70569 Stuttgart, Germany. christoph.meier@iao.fhg.de
[3] Fraunhofer Institute for Secure Telecooperation (Fhg-SIT),COR, Rheinstrasse 75 Darmstadt, D-64295, Germany. reinema@sit.fraunhofer.de
[4] Marjolein van Bodegom, Pentascope, Javastraat 1, Den Haag, Holland. vanbodegom@pentascope.nl

Developments in information technology, globalisation, organizational patterns, economic climate and movement away from manufacturing have contributed to new working practices, with a rise in task based temporary teams, distance working, and emphasis on knowledge management. Current systems do not necessarily support task based, distributed working. This paper reflects, from a user's standpoint, on the approach and the results of the first year of the UNITE project, which aims to produce an integrated system to address the challenges posed by mobile teamwork.

Introduction

Today's 'global village' is populated with international companies and employees, who work on projects in virtual teams. They may be distributed within one or many geographically dispersed organisations. Such teams consist of domain experts, who are relative strangers, coming together for a specific purpose. Traditional offices, and their associated infrastructure, do not allow smooth working on such projects. For example, systems may be incompatible, international phone calls may be difficult to co-ordinate, and any form of conferencing may require movement to different rooms, and the scheduling of shared resources, such as technical support staff. A new type of working environment needs to be created to allow geographically separated people to collaborate naturally and effectively (Reinema et al 1998 and 2000). This should reflect their shared context of work, that of a joint project. This should:

- allow geographically separated people to collaborate naturally and effectively.
- support the simultaneous working on and switching between multiple projects.
- appropriately reflect the shared work context – that of a joint project. When teams are co-located it is relatively simple to gain knowledge of one's colleagues, their working

habits and preferences, project status, current activities etc. When team members are separated such information has to be made explicit.

- provide an environment that seems familiar by exploiting people's skill in navigating and working in the real world, so enabling smooth transition between the real and virtual world
- reveal the resources available that team members are working from or have available.

Rather than develop more CSCW (computer supported co-operative working) applications (see Figure 1) to support geographically dispersed teams and their mobile members, the UNITE solution is to take existing products as a reservoir of building blocks and encapsulate them in a unified, integrated system, which meets these new user requirements. Such a system would allow people to use their preferred applications but embed them in an integrated system where applications seamlessly interweave with each other. This will remove the technological incompatibilities that blight communication especially in projects of short duration (where team members may spend longer trying to set up a workable communication infrastructure than they do in task related activities). With the mushrooming of applications and communication devices the only valid solution is dynamic integration of technology.

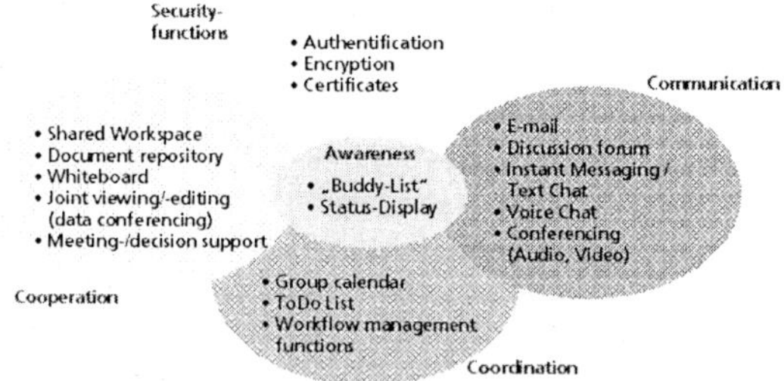

Figure 1 Functional Blocks/Modules of co-operative systems (Meier, 2001)

To this end, the UNITE development team (with members from Steria, IBM, the Fraunhofer Institutes SIT and IAO, ADETTI and Coventry University) are developing a web based, co-operative workplace that can enable people to switch from one project to another in a highly flexible working environment. This means that at any time, wherever they are working, be it on company premises, at home, or on the move, UNITE users will be able to focus on the context of one particular project and task, and be freed from physical locations. For companies, such a system will bring increases in flexibility, efficiency and effectiveness, e.g. the amount of time spent travelling and trying to make contact with colleagues will be reduced, the flow of information around the organisation will be increased. For individuals, problems of mobile working (Woodcock, 1999) such as isolation, lack of informal communication, removal from the locus of control, poor support by central management, inability to create affective ties will be reduced. Therest of the paper describes the user requirements for such a system.

User Requirements Capture

At the start of the project, four target customers for UNITE were identified: Start-Ups, especially in the field of new service companies (e.g. consultancies, multimedia agencies); Networks, including multiple different small and micro firms especially from the service sector; Medium sized enterprises with growing geographical distribution and increasing internationalisation requirements; and large companies, where functional units and teams are increasingly geographically distributed.

Our intention is to develop a collaboration platform (the UNITE platform) for these target customers. Given the duration of the project (2 years), we needed to arrive at a good understanding of the work processes of potential users. To do this we worked with a user company (Pentascope), considered to be typical of the target customers identified. Once the user requirements were gathered, we abstracted from these, using the collective knowledge of the consortium, to arrive at a collection of general user requirements, features and functions that would satisfy the needs of all identified target customers.

Pentascope is a knowledge based network organisation (KBNO) of highly mobile workers, who require flexibility and control over their work environments. Pentascope consultants usually work simultaneously on 3-4 internal and 2-3 external projects, of different duration. They occasionally need to share physical office resources and often need access to the expertise of others in the organisation when working with clients. It is the requirements of such organisations that form the basis of the cooperative working solution we are developing. They have the following characteristics:

- Their employees work at multiple places: at company or customer premises, at home, or on the move. This requires connection to tele-workers at different physical locations;
- The number of their employees in such organisations often larger than 250. If these employees are not co-located it is hard to establish a shared understanding of what is happening at individual, team, or organisational level;
- Their business is highly predicated by teamwork, i.e. project-driven, timely access to experts, close communications between team members, document sharing;
- The organisation is matrix-based, with tasks executed by dynamically-formed subgroups;
- They rely on responsibility and discipline at the level of their employees;
- They form multidisciplinary organisations, in which different domain experts have to interact and depend on each other. This frequently gives rise to communication breakdowns.

A three-step approach was taken to capturing the user requirements.

1. Familiarisation with users - Initial two day visit to trial users (Pentascope) during which interviews were conducted in order to familiarize the UNITE consortium with the users, their organization and ways of working.

2. Capturing user requirements - An intense two day workshop with members of Pentascope to gather in-depth information on work contexts and procedures. This included exercises such as drawing images relating to aspects of work (e.g., "What do you enjoy about working here?", "What is teamwork like?", "Yesterday what was your

work environment like?"). In asking the participants to first draw and then verbally comment on their work experience, experientially rich accounts were produced that went beyond practiced verbal descriptions. In later sessions, the focus narrowed down on project related work. Accounts of steps and activities over the life-cycle of a project were collected and subsequently worked through and detailed. Of especial importance were "doing work in the course of meetings" and "communication issues."
3. From requirements to specification - The information was reflected on by the project partners, in the light of their experience with supporting cooperation at a distance. The outcome of this was a detailed documentation of usage scenarios, user requirements and functionality iteratively co-produced by developers, users and evaluators.

User Requirements

The user requirements capture process (led by FhG-IAO) produced insights into the work processes, the particular culture of this company and led, finally, to an appreciation of particular problems members experience (e.g. with respect to internal communication). The outcome provided three sets of requirements for the UNITE platform.

General user requirements
These requirements refer to the global needs of users. They include: firstly, the integration of work contexts. Users can be in one of multiple work contexts: 1) an individual context, when they work in isolation; 2) a project context, when they work with the other team members of a given project. If they work on multiple projects, they are in one of many project contexts as well as an organisation context, when they behave as an employee of an organisation. Secondly, specific support of project style of work. Users need project and task management, document management, and communication with others in either the aforementioned work context, and thirdly, shielding from the heterogeneity of the underlying IT infrastructures and the devices used by the team members, e.g. fixed workstations and phones, notebook-computers, mobile phones.

Business processes requirements
The following are indicative of the requirements the UNITE platform system should support for work tasks to be completed. These include bulletin boards and discussion forums, personal address book and calendar, a repository for sharing documents and bookmarks, an electronic whiteboard for text, drawings, pictures, audio/video clips that can be used for socialising (e.g., as "fun corner"), a single point of entry with easy log-in for all uses of the platform (e.g., an individual portal leading to those project portals a particular member is involved in)

If we take the scenario of trying to arrange a meeting users need status displays indicating availability of other team members and physical resources (e.g., equipment and rooms) where the meeting could be held, a means of finding dates and times for joint meetings (by comparing calendar information of different team members), a unified messaging service (e.g., forwarding of e-mails to mobile and transformation into voice-message) to facilitate contact, easy publication of meeting related information on a website available to all members, a way of linking meeting-related materials to one "meeting object" and automated notification of changes for all participants.

Once the meeting is held, the users require automated set up of synchronous communication (e.g., phone/mobile, voice chat, video conferencing) according to availability of other partner(s) by means of a single click on an avatar. This was seen as important as not all team members may be able to meet at the same time and place. To support this, configurable teleconferencing support (e.g., multi-point audio/video, whiteboard, application sharing, polling) is required.

Implementation requirements
These requirements relate to the way that the UNITE platform will need to be implemented in an organisation to ensure it's success. Two issues have been identified, firstly, an incremental implementation path, starting with a few users working on a single project and then extending its availability and secondly, easy communication between organisation members employing the UNITE platform and those who do not, thus providing users with benefits even while the platform has not been extended to the entire company.

The UNITE Collaborative Platform

The requirements were encapsulated in a set of task-based scenarios e.g. the steps undertaken to arrange a project meeting, and were translated into groups of functionalities for the developers to incorporate firstly in a Basic Platform (with limited functionality and services) and then in an Enhanced Platform that will, for example, support videoconferencing.

At the heart of UNITE is an html interface integrating project based resources (such as libraries, communication tools). It is intended that the interface is presented to the user as a co-operative workplace analogous to real offices where space is distributed according to activities such as individual work, joint work, conferences, presentations, informal exchange, etc. As individual projects have different resource requirements, so project spaces can be flexibly created and adapted and dissolved on completion. The project resources (e.g. project schedule and current status, organisational directives) are visible and can be acted on by team members wherever they are located.

Each team member will have an avatar and be able to move their cursor, for example, to approach a colleague working at their desk, inspect the project calendar, or enter the conference area. Automatically, and without the need to explicitly open software, the user can 'talk' to a colleague, inspect documents, or participate in a conference session. The required communication and processing tools, together with security measures for access control and authorisation are implicitly initiated and terminated upon entering and leaving specific areas.

The fully group-aware interface allows users to see the contents of repositories, providing access to services, and current status of team members. Any work done by team members in the co-operative workplace is considered project related. The workplace will keep track of the context of actions, and make sure items are stored in and restored to their proper context. Context switching between projects will be supported.

At the user interface, a virtual office will be structured like a real team office. This preserves some of the advantages of a local shared office, such as transparency, ease of interaction, availability of project data. Team members are present as avatars in the virtual office as if they were physically present in a single office. The virtual office captures the project's context, by including its input and output, resources, tools and services.

The virtual office will present its users a context oriented view of its contents and members. This will be achieved by importing the necessary basic tools, services, and resources into the co-operation platform, relating them to each other where meaningful, embedding them in the project context, surrounding them with a context oriented user interface and thus creating value-added project services. Phone and fax calls in a virtual office for example, would only be those related to the project, conference sessions would automatically be tied to the context of project data and cost, any collaboration input and output would be matched against the project database, cost file, security rules etc.

Current and Future Work

At the time of writing, the Basic Platform is undergoing acceptance tests and usability inspection in preparation for the usability trials at Pentascope in 2002. The evaluation will take the form of a 6 week trial in which a team will use UNITE to support all synchronous and asynchronous project related activity. It will include pre and post trial reflections, and an intensive 2 day, mid project video-ed trial to gain additional insights into usability problems (e.g. learning, errors, memorability, efficiency and satisfaction). Evaluation will focus on, overall system usability (using SUMI (Software Usability Measurement Instrument) and in-house questionnaires), the manner in which the system was used during the project (using diary studies), the extent to which UNITE aided or changed working practice (as compared to current modes of working) and the extent to which UNITE offered users valuable support.

Acknowledgements

This project is funded by the EU under the Information Society Technologies (IST) Programme (Project No. IST-2000-25436). It represents the view of the authors only.

References

Klöckner, K.,Pankoke-Babatz, U. and Prinz, W. 1999, Experiences with a cooperative design process in developing a telecooperation system for collaborative document production. Behaviour & Information Technology, 18 (5), 373-383

Meier, C. 2001, Virtuelle Teamarbeitsräume im WWW. Wirtschaftspsychologie, Heft 4

Reinema, R., Bahr, K. Baukloh, M., Burkard, H.-J., Schulze, G. 1998, Cooperative Buildings–Workspaces of the Future. In: Callaos, C., Omolayole, O., Wang, L. (Eds.), Proceedings of the World Multiconference on Systemics Cybernetics and Informatics (SCI'98), Orlando, Florida, Vol. 1, 121-128

Reinema, R., Bahr, K., Burkhardt, H.-J., Hovestadt, L. 2000, Cooperative Rooms -- Symbiosis of Real and Virtual Worlds; In: Proceedings of 8th International Conference on Telecommunication Systems, Modelling and Analysis, Nashville, Texas, March

Woodcock, A. (1999), Human factors perspectives on information technology and globalisation. *Information Society '99*, Vilnius, Lithuania, 166-177

USER REQUIREMENTS ANALYSIS APPLIED TO ENVIRONMENTAL SYSTEMS

Martin C. Maguire

Loughborough University
Research School in Ergonomics and Human Factors
(formerly HUSAT and ICE)
The Elms, Elms Grove, Loughborough, LE11 1RG, UK
m.c.maguire@lboro.ac.uk

The analysis of user requirements for a new system is a key stage in system development. This paper reviews the approach adopted to elicit user requirements for two EC projects: FOREMMS (forest monitoring and management) and EUROCLIM (climate change monitoring and prediction), providing data on the natural environment in Europe. The paper describes the survey approach used to analyse the user requirements for these systems, and highlights techniques employed and lessons learned.

Introduction

For any system to be successful, it is necessary to understand what the future users need and would like from the system. Loughborough University is a partner in two related EC funded projects to develop monitoring systems to support environmental work and research across Europe. One project is FOREMMS (Forest Environmental Monitoring and Management System) and the other is EUROCLIM (European Climate Change Monitoring and Prediction System).

The EU FOREMMS project has the goal of developing and demonstrating an advanced forest environmental and management prototype system (http://www.nr.no/foremms). The functioning of forest ecosystems is endangered, biodiversity is decreasing, and the quality of human life is reduced in many places in Europe due to the high density of human population, intensive utilisation of forests, accumulation of poisonous emissions and the increasing amount of green house gases in the atmosphere. Thus the FOREMMS project, co-ordinated by the Norwegian Computer Center in Oslo, uses remote sensing techniques including satellite monitoring, airborne measurement and automatic ground stations, to gather relevant data to support forest health and good forest management.

The EU EUROCLIM project has the aim of developing an advanced climate monitoring and prediction system for Europe (http://euroclim.nr.no). One of the greatest threats to human beings is climate change and, according to the Intergovermental Panel on Climate Change (IPCC), it is not known whether northern Europe will experience regional cooling or warming in a future warmer world. However, most likely, global warming will change the living conditions in Europe significantly and the weather will show more extreme conditions, like flooding and hurricanes. The best natural indicator of global warming is the cryosphere, i.e. masses of sea-ice, snow, and glaciers. The EUROCLIM project is developing an online resource of satellite and field measurement data from a network of sites across Europe. The aim is to establish a 20-year database of climatic data (starting from 1980) and to update it each week into the future. There will also be modelling and statistical facilities to estimate future climate change trends and extremes etc. For both FOREMMS and EUROCLIM, data will be stored in digital form so that maps and charts can be produced for professional users and the public to access via the Web. Professional users will also be able to download datasets for use in their own research work.

One of the author's roles in both projects has been to analyse user requirements for both of these systems. This paper highlights some of the issues and problems associated with user requirements analysis common to both of the projects and discusses solutions adopted. This may be useful for user requirements activities in the future.

Survey approach

As with many projects, timescales and resources were limited and a cost effective method was required to gather user requirements for the FOREMMS and EUROCLIM systems. For both systems there were a diverse range of potential user groups, including:

- commercial companies (e.g. forest owners, climate consultants)
- research institutes (concerned with forest ecosystems and climate change)
- campaign groups
- government and EU representatives
- journalists
- teachers and lecturers
- the general public.

It was therefore decided to develop the survey as a detailed questionnaire, which would gather as much relevant information as possible. Project partners were asked to disseminate it to potential users in their own countries, to be returned to the author for analysis. This was found to be a practical approach, obtaining consistent data from diverse users across Europe.

The need to understand the technical area

One of the problems for ergonomists within system design is to gain sufficient understanding of the technical area that the system supports. It is only after gaining a reasonable comprehension that it becomes possible to plan the questions to be asked within the user requirements survey. The process does however help to develop an understandable questionnaire for non-scientific users which explains terms such as cryospheric, biosphere, precipitation, aforestation, biomass, LAI (leaf area index) etc. Also the discussion of potential topics to explore enables the issues become clear to the technical developers and user representatives. For instance, the EUROCLIM system will provide datasets for cryospheric variables (snow cover, glaciers, sea-ice). The need was identified to investigate specific *quality limits* for the data, if it was to be useful to the end users. Thus for each data variable (e.g. snow cover) the survey asked the professional users for their requirements for data resolution (how much ground one data point had to cover), update frequency from the satellite, accuracy of the data, and importance for the task. Once these characteristics were identified, they were fed directly into the system specification discussions.

Contacting potential users and survey distribution

The task of contacting potential users was distributed across the FOREMMS and EUROCLIM consortia. This allowed technical partners to use their own contacts to distribute survey forms to relevant persons e.g. in other scientific or public organisations they had collaborated with in the past. However new contacts also had to be established. Although cold calling new organisations can be daunting, it was found that contact by phone followed by email worked effectively. Patience was, of course, required to explain the project to different persons within an organisation before reaching the right person. However by explaining the project objectives and describing its relevance to the organisation, the potential user normally provided a positive and helpful response. Some respondents, unable to participate offered to distribute the questionnaire to colleagues or even to people in other organisations.

It was found that the professional users were normally happy to receive the questionnaire by email as a 'soft' document which they could either fill-in and send back electronically, or print off to complete by hand and return by post.

To obtain general public reactions, it was decided to impose minimal constraints on the sample of public users and to simply obtain an even balance of males and females across different age groups. The aim was to obtain general reactions to the proposed FOREMMS and EUROCLIM systems, and to find out what kinds of information they would require. Since one of the aims within both FOREMMS and EUROCLIM is to raise public awareness and interest in environmental topics, the views of those who are relatively apathetic is as important to record as those who are enthusiastic.

Grouping diverse user types

Both FOREMMS and EUROCLIM are attempting to provide resources for a wide range of scientific and professional users, as well as to offer general information for teachers and the public. For both projects, between 30 and 40 scientific users were surveyed as a basis for user needs analysis. Since the users expressed a wide range of needs, particularly in terms of the types and quality of datasets they required, it was necessary to select a subset of data variables while trying to satisfy as high a proportion of users as possible. It was found useful to group the scientific/professional users into the different types of task or topic area they were interested in, to see for which areas it was feasible to support. The tables below show the different groups that were identified for both projects:

Table 1. Main professional users grouped by task

FOREMMS project	EUROCLIM project
Forest detection (particularly ancient forests), classification and general characteristics	Natural hazards and extreme weather.
Studying of forest health	Agricultural practice and management.
Forest land cover and activities.	Socio-economic effects of climate change.
Forest management information	Effect on ecology, species and habitat.
Level of biodiversity	Needs of the building industry.
History of forests and future trends.	Arctic and mountain environment change.
Actions required to preserve/rejuvenate forests	Water and energy resource management.
Forest state monitoring	Climate change monitoring.
Policy making	

Questions to ask

The FOREMMS and EUROCLIM projects were based on well developed system concepts but needed the detailed user requirements to be specified. As is typical with user requirements elicitation activities, it is not a simple process to ask users what they require as they will not always be able to be specific and will have a limited understanding of what is possible. It was seen as necessary to ask a wide range of questions probing different aspects of users' current tasks and their desired system characteristics. These covered the *system functions* - the basic facilities provided by the system, and the *system qualities* such as ease of use, ability to work with different Internet browsers, response times, etc, that determine the quality of the interaction (Robertson and Robertson, 1999). There are also other aspects of the system that will be important to the user such as whether it contains the appropriate *data content* to support their particular needs, and whether data is presented appropriately via screen layouts. It is necessary to investigate all these different aspects in order to obtain a comprehensive view of the user needs and preferences.

At a more fundamental level, it is important to understand the tasks that users perform and, from a brief description of the system, to ask users how they would like to use it (ideally) to support their work. These first impressions are useful in highlighting user expectations without the user being diverted by more detailed questions.

A helpful general structure for the professional user survey forms was as follows:

- Background information on the user/respondent.
- Main interests in relation to forest environment/climate research.
- What would the system provide in an ideal world.
- Specific needs based on a list of data types potentially available.
- Other kinds of data required.
- System interaction and presentation requirements.
- Comment on example scenarios of use.
- Needs for statistical/trend information and modelling output
- Additional comments.

The general public were presented with simpler versions of the professional surveys. They were asked were about their interests in relation to the environment, to describe any information they have read about environmental issues and how good these sources were. They were asked what would be the desirable characteristics for the system(s) as either members of the public or as teachers. They were given an outline structure of the system content, asked to comment on it and to name any missing items of information they would like to see. They were also asked to say what they thought would be the benefits of the system to themselves or their families.

Analysis of results

The surveys for FOREMMS and EUROCLIM generated much data. In order to manage this, the responses were assembled into a spreadsheet (one row per user). For long comments, it was convenient to copy and paste these from the survey form into the spreadsheet. Once the spreadsheet was complete, it was then possible to perform frequency counts on answers, compile bar charts, and pick out relevant quotations. Tables were also compiled to show what kinds of data are required for each type of activity (listed in Table 1), enabling the system to be targeted at specific types of user activity. With large amounts of user requirements data, it is essential to structure and summarise it as much as possible so that others in the design team may read it conveniently. However it is also useful to relate particular responses to the scientific organisations that made them allowing the designers to understand the basis for certain data requirements. Recording contact details also allows for follow-up clarification with the user.

User requirements specification

In *FOREMMS*, the professional users were asked what questions they would like the system to help them answer. A wide range were specified e.g. What is the best way to reforest forest areas after tree cutting?; How can lumber value be increased while maintaining biodiversity?; What is the best method of reforestation (planting or seeding)?, What are the historical trends and the effect of human involvement on forests?; Where are areas of original forest that need preserving? (Maguire, 2001). Based on these questions, the design team identified general scenarios of use that FOREMMS could support, such as:

- *Display forest coverage and forest type for a particular region.*
- *Display areas of forest health concerns indicated by discolouration and defoliation for a particular region.*

The scenarios were specified in a system design document in terms of 'use cases'. These were described in terms of: importance to users, user inputs required, presentation of results, data and algorithms needed to generate the results, comments and open issues. This approach was helpful in capturing functional needs in a user-oriented way. The design team then identified the data parameters that could be acquired by satellite to satisfy each use case i.e. forest cover, forest type, carbon stock/biomass, forest patchiness, discolouration and defoliation.

In *EUROCLIM,* user needs for the different cryospheric variables (snow cover, glaciers and sea ice) were investigated, as well as the need for prediction and modelling facilities (Maguire, 2002). Again general scenarios were identified, such as:

- *A water authority wants to know future rainfall levels to help plan reservoirs*
- *An administrator wishes to find out what will be the future frequency of snow melt floods that may cause damage to an area.*
- *A farm ministry wishes to know the length of the summer and growing season in past and in the future.*
- *A ski-resort planner wishes to know how snow cover will change in a specific region.*

Again relevant data parameters were prioritised e.g. snow cover, sea ice thickness, glacier volume, etc. It was found that the professional users in FOREMMS had an interest in accessing many datasets at a high level of accuracy. In EUROCLIM the approach was adapted to ask respondents to prioritise specific requirements to help focus on the most important requirements for them. A software utility was also developed allowing different potential system datasets to be specified to show what proportion of users this would satisfy.

In FOREMMS, the *general public* wanted information that would improve their understanding of healthy forests. They also wanted to know what they could do personally to help preserve forest ecosystems, and how to visit and enjoy forests and avoid damaging them. For climate change (EUROCLIM), users also wanted assistance to improve their basic understanding of climate issues. They wanted to know, Is global warming happening?; What are the causes?; What are the effects going to be in my region?; ..on peoples' living conditions?; .. on the climate generally, and on tourism and leisure activities?. They also wanted to know about the timescale for change, the severity of effects, what actions are being taken by governments and other organisations, and what they could do personally. Thus while there is a public appetite for environmental information, they seem to require material that improves their understanding and information of practical use to them.

Simulation to obtain further feedback

In order to summarise the functions, qualities and data requirements for the systems, user interface simulations and prototypes have been developed. This provides a good way to communicate the user requirements within the design team, and to obtain confirmation from the users themselves. An example FOREMMS screen, developed in Powerpoint™, is shown in Figure 1.

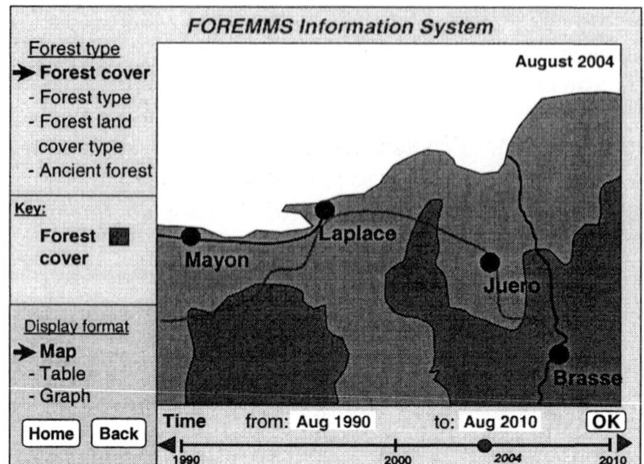

Figure 1. Example screen simulation for FOREMMS user interface

In EUROCLIM, a simulation will also be developed to obtain early feedback from end users on an individual basis and in panel sessions. Later on, in both FOREMMS and EUROCLIM, user trials will be organised where user representatives will test operational versions of the systems by performing representative tasks, based on the usage scenarios.

Conclusion

This paper has reported a survey approach for the analysis of user requirements for two environmental data resource systems. This method provides a good overview of the needs of a wide range of users. However the user requirements and early designs need to be further investigated and refined, using system prototypes and simulations. They provide a useful evolving model for the project team (e.g. developers, user representatives and ergonomists) to refer to during the system development process.

References

Maguire, M. C. 2001, Deliverable 1 - User needs report, *FOREMMS IST-1999-11228 project*, **October**, HUSAT/RSEHF, Loughborough University, Loughborough, UK.

Maguire, M. C. 2002, Deliverable 1 - User requirements report, *EUROCLIM IST-2000-28766 project*, **January**, HUSAT/RSEHF, Loughborough University, Loughborough, UK.

Robertson, S. and Robertson, J. 1999, *Mastering the Requirements Process.* (Addison-Wesley, Harlow, UK). ISBN 0-201-36046-2.

DRIVER EDUCATION FOR PRE-TEENS: A NEW APPROACH TOWARD REDUCING CRASHES AMONG YOUNG NOVICE DRIVERS

Jerry Wachtel

The Veridian Group, Inc.
567 Panoramic Way, East Cottage
Berkeley, California 94704, USA

This paper reports on a study performed to evaluate the feasibility of educating grade-school age children about road safety. We created a series of integrated classroom programs on topics of speeding, aggressive driving, and sharing the road with bicycles. Lessons, a teacher's guide, and mastery tests were prepared in support of an age-appropriate curriculum that was presented to children (ages 7-9 and 9-11) at five different schools. Using a "before and after with controls" protocol, the experimental groups were taught the road safety curriculum while the control groups received unrelated civics lessons. In general, our hypotheses were confirmed. The extent of the superior achievement on the post-test by experimental groups ranged from 1.4 to more than 5 times that of the control groups. Participant feedback was positive, and lessons were learned to enhance future expansion of the program and its development as an internet-based curriculum.

Introduction

Why A Program Of Road Safety In Grade Schools?

Motor vehicle crashes are the leading cause of death for children of every age from 6 to 14 years in the United States and much of the western world (National Center for Statistics & Analysis, 1996; European Transport Safety Council, 1995). By the time teenagers become licensed drivers the statistics become more alarming. Although young drivers between 15 and 20 years old represented only 6.7 percent of all licensed drivers in the U.S. in 1995, they accounted for 14 percent of all fatal crashes. When driver fatality rates are calculated on the basis of estimated annual travel, the rate for teenage drivers is about four times as high as the rate for drivers 25 to 65 years old. These data represent a serious public health issue.

A look behind these statistics points out additional disturbing information. Nearly one-third of the 15-20 year-old drivers involved in fatal crashes who had an invalid driver's license at the time of the crash also had a previous license suspension or revocation; and nearly one-third of those 15-20 year-olds killed in crashes had been drinking. Of young drivers killed in motorcycle crashes in 1994, 47 percent were not wearing helmets, 24 percent were unlicensed and 20 percent had been drinking. These data point to underlying factors that may contribute to crashes among young, novice drivers. Such factors involve risky behaviors that may be mediated by attitudes toward driving which may be representative of other measurable attitudes held by subsets of young drivers. A decision to drink and drive, to drive without a license, or to ride a motorcycle without a helmet, is not likely to be spontaneous, but rather the result of some deliberative thought process. Findings by Canada's Traffic Injury Research Foundation indicate that those youths who engage in risk-taking behavior while driving are quite likely to engage in risky behaviors in other aspects of their lives as well (Simpson, 1993). In short, factors associated with risky behaviors in one's overall lifestyle are significantly related to vehicle crashes among young drivers.

Although children may not legally drive until they are teenagers, there are compelling reasons why issues of road safety are important and suitable subjects for an educational curriculum in grade schools. These include:

1. Children are passengers in cars driven by their parents or older siblings. Evidence suggests that expressions of concern by child passengers about road safety issues can influence the behaviors of drivers of vehicles in which those children are riding.

2. Learning about road safety can help children become safer users of the road themselves, long before they begin driving. All children are pedestrians who must share the road with motor vehicles, and most are also bicyclists, roller-skaters, skate-boarders, etc.

3. Road user behavior does not occur in isolation of other life events. The behavior of drivers on the road is a microcosm of their behavior when interacting with others in other domains. Principles of road safety and the civil treatment of others should be considered in the larger context of behavior in society at large. Educational curricula that address broad issues of civic behavior such as violence or discrimination are a natural setting for the inclusion of lessons about road safety, because road safety is inherently part of this larger societal milieu.

4. The teenage years are those in which children rebel, respond more to peers than to elders, take risks, and test limits. These behaviors are particularly dangerous in the context of learning to drive. One possible approach to mediate these "rites of passage" behaviors, essentially untried and untested, is to begin the education of road safety early, thus providing a more stable foundation which may assist the individual during the turbulent teen years.

Is A Road Safety Curriculum Suitable For Young Children?

Today, formal education is begun far earlier than in the past. We once taught about nutrition only late in primary school - now we do it in kindergarten; we once taught foreign languages in intermediate school - now we debate whether this should be done at age 2-3 or 5-6. There is ample evidence that grade school children can be successfully educated about important social and societal issues. We found numerous programs of instruction that addressed topics as diverse as injury prevention, poison control, disease education, and ethics.

In the U.S. little systematic attention has been given to programs of road safety education in primary schools. Driver education programs are typically aimed at teenagers who are approaching legal driving age - far later than education professionals argue such learning could effectively take place. Although there have been some exceptions, for example, the "Kids Teaching Kids" (KTK) program sponsored by the Minnesota Highway Safety Center at St. Cloud State University (Isberner, 1999), these have not generally been subjected to comprehensive tests of their effectiveness.

We believe that grade school is an appropriate time and place to begin a curriculum on road safety; that an age-appropriate curriculum could, if properly designed, implemented and tested, lead to a positive change in children's knowledge about, and attitudes toward, issues of road safety. Further, we believe that these knowledges and attitudes, if reinforced over time, could contribute to a meaningful reduction in road user injury and death. The purpose of this study was to take the principles of early childhood education, couple them with the advances in educational delivery technology, and initiate a pilot test of road safety education at the primary school level.

Technical Approach

Subjects

Students in two age groups participated in this study. Older students (CM) included children aged 9-11, and younger students (CE) included children aged 7-9. Within each age group, classes were subdivided into a younger (CM1, CE1) and an older (CM2, CE2) subgroup. Teachers, administrators and students at one urban and four rural schools in and around the city of Lyon, France took part in the study. In all, more than 400 children participated.

Methods and Materials

We developed lessons about three topics - speeding, aggressive driving, and road sharing with bicycles, all within the broader context of civics. We also prepared a comprehensive "Facilitator's Guide." The curriculum and Guide were developed based on the work of educators who have written extensively about teaching "critical thinking" skills (Baron and Sternberg, 1987; Wallerstein and Bernstein, 1988). We created and pilot-tested an engaging story around the three lessons, integrated them into an educational video, and used the video as the stimulus to "trigger" classroom discussion and activities.

Experimental Design

The experimental design was that of a "before-and-after study with controls." We used a repeated measures design in which the repeated measures were three evaluation tests. Test results were compared across several "between subjects" variables, specifically: (a) group membership (Experimental vs. Control); (b) gender; (c) age, and (d) school.

Hypotheses

We posed several hypotheses and research questions. However, only the main hypothesis is discussed in this report: Those children who receive the experimental program of instruction will improve their understanding of the key road safety issues when compared to those children who do not receive the instruction.

Program Evaluation

We developed a series of three tests to evaluate students' understanding of the material. The tests were developed and pilot-tested, then modified as a result of suggestions from our educational advisors. The younger and older groups received different tests. The pretest (T1) was administered just prior to the introduction of the curriculum. It established a baseline of the children's knowledge and understanding of the material, and enabled us to determine whether there were significant differences between the groups at the start of the project. Following the pretest the Experimental groups were presented with the road safety curriculum while the Control groups received education in unrelated topics of civics. Immediately upon completion of these lessons both groups were given the immediate posttest (T2) to measure the extent to which they had mastered the material. After three weeks in which no instruction was given in the subjects of interest, the follow-up posttest (T3) was administered. This served to measure the degree to which the learned material was retained.

Results

The data for the younger (CE) and older (CM) children were analyzed separately.

Younger Children

At T1, the Control group performed better than the Experimental group. However, at T2, the Experimental group mean was higher than that of the Control group, showing an improvement factor of 2.85:1 in the number of test questions answered correctly.

Overall, there were essentially no changes in the Experimental students' scores between Tests 2 and 3, indicating that there was little, if any, forgetting of the material learned. When separated by gender, however, the results differ. The girls retained the material quite well. The boys' scores fell from an average of 14.2 correct answers to 13.8 (out of 20).

Because of class structure in French primary schools, we further subdivided the group into older and younger classes. Not surprisingly the older students had consistently higher scores than their one-year younger cohort. The data show that the program was successful at both age levels. For the older young children, the Experimental Group improvement from T1 to T2 was more than four times greater than for the Control Group; for the younger young children, the Experimental Group improvement from T1 to T2 was twice as great as that of the Control Group. This is shown in Figure 1.

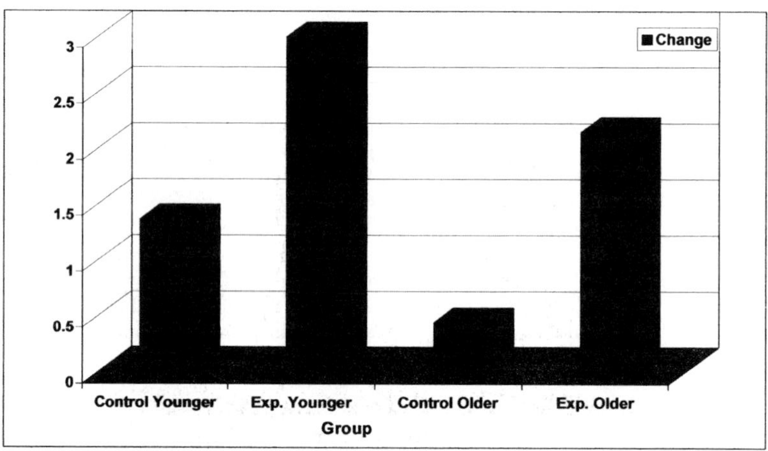

Figure 1. Younger Children by Age – Change from Test 1 to Test 2 Questions Answered Correctly

Older Children

When comparing older students, the Experimental group scored higher than the Control group on both T1 and T2. Although both groups improved their average overall scores from T1 to T2, the Experimental group improved more than the Control group.

Within gender groups, the Experimental girls started out with higher scores than the Control girls, and improved more than did the Control girls to the point where, at the time of T2, the Experimental girls had higher average scores than the Control girls by nearly two full questions (out of 22). For the boys, the situation was somewhat different. The Experimental boys started out with higher scores than the Control boys. But then both groups improved almost equally from T1 to T2, gaining an average of one question.

Between gender groups, we found that Experimental girls had higher scores than Experimental boys at both T1 and T2, and the Experimental girls improved their scores more than did the Experimental boys between these two exams. Further, if we combine the Experimental and Control groups, girls scored higher than boys at both T1 and T2. In short, both girls and boys improved from T1 to T2, but the girls benefited slightly more from the instruction than did the boys.

As expected, the older group of older students (CM2) had higher scores than the younger group (CM1) both before the educational material was presented and after the immediate posttest. As predicted, the Experimental group of younger old children improved between T1 and T2 by twice the extent of the Control group. Surprisingly, however, this was not the case for the older group of old children. Here, the Control group improved to a degree essentially equal to the Experimental group. Thus, although the performance improvement of the Experimental group was in accordance with our hypothesis, the performance improvement for the Control group was not anticipated. Taking the average of comparative improvement of the Experimental vs. the Control groups by combining the CM1 and CM2 groups, the results support the hypothesis, with the Experimental groups showing an improvement from T1 to T2 of about 1.5 times that shown by the control groups. Figure 2 shows the change scores from Test 1 to Test 2 for the older children by age.

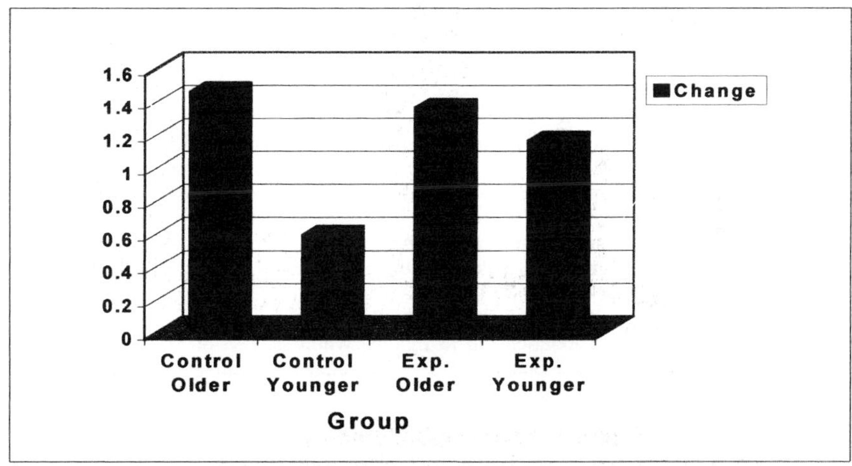

Figure 2. Older Children By Gender – Change from Test 1 to Test 2 Questions Answered Correctly

Discussion And Interpretation Of Findings

This project was conceived as a test of both the concept (road safety education for grade school age children) and the approach (the use of a trigger videotape as a surrogate for more immersive educational media, followed by interactive classroom exercises and home assignments). Certain trade-offs had to be made in experimental design to ensure teacher and administrator support for the program. Some of these trade-offs weakened the conclusions that could be drawn from the data. We also learned, after the fact, that one of the control group teachers improperly corrected the evaluation tests together with her class. In this, and all other cases, the effects of these trade-offs were the same; to artificially improve control group scores at the expense of experimental group scores. This leads us to believe that our findings would have been considerably stronger had these constraints not existed. Nonetheless, with few exceptions, our hypotheses were supported.

Conclusion

A study was performed to assess the feasibility of providing an educational curriculum on road safety to grade school age children. A series of integrated lessons on subjects of speeding, aggressive driving, and sharing the road with bicycles, triggered by a videotaped story, was presented to more than 400 students in five schools in France. Despite some administrative constraints that artificially reduced the differences in measured performance between Experimental and Control groups, the study's major hypothesis was confirmed. Both the nature and the level of the educational material, and the method of delivery, were demonstrated to be successful in improving the knowledge of those students who received the curriculum. Further, the inherent interest of the subject matter, coupled with the specialized delivery method, helped make the learned lessons resistant to early forgetting. The results demonstrated the usefulness of this course of instruction and its value to grade school children. Useful insights were gained in support of further development and expansion of the program.

Acknowledgements

This study was funded through a research grant provided by Fondation MAIF, Niort, France. The author gratefully acknowledges Dr. Charles Berthet, Professor Liliane Gallet-Blanchard, Ms. Catherine LeGuen, and Ms. Sylvie Audelan-Talon of the Fondation for their support and assistance throughout this project.

References

National Center for Statistics & Analysis. (1996). *Traffic Safety Facts 1996: Children.* Washington, DC: U.S. Department of Transportation, National Highway Traffic Safety Administration.

European Transport Safety Council (ETSC) (1995). *Young Driver and Rider Casualties: What Can We Do?*

Simpson, H.M. (1993). "Implications of current research for educating novice drivers." *Proceedings: Working Conference on Novice Driver Education.* Edmonton, Alberta: Alberta Motor Association, April.

Isberner, B.J. (1999). "A Study of the Effects of the Kids Teaching Kids Peer Led Elementary Traffic Safety Program in Relationship to Health, Wellness, Use of Seat Belts and Safe Novice Drivers in Minnesota." *Doctoral Dissertation Submitted to the Graduate Faculty of LaSalle University in Partial Fulfillment of the Requirements for the Doctor of Philosophy in Education, Ph.D. Degree Program.*

Baron, J. & Sternberg, R. (1987). *Teaching Thinking Skills: Theory and Practice.* W.H. Freeman Co., New York, NY.

Wallerstein, N., & Bernstein, E. (1988). "Empowerment Education: Freire's Ideas Adapted to Health Education." *Health Education Quarterly*, 15(4), 379-394.

AUTHOR INDEX

SUBJECT INDEX